Birkhäuser

Progress in Computer Science and Applied Logic

Volume 26

For further volumes:
http://www.springer.com/series/4814

Manindra Agrawal • Vikraman Arvind
Editors

Perspectives in Computational Complexity

The Somenath Biswas Anniversary Volume

Birkhäuser

Editors
Manindra Agrawal
Department of Computer Science
and Engineering
Indian Institute of Technology
Kanpur
India

Vikraman Arvind
CIT Campus
Institute of Mathematical Sciences
Chennai
India

ISBN 978-3-319-35058-5 ISBN 978-3-319-05446-9 (eBook)
DOI 10.1007/978-3-319-05446-9
Springer Cham Heidelberg New York Dordrecht London

Mathematics Subject Classification (2010): 03-XX, 03D15, 68-XX, 68Q15

Printed on acid-free paper

Springer is part of Springer Science+Business Media (www.birkhauser-science.com)

Contributors

Eric Allender Department of Computer Science, Rutgers University, New Brunswick, NJ, USA

Vikraman Arvind Institute of Mathematical Sciences, Chennai, India

S. Ajesh Babu Microsoft Research India, Bangalore, India

Markus Bläser Computer Science, Saarland University, Saarbrücken, Germany

Sumanta Ghosh Department of Computer Science and Engineering, Indian Institute of Technology Kanpur, Kanpur, Uttar Pradesh, India

Neeraj Kayal Microsoft Research, Bangalore, India

Piyush P. Kurur Department of Computer Science and Engineering, Indian Institute of Technology Kanpur, Kanpur, Uttar Pradesh, India

Meena Mahajan The Institute of Mathematical Sciences, Chennai, India

Bruno Poizat Institut Camille Jordan, Université Claude Bernard, Villeurbanne-cedex, France

Jaikumar Radhakrishnan School of Technology and Computer Science, Tata Institute of Fundamental Research, Mumbai, India

Ramprasad Saptharishi Microsoft Research, Bangalore, India

Nitin Saxena Department of CSE, IIT Kanpur, Kanpur, India

Jacobo Torán Department of Theoretical Computer Science, University of Ulm, Ulm, Germany

N. Variyam Vinodchandran Department of Computer Science and Engineering, University of Nebraska-Lincoln, Lincoln, NE, USA

Somenath Biswas 2013

Preface

In the summer of 2012, we organized a three day "Computational Complexity" workshop at the Indian Institute of Technology, Kanpur, India, in honor of Professor Somenath Biswas to celebrate his 60th birthday. It was a fitting event for the occasion, well-attended by several well-known experts in the field from different parts of the world.

Professor Biswas is one of the first complexity theorists from India. In his teaching and research career spanning over 30 years, apart from doing quality research, he has contributed immensely to the development of the field in India. We felt that to bring out a *festschrift* volume of articles in complexity theory, based partly on the talks at the workshop, would be a lasting tribute. We are deeply grateful to the eminent researchers who enthusiastically agreed to contribute articles for this project and also helped, in a cross-refereeing process, with refereeing each other's contributed articles. These articles span different aspects of recent complexity theory research, including the isomorphism conjecture, arithmetic circuit complexity, space-bounded complexity classes, proof complexity, applications of entropy, and the complexity of graph isomorphism. As former students of Somenath Biswas, we feel privileged to edit this volume. It is an expression, as it were, of our affection and regard for him.

We are grateful to Eric Allender for his valuable advice and support from the early stages of this book project, and for suggesting the "Progress in Computer Science and Applied Logic" Springer-Birkhäuser series. We would also like to thank Erich Grädel, the chief editor of the series for enthusiastically supporting the project.

March 2014

Manindra Agrawal
Vikraman Arvind

Contents

Chapter 1
Complexity Theory Basics: NP and NL

Vikraman Arvind

Abstract We introduce basic concepts and results in computational complexity as background for some of the articles in this volume. Our focus is on the complexity classes nondeterministic polynomial time (NP) and nondeterministic logarithmic space (NL). The presentation is aimed at computer science students at a senior undergraduate level, and assumes some familiarity with algorithm design and theory of computation. The material is covered at a fairly brisk pace. Several results and proof details are incorporated in exercises which the reader is urged to solve or look up in textbooks such as [BDG88, Pap94, AB09].

Keywords Nondeterministic polynomial time · NP-completeness · Nondeterministic logarithmic space

Mathematics Subject Classification (2010) Primary 68Q15.

1.1 NP-completeness

The story of modern computational complexity begins with the advent of stored program computers in the 1940s and the need for efficiently solving optimization problems by programming the computer. It was soon discovered that a brute-force enumerative search for the optimal solution yields only algorithms that take exponential time, since the number of candidate solutions for optimization problems is typically exponential in the input size. Such solutions were not practical even for inputs of moderate size. Cobham [Cob64] and Edmonds [Edm65], around 1965, independently suggested polynomial-time boundedness as an appropriate theoretical criterion for efficient computation. This was an important conceptual contribution.

V. Arvind (✉)
Institute of Mathematical Sciences, CIT Campus, Taramani, Chennai 600113, India
e-mail: arvind@imsc.res.in

M. Agrawal and V. Arvind (eds.), *Perspectives in Computational Complexity*,
Progress in Computer Science and Applied Logic 26, DOI: 10.1007/978-3-319-05446-9_1,

Although algorithms with linear or quadratic time bounds are desirable in practice, polynomial time computation is theoretically satisfactory for several reasons. First, it rules out exhaustive search solutions which are typically exponential time. Also, since polynomials are closed under composition, it makes polynomial-time bounded computation closed under procedure calls. Thus, a polynomial-time bounded program making calls to a library of polynomial-time subroutines is still polynomial time. Furthermore, as reasonable models of computation can simulate each other with at most a polynomial-time slowdown,[1] it makes the notion of polynomial-time solvability independent of the computation model.

The next big step was the pioneering research of Cook [Coo71], Levin [Lev73], and Karp [Kar72]. The class NP, consisting of decision problems that have nondeterministic polynomial-time decision procedures, was identified as the class that captures most natural optimization problems of interest. The notion of polynomial-time reductions was used to compare the relative difficulty of problems. Propositional formula satisfiability was shown to be NP-complete under polynomial-time reductions [Coo71, Lev73]. Then several decision problems, arising from optimization problems, were shown NP-complete [Kar72]. NP-complete problems are the hardest problems in the class NP as opposed to decision problems that are polynomial-time solvable.

Let Σ denote a fixed finite alphabet $|\Sigma| \geq 2$. Input instances of decision problems are encoded as finite strings over Σ. Thus, Σ^* comprises of all input instances for a decision problem and the "yes" instances $A \subseteq \Sigma^*$ form a language. Therefore, we can identify decision problems with languages and we use the terms interchangeably. As is customary in computation theory, we will use the standard Turing machine model as the model of computation (see e.g. [HU79]).

Definition 1.1 Let $A, B \subseteq \Sigma^*$ be languages.

1. Then A is said to be polynomial-time *many-one reducible* to B, denoted $A \leq_m^p B$, if there is a polynomial-time computable function f such that for all $x \in \Sigma^*$

$$x \in A \text{ if and only if } f(x) \in B.$$

2. More generally, A is said to be polynomial-time *Turing reducible* to B if there is a polynomial-time bounded oracle Turing machine M such that M^B accepts the language A.

A language $L \subseteq \Sigma^*$ is in the complexity class P if there is a polynomial-time bounded deterministic Turing machine (equivalently, a polynomial-time algorithm) for checking membership in L.

A language $L \subseteq \Sigma^*$ is in the complexity class NP if there is a language $A \in \mathrm{P}$ and a polynomial p such that for all $x \in \Sigma^*$

$$x \in L \text{ if and only if } \exists y \in \Sigma^{\leq p(|x|)} : \langle x, y \rangle \in A,$$

[1] This is a polynomial-time version of the Church-Turing thesis known as the feasibility thesis [vEB90].

where $\Sigma^{\le p(|x|)}$ denotes strings of length at most $p(|x|)$ over Σ. In other words, the complexity class NP consists of languages L such $x \in L$ has a polynomial-size certificate y of membership in L, and the certificate is polynomial-time verifiable by checking if $\langle x, y\rangle \in A$.

Exercise Show that the following decision problems are in the class NP:

1. Given an undirected graph G and a number k as input, decide if G has a clique of size at least k (known as the CLIQUE problem). The graph G is given by either its adjacency list or adjacency matrix.
2. Given a positive integer m (encoded in binary) decide if it is composite.
3. Given a system of linear equations $Ax = b$ over rationals, where the rational entries of the matrix A and column vector b are encoded in binary, decide if it has a solution.

We now formally define NP-completeness. A language $L \subseteq \Sigma^*$ is said to be NP-complete if L is in NP and each $L' \in$ NP is polynomial-time many-one reducible to L.

Since polynomial-time reducibility between languages is a transitive relation, once a language L' is shown NP-complete, it suffices to show that L' is polynomial-time many-one reducible to L in order to prove that the language L in NP is also NP-complete. A problem that can be directly shown NP-complete is the following:

$$K = \{\langle M, x, 1^t\rangle \mid M \text{ accepts } x \text{ in at most } t \text{ steps}\},$$

where M in the above definition denotes the Turing machine code (as a list of quintuples) of a *nondeterministic* Turing machine.

Exercise Show that K is NP-complete. (Hint: In order to show K is in NP you will need to use a suitable universal Turing machine).

But the first problem shown NP-complete by Cook and Levin was a natural problem, known as the satisfiability problem for propositional formulas, which opened the floodgate to NP-complete problems and the subject of computational complexity.

Theorem 1.2 (Cook-Levin theorem) [Coo71, Lev73] *The satisfiability problem for propositional formulas is NP-complete.*

A propositional formula F is in *conjunctive normal form* (CNF in short) if $F = C_1 \wedge C_2 \wedge \cdots \wedge C_m$ where each C_i is an OR of variables or their negations. The proof of the Cook-Levin theorem actually shows the stronger result that the satisfiability problem for propositional CNF formulas is NP-complete.

Clearly, NP-complete problems are the hardest problems in NP and all NP-complete problems are polynomial-time equivalent. That is to say, if a polynomial-time algorithm is discovered for any NP-complete problem then we have a polynomial-time algorithm for any problem in NP. Whether P equals NP or not is the central open problem in computational complexity.

Exercise

1. Show that the CLIQUE problem (defined in Exercise 2) isNP-complete by giving a reduction to it from propositional CNF formula satisfiability.
2. A *vertex cover* for an undirected graph G is a subset S of vertices such that for each edge (u, v) of G we have $\{u, v\} \cap S \neq \emptyset$. Given an undirected graph G and a number k as input the VC problem is to decide if G has a vertex cover of size at most k. Show that VC is NP-complete. The graph G is given by either its adjacency list or adjacency matrix.

1.2 Inside NP

Thus, within the class NP we have polynomial-time solvable problems on the one hand, which is the subclass P. At the other extreme, we have NP-complete problems, of which there are abundantly many [GJ79], because most natural optimization problems that arise in practice and we wish to solve efficiently turn out to be NP-complete. A natural question that arises is whether NP contains other problems. The answer to this question is given by Ladner's theorem which states that if P $\neq$ NP then there are problems in NP that are neither in P nor NP-complete. We discuss a proof attributed to Russell Impagliazzo [DF03].

Theorem 2.1 (Ladner's theorem)[Lad75] *If* P $\neq$ NP *there is a problem* $A \in$ NP *that is neither in* P *nor* NP*-complete.*

Proof By assumption the NP-complete problem SAT is not in P. Following standard notation [BDG88, Pap94], let DTIME$[g(n)]$ denote the class of languages accepted by deterministic Turing machines that halt in time bounded by $g(n)$ on inputs of length n. Now, if we knew that SAT $\notin$ DTIME$[g(n)]$ for some fixed superpolynomial function $g(n)$, for example $g(n) = n^{\log n}$, then we can easily find a language $A \in$ NP that is neither in P nor NP-complete. Indeed, let

$$A := \{x01^{|x|^{\log\log|x|}} \mid x \in \text{SAT}\}.$$

Clearly, $A \in$ NP because given a string of the form $x01^k$ we can guess and verify a satisfying assignment for the SAT instance x and check in polynomial time that the pad 1^k is of length $|x|^{\log\log|x|}$. Suppose A is NP-complete. Then SAT $\leq_m^p A$ via some polynomial-time computable reduction f. Notice that

$$x \in \text{SAT} \iff f(x) = x'01^{|x'|^{\log\log|x'|}} \in A \iff x' \in \text{SAT}.$$

Since f is polynomial-time computable, $|f(x)| \leq |x|^c$ for some constant c and hence $|x'| < |x|$ for all but finitely many instances $x \in$ SAT. Thus, it suffices to check if the smaller instance $x' \in$ SAT. Repeatedly applying this argument gives a polynomial-time procedure for SAT contradicting P $\neq$ NP. Hence A cannot be NP-complete.

On the other hand, we claim that $A \notin$ P. For, if A were in P that would give a $|x|^{O(\log\log|x|)}$ time algorithm to decide if $x \in$ SAT, contradicting the assumption that SAT is not in DTIME$[n^{\log n}]$.

Unfortunately, we can only assume that SAT is not in P and cannot make the stronger assumption that SAT is not in DTIME$[n^{\log n}]$. So we need to define the padded language A more carefully to make the above argument work. Let

$$A := \{x01^k \mid x \in \text{SAT}, \ \ k = f(|x|)\},$$

where the padding function $f(n)$ will be computable in time polynomial in n and constructed by diagonalization. Let $\{M_i\}_{i>0}$ be a recursive enumeration of polynomial-time clocked, deterministic Turing machines; more precisely, let M_i be clocked to run for $n^i + i$ steps on length n inputs. The function f is defined as follows:

1. $i := 1$.
2. For $n := 1$ to ∞ do
3. Let $f(n) = n^i$.
4. If there is an input x of length at most $\log n$ such that $M_i(x)$ accepts and $x \notin A$ or $M_i(x)$ rejects and $x \in A$ then $i := i + 1$.
5. endfor

Notice that at the nth iteration of the for-loop, in which $f(n)$ gets defined, the function $f(m)$ is already defined for $m < n$ and hence checking $x \in A$ for $\log n$ length inputs is well defined. Moreover, checking membership of $x \in A$ can be done in polynomial in n time since $|x| \leq \log n$. Hence, f is defined by the above procedure for all n and is computable in time polynomial in n. This defines the set A.

Suppose $A \in$ P. Then $A = L(M_i)$ for some machine M_i. By construction, there are constants n_0 and $k > i$ such that $f(n) = n^k$ for all $n > n_0$. That means, for all but finitely many input lengths, SAT is polynomial-time reducible to A by the map $x \mapsto x01^{|x|^k}$ which contradicts the assumption P $\neq$ NP. It follows that for each constant i we have $f(n) > n^i$ for all but finitely many n.

Suppose A is NP-complete and g is a polynomial-time reduction from SAT to A. We will give a polynomial-time algorithm for SAT contradicting the assumption. Since g is polynomial-time computable, we have $|g(x)| \leq |x|^c$ for some constant $c > 0$ and all but finitely many inputs x. If $g(x)$ is not of the form $y01^{f(|y|)}$ we can reject x. Also, if $g(\text{SAT})$ is finite then SAT is trivially in P by table look-up. Suppose $g(\text{SAT})$ is infinite. Then for all but finitely many $x \in$ SAT we have $g(x) = y01^{f(|y|)}$ where $f(|y|) > |y|^c$. The finitely many exceptions we can keep in a table. Thus, given a SAT instance x we first compute $g(x) = y01^{f(|y|)}$. If x is not in the look-up table, since $f(|y|) < |g(x)| \leq |x|^c$, it follows that $|y| < |x|$ and $x \in$ SAT if and only if $y \in$ SAT. We can now recurse on the instance y. Overall this gives a polynomial-time SAT algorithm contradicting the assumption. This concludes the proof. □

Exercise Suitably adapt the above proof to show for any language $A \notin$ P that there is a language $B \notin$ P such that $B \leq_m^p A$ but $A \not\leq_m^p B$.

1.2.1 The Class NP ∩ coNP

The class coNP consists of languages L such that $\Sigma^* \setminus L$ is in NP.

Indeed, for any class of languages $\mathcal{C}$ we can define co$\mathcal{C}$:

$$\text{co}\mathcal{C} = \{L \subseteq \Sigma^* \mid \Sigma^* \setminus L \in \mathcal{C}\}.$$

Remark 2.2 The class coNP consists of all languages whose complements are in NP. For example $\overline{\text{SAT}}$ is in coNP and, by virtue of SAT being NP-complete, $\overline{\text{SAT}}$ is coNP-complete under polynomial-time many-one reductions. The set TAUT consisting of all propositional tautologies is also coNP-complete (exercise: verify this). Whether NP equals coNP is a major open problem. We can view the NP versus coNP question from the logical perspective of propositional proof systems. For any language $L \in$ NP, by definition for each $x \in L$ there is a polynomial-length proof of membership that can be checked in polynomial time. This can be thought of as a "sound and complete proof system" for L. Thus, the question whether NP = coNP amounts to asking if there is a proof system for propositional tautologies in which all tautologies have polynomial length proofs. This leads to a study of propositional proof systems of different strengths with the aim of proving lower bounds for proof lengths in the proof systems. The article by Jacobo Torán in this volume presents aspects of this fascinating topic with pointers to current research and open problems.

We will now discuss decision problems that are in NP ∩ coNP. For any $L \in$ NP ∩ coNP there are languages A and B in P and a polynomial p such that for each $x \in \Sigma^*$

$$\begin{aligned} x \in L \quad &\leftrightarrow \quad \exists y \in \Sigma^{p(|x|)} \langle x, y\rangle \in A \\ &\leftrightarrow \quad \forall z \in \Sigma^{p(|x|)} \langle x, y\rangle \in B \end{aligned}$$

These are the so-called well-characterized problems. They are well characterized in the sense that membership of x in L can be characterized using an existentially quantified predicate, and can also be characterized using a universally quantified predicate. It is a remarkable phenomenon in complexity theory that the discovery of such characterization often precedes (even anticipates) the discovery of a polynomial-time algorithm for the problem. We discuss a few well-known examples.

The language PM = $\{G \mid G$ has a perfect matching$\}$ has a polynomial-time algorithm as shown in the already mentioned famous paper of Edmonds [Edm65]. However, in the 1940s Tutte, in his well-known theorem stated below, had anticipated this by "well-characterizing" the PM problem.

Theorem 2.3 (Tutte's 1-factor theorem) [Tut47] *An undirected graph G has a perfect matching if and only if for every subset S of the vertex set the number of odd-size components in the graph $G \setminus S$ is bounded by $|S|$.*

Similarly, Farkas' lemma [Sch98, Sect. 7.3] stated below "well-characterizes" linear programming which was shown to be in polynomial time many decades later by Kachiyan [Sch98, Chap. 13].

Theorem 2.4 (Farkas' Lemma) *The system of linear inequalities $Ax \leq b$ has a solution if and only if for all $y^T \geq 0$ if $y^T A = 0$ then $y^T b \geq 0$.*

Exercise Use the characterizations in Theorems 2.3 and 2.4 to show that the perfect matching problem and feasibility of linear inequalities problem are in NP ∩ coNP.

Exercise Show that NP = coNP if and only if some problem in NP ∩ coNP is NP-complete.

The above discussion might lead one to believe that perhaps NP ∩ coNP equals P. However, there are problems in NP ∩ coNP that have defied all attempted polynomial-time solutions. The decision version of the Integer factoring problem is a noteworthy example. Each positive integer n can be uniquely factorized as $n = p_1^{e_1} p_2^{e_2}, \ldots, p_k^{e_k}$ where $p_1 < p_2 < \cdots < p_k$ are distinct primes. Let $enc(n)$ denote an encoding in binary of this factorization. Consider the language

$$\text{FACT} = \{\langle n, i, b\rangle \mid \text{ the } i^{th} \text{ bit of } enc(n) \text{ is } b\}.$$

Exercise

1. Assuming primality testing is in P show that FACT is in NP ∩ coNP.
2. Show that the integer factoring problem can be solved in polynomial time with calls to a decision procedure for FACT.

1.3 The Berman-Hartmanis Conjecture

Polynomial-time reductions define a natural order $\leq_m^p$ on NP languages that raises some fundamental questions about the structure of NP languages.

Exercise Show that the order $\leq_m^p$ on NP languages is a binary relation that is reflexive and transitive but not symmetric.

In order to obtain a partial order from $\leq_m^p$ we define the equivalence relation $A \equiv_m^p B$ if and only if $A \leq_m^p B$ and $B \leq_m^p A$.

Exercise

1. Show that $\equiv_m^p$ is an equivalence relation on NP.
2. For $A \in$ NP let $[A]$ denote the equivalence class containing A for the equivalence relation $\equiv_m^p$. Show that $\leq_m^p$, suitably defined on the equivalences classes $[A]$ for $A \in$ NP, yields a partial order.

This partial order has its top element as [SAT], the equivalence class of all NP-complete languages. Its bottom element is P. Notice that class P contains, among all polynomial-time solvable decision problems, all finite languages. In contrast, assuming P $\neq$ NP, all sets in [SAT], being NP-complete, are infinite.

Definition 3.1 [BH77] Let $A, B \subseteq \Sigma^*$. A polynomial-time many-one reduction f from A to B is a *polynomial-time isomorphism* if $f : \Sigma^* \rightarrow \Sigma^*$ is a bijection and f^{-1} is also polynomial-time computable.

Berman and Hartmanis [BH77], in 1977 conjectured that all NP-complete sets are polynomial-time isomorphic to each other. Since they could show [BH77] many natural NP-complete problems to be isomorphic, empirically this appears plausible. Although the conjecture is not currently believed to be true, it gave impetus and direction to a lot of interesting complexity theory research. The article by Eric Allender in this volume surveys the interesting complexity theory research related to the Berman-Hartmanis conjecture over the last two decades. Our aim here is to provide some useful background.

A polynomial-time computable function $f : \Sigma^* \rightarrow \Sigma^*$ is *1-invertible* if f is injective and f^{-1} is also polynomial-time computable. More precisely, there is a polynomial-time algorithm that on input $y \in \Sigma^*$ computes $f^{-1}(y)$ if y is in the range of f and outputs $\perp$ otherwise.

Now, suppose A and B are NP-complete languages such that A is reducible to B via a 1-invertible function f and B is reducible to A via a 1-invertible function g, can we then conclude that A and B are polynomial-time isomorphic. The motivation for this approach is its analogy to the setting of the Schröder-Bernstein theorem in set theory which we recall with a quick proof in order to generalize it to the isomorphism setting.

Theorem 3.2 (Schröder-Bernstein theorem) *Let A and B be sets and $f : A \rightarrow B$ and $g : B \rightarrow A$ be injective functions. Then there is a bijection between A and B.*

Proof The proof idea involves examining the alternating *preimage sequence* of points $x, g^{-1}(x), f^{-}(g^{-1}(x)), \ldots$ for each $x \in A$. Since both f and g are injective, notice that for any $x \in A$ and $y \in B$ the preimages $g^{-1}(x)$ and $f^{-1}(y)$, if they exist, are unique. We can partition A into three parts A_1, A_2, and A_3. The part A_1 consists of $x \in A$ such that the preimage sequence is finite and ends in A. The part A_2 consists of $x \in A$ such that it has a finite preimage sequence ending in B, and A_3 consist of the elements $x \in A$ whose preimage sequence is infinite. Likewise, B is partitioned into three parts B_1, B_2, and B_3 of elements $y \in B$ whose pre-image sequence either ends in A or in B or is infinite, respectively. Define a function $h : A \rightarrow B$ as follows:

$$\begin{aligned} \forall\, x \in A_1 \quad & h(x) = f(x), \\ \forall\, x \in A_2 \quad & h(x) = g^{-1}(x), \\ \forall\, x \in A_1 \quad & h(x) = f(x). \end{aligned}$$

It is easy to see that h is a bijection. □

Exercise Verify that h defined in the proof is indeed a bijection from A to B.

A function $f : \Sigma^* \rightarrow \Sigma^*$ is called *length increasing* if $|f(x)| > |x|$ for all $x \in \Sigma^*$. In order to adapt the Schröder-Bernstein proof strategy for showing polynomial-time isomorphisms between NP-complete sets, it turns out that length-increasing 1-invertible reductions is a suitable notion.

Theorem 3.3 [BH77] *Let $A, B \subseteq \Sigma^*$ be two languages such that there are length-increasing 1-invertible reductions from A to B and from B to A. Then A and B are polynomial-time isomorphic.*

The proof of this theorem is in the following exercise.

Exercise

1. Suppose f and g are length-increasing 1-invertible reductions from A to B and from B to A respectively. Show that the partition corresponding to infinite preimage sequences is empty for both f and g.
2. Verify that the bijection $h : \Sigma^* \rightarrow \Sigma^*$ defined in the proof of Theorem 3.2 is a polynomial-time isomorphism between A and B.

The question is how do we get hold of length increasing 1-invertible reductions? Berman and Hartmanis [BH77] discovered another natural property that several NP-complete languages are endowed with. A language $A \subseteq \Sigma^*$ is said to be *paddable* if there is a 1-invertible reduction *pad* from $A \times \Sigma^*$ to A. It turns out that many natural NP-complete problems are paddable.

Exercise Show that SAT, 3-SAT, CLIQUE, VC are all paddable languages.

Now, it turns out that paddable NP-complete languages are all isomorphic [BH77]. Since many NP-complete problems are paddable, it lead Berman and Hartmanis to make their conjecture.

Exercise

1. If $A \leq_m^p B$ and B is paddable then show that A is polynomial-time reducible to B via a 1-invertible length increasing function.
2. Conclude that if A and B are paddable NP-complete languages then they are polynomial-time isomorphic.

1.4 Are There Sparse NP-Complete Sets?

Let $A \subseteq \Sigma^*$ be any language. The *density* of A at length n is defined to be the number $|A^{=n}|$ of length n strings in A. The language A is exponentially dense if $|A^{=n}|$ is at least 2^{n^ϵ} for some constant $\epsilon > 0$. Natural NP-complete problems have exponential density.

Exercise Show that SAT, CLIQUE, VC have exponential density.

A consequence of the Berman-Hartmanis conjecture is that all NP-complete languages have exponential density. Can we show that languages with subexponential density cannot be NP-hard?

A language $A \subseteq \Sigma^*$ is said to be *sparse* if there is a polynomial $p(n)$ such that $|A^{=n}| \leq p(n)$ for all n.

Theorem 4.1 (Mahaney) [Mah82] *No sparse language is* NP-*hard unless* P = NP.

Proof We present a more recent proof [Agr11]. Suppose SAT $\leq_m^p A$ for some arbitrary sparse language A via a polynomial-time reduction f. For some polynomial $p(n)$ we have $|f(x)| \leq p(|x|)$.

We give a polynomial-time algorithm for SAT. Given a formula F of size s in boolean variables $x_1, x_2, \ldots, x_n$ as input, the algorithm proceeds in stages $0 \leq i \leq n$. At stage i, the algorithm maintains a list of formulas $\{F_a \mid a \in I\}$ such that each $F_a, a \in I$ is obtained from F by the truth assignment a to the first i variables $x_1, x_2, \ldots, x_i$ with the property that

$$F \in \text{SAT if and only if } F_a \in \text{SAT for some } a \in I. \tag{1.1}$$

If $|I| > q(s) + 1$, for a suitable polynomial $q(s)$ which will be defined in the course of the proof, then the algorithm applies a pruning operation to discard some a from the list I such that Property (1.1) holds for $I \setminus \{a\}$ as well. We now describe the pruning operation:

Consider all pairs of disjunctions $F_{ab} = F_a \vee F_b$ for $a, b \in I$. If s is the size of formula F then clearly $2s$ bounds the size of F_{ab} for all $a, b \in I$. Fix an $a \in I$ and consider $f(F_{ab})$ for all $b \in I \setminus \{a\}$. If $F_a \in$ SAT then clearly $F_{ab} \in$ SAT for all $b \in I \setminus \{a\}$. Hence, $\mathbb{F}_a \in$ SAT implies $f(F_{ab}) \in A^{\leq p(2s)}$. Since A is sparse we know that $|A^{\leq p(2s)}| \leq q(s)$ for some polynomial q. The algorithm computes $f(F_{ab})$ for all $b \in I \setminus \{a\}$ and does the following:

1. If $f(F_{ab})$ are all distinct for $b \in A \setminus \{a\}$ then F_a cannot be in SAT and a can be discarded from I.
2. Otherwise, for some $b \neq c \in I \setminus \{a\}$ we have $f(F_{ab}) = f(F_{ac})$. The algorithm can discard either b or c from I.

After pruning the list to get $\{F_a \mid a \in I\}$ such that $|I| \leq q(s) + 1$ the algorithm proceeds to the next stage where it first doubles the list of formulas by setting x_{i+1} to 0 and 1 to get $\{F_b \mid b \in J\}$ where J consists of all extensions of assignments in I obtained by setting x_{i+1} to 0 and 1.

Continuing thus, at the nth stage the algorithm accepts SAT if the set I at that stage includes a satisfying assignment. □

Let $A \in$ NP. By definition there are a polynomial-time computable language $B \subseteq \Sigma^* \times \Sigma^*$ and a polynomial bound $p(n)$ such that $x \in A$ if and only if

$$\exists y \in \Sigma^{p(|x|)} \langle x, y \rangle \in B.$$

Define the language $\text{pre}(A) = \{\langle x, w\rangle \mid \exists u \in \Sigma^{p(|x|)-|w|} : \langle x, wu\rangle \in B\}$.

Exercise Show that pre(A) is in NP and $A \leq_m^p \text{pre}(A)$.

A *tally* language is a subset of 0^*. The next exercise is about tally languages and is analogous to Theorem 4.1.

Exercise For a language $A \in \text{NP}$ if pre(A) is polynomial-time reducible to a tally language then show that pre(A), and hence A, is in P.

1.4.1 Subexponentially Dense Languages

A subset $S \subseteq \Sigma^*$ is of *subexponential density* if for every $\epsilon > 0$ there is an $n_0 \in \mathbb{N}$ such that for all $n > n_0$ we have $|S^{=n}| \leq 2^{n^\epsilon}$. A natural question is whether Mahaney's theorem can be generalized to sets of subexponential density.

The class of languages $\text{SUBEXP} = \cap_{\epsilon>0}\text{DTIME}[2^{n^\epsilon}]$ consists of languages that have subexponential time decision procedures.

In the proof of Theorem 4.1 we gave a polynomial-time algorithm for SAT assuming that it is reducible to some sparse language. Modify the proof to show the following.

Exercise If SAT is polynomial-time many-one reducible to a language of subexponential density then show that NP $\subseteq$ SUBEXP.

The inclusion NP $\subseteq$ SUBEXP is believed unlikely. For instance, it would imply a $2^{o(n)}$ time algorithm for the satisfiability of 3CNF formulas, where n is the number of boolean variables in the input formula. This would, in turn, imply similar subexponential algorithms for certain other NP-complete problems [IPZ01]. For a complexity theory tailored to the problem of designing faster exponential-time algorithms for NP-complete problems see [IPZ01] and related papers.

For the rest of this section we briefly discuss another unlikely complexity-theoretic consequence, assuming SAT is reducible to a set of subexponential density. In the process we will introduce some more basic complexity theory.

We start with the definition of the *polynomial-time hierarchy*. A language L is in the class Σ_k^p if there are a polynomial $p(n)$ and language $A \in \text{P}$ such that

$$L = \{x \mid \exists y_1 \forall y_2 \cdots Qy_k : |y_i| \leq p(|x|) \text{ for all } i \\ \text{and } \langle x, y_1, y_2, \ldots, y_k\rangle \in A\},$$

where the quantifier "Q" is "$\exists$" if k is odd and "$\forall$" if k is even. The class Π_k^p consists of all languages L such that $\Sigma^* \setminus L$ is in Σ_k^p. The following observations are immediate from the definition:

$$\Sigma_0^p = \Pi_0^p = \mathrm{P},$$
$$\Sigma_1^p = \mathrm{NP},\ \Pi_1^p = \mathrm{coNP},$$
$$\Sigma_k^p \subseteq \Pi_{k+1}^p \text{ and } \Pi_k^p \subseteq \Sigma_{k+1}^p.$$

The entire polynomial-time hierarchy is defined as the union $\mathrm{PH} = \cup_{k \geq 0} \Sigma_k^p$.

Exercise Show that $\Sigma_k^p = \Pi_k^p$ implies $\mathrm{PH} = \Sigma_k^p$.

We now introduce the *nonuniform* analog of P and NP. A polynomially bounded *advice* function is any function $a : \mathbb{N} \to \Sigma^*$, mapping positive integers to strings, such that $|a(n)| \leq p(n)$ for some polynomial p and all $n \in \mathbb{N}$

Let $\mathcal{C}$ be any class of languages. A language L is in the class $\mathcal{C}$/poly if there are a polynomially bounded advice function $a(n)$ and a language $A \in \mathcal{C}$ such that for all $n \in \mathbb{N}$ and $\mathbf{x} \in \Sigma^*$:

$$x \in L \text{ if and only if } \langle x, a(|x|)\rangle \in A.$$

Notice that the function $a(n)$ is arbitrary and could even be noncomputable. On the other hand, the advice string $a(|x|)$ is polynomially bounded and is the same for all length n inputs.

The classes P/poly and NP/poly are the usual nonuniform analogues of P and NP respectively. These complexity classes have alternative definitions in terms of boolean circuits which we now introduce in a small digression.

An n-variate *boolean function* is a function $f : \{0, 1\}^n \to \{0, 1\}$. Boolean circuits are a natural computational model for computing boolean functions. A *boolean circuit* C is a directed acyclic graph whose vertices of indegree 0 are labeled either by a boolean constant (either 0 or 1) or by one of the n input boolean variables $x_1, x_2, \ldots, x_n$. Every other vertex of the graph has indegree either one or two. The vertices of indegree one are labeled as NOT gates and each vertex of indegree two is labeled either as an AND gate or an OR gate. A special vertex of the circuit is designated as the *output* gate. The *size* of the circuit C, denoted $\mathrm{size}(C)$ is the number of gates in C.

The circuit is evaluated in any topologically sorted order of the gates, following the usual semantics of NOT, AND, and OR at each gate. Each gate computes a boolean function and the n-variate boolean function computed at the output gate is defined as the function computed by the circuit.

Let $\Sigma = \{0, 1\}$ and $L \subseteq \Sigma^*$ be a language. Consider a family of circuits $\{C_n\}_{n>0}$, where C_n computes an n-variate boolean function for each n. We say that the circuit family $\{C_n\}_{n>0}$ computes the language L if for each n we have

$$L^{=n} = \{x \in \Sigma^n \mid C_n(x) = 1\}.$$

We call $\{C_n\}_{n>0}$ a *polynomial-size* circuit family if $\mathrm{size}(C_n) \leq p(n)$ for some polynomial $p(n)$ and all n.

Exercise Show that a language L is in P/poly if and only if there is a polynomial-size circuit family that computes L. Similarly, characterize NP/poly using circuit families.

As there is no bound on the computational complexity of advice functions, the class P/poly contains even noncomputable languages (exercise: prove this). Are there NP-complete languages in P/poly? This could be useful for efficiently solving NP-complete languages for fixed input lengths. The advice function, computed as a preprocessing step, serves as a small table that can be looked up to efficiently solve instances of that fixed length. However, this seems to be unlikely because it implies that the polynomial hierarchy will collapse to the second level.

Theorem 4.2

1. *If an* NP-*complete language is in* P/poly *then* $\mathrm{PH} = \Sigma_2^p$ [KL80].
2. *If there is a* coNP-*complete language in* NP/poly *then* $\mathrm{PH} = \Sigma_3^p$ [Yap83].

Coming back to sets of subexponential density, it is shown in [BH08], using a nice counting argument from [FS08], that if SAT is polynomial-time many-one reducible to a set S of subexponential density then $\mathrm{coNP} \subset \mathrm{NP/poly}$ which in turn would imply $\mathrm{PH} = \Sigma_3^p$ by the above theorem.

1.5 Nondeterministic Logspace

We now turn to logarithmic space bounded complexity classes. A deterministic logarithmic space-bounded Turing machine is a deterministic Turing machine that has a read-only input tape and one or more worktapes, where n denotes the input size and the workspace on each worktape is bounded by $O(\log n)$. By the tape-compression theorem the constant factor in the space bound is not important. It can be reduced by suitably increasing the tape alphabet size. Furthermore, multiple worktapes can be replaced by a single worktape in space-bounded computation. We denote by DSPACE[$\log n$] the class of decision problems that can be solved in deterministic logspace bounded Turing machines. Likewise, NSPACE[$\log n$] denotes the class of languages accepted by *nondeterministic* logspace bounded Turing machines, where acceptance is defined as usual: the machine accepts an input if there is some accepting computation path for the machine on that input. The definitions and results go back to the seminal work of Hartmanis and Stearns [HS65] and are well treated in the classic textbook [HU79, Chaps. 12 and 13].

The class DSPACE[$\log n$] is denoted as L. Likewise, the nondeterministic class NSPACE[$\log n$] is denoted as NL.

Proposition 5.1 $\mathrm{L} \subseteq \mathrm{NL} \subseteq \mathrm{P}$.

Proof The first containment is obvious. Consider the nondeterministic computation of an NL machine M on an input x of length n. Notice that the configurations of M

have $O(\log n)$ size descriptions. More precisely, a configuration is described by the current state, the head position on the input tape (which requires $\log n$ bits), the head position on the work tape (which requires $\log \log n$ bits), along with the contents of the work tape which is $O(\log n)$ bits. Furthermore, given two configurations I and I' we can easily check from the description of the machine M if M can go from I to I' in one step. Thus, in polynomial in n time we can define a directed graph whose nodes are the configurations and edges are (I, I') if M can move from I to I' in one step. Clearly, M accepts x if and only if there is a directed path from the initial configuration to an accepting configuration. This can be solved in polynomial time by a DFS-based algorithm. Thus, there is a polynomial time algorithm to determine if M accepts x. □

It is an open problem whether L equals NL. The L versus NL problem makes an interesting study in contrast with the P versus NP problem as the results explained in this section will show.

In order to talk about NL-complete problems, we can define logspace many-one reductions: $f : \Sigma^* \rightarrow \Sigma^*$ is logspace computable if there is a deterministic logspace-bounded Turing machine that on input $\langle x, i\rangle$ computes the ith bit of $f(x)$ (if $i > |f(x)|$ the machine indicates that the ith bit is undefined). It is easy to check that the composition of logspace computable functions is logspace computable. A language $A \subseteq \Sigma^*$ is NL-complete if $A \in$ NL and for every language $B \in$ NL, B is logspace many-one reducible to A.

Let REACH $= \{\langle G, s, t\rangle \mid G$ is a directed graph with a directed path from s to $t\}$.

Exercise Show that *REACH* is NL-complete (Hint: use the configuration graph from the proof of Proposition 5.1 for the reduction).

Unlike NP and coNP which are believed to be different, the classes NL and coNL are equal. This result was shown independently by Immerman and Szelepcenyi [Imm88, Sze88] in 1987. The Immerman-Szelepcenyi theorem was an important breakthrough in our understanding of nondeterministic complexity classes.

Theorem 5.2 (Immerman-Szelepcenyi) [Imm88, Sze88] *For any space-constructible function $s(n) \geq \log n$ the classes* NSPACE[$s(n)$] *and co*NSPACE[$s(n)$] *are equal. In particular,* NL = coNL.

A formal detailed proof can be found in a textbook [Pap94, Theorem 7.6]. The following exercise outlines the proof with hints for the ambitious reader. Let $\overline{\text{REACH}} = \{\langle G, s, t\rangle \mid G$ is a digraph with no s-to-t directed path$\}$.

Exercise

1. Show that in order to prove NL = coNL it suffices to give an NL algorithm for the problem $\overline{\text{REACH}}$.
2. Let (G, s, t) be an input instance of $\overline{\text{REACH}}$. Suppose the number N of vertices in G that are reachable from s is given to the algorithm as additional input. Then design an NL algorithm that accepts (G, s, t) precisely when there is no directed s-to-t path in G.

3. Let (G, s, t) be an input instance of $\overline{\text{REACH}}$. Given as additional input the number N_i of vertices in G that are reachable from s by directed paths of length at most i. Design an NL algorithm that on each accepting computation path halts with the number N_{i+1} as the contents of the worktape.
4. Put these parts together to obtain an NL algorithm for $\overline{\text{REACH}}$.

In the 1970s Savitch's theorem [Pap94, Theorem 7.5] had already shown that nondeterministic space is quite different from nondeterministic time.

Theorem 5.3 (Savitch) *For any space-constructible function* $s(n) \geq \log n$ *we have* $\text{NSPACE}[s(n)] \subseteq \text{DSPACE}[s^2(n)]$.

Exercise Show that REACH is in $\text{DSPACE}[\log^2 n]$ using the following divide and conquer strategy: there is an s-to-t directed path in G of length k if and only if for some vertex r of G there is a directed s-to-r path of length $\lceil k/2 \rceil$ and a directed r-to-t path of length $\lfloor k/2 \rfloor$ in G.

The above results definitely encourage the optimist to believe that it is possible to design a deterministic logspace algorithm for REACH and hence show NL = L. Vinodchandran's article surveys recent progress on space-bounded algorithms for REACH. A big breakthrough was obtained by Reingold [Rei08] in 2004 by showing that *undirected* graph reachability is in L. We discuss some aspects of Reingold's result in the Sect. 1.6.

1.6 Undirected Graph Reachability

In this section we develop the background and outline some of the ideas that are involved in the proof of Reingold's theorem [Rei08] which states that undirected graph reachability is in L.

The corresponding language we denote as UREACH: UREACH $= \{\langle G, s, t \rangle \mid G$ is an undirected graph containing a path from s to $t\}$. Clearly, UREACH is a special case of REACH and hence is in NL.

First, we note that without loss of generality we can assume the undirected graph G is 3-regular. That is to say, every vertex in the graph G has degree exactly 3. More precisely, given an instance $\langle G, s, t \rangle$ of UREACH, there is a deterministic logspace computation that we can apply to transform it into another instance $\langle G', s', t' \rangle$ in which G' is 3-regular such that $\langle G', s', t' \rangle \in$ UREACH if and only if $\langle G, s, t \rangle \in$ UREACH. The main idea involved in this transformation is that a vertex v in G of degree $d > 3$ can be replaced by a cycle of length d consisting of vertices $v_1, v_2, \ldots, v_d$, and the ith neighbor of v (say, in lexicographic order) made adjacent to v_i. This transformation can be carried out in logspace and in the transformed graph there is a path between s and t if and only if there is one in G. Repeated application of this step takes care of all vertices in G of degree more than 3. Vertices of degree 1 or 2 can be contracted (taking appropriate care if s or t is encountered in the contraction process).

Exercise Verify the details of the logspace transformation sketched above.

1.6.1 Reachability in Graphs of Logarithmic Diameter

Let G be a connected undirected graph. For every pair of vertices u, v of G let $d_G(u, v)$ denote the length of the shortest u to v path in G. The *diameter* of G is the maximum distance $\max_{u,v} d_G(u, v)$ between vertices in G. Suppose $c, d > 0$ are constants and $\langle G, s, t\rangle$ is an input instance of UREACH such that the maximum degree of G is d and the diameter of G is bounded by $c \log n$. There is a simple deterministic logspace algorithm to check if there is an s to t path in G. Since each vertex u of G has at most d neighbors, its neighborhood $N(u)$ can be indexed by $\{1, 2, \ldots, d\}$, say in increasing order of vertex names. Furthermore, we are looking for a path of length at most $c \log n$ from s to t. We can encode all paths of length $c \log n$ starting from s by sequences $d_1, d_2, \ldots, d_{c \log n}$, where $1 \leq d_i \leq d$ for each i. This sequence can be written down on the worktape using at most $O(c \log d \log n)$ space and the algorithm can cycle through all such sequences one by one in lexicographic order. For a sequence $d_1, d_2, \ldots, d_{c \log n}$ the path in G defined by it is $s = u_0, u_1, \ldots, u_{c \log n}$, where u_i is the d_ith neighbor of u_{i-1} for each i. The algorithm can generate the vertices of this path one by one. In order to generate u_i it only needs to store i and u_{i-1} which is $O(\log n)$ space. The algorithm accepts if $u_i = t$ for some i in some sequence.

The diameter of n-vertex graphs can be $\Omega(n)$ in general. However, it turns out, as shown by Reingold [Rei08] that there is a deterministic logspace computation that transforms an instance $\langle G, s, t\rangle$ of UREACH into an instance $\langle G', s', t'\rangle$ such that

(i) s' and t' are in the same connected component of G' if and only if s and t are connected in G.
(ii) G' is a d-regular graph for some constant d and has diameter bounded by $c \log n$ for some constant c.

Reingold's construction of G' [Rei08] is based on properties of expander graphs and the explicit construction of expanders. We will not describe the details of his algorithm. Instead, we will present a randomized logspace algorithm for UREACH and in the process explain some basic properties of expander graphs.

1.6.2 A Randomized Logspace Algorithm for UREACH

A randomized logspace bounded Turing machine is a Turing machine M that has, in addition to a read-only input tape and a logspace bounded worktape, one-way access to a random tape on which is written the outcome of a sequence of unbiased and independent random bits and in each time instant the random tape head moves one step to the right reading the next random bit.

Definition 6.1 A language $L \subseteq \Sigma^*$ is said to be in the class RL if there is a randomized logspace bounded Turing machine that runs in polynomial time such that

$$x \in L \longrightarrow \text{Prob}[M(x) \text{ accepts}] \geq 1/2$$
$$x \notin L \longrightarrow \text{Prob}[M(x) \text{ accepts}] = 0$$

The class RL stands for randomized logspace with 1-sided error. We can analogously define the more general class BPL which allows two-sided error.

Exercise L $\subseteq$ RL $\subseteq$ NL.

An interesting point in Definition 6.1 is that we need to insist on a polynomial time bound for the randomized machine in order to get a subclass of NL. This is in contrast to the definitions of L and NL where the polynomial time bounds is enforced by detecting repetition of configuration. In the case of randomized logspace computation, allowing long computation paths can influence acceptance probabilities in nontrivial ways. The following exercise explains this aspect.

Exercise Show that randomized logspace machines with one-sided error (and error probability bounded by $1/2$ as in Definition 6.1) that are allowed to run for exponential time accept precisely the class NL.

Our aim is to show that UREACH is in RL. As explained earlier it suffices to consider undirected graphs that are d-regular for some constant d.

Let $G = (V, E)$ be a d-regular graph on n vertices and let A denote its *normalized adjacency matrix*: $A_{ij} = 1/d$ if $ij \in E$ and $A_{ij} = 0$ otherwise.

Since G is undirected, note that A is a real symmetric matrix. An eigenvalue of A is a complex number λ such that $Ax = \lambda x$ for some nonzero vector x. Since A is a symmetric matrix, by standard linear algebra [HK71] it follows that A has all real eigenvalues. Let $\lambda_1 \geq \lambda_2 \geq \cdots \geq \lambda_n$ denote its n eigenvalues. It is easy to see that $-1 \leq \lambda_i \leq 1$ for all i.

The largest eigenvalue of A is 1 and an eigenvector corresponding to eigenvalue 1 is the all 1's vector. We can also think of the uniform distribution $(1/n, 1/n, \ldots, 1/n)^T$ on the vertex set as the eigenvector corresponding to 1.

Exercise

1. Prove that G has k connected components if and only if $1 = \lambda_1 = \lambda_2 = \cdots = \lambda_k > \lambda_{k+1} \geq \cdots \geq \lambda_n$, where $\lambda_1 \geq \lambda_2 \geq \cdots \geq \lambda_n$ denote the n eigenvalues of the normalized adjacency matrix A.
2. Prove that G is bipartite if and only if -1 is an eigenvalue of A.

As a consequence of the statements in the above exercise, if G is connected then $\lambda_2 < 1 = \lambda_1$. Indeed, when G is connected it is possible to give a better upper bound than 1 for λ_2. We will explain this bound and also use it to design a randomized logspace algorithm for UREACH.

Suppose λ and λ' are distinct eigenvalues of A, and x and x' be corresponding eigenvectors. Then we claim x and x' are mutually orthogonal. That is to say, $x^T x' = 0$. The claim holds because $x^T A x' = \lambda x^T x'$ as well as $x^T A x' = \lambda' x^T x'$ and $\lambda \neq \lambda'$.

By the Courant-Fisher-Weyl minmax theorem [CH53, Chap. 1, Sect. 4], we can write $\lambda_2 = \max_{x \perp \mathbf{1}, ||x||=1} x^T A x$, where $\mathbf{1}$ stands for the all 1's vector. The condition $x \perp \mathbf{1}$ is equivalent to $\sum_i x_i = 0$. Since $x \neq 0$ there are some negative and some positive entries in x. Let $x_s < 0$ be the smallest entry of x and $x_t > 0$ be the largest. Since G is connected, there is a path P of length at most $n-1$ from s to t in G. Simplifying $\lambda_2 = x^T A x$ we obtain

$$\begin{aligned}\lambda_2 &= \frac{1}{d}\sum_{ij \in E} 2x_i x_j \\ &= 1 - \frac{1}{d}\sum_{ij \in E}(x_i - x_j)^2 \\ &\leq 1 - \frac{1}{d}\sum_{ij \in P}(x_i - x_j)^2.\end{aligned}$$

Now, by the triangle inequality we have $|x_t - x_s| \leq \sum_{ij \in P} |x_i - x_j|$, and by the Cauchy-Schwartz inequality we have $\sum_{ij \in P} |x_i - x_j| \leq \sqrt{n}\sqrt{\sum_{ij \in P} |x_i - x_j|^2}$. Squaring both sides, we obtain $\sum_{ij \in P} |x_i - x_j|^2 \geq (x_t - x_s)^2/n$, and combined with the above inequality for λ_2 this yields

$$\lambda_2 \leq 1 - \frac{1}{dn}(x_t - x_s)^2.$$

Now, since $\sum_i x_i^2 = 1$, it follows that $x_t - x_s \geq 2/\sqrt{n}$. Putting it together we have shown the following:

Lemma 6.2 *For a connected graph G, $\lambda_2(G) \leq 1 - \frac{4}{dn^2}$.*

We now turn to the randomized logspace algorithm for UREACH [AKL+79]. Let $\langle G, s, t\rangle$ be an instance of UREACH where G is a d-regular graph. The algorithm does a simple random walk on G starting at the vertex s. At the ith step of the random walk if the current vertex is u, then in the $(i+1)^{st}$ step the walk picks one of the d neighbors of u uniformly at random and moves to it. Because of the bound on $\lambda_2(G)$, we can show that if t lies in the same connected component as s then with high probability the random walk will visit t within polynomially many steps. Furthermore, this algorithm can be implemented by a randomized logspace Turing machine in the following manner. The current vertex u is stored in the worktape. The next $\log d$ random bits that pick a neighbor of u are read from the random tape by the tapehead moving right and reading a bit for $\log d$ steps.

Theorem 6.3 [AKL+79] UREACH *is in* RL.

Proof Let $\langle G, s, t\rangle$ be an instance of UREACH. We can assume that G is an n-vertex, d-regular graph for some $d \geq 3$. If there is no path from s to t then the random walk starting at s will never visit t. When t is reachable from s, we need to analyze the random walk and lower bound the probability of visiting t in a given number of steps.

At the ith step of the random walk, the state can be described by a probability distribution vector $\mathbf{x} = (x_1, x_2, \ldots, x_n)^T$, where $x_k \geq 0$ for each k and $\sum_k x_k = 1$. Each x_k in this vector denotes the probability of the random walk being at vertex k at the ith step. We note that any probability vector x can be written as $\mathbf{x} = \mathbf{u} + \mathbf{u}^{\perp}$, where $\mathbf{u} = (1/n, 1/n, \ldots, 1/n)^T$ is the uniform distribution and their difference $\mathbf{u}^{\perp} = \mathbf{x} - \mathbf{u}$ is clearly orthogonal to $\mathbf{u}$ because the inner product $(\mathbf{u}, \mathbf{x} - \mathbf{u}) = 0$. The normalized adjacency matrix A is the stochastic matrix that governs this random walk. If the probability distribution vector at the ith step is $\mathbf{x}$ then at the $(i+1)^{st}$ step the distribution vector is $A\mathbf{x}$.

Now, the initial distribution vector is $\mathbf{x} = (e_1, e_2, \ldots, e_n)^T$, where $e_s = 1$ and $e_j = 0$ for all $j \neq s$. Thus, after i steps of the random walk the probability distribution is $A^i\mathbf{x}$. Therefore, writing $\mathbf{x} = \mathbf{u} + \mathbf{u}^{\perp}$ we get

$$A^i\mathbf{x} = \mathbf{u} + A^i\mathbf{u}^{\perp}.$$

Since $\lambda_2 = \max_{x \perp 1, ||x||=1} x^T A x$, it follows that

$$||A^i\mathbf{u}^{\perp}||^2 = |(u^{\perp})^T A^{2i} u^{\perp})| \leq |\lambda_2(G)|^{2i} ||u^{\perp}||^2.$$

We can bound $||u^{\perp}||^2 \leq 2$, since $||u^{\perp}|| \leq ||\mathbf{x}|| + ||\mathbf{u}||$, $||\mathbf{x}|| = 1$ and $||\mathbf{u}|| = 1/\sqrt{n}$. Letting $i = 2dn^3$ in Lemma 6.2 gives the bound $|\lambda_2(G)|^{4dn^3} \leq e^{-n}$. It follows that

$$||A^i\mathbf{x} - \mathbf{u}||^2 = ||A^i u^{\perp})||^2 \leq e^{-n}.$$

Since each entry of $\mathbf{u}$ is $1/n$, the above bound implies that each entry of $A^i\mathbf{x}$ is at least $1/2n$. We can conclude that if t reachable from s then with probability at least $1/2n$ the random walk starting at s visits t within $i = 2dn^3$ steps. Consequently, the probability that the random walk visits t within $4dn^4$ steps is at least $1-(1-1/2n)^{2n} > 1/2$ for sufficiently large n. This bound on the success probability proves that the random walk algorithm is indeed an RL algorithm. □

1.7 Logspace Counting Classes

We now briefly introduce logspace counting classes. These complexity classes provide a tight classification of many natural computational problems. Formal definitions of these classes and their basic properties can be found in [BDHM92, AO96]. The central complexity class here is denoted GapL and captures the complexity of computing the integer determinant.

Definition 7.1 GapL is the class of functions $f : \Sigma^* \to \mathbb{Z}$, for which there is a logspace bounded nondeterministic Turing machine M such that on input $x \in \Sigma^*$, we have $f(x) = \mathrm{acc}_M(x) - \mathrm{rej}_M(x)$, where $\mathrm{acc}_M(x)$ and $\mathrm{rej}_M(x)$ denote the number of accepting paths of M and the number of rejecting paths of M, respectively, on input x.

GapL is a function class and we have an appropriate notion of logspace computable reductions for this class. Given two functions $f, g : \Sigma^* \to \mathbb{Z}$ we say that f is logspace many-one reducible to g (as usual, denoted $f \leq_m^L g$) if there are two logspace computable functions e and v such that for every $x \in \Sigma^*$

$$f(x) = v(g(e(x))).$$

Notice that if g is computable in logspace then f is also logspace computable. A function $g \in$ GapL is GapL-*complete* under logspace reductions if $f \leq_m^L g$ for every $f \in$ GapL. A fundamental result due Toda [Tod91] is that the integer determinant is GapL-complete. The result has an elegant proof by Mahajan and Vinay [MV97]. A problem analogous to REACH that is easily shown to be GapL-complete is in the following exercise.

Exercise Given a directed acyclic graph G with a source vertex s and two sink vertices t_1 and t_2, let $g(G, s, t_1, t_2)$ denote the difference of the number of directed paths from s to t_1 from the number of directed paths from s to t_2. Show that this function is GapL-complete.

The class GapL is analogous to the complexity class #P [Val79] introduced by Valiant to study the complexity of computing the integer permanent. A function $f : \Sigma^* \to \mathbb{N}$ is said to be in #P if there is an NP machine M such that $f(x) = acc_M(x)$ for all $x \in \Sigma^*$. A celebrated result due to Valiant [Val79] is the #P-completeness of the integer permanent (for an appropriate notion of reductions). The reader can find details of the proof in the textbook [AB09].

Remark 7.2 As an aside, comparing the computational complexity of the determinant and the permanent is a fascinating topic that is central to the area of arithmetic circuits. This book includes several chapters on the topic of arithmetic circuits. Meena Mahajan's article presents a detailed survey on Valiant's algebraic complexity classes.

The complexity of computing the determinant of matrices over the field $\mathbb{F}_2$ is captured by the language class $\oplus$L defined below.

Definition 7.3 A language $L \subseteq \Sigma^*$ is in the complexity class $\oplus$L (called "parity-L") if there is a logspace bounded nondeterministic Turing machine M such that $x \in L \iff acc_M(x)$ is odd.

Exercise Show that the following language is $\oplus$L complete under logspace many-one reductions:

$$\text{ODD-REACH} = \{\langle G, s, t\rangle \mid G \text{ is a DAG such that there is an odd number of directed paths from } s \text{ to } t\}.$$

Exercise Assuming that the integer determinant is GapL-complete show that

$$\text{NONSINGULAR} = \{M \mid M \text{ is a square integer matrix such that } \det(M) \equiv 1 (\text{mod } 2)\}$$

is complete for ⊕L under logspace many-one reductions.

The last complexity class we will introduce in this chapter is *unambiguous logspace* denoted UL. A logspace bounded nondeterministic Turing machine M is said to be *unambiguous* if on every input $x \in \Sigma^*$ the machine M has *at most* one accepting path. A language L is in the complexity class UL if there is an unambiguous logspace-bounded nondeterministic Turing machine M that accepts the language L.

A surprising inclusion of complexity classes shown in [RA00] is that NL $\subseteq$ UL/poly.[2] Whether NL equals UL or not is an intriguing open problem and is discussed in Vinodchandran's article in this volume.

Acknowledgments I am grateful to Somenath Biswas for his comments and suggestions.

References

[AB09] S. Arora, B. Barak, *Computational Complexity, a Modern Approach* (Cambridge University Press, Cambridge, 2009)

[Agr11] M. Agrawal, *The Isomorphism Conjecture for NP* (World Scientific Press, Singapore, 2011)

[AKL+79] R. Aleliunas, R.M. Karp, R. J. Lipton, L. Lovász, C. Rackoff, Random walks, universal traversal sequences, and the complexity of maze problems, in *FOCS* (1979), pp. 218–223. IEEE

[AO96] E. Allender, M. Ogihara, Relationships among PL, # and the determinant. RAIRO Theor. Inf. Appl. **30**(1), 1–21 (1996)

[BDG88] J.L. Balcázar, J. Diaz, J. Gabarro, *Structural Complexity I and II* (Springer, Heidelberg, 1988)

[BDHM92] G. Buntrock, C. Damm, U. Hertrampf, C. Meinel, Structure and importance of logspace-MOD class. Theory Comput. Syst. **25**(3), 223–237 (1992)

[BH77] L. Berman, J. Hartmanis, On isomorphisms and density of np and other complete sets. SIAM J. Comput. **6**(2), 305–322 (1977)

[BH08] H. Buhrman, J. Hitchcock, Np-hard sets are exponentially dense unless conp⊆p/poly, in *IEEE Conference on Computational Complexity* (2008), pp. 1–7

[CH53] R. Courant, D. Hilbert, *Methods in Mathematical Physics*, vol. I (Interscience Publishers, Hoboken, 1953)

[Cob64] A. Cobham, The intrinsic computational difficulty of functions, in *International Congress for Logic, Methodology and Philosophy of Science (ICLMPS)* (1964)

[Coo71] S.A. Cook, The complexity of theorem proving procedures, in *stoc71* (1971), pp. 151–158

[DF03] R.G. Downey, L. Fortnow, Uniformly hard languages. Theor. Comput. Sci. **2**(298), 303–315 (2003)

---

[2] This inclusion is surprising because the definition of UL/poly is quite stringent: the nondeterministic logspace machine is required to be unambiguous for every advice string.

[Edm65] J. Edmonds, Paths, trees, and flowers. Canad. J. Math **17**, 449–467 (1965)
[FS08] L. Fortnow, R. Santhanam, Infeasibility of instance compression and succinct PCP's for NP, in *Proceedings 40th Annual Symposium on Theory of Computing* (ACM, 2008), pp. 133–142
[GJ79] M.R. Garey, D.S. Johnson, *Computers and Intractability: A Guide to the Theory of NP-Completeness* (W.H. Freeman, New York, 1979)
[HK71] K. Hoffman, R. Kunze, *Linear Algebra*, 2nd edn. (Prentice Hall, Upper Saddle River, 1971)
[HS65] J. Hartmanis, R.E. Stearns, On the computational complexity of algorithms. Trans. Am. Math. Soc. **117**, 285–306 (1965)
[HU79] J. Hopcroft, J. Ullman. *Introduction to Automata Theory, Anguages, and Computation* (Addison Wesley, Boston, 1979)
[Imm88] N. Immerman, Nondeterministic space is closed under complementation. SIAM J. Comput. **17**, 935–939 (1988)
[IPZ01] R. Impagliazzo, R. Paturi, F. Zane, Which problems have strongly exponential complexity? J. Comput. Syst. Sci. **63**(4), 512–530 (2001)
[Kar72] R.M. Karp, Reducibility among combinatorial problems, in *Complexity of Computer Computations* (1972), pp. 85–103
[KL80] R. Karp, R.J. Lipton, Some connections between nonuniform and uniform complexity classes, in *Proceedings 12th Annual Symposium on Theory of Computing* (ACM, 1980), pp. 302–309
[Lad75] R.E. Ladner, On the structure of polynomial time reducibility. J. ACM **22**(1), 155–171 (1975)
[Lev73] L.A. Levin, Universal sorting problems. Probl. Peredachi Informatsii **9**(3), 265–266 (1973). in Russian
[Mah82] S.R. Mahaney, Sparse complete sets of NP: Solution of a conjecture of berman and hartmanis. J. Comput. Syst. Sci. **25**(2), 130–143 (1982)
[MV97] M. Mahajan, V. Vinay, Determinant: Combinatorics, algorithms, and complexity. Chicago J. Theor. Comput. Sci. 5 (1997)
[Pap94] C.H. Papadimitriou, *Computational Complexity* (Addison-Wesley, Boston, 1994)
[RA00] K. Reinhardt, E. Allender, Making nondeterminism ambiguous. SIAM J. Comput. **29**(4), 1118–1131 (2000)
[Rei08] O. Reingold, Undirected connectivity in log-space. J. ACM **55**(4), 1–24 (2008)
[Sch98] A. Schrijver, *Theory of integer and linear programming* (John Wiley and Sons, Inc. New York, 1998)
[Sze88] R. Szelepcsényi, The method of forced enumeration for nondeterministic automata. Acta Informatica **26**, 279–284 (1988)
[Tod91] S. Toda, *Counting problems computationally equivalent to the determinant*. Technical Report CSIM 91-07, Dept of Computer Sc. and Inf. Math., Univ. of Electro-Communication, Tokyo, Japan (1991)
[Tut47] W.T. Tutte, The factorization of linear graphs. The J. Lond. Mathe. Soc. **22**, 107–111 (1947)
[Val79] L.G. Valiant, The complexity of computing the permanent. Theor. Comput. Sci. **8**, 189–201 (1979)
[vEB90] P. van Emde Boas, Machine models and simulation, in *Handbook of Theoretical Computer Science, Volume A: Algorithms and Complexity* (Elsevier, Amsterdam, 1990), pp. 1–66
[Yap83] C.K. Yap, Some consequences of nonuniform conditions on uniform classes. Theor. Comput. Sci. **26**(3), 287–300 (1983)

Chapter 2
Investigations Concerning the Structure of Complete Sets

Eric Allender

Abstract This chapter will discuss developments bearing on three related research directions where Somenath Biswas has made pioneering contributions:

- Isomorphism of Complete Sets
- Creative Sets
- Universal Relations

Some open questions in each of these directions will be highlighted.

Keywords Berman-Hartmanis conjecture · Isomorphism · NP-completeness · Creativity · Universality

Mathematics Subject Classification (2010) Primary 68Q15

2.1 Introduction

How many NP-complete sets are there?

Although there is a trivial and uninteresting answer to this question (namely: there is a countably infinite number of NP-complete sets), there is a large body of work investigating the proposition that in actuality there is *precisely one* NP-complete set (modulo minor encoding details).

Let us clarify what is meant by "minor encoding details": When we consider the set SAT of satisfiable Boolean formulae, it is irrelevant if we encode formulae using round parentheses () or square ones [], or if we write variables in italic font or in bold face. Any of these choices would lead to a reasonable encoding of SAT; they all yield encodings of SAT that are equivalent in some sense.

E. Allender (✉)
Department of Computer Science, Rutgers University, New Brunswick, NJ 08855, USA
e-mail: allender@cs.rutgers.edu

M. Agrawal and V. Arvind (eds.), *Perspectives in Computational Complexity*, Progress in Computer Science and Applied Logic 26, DOI: 10.1007/978-3-319-05446-9_2,

One way to attempt to formalize this notion of "equivalence" is to say that two sets A and B, $A, B \subseteq \{0, 1\}^*$, are p-isomorphic if there is a bijection f defined on $\{0, 1\}^*$ computable and invertible in polynomial time, such that $f(A) = B$. This approach leads to the famous Berman-Hartmanis conjecture [BH77], which asserts that all of the sets that are NP-complete under $\leq_m^{\mathrm{p}}$ reductions are p-isomorphic.

The isomorphism conjecture(s) will be discussed in more detail in Sect. 2.2. However, a bit of background about isomorphism of complete sets is necessary here, in order to provide a coherent overview of the current paper. The Berman-Hartmanis conjecture arose, at least in part, because of a cultural inheritance from the study of computability theory. If we accept the rough idea that NP is analogous to the class of computably-enumerable sets, and polynomial time is analogous to the class of computable functions, then the Berman-Hartmanis conjecture is analogous to the Myhill Isomorphism Theorem in computability theory, which states: All of the sets that are complete for the class of computably-enumerable sets under $\leq_m$-reductions are computably-isomorphic to the Halting Problem. (For expositions of this work, see [Rog67] or [Soa87].)

Central to the proof of the Myhill Isomorphism Theorem is the notion of a *creative* set. We postpone until Sect. 2.3 the precise definition of "creativity," but this is an appropriate time to mention that the name was coined by Emil Post [Pos44], who was profoundly influenced by certain consequences of Gödel's incompleteness theorems. Post believed that there was a link between the notion of "mathematical creativity" and the fact that there is a computable function that, given a set of consistent axioms for arithmetic, will produce a true statement that cannot be proved from those axioms. Post's definition of creativity abstracts out this property of the set of theorems provable from a list of axioms.

In the setting of recursion theory, the creative sets turn out to exactly coincide with the sets that are complete for the class of computably enumerable sets under $\leq_m$-reducibility, and this is useful in proving Myhill's Isomorphism Theorem. Thus it was natural for researchers to try to define a resource-bounded analog of creativity. But it is not entirely clear what is the best way to define such an analog. Different definitions were presented by various authors [JY85, Wan91], but the definition of *NP-creative* sets by Agrawal and Biswas [AB96] provides several advantages over other definitions. (For instance, all NP-creative sets are NP-complete; this is not known for some other notions.) Just as all of the NP-complete sets in Garey and Johnson [GJ79] are p-isomorphic to SAT, so also are they all NP-creative. Although Agrawal and Biswas refrain from conjecturing that all NP-complete sets are NP-creative, we may as well consider a "creativity" version of the Berman-Hartmanis conjecture:

The **Creativity Hypothesis**: The class of NP-creative sets coincides with the class of sets that are NP-complete under $\leq_m^{\mathrm{p}}$ reductions.

One might at first guess that, since SAT is NP-creative, then *everything* that is p-isomorphic to SAT would also be NP-creative—but Agrawal and Biswas showed that, if this is true, then the Creativity Hypothesis is true (and hence $\mathrm{P} \neq \mathrm{NP}$).

Thus, NP-creativity and p-isomorphism yield two possibly different subclasses of the NP-complete sets, each of which captures a notion of "naturalness" (in the sense that all of the currently-known "natural" examples of NP-complete sets are both creative and p-isomorphic to SAT). Neither the Creativity Hypothesis nor the Berman-Hartmanis Conjecture is known to imply the other. Section 2.3 discusses creativity in more detail.

The final section of this chapter highlights one additional direction in which Somenath Biswas has pushed, in order to give additional insight into the structure underlying completeness. Although the Berman-Hartmanis conjecture focuses on the NP-complete *sets*, let us not forget that much of the practical interest in NP-completeness derives from the desire to find *witnesses* for membership in an NP-complete set. That is, at a fundamental level, it is not a *set*, such as HAMILTONIAN-CIRCUIT, that is of primary interest, but rather the corresponding *relation* consisting of pairs (G, C) such that C is a Hamiltonian cycle in the graph G.

Is it possible that the (string, witness) relations for every NP-complete set are all "the same" in some sense? Note that it is not at all obvious how to formulate this sense of "sameness." For instance, if there is a polynomial-time relation $W(x, y)$ consisting of witnesses y for string x, then there is a relation $W'(x, z)$ such that $W'(x, z)$ is true if and only if $W(x, y)$ holds, where y is the string that results by deleting every second symbol of z. These two relations both serve as witness relations for the same set in NP, but they do give different numbers of witnesses for the same string, and thus they fail to be "the same" on a fairly basic level. And yet, they do contain exactly the same information, in some intuitive sense. Agrawal and Biswas succeeded [AB92] in giving a useful definition of "universal relations", in order to capture the sense in which the defining relations for all known NP-complete sets seem to be "the same." More recently, Chaudhary, Sinha, and Biswas have adapted this notion for nondeterministic logspace [CSB07]. This topic is explored in Sect. 2.4.

2.2 The Isomorphism Conjecture(s)

An outstanding survey of recent developments related to isomorphisms of complete sets is now available [Agr11], and the reader is urged to consult that source for a more complete introduction to the topic and an in-depth discussion of the current state of the field. The discussion here will focus on describing aspects of the topic that are (1) related to the work of Somenath Biswas, or (2) related to some open questions or developments that are not mentioned in [Agr11].

The winds of public opinion have blown back and forth, regarding the Berman-Hartmanis Conjecture. It appears to have initially been viewed as fairly plausible. One of the first published accounts questioning whether the isomorphism conjecture is true appears in the work of Joseph and Young [JY85]. They defined a class of NP-complete sets they named the *k-creative sets* (which will also be discussed in Sect. 2.3), and they explicitly conjectured that some k-creative sets are not p-isomorphic to SAT. In particular, for any one-one length-increasing function f computable in polynomial time, they defined a set K_f, and they pointed out that, if f is suitably hard to invert, then it is hard to see how K_f can be p-isomorphic to SAT. Kurtz, Mahaney, and Royer subsequently elaborated on this intuition, and formulated the *Encrypted Complete Set Conjecture*, which states that there is a one-one, length-increasing, one-way function f such that SAT and $f(\mathrm{SAT})$ are not p-isomorphic.

Several papers were then written, all of which tended to buttress support for the Encrypted Complete Set Conjecture (all of which are discussed in the survey [Agr11]). But then attention shifted to some interesting classes of *restricted* $\leq_m^{\mathrm{p}}$-reductions; we will discuss some of these developments in more detail below—but the general trend of these investigations has been to weaken our confidence in the Encrypted Complete Set Conjecture. More recently, there has been a productive series of investigations of *more powerful* classes of reductions, notably including m-reductions computed in P/poly [AW09, Agr02] and in NP $\cap$ coNP [HHP07] (this latter class of reductions is known as SNP-reductions). As a consequence, we now know that some fairly plausible hypotheses imply that all sets complete for NP under P/poly-reductions and SNP-reductions are P/poly-isomorphic and SNP-isomorphic, respectively. Combined with the results about restricted $\leq_m^{\mathrm{p}}$ reductions that will be discussed below, the picture that emerges is that NP-complete sets are either provably isomorphic or at least are reasonably likely to be isomorphic, both when one considers reductions strictly *less* powerful and *more* powerful than $\leq_m^{\mathrm{p}}$ reductions. It remains to be seen whether any of these lessons ultimately shed much light on the case of $\leq_m^{\mathrm{p}}$ reductions themselves.

2.2.1 Restricted Reductions

It is debatable whether $\leq_m^{\mathrm{p}}$ reductions really constitute the most important class of reductions. There is a rich structure of complexity classes within P, and $\leq_m^{\mathrm{p}}$-reducibility is essentially useless in elucidating this structure. This was the motivation for Jones et al. to introduce log-space reductions [Jon75]—but even in that pioneering work, it was realized that a higher precision tool was necessary, in order to investigate the structure of logspace, which is why Jones introduced what he called *log-bounded rudimentary reductions* [Jon75]. This was several years before the modern study of circuit complexity got under way, and it took a while before it was noticed that log-bounded rudimentary reductions actually correspond to many-one reductions computed by uniform AC^0 circuits (that is, constant-depth polynomial-size families of circuits of AND and OR gates) [AG91]. Ultimately, AC^0 reductions have proved to be the most useful notion of reducibility for investigating subclasses of P, surpassing both NC^1 reducibility [CM87] and 1-L reducibility (discussed below). However, AC^0 reducibility posed greater challenges initially, and thus progress was made first with 1-L reducibility.

2.2.2 1-L Reductions

1-L reductions are functions computed by logspace-bounded Turing machines that make a single pass over their input tape (from left to right). They were introduced by Hartmanis, Immerman, and Mahaney [HIM78, HM81] for many of the same reasons that had led Jones to introduce log-bounded rudimentary reductions. 1-L reductions offered the advantages of being significantly more convenient and intuitive (since the original formulation of log-bounded rudimentary reductions lacked the intuitive appeal of the AC^0 formulation). In this brief overview, we avoid giving more detailed definitions of 1-L reductions, but it is appropriate to note that there are some differences in the formulations as presented in [HIM78] and [HM81], and that some of these formulations result in a class of reductions that is not closed under composition (see [All88]).

1-L reductions are easy to invert, and this fact, combined with some diagonalization techniques, enabled a proof that all sets complete for PSPACE under 1-L reductions are p-isomorphic [All88]. They are not isomorphic under 1-L isomorphisms [BH92], but they are complete under isomorphisms computable in nondeterministic logspace [HH93].

Agrawal and Biswas [AB96] succeeded in showing that, even for classes as small as deterministic logspace (and indeed, for any class that is closed under logspace reductions that produce output at most linearly longer than the input) the sets complete under 1-L reductions are complete under one-one, length-increasing, polynomial-time invertible reductions. (Thus by [BH77] all such sets are p-isomorphic.) Finally, Agrawal proved that all such sets are isomorphic via reductions computed by one-way nondeterministic logspace (1-NL) machines [Agr96] (and the same paper proves an analog of the Berman-Hartmanis conjecture for 1-NL reductions).

At this point, study of the structure of sets complete under 1-L reductions effectively stopped.[1] The major open questions had been solved. But this was merely a prelude to a much more exciting and significant development in the history of work on the isomorphism problem, focusing on reductions that are computable by constant-depth circuits. Indeed, although there are problems (such as the PARITY problem) that are computable by 1-L machines but are not computed by AC^0 circuits [FSS84], Agrawal had shown that, for essentially all complexity classes of interest, all sets complete under 1-L reductions *are* complete under reductions computable in AC^0 [Agr96]. And whereas all functions computable by 1-L machines are easy to invert, this is not the case for AC^0. Thus, by considering AC^0 reductions, the research community was moving on to a richer class of complete sets, and was confronting some of the essential issues raised by the Encrypted Complete Set Conjecture.

---

[1] This is not to suggest that work on 1-L computation stopped. Indeed, much of the very large body of work on *streaming algorithms* consists of the study of 1-L computation.

2.2.3 Constant-Depth Circuits

Attention was first focused on AC^0 isomorphisms by considering a very restricted class of AC^0 reductions: projections (which are reductions computed by circuits with no gates, other than negation gates). Sets complete for NP (and other classes) under uniform projections are isomorphic under uniform AC^0 isomorphisms [ABI97].

The logical next step was to work on extending this result from projections to NC^0 functions (that is, functions computed by constant-depth circuits with fan-in $O(1)$, so that each output bit depends on only $O(1)$ input bits—as contrasted with projections, where each output bit depends on either zero or one input bit). As part of this investigation, it was also discovered that, at least for the class NC^1, the sets complete under AC^0 reductions are also complete under NC^0 reductions, thereby obtaining the first theorem showing that the sets complete under AC^0 reductions are all AC^0 isomorphic [AA96]. Subsequently, the authors were joined by Rudich, in showing that this holds not only for NC^1, but also for NP and for most other complexity classes of interest [AAR98].

These initial AC^0 isomorphism theorems were proved only in the nonuniform setting. After a series of intermediate steps improving the uniformity condition [AAIPR01, Agr01], Agrawal succeeded in overcoming some daunting technical difficulties, in presenting a Dlogtime-Uniform version of the isomorphism theorem [Agr11], which stands as one of the crowning achievements of the study of the structure of complete sets. This work not only shows that a natural re-phrasing of the Berman-Hartmanis Conjecture (in terms of AC^0 reductions and isomorphisms) is true, but also gives a convincing setting where the Encrypted Complete Set Conjecture fails (since even when f is an AC^0 function that provably cannot be inverted in AC^0, $f(\text{SAT})$ is still AC^0-isomorphic to SAT).

2.2.4 Open Questions

Again, please refer to [Agr11] for several interesting open questions. Here are a few additional questions relating to isomorphisms, that are not discussed there.

Two important problems that are not believed to be NP-complete are Factoring and the Minimum Circuit Size Problem:

$$\text{FACT} = \{(x, i, b) : \text{the } i\text{th bit of the prime factorization of } x \text{ is } b\}.$$

$$\text{MCSP} = \{(\chi_f, s) : \chi_f \text{ denotes a string of length } 2^m \text{ (for some } m) \text{ that is the truth-table of a Boolean function } f \text{ on } m \text{ variables and } s \text{ denotes an integer such that } f \text{ can be computed by a Boolean circuit of size at most } s.\}$$

(In the definition of FACT, the prime factorization is presented as $p_1^{e_1}, \ldots, p_k^{e_k}$, where each exponent $e_i > 0$, and $p_i < p_{i+1}$, so that each number has a unique prime factorization.)

I suspect that FACT is probably not complete for *any* reasonable complexity class under AC^0 reductions. In an earlier survey [All01] I outlined a possible approach toward proving that this is the case. Namely, I noted that it would suffice to show that there is no one-one length-increasing NC^0 reduction from FACT$\times\{0,1\}^*$ to FACT (or no isomorphism between these sets, computable and invertible in depth-three AC^0). All of my attempts to construct such a reduction have involved multiplication in some form, and this is not computable in AC^0. Perhaps, I suggested, one could show that multiplication is inherent, in computing such a reduction. Now, however, after some illuminating discussions with Michal Koucký, I no longer think that this is a promising approach. One way to build a padding function would be to map the pair $((x,i,b),y)$ to the triple (z,i,b), where $z = xy'$, where y' has binary representation $10^\ell y_1 0^\ell y_2 0^\ell \ldots y_{n-1} 0^\ell y_n 0^\ell z'$ where ℓ is suitably large, and where z' has $\log^{O(1)} n$ bits. If y' is prime, then it will be the largest prime factor of z, and thus the initial part of the prime factorizations of x and of z will be the same. The product xy' can be computed in AC^0, because of the padding by 0^ℓ and because z' is small. It is reasonably likely that a value of z' exists so that y' will be prime, although number theorists have not yet established that this holds. It would be *very* hard to show that *no* such z' can be found in uniform AC^0. Thus it is reasonably likely that a padding function for (a suitable encoding of) FACT does exist in AC^0. This does not guarantee that such a padding function can be found in NC^0, but it does illustrate some of the difficulties of pursuing this approach.

Kabanets and Cai have presented some arguments, suggesting that MCSP is not complete for NP under $\leq_m^p$ reductions [KC00]. Can one obtain even stronger evidence, suggesting that MCSP is not p-isomorphic to SAT? It is certainly not clear that MCSP should have a padding function (i.e., a polynomial-time computable and invertible function f mapping MCSP$\times\{0,1\}^*$ onto MCSP). It is even harder to see how to construct a padding function if one fixes the circuit size s to be something exponentially large, but still much smaller than 2^m, such as this set:

MCSP2 $= \{\chi_f : \chi_f$ denotes a string of length 2^m that is the truth-table of a Boolean function f on m variables such that f can be computed by a Boolean circuit of size at most $2^{m/2}$.$\}$

As Kabanets and Cai observe [KC00], if MCSP2 has a padding function computed in polynomial time, then $BPP = P$. The connection between the paddability of MCSP2 and the BPP versus P problem arises through the easy observation that any set C isomorphic to SAT has a P-printable sets contained both in C and in $\overline{C}$, combined with the following equivalence (where the first condition listed is the well-studied Impagliazzo-Wigderson derandomization hypothesis [IW97]):

- There is a set $A \in$ E that requires circuits of size greater than $2^{n/2}$ for all large n iff
- There is a P-printable set B contained in the complement of MCSP2 of the form $B = \{\chi_f : \chi_f$ denotes a string of length 2^m that is the truth-table of $A^{=m}\}$.

The observations above do *not* provide much evidence against MCSP2 being isomorphic to SAT; rather, they merely indicate that it will not be easy to prove that it *is* isomorphic to SAT. What unlikely consequences would follow if MCSP2 (or MCSP) turned out to be isomorphic to SAT?

2.3 Creative Sets

A set is defined to be *creative* if its complement is *productive* (and this holds for all of the variants of "creativity" and "productivity" that have been considered). Thus, in order to discuss creative sets, we must first define productive sets.

A set A is *productive over a class of languages* $\mathcal{C}$ if there is a function (a so-called "*productive function*") witnessing that $A \notin \mathcal{C}$, in some sense. In order to make this definition precise, we must be explicit what notion of Turing machine indices I we are using to represent elements of $\mathcal{C}$, and what class of functions F we will allow as productive functions. Thus we can say that A is (F, I)-productive if there is a function $f \in F$ such that, for every $i \in I$, $f(i) \in A$ if and only if $f(i)$ is not in the language accepted by machine i. That is, given i, f finds an input on which A differs from the i-th element of $\mathcal{C}$.

Agrawal and Biswas define a set A to be NP-creative if its complement is (polynomial-time,I)-productive, where I is an indexing of nondeterministic polynomial-time Turing machines, where machine M_i has the property that, on all inputs, it runs in time $|i|$ [AB96]. It is far from obvious that this is an appropriate definition, since these machines all run in $O(1)$ time! Thus, in particular, $\{L(M_i) : i \in I\}$ is *not* equal to NP, and does not even contain all of the sets in, say, AC^0! Nevertheless, Agrawal and Biswas are able to demonstrate that this definition yields at least as many sets as an earlier notion of creativity (the "k-creative" sets of [JY85], which were re-dubbed "k-completely-creative" sets by Wang [Wan91] to distinguish them from another creativity notion he introduced), and they also show that all NP-creative sets are NP-complete (in contrast to the situation for the "k-creative" sets of [Wan91], which are neither known to include sets such as SAT, nor to be contained in the class of NP-complete sets). Figure 2.1 indicates the inclusion relations among these various classes of "creative" sets for NP.

However, when these creativeness definitions are adapted to larger complexity classes (such as EXP), they all coincide exactly with the class of sets complete under $\leq_m^p$ reductions. (The issue boils down to a question of whether the complexity class $\mathcal{C}$ can diagonalize over the class F of productive functions. This is true when $\mathcal{C} =$ EXP, but is not known to be true for $\mathcal{C} =$ NP.)

Even though the definition of NP-creative sets is less intuitive than the definition of the class of sets that are p-isomorphic to SAT, Agrawal and Biswas make a convincing argument that all "natural" NP-complete sets (including all of the NP-complete sets listed in [GJ79]) are NP-creative. Thus there is some merit in investigating the "Creativity Hypothesis" mentioned in the introduction—the hypothesis that all NP-complete sets are NP-creative—as an alternative to the Berman-Hartmanis

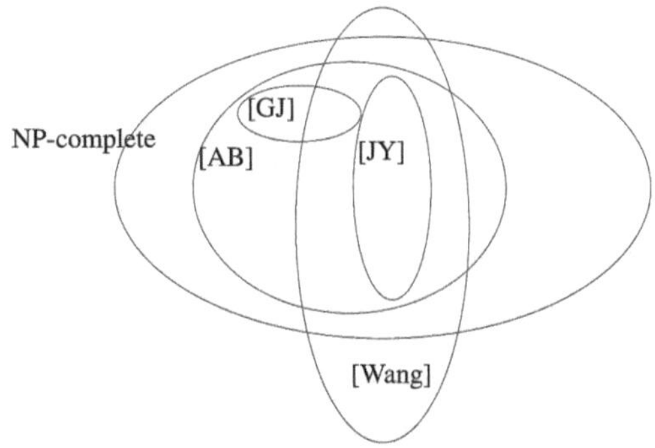

Fig. 2.1 Diagram, showing (likely) inclusions among classes of "creative" sets for *NP*. The region labeled "*[GJ]*" indicates the list of "natural" NP-complete problems catalogued in [GJ79]. It is not known to contain any of the k-creative sets defined by Joseph and Young [JY85], indicated by the region labeled *[JY]*. This same class was called "k-completely-creative" by Wang [Wan91], who also introduced another class of k-creative sets, indicated by the region labeled [Wang]; it is not known whether all of those sets are NP-complete. The region labeled *[AB]* indicates the NP-creative sets of Agrawal and Biswas [AB96]

conjecture. Proving that either of these conditions hold would entail proving P $\neq$ NP. (In the case of the Creativity Hypothesis, Agrawal and Biswas show that any NP-creative set is complete for NP under reductions that are "exponentially honest," in the sense that, for some constant c, $2^{c|f(x)|} > |x|$ for all x [AB96]. Thus, in particular, no finite set can be NP-creative.) It is particularly interesting that Agrawal and Biswas show that, if all of the sets that are p-isomorphic to SAT are NP-creative, then the Creativity Hypothesis holds.

The Creativity Hypothesis has not received much attention. Here are some questions that might yield some interesting insights:

- Are all of the sets that are complete for NP under AC^0 reductions NP-creative? How about the sets that are complete under first-order projections? Or the sets that are complete for NP under 1-L reductions?
- If one assumes that FACT or MCSP are NP-creative, can one derive stronger conclusions than if one merely assumes that these sets are NP-complete?
- Agrawal and Biswas have shown that the complement of any NP-creative set contains an infinite subset in NP. Consider a set such as $\{x$: the time-n^2-bounded Kolmogorov complexity of x is greater than $|x|/2\}$. Would we expect this set to have an infinite NP-subset? (Actually, the answer is probably *Yes*! It is observed in Sect. 2.2.4 that, under the Impagliazzo-Wigderson derandomization hypothesis, this set even has a P-printable subset.) Can one derive strong and unlikely conclusions from the assumption that this set is the complement of an NP-creative set?

2.4 Universal Relations

All of the NP-complete sets that are p-isomorphic to SAT have a padding function, and even the NP-complete sets that are not known to be p-isomorphic to SAT (such as certain k-creative sets, and sets of the form $f(\mathrm{SAT})$ where f is one-way) have "padding" functions if we drop the requirement of invertibility (i.e., a reduction from $A \times \Sigma^*$ to A that is one–one and length-increasing, but is not necessarily invertible). Similarly, all known NP-complete sets are disjunctive-self-reducible. (A set A is called "disjunctive-self-reducible" if there is a polynomial-time-computable function that takes a string x as input, and produces a list $y_1, \ldots, y_{|x|^{O(1)}}$ as output, such that $x \in A$ iff $\exists i\ y_i \in A$.)

Agrawal and Biswas defined two operators on relations (which they name the *join* and *equivalence* operators) that are related to paddability (without invertibility) and disjunctive-self-reducibility, respectively (in the sense that if the witness relation for a set A has the given operator computable in polynomial time, then A is paddable or disjunctive-self-reducible, respectively). Remarkably, they were able to show that the relations with these two operators are *precisely* the relations from which any other NP-witness relation can be "recovered" in a fairly natural sense. (The details of these definitions will not be repeated here; see [AB92]).

One thing that I particularly like about [AB92] is their presentation of a new class of NP-complete sets. Let f be *any* one–one and size-increasing polynomial function. They define a relation R_f as follows: $(z, w) \in R_f$ if $|w| = 4|z|^3$ and one of the following three conditions hold:

1. $|z| = 1$ and $w \in \{0100, 0101, 0110, 0111, 1001, 1010, 1011\}$.
2. For some $r > 0$, $w = \#^r x_1\#w_1\#\#x_2\#w_2\#\# \ldots \#\#x_n\#w_n$, where $f(1\#x_1\#x_2\# \ldots \#x_n) = z$ and for all $i \leq n$, $(x_i, w_i) \in R_f$.
3. For some $r > 0$, $w = \#^r x\#i_1\#i_2\# \ldots \#i_n\#j_1\# \ldots \#j_n\#\#w'$, where $f(1\#x\#i_1\# \ldots \#i_n\#j_1\# \ldots \#j_n) = z$ and for each $k \leq n$, bits number i_k and j_k of w' are the same.

Agrawal and Biswas show that $\{x : \exists y(x, y) \in R_f\}$ is NP-complete. I know of no *direct* way to see that this set is NP-complete; the proof of completeness presented by Agrawal and Biswas follows because the relation R_f is universal (because it has the required join and equivalence operators). It would be interesting to know if there is any example of a natural NP-complete problem, for which it is easier to prove NP-completeness by presenting the join and equivalence operators, than to present a traditional $\leq_m^{\mathrm{p}}$ reduction.

The theory of Probabilistically-Checkable Proofs tells us that problems in NP have witness relations with very special encoding structure. It would be interesting to know if this body of knowledge can be merged with the theory of universal relations, to obtain any new insights.

Figures 2.2, 2.3 and 2.4 shows inclusion relations among the different notions considered in this survey, all of which present ways to give a mathematically precise definition that can serve as a proxy for the vague concept of what it means to be a "natural" NP-complete set:

Fig. 2.2 Diagram, showing inclusions among notions of “natural” NP-complete sets discussed in this survey

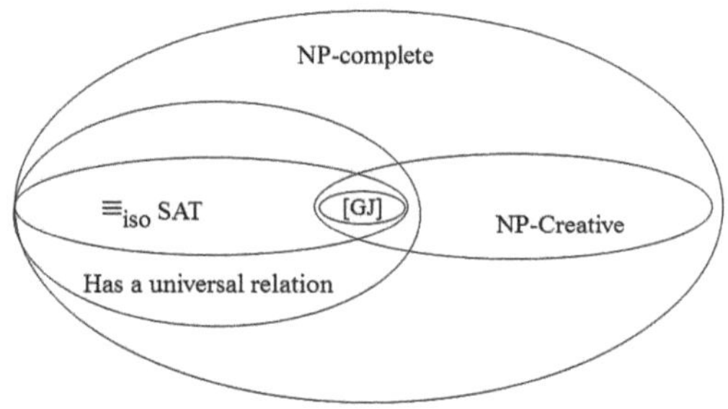

Fig. 2.3 Diagram, showing inclusions among notions of “natural” NP-complete sets, if p-isomorphism preserves NP-Creativity

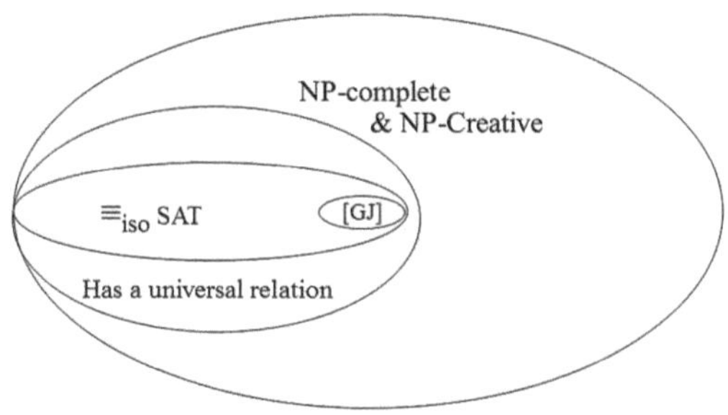

Fig. 2.4 Diagram, showing inclusions among notions of “natural” NP-complete sets, if all NP-Creative sets are p-isomorphic

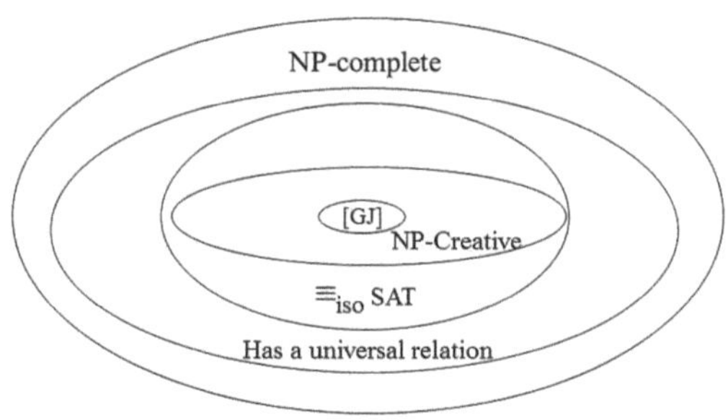

- being p-isomorphic to SAT.
- being NP-creative.
- having a universal relation.

Do all NP-creative sets have universal relations? (Note in this regard that Agrawal and Biswas show that sets with universal relations are all complete for NP under polynomially honest reductions (i.e., reductions f where there is a polynomial p such that $p(|f(x)|) \geq |x|$ for all x [AB92]), whereas the NP-creative sets are only known to be complete under exponentially honest reductions [AB96]. Thus it might be better to ask first whether all NP-creative sets that are complete under polynomially honest reductions have universal relations.)

Are the standard witness relations for MCSP and FACT universal? (Possibly this question is easy to answer …) Note in this regard that Agrawal and Biswas show that the standard witness relation for Graph Isomorphism is not universal—as well as a (somewhat nonstandard) witness relation for Simple Max Cut. They introduce a more general notion of universal relations, which they call "generalized universal relations," and show that the Simple Max Cut witness relation is generalized universal. If one is able to show that the standard witness relations for MCSP and FACT are not universal, then perhaps one can show that they are also not generalized universal. This would provide some additional evidence that these problems are not NP-complete.

Are there any additional implications that one can prove, regarding the Berman-Hartmanis conjecture, the Creativity Hypothesis, and the question of whether all NP-complete sets have universal witness relations?

(There has been some additional work by other authors, regarding universal relations. The reader is referred to [CSB07] for a discussion of this work.)

2.5 Conclusions

The notions of p-isomorphism, NP-creativity, and universality provide three ways to identify properties that are shared by all of the "natural" NP-complete sets. Although the work of Somenath Biswas and others has given us a body of interesting results regarding these notions, a number of intriguing open questions remain.

Acknowledgments Supported in part by NSF Grants CCF-0832787 and CCF-1064785.

References

[Agr96] M. Agrawal, On the isomorphism conjecture for weak reducibilities. J. Comput. Syst. Sci. **53**(2), 267–282 (1996)
[Agr01] M. Agrawal, Towards uniform AC^0–isomorphisms. in *Proceedings IEEE Conference on Computational Complexity* (2001). pp. 13–20
[Agr02] M. Agrawal, Pseudo-random generators and structure of complete degrees. in *IEEE Conference on Computational Complexity* (2002). pp. 139–147
[Agr11] M. Agrawal, The isomorphism conjecture for constant depth reductions. J. Comput. Syst. Sci. **77**(1), 3–13 (2011)
[Agr11] M. Agrawal, in *The Isomorphism Conjecture for NP*, ed. by S.B. Cooper, A. Sorbi. Computability in Context: Computation and Logic in the Real World (World Scientific Press, 2011), pp. 19–48
[AA96] M. Agrawal, E. Allender, An isomorphism theorem for circuit complexity. in *Proceedings IEEE Conference on Computational Complexity* (1996), pp. 2–11.
[AAIPR01] M. Agrawal, E. Allender, R. Impagliazzo, T. Pitassi, S. Rudich, Reducing the complexity of reductions. Comput. Complex. **10**(2), 117–138 (2001)
[AAR98] M. Agrawal, E. Allender, S. Rudich, Reductions in circuit complexity: an isomorphism theorem and a gap theorem. J. Comput. Syst. Sci. **57**(2), 127–143 (1998)
[AB92] M. Agrawal, S. Biswas, Universal relations. in *Proceedings IEEE Conference on Structure in Complexity Theory* (1992), pp. 207–220.
[AB96] M. Agrawal, S. Biswas, NP-creative sets: a new class of creative sets in NP. Math. Syst. Theor. **29**(5), 487–505 (1996)

[AB96] M. Agrawal, S. Biswas, Polynomial-time isomorphism of 1-L-complete sets. J. Comput. Syst. Sci. **53**(2), 155–160 (1996)

[AW09] M. Agrawal, O. Watanabe. One-way functions and the Berman-Hartmanis conjecture. in *Proceedings IEEE Conference on Computational Complexity* (2009). pp. 194–202

[All88] E. Allender, Isomorphisms and 1-L reductions. J. Comput. Syst. Sci. **36**(3), 336–350 (1988)

[All01] E. Allender, in *Some Pointed Questions Concerning Asymptotic Lower Bounds, and News From the Isomorphism Front*, ed. by G. Paun, G. Rozenberg, A. Salomaa. Current Trends in Theoretical Computer Science: Entering the 21st Century (World Scientific Press, 2001), pp. 25–41

[ABI97] E. Allender, J.L. Balcázar, N. Immerman, A first-order isomorphism theorem. SIAM J. Comput. **26**(2), 557–567 (1997)

[AG91] E. Allender, V. Gore, Rudimentary reductions revisited. Inf. Process. Lett. **40**(2), 89–95 (1991)

[BH77] L. Berman, J. Hartmanis, On isomorphism and density of NP and other complete sets. SIAM J. Comput. **6**, 305–322 (1977)

[BH92] H.-J. Burtschick, A. Hoene, in *The Degree Structure of 1-L Reductions*, ed. by I.M. Havel, V. Koubek. MFCS, Lecture Notes in Computer Science vol. 629 (Springer, 1992), pp. 153–161

[CSB07] V. Choudhary, A.K. Sinha, S. Biswas. Universality for nondeterministic logspace. in *Proceedings 1st International Conference on Language and Automata Theory and Applications (LATA)* (2007), pp. 103–114

[CM87] S.A. Cook, P. McKenzie, Problems complete for deterministic logarithmic space. J. Algorithms **8**(3), 385–394 (1987)

[FSS84] M.L. Furst, J.B. Saxe, M. Sipser, Parity, circuits, and the polynomial-time hierarchy. Math. Syst. Theor. **17**(1), 13–27 (1984)

[GJ79] M.R. Garey, D.S. Johnson, *Computers and Intractability: A Guide to the theory of NP-completeness* (W.H. Freeman and Company, New York, 1979)

[HHP07] R.C. Harkins, J.M. Hitchcock, A. Pavan, Strong reductions and isomorphism of complete sets. in *Proceedings Conference on Foundations of Software Technology and Theoretical Computer Science (FST&TCS)*. Lecture Notes in Computer Science vol. 4855 (Springer, 2007), pp. 168–178

[HIM78] J. Hartmanis, N. Immerman, S.R. Mahaney, One-way log-tape reductions. in *Proceedings IEEE Symposium on Foundation of Computer Science (FOCS)* (1978). pp. 65–72

[HM81] J. Hartmanis, S.R. Mahaney, Languages simultaneously complete for one-way and two-way log-tape automata. SIAM J. Comput. **10**(2), 383–390 (1981)

[HH93] L.A. Hemachandra, A. Hoene, Collapsing degrees via strong computation. J. Comput. Syst. Sci. **46**(3), 363–380 (1993)

[IW97] R. Impagliazzo, A. Wigderson, $P = BPP$ if E requires exponential circuits: derandomizing the XOR lemma. in *Proceedings ACM Symposium on Theory of Computing (STOC)* (1997), pp. 220–229

[Jon75] N.D. Jones, Space-bounded reducibility among combinatorial problems. J. Comput. Syst. Sci. **11**(1), 68–85 (1975)

[JY85] D. Joseph, P. Young, Some remarks on witness functions for nonpolynomial and noncomplete sets in NP. Theor. Comput. Sci. **39**, 225–237 (1985)

[KC00] V. Kabanets, J.-Y. Cai, Circuit minimization problem. in *Proceedings ACM Symposium on Theory of Computing 456 (STOC)* (2000), pp. 73–79

[Pos44] E.L. Post, Recursively enumerable sets of positive integers and their decision problems. Bull. Am. Math. Soc. **50**, 284–316 (1944)

[Rog67] H. Rogers, *Theory of Recursive Functions and Effective Computability*. (McGraw-Hill, New York, 1967)

[Soa87] R.I. Soare, *Recursively-Enumerable Sets and Degrees* (Springer, Berlin, 1987)

[Wan91] J. Wang, On p-creative sets and p-completely creative sets. Theor. Comput. Sci. **85**(1), 1–31 (1991)

Chapter 3
Space Complexity of the Directed Reachability Problem over Surface-Embedded Graphs

N. Variyam Vinodchandran

Abstract The graph reachability problem, the computational task of deciding whether there is a path between two given nodes in a graph, is the canonical problem for studying space-bounded computations. Three central open questions regarding the space complexity of the reachability problem over directed graphs are: (1) improving Savitch's $O(\log^2 n)$ space bound, (2) designing a polynomial-time algorithm with $O(n^{1-\epsilon})$ space bound, and (3) designing an *unambiguous* non-deterministic log-space (UL) algorithm. These are well-known open questions in complexity theory, and solving any one of them will be a major breakthrough. We discuss some of the recent progress reported on these questions for certain subclasses of surface-embedded directed graphs.

Keywords Graph reachability · Space-bounded computation · Computational complexity

Mathematics Subject Classification (2010) Primary 68Q15

3.1 Introduction

The graph reachability problem, the problem of deciding whether there is a path from a given vertex s to a vertex t in a given graph, is central to space-bounded computations. Various versions of this problem characterize several important space complexity classes. Over directed graphs, it is the canonical complete problem for

---

Supported in part by the NSF grant CCF-0916525.

N. V. Vinodchandran (✉)
Department of Computer Science and Engineering, University of Nebraska-Lincoln, Lincoln, NE, USA
e-mail: vinod@cse.unl.edu

M. Agrawal and V. Arvind (eds.), *Perspectives in Computational Complexity*, Progress in Computer Science and Applied Logic 26, DOI: 10.1007/978-3-319-05446-9_3,

non-deterministic log-space (NL). The breakthrough result of Reingold implies that the undirected reachability problem characterizes the complexity of deterministic log-space (L) [Rei08]. It is also known that a certain restricted promise version of the reachability problem over directed graphs characterizes randomized log-space computations (RL) [RTV06]. Clearly, progress in space complexity studies is directly related to progress in understanding graph reachability problems. We refer the readers to an excellent (albeit two decades old) survey by Avi Wigderson [Wig92] to further understand the significance of reachability problems in complexity theory.

This article is far from an exhaustive survey on the space complexity of the graph reachability problem. In particular, some of the major progress (such as Reingold's algorithm for undirected graph reachability and Saks and Zhou's deterministic simulation of randomized log-space) are not discussed here. Instead, we limit our discussion to some recent progress that the author and his collaborators reported on these questions for certain subclasses of surface-embedded directed graphs. It is mostly an elaboration of the talk that the author gave on Prof. Somenath Biswas's 60th birthday celebrations at IIT Kanpur in August of 2012.

3.1.1 Three Central Questions

We first discuss three central questions concerning the space complexity of the directed graph reachability problem. These are well-known and difficult open questions in the area, and progress on any of these is a step towards the much bigger NL versus L question (the first two problems are discussed in Wigderson's 1992 survey [Wig92]). However, the author feels that the known barriers for attacking these problems are much less severe than those known for many difficult open problems in time-bounded complexity classes and circuit lower bounds, and believes that breakthrough progress on these problems can be made in the near future.

Problem 1: Improving Savitch's bound. About four decades ago Savitch proved that the reachability problem over directed graphs with n vertices can be solved in space $O(\log^2 n)$ deterministically [Sav70]. This implies that problems that can be solved nondeterministically in space $s(n)$ have *deterministic* algorithms with $O(s^2(n))$ space bound. Thus, for polynomial space bounds, nondeterminism does not add any additional power to determinism. For the important case when the space bound is $O(\log n)$, Savitch's theorem implies that all problems in NL can be solved deterministically in $O(\log^2 n)$ space. *Is this quadratic blow-up necessary?* This is one of the most important open problems in this topic.

Problem 2: Improving the BBRS bound. The time complexity of Savitch's algorithm is $\Theta(n^{\log n})$—a super-polynomial bound. The standard breadth first search algorithm (BFS) is another fundamental algorithm for solving graph reachability. BFS can be implemented in linear time but requires linear space. BFS is efficient in time but not in space, and Savitch's algorithm is efficient in space but takes super-polynomial time. Hence a natural and significant question that researchers have considered is whether we can design an algorithm for the directed graph reachability problem that is efficient in both time and space. In particular, can we design a polynomial-time algorithm for the directed graph reachability problem that uses only $O(n^{1-\epsilon})$ space for some constant ϵ? The best known result in this direction is the two decades old bound due to Barnes et al. [BBRS98] (which we call the BBRS bound). By cleverly combining BFS and Savitch's algorithm, Barnes et al. designed a polynomial-time algorithm for reachability that uses space $O(n/2^{\sqrt{\log n}})$—a slightly sub-linear function. Improving the BBRS bound remains another significant open question regarding the space complexity of the directed reachability problem.

Problem 3: NL versus UL Problem. UL denotes an unambiguous subclass of NL. A decision problem L is in UL if and only if there exists a nondeterministic log-space machine M deciding L such that, for every instance x, M has at most one accepting computation on input x [BJLR91, ÀJ93]. Thus UL is a complexity class that is sandwiched between NL and L. Is NL = UL? While typically such collapse results are unlikely in complexity theory (and even if they are likely, they are nearly impossible to prove), there is an increasing body of evidence that in this specific case the answer is yes, and the author believes that we might be able to prove this equality using known techniques. Reinhardt and Allender showed using the isolation lemma that in the nonuniform setting NL coincides with UL; that is NL/poly = UL/poly [RA00]. Further, in a subsequent paper, Allender, Reinhardt, and Zhou showed that, under a certain hardness assumption the construction given in [RA00] can be *derandomized* to show that NL = UL [ARZ99]. Thus it is very likely (at least to the author) that NL = UL, though we do not yet have a proof. Can we show NL = UL unconditionally?

3.1.2 Outline

In the next two sections we discuss some progress that we have made towards these three open questions—Sect. 3.2 on Problems 1 and 2, and Sect. 3.3 on the NL versus UL problem. All the results discussed in these sections are for directed graphs embedded on topological surfaces. As an aside, in Sect. 3.4 we reproduce the proof of the BBRS bound from [BBRS98], partly to bring more attention to this nice algorithm.

3.2 Space Efficient Reachability Algorithms for Graphs with Topological Structure

An important sub-case of Problem 1 (and Problem 2) is to design reachability algorithms that beat Savitch's bound (respectively, the BBRS bound) for directed graphs with some topological structure (graphs that are embedded on topological surfaces). We discuss some recent progress reported along this direction. The main results are

(1) algorithms that beat both Savitch's bound and the BBRS bound for a subclass of directed *acyclic* graphs parameterized by the number of sources and the genus of the embedding [SBV10, SV12] (2) an algorithm for directed planar reachability that improves on the BBRS bound [INP+13]. The main approach in both these results is that of space-efficient "kernelization."

Kernelization is a known preprocessing technique in designing algorithms (for example in the area of fixed parameter tractability). Kernelization algorithms are reductions from a problem to itself so that the easy part of the instance is abstracted out and the core part is retained in the reduced instance. The hope is that the core part will be of smaller size and hence known algorithms can be applied to this compressed instance yielding algorithms with better complexity. We first illustrate this in a simple scenario.

Consider a reachability instance $\langle G, s, t\rangle$ where $G = (V, E)$ is a n-vertex graph with the guarantee that it has at most k directed edges (the remaining edges are undirected). Let G_{un} be the undirected graph we get by removing all the directed edges from G. For a directed edge $e = (u, v)$ let $tail(e) = u$ and $head(e) = v$. We show a simple log-space reduction that takes $\langle G, s, t\rangle$ and produces a reachability instance $\langle G', s', t'\rangle$ where G' is a directed graph with $O(k)$ vertices.

In the reduced graph $G' = (V', E')$, $V' = \{s', t'\} \cup \{v_e \mid e \text{ is a directed edge in } G\}$. The pair $(v_{e_1}, v_{e_2}) \in E'$ if $tail(e_2)$ is in the same connected component as $head(e_1)$ in G_{un}. For a $v_e \in V'$, $(s', v_e) \in E'$ if $tail(e)$ is in the same connected component of s in G_{un} and (v_e, t') if $head(e)$ is in the same connected component of t in G_{un}. Notice that this reduction is log-space since for checking whether two vertices u, v are in the same connected component of G_{un}, we can use Reingold's log-space algorithm for undirected reachability. It is clear that there is a s-t path in G if and only if there is a s'-t' path in G'. Using this reduction together with Savitch's algorithm we get that reachability in graphs with $n^{o(1)}$ directed edges can be solved in $o(\log^2 n)$. Also, by applying BFS to the reduced graph, we get that for any $\epsilon > 0$, reachability in graphs with $O(n^{1-\epsilon})$ directed edges can be solved in polynomial time and $O(n^{1-\epsilon})$ space.

We now describe the main kernelization result of [SBV10, SV12] and its application. Let $\mathcal{G}(m, g)$ denote the class of DAGs with at most $m = m(n)$ source vertices embedded on a surface (orientable or non-orientable) of genus at most $g = g(n)$, where n is the number of vertices. Building on [SBV10], in [SV12] we show the following reduction for graphs in $\mathcal{G}(m, g)$.

Theorem 1 [SV12] *There is a log-space reduction that, given an instance $\langle G, s, t\rangle$ (presented as a combinatorial embedding) where $G \in \mathcal{G}(m, g)$ and s, t are vertices of G, outputs an instance $\langle G', s', t'\rangle$ where G' is a directed graph and s', t' vertices of G', so that (a) there is a directed path from s to t in G if and only if there is a directed path from s' to t' in G', (b) G' has $O(m + g)$ vertices.*

By combining the above reduction with Savitch's theorem (with $m = g = 2^{O(\sqrt{\log n})}$) we get that reachability over graphs with $2^{O(\sqrt{\log n})}$ sources embedded on a surface of genus $2^{O(\sqrt{\log n})}$ can be decided in deterministic log-space. For $m = g = n^{o(1)}$ we get $o(\log^2 n)$ space algorithm for reachability that beats Savitch's bound. For $m = g = O(n^{1-\epsilon})$, we get $O(n^{1-\epsilon})$ space algorithm with polynomial running time for reachability, for any small constant ϵ, improving the BBRS bound.

One of the motivations for investigating the reachability problem for this class of surface-embedded graphs comes from earlier work due to Allender, Barrington, Chakraborthy, Datta, and Roy [ABC+09]. In [ABC+09], the authors considered reachability in planar DAGs with a *single* source vertex. They called this class of graphs Single source Multiple sink Planar Dags (SMPD). SMPD generalizes Single source Single sink Planar Dags (SSPD). SSPDs are interesting since they generalize *series parallel graphs* which is a well-studied restriction of directed acyclic graphs. Allender et al. presented a log-space algorithm for reachability in SMPD and left open whether reachability can be solved using logarithmic space over planar DAGs with multiple source nodes. In [SBV10], building on the SMPD algorithm, we present a log-space algorithm for planar dags with logarithmic number of sources. In the subsequent paper [SV12], via a careful use of techniques developed in [SBV10], we proved the log-space kernelization theorem that in particular implied a log-space algorithm for reachability in graphs with $2^{O(\sqrt{\log n})}$ sources, embedded on a surface of genus $2^{O(\sqrt{\log n})}$. The proof of this theorem is technically involved and we do not discuss it here. It remains a significant open question whether reachability for planar Dags (without any restriction on the number of sources) can be solved deterministically in $o(\log^2 n)$ space.

While improving Savitch's bound even for planar graphs remains open, the question of improving the BBRS bound for planar graphs was settled recently. Using a kernelization approach, in [INP+13], we show that the directed planar reachability problem can be solved in polynomial time using roughly $O(n^{1/2})$ space. This result extends a similar bound for the reachability problem over *grid graphs* due to Asano and Doerr [AD11].

Theorem 2 [INP+13] *For any constant $0 < \epsilon < 1/2$, there is an algorithm that, given a directed planar graph G and two vertices s and t, decides whether there is a path from s to t. This algorithm runs in time $n^{O(1/\epsilon)}$ and uses $O(n^{1/2+\epsilon})$ space, where n is the number of vertices of G.*

For showing this result, we first design a polynomial-time and $\tilde{O}(\sqrt{n})$-space algorithm for computing a "separator" of $O(\sqrt{n})$ size for an undirected planar graph. (For any undirected graph G and for any constant ρ, $0 < \rho < 1$, a *ρ-separator* of G is a a subset of vertices S whose removal disconnects G into two subgraphs A and B, such that $|A|$ and $|B|$ is at most ρn). This algorithm is based on a parallel algorithm for constructing a planar separator due to Gazit and Miller [GM87].

Theorem 3 [INP+13] *There is an algorithm that takes an undirected planar graph G with n vertices as input and outputs a (8/9)-separator of G of size $O(\sqrt{n})$. This algorithm runs in polynomial time and uses $\tilde{O}(\sqrt{n})$ space. (Here $\tilde{O}(s(n))$ denotes $O(s(n)(\log n)^{O(1)})$).*

Proof Sketch While for obtaining the $O(n^{1/2+\epsilon})$ space bound we need a recursive approach, it is easy to illustrate the idea for the case when the space bound is $O(n^{2/3})$. Let $G = (V, E)$ be the input directed planar graph. Let G_u be the underlying undirected graph. The first step is to apply the planar separator algorithm repeatedly k times on the connected components of G_u that are bigger than $n^{2/3}$ to further partition the graph until every component is of size $\leq n^{2/3}$. It is easy to see that after $k = \lceil \frac{2}{3} \times \frac{\log n}{\log(9/8)} \rceil$ applications we get a collection $\mathcal{S}$ of separators with total size $O(n^{2/3})$ so that removing $\mathcal{S}$ partitions the graph into disconnected components where each component is of size $\leq n^{2/3}$. (This is a standard trick used in many separator-based algorithms). Let $C_1, C_2, \ldots, C_l$ be the set of vertices in these components.

Now consider the kernel graph $\mathcal{G} = (\mathcal{V}, \mathcal{E})$ where $\mathcal{V} = \mathcal{S} \cup \{s, t\}$. For any two nodes x and y in $\mathcal{V}$, (x, y) is a directed edge if and only if there is a directed path from x to y in the subgraph of G that is induced by $\mathcal{V} \cup C_i$ (call this G_i), for some connected component C_i in the partition. Clearly, the number of nodes in $\mathcal{G}$ is $O(n^{2/3})$. Now consider the algorithm that decides whether there is a directed path from s to t in G by performing a BFS on $\mathcal{G}$ starting at s. Whenever BFS queries $(x, y) \in \mathcal{E}$?, the algorithm performs BFSs for each of the graphs G_i starting at x looking for a path from x to y, and returns YES if for some G_i this inner BFS returns true. Notice that since $|\mathcal{V} \cup C_i|$ is at most $O(n^{2/3})$, each of this BFSs can be implemented in $O(n^{2/3})$ space and polynomial time. Hence, overall the algorithm takes $O(n^{2/3})$ space and polynomial time.

For extending this proof to the $O(n^{1/2+\epsilon})$ space bound, we need $|\mathcal{S}| = O(n^{1/2+\epsilon})$. But that will result in large components and a simple application of the inner BFSs will not give the required space bound. Instead, we can apply the algorithm recursively. By limiting the number of recursive applications to a constant, we can make sure that the running time remains polynomial. We omit the details. □

Before we move on to the next section we mention that there is a certain computational model known as NNJAG model in which it is possible to prove lower bounds those match both Savitch's bound and the BBRS bound [Poo93, CR80, EPA99]. The NNJAG model is a branching program model tailored towards the reachability problem with restricted access to the input graph. While all the known algorithms for general reachability (such as BFS, DFS, Savitch's algorithm, BBRS algorithm) can be implemented in the NNJAG without substantial increase in time and space (in comparison to implementations on a random access machine), it is not clear to the author how a general approach such as kernelization can be handled in these models. It is conceivable that algorithms based on kernelization can overcome NNJAG lower bounds and help solve these open problems.

3.3 NL Versus UL Problem

The main progress on this problem has also been on graphs with some topological structure. We first discuss a technique developed by Reinhardt and Allender [RA00] since all the known proofs on this problem use their technique.

A positively and polynomially weighted graph is said to be *min-unique* if, between any two nodes the minimum weight path (if it exists) is unique. Here the weight of a path is the sum of the weights of its edges. Reinhardt and Allender [RA00] showed, using an adaptation of the inductive counting technique of Immerman [Imm88] and Szelepcsényi [Sze88], that the reachability question in min-unique graphs can be decided in UL. They combine this construction with an observation due to Wigderson [Wig94] that the isolation lemma of Mulmuley et al. [MVV87] can be used to nonuniformly assign weights to make a given graph min-unique. These two steps imply the collapse result that NL is in nonuniform UL.

Thus a space-efficient derandomization of the isolation lemma will show that NL = UL. However, derandomizing isolation lemma in its generality will have much deeper consequences and is a well-known and difficult open problem [AM08]. Instead, a viable and concrete approach for showing NL = UL is to first consider a class of graphs over which the reachability problem is NL-complete, and prescribe a *deterministic* log-space computable weight function which will make graphs in this class min-unique.

In [ABC+09], the authors solve this min-unique weight assignment problem for the class of *layered grid graphs*. Layered grid graphs are graphs with vertices on a $n \times n$ grid and the edges that go west-to-east and south-to-north. Subsequently in [BTV09], we show how to extend this weight function to general grid graphs (without restriction on the direction of edges). This implied that directed planar reachability is in UL since the directed planar reachability problem is known to be reducible to the grid graph reachability problem [ADR05]. In fact this even implied that the reachability problem for graphs embedded on constant genus surfaces and graphs that are $K_{3,3}$-free and K_5-free are in UL since the reachability problem for these classes of graphs reduces to the directed planar reachability problem [KV10, TW09] in log-space.

While, in [BTV09] we showed that directed planar reachability is in UL, it was not clear then how to solve the min-unique weight assignment problem for planar graphs. In a subsequent chapter, we solve this problem using *Green's Theorem*, a celebrated result from multivariate calculus [TV12]. Since it is a slightly nonstandard approach to use an analytical result to solve discrete problems, we believe this approach has the potential to solve the general NL versus UL problem. We next present the proof of the min-unique weight assignment problem for directed planar graphs based on Green's theorem.

Green's theorem, stated below, relates a certain curve integral over a closed curve on the plane to a related double integral over the region enclosed by this curve.

Theorem 4 Green's Theorem *Let C be a closed, piecewise smooth, simple curve on the plane which is oriented counterclockwise. Let R_C be the region bounded by C. Let P and Q be functions of (x, y) defined on a region containing R_C that have continuous partial derivatives in the region. Then*

$$\oint_C (P\,dx + Q\,dy) = \iint_{R_C} \left(\frac{\partial Q}{\partial x} - \frac{\partial P}{\partial y} \right) dA$$

We use the following corollary that we get if we substitute $Q(x, y) = x$ and $P(x, y) = 0$ in Green's theorem.

Corollary 5 (Area by line integrals) *Let C be a closed, piecewise smooth, simple curve on the plane that is oriented counterclockwise. Let R_C be the region bounded by C. Then,*

$$\text{Area}(R_C) = \oint_C x\, dy.$$

Theorem 6 [TV12] *There is a log-space algorithm that, given any planar graph G, assigns weights to the edges so that the resulting weighted graph is min-unique.*

Let us assume that the planar graph $G = (V, E)$ is presented as a straight-line drawing. That is, each vertex v is represented as a point (x_v, y_v) in the plane so that an edge (u, v) is a line between points (x_u, y_u) and (x_v, y_v). In addition, no such lines intersect other than at the vertices. Moreover, we assume that the coordinates are integer points with values bounded by poly(n) (n is the number of vertices). Typically, planar graphs are presented as a combinatorial embedding and it is not clear how such line drawings can be computed in log-space from a combinatorial embedding. However, this is not critical and in [TV12] we show how to handle this presentation issue.

Let $e = (u, v)$ be a directed edge directed from u to v where u is identified with the point (x_u, y_u) and v is identified with (x_v, y_v). For such a directed edge, define a weight function w as follows:

$$w_{gt}(e) = 2 \times \oint_e x\, dy = (y_v - y_u)(x_v + x_u)$$

The required isolation property of the weight function is proved using the following crucial lemma.

Lemma 7 *Let G be a directed planar graph and let C be any directed simple cycle in G. Let R_C be the region enclosed by C. Then the weight of the cycle C, $|w_{gt}(C)| = 2 \times \text{Area}(\text{R}_\text{c})$. In particular, $w_{gt}(C)$ is nonzero.*

Proof Let $C = (e_1, e_2, \ldots, e_l)$ be a directed cycle-oriented counterclockwise. Then we have

$$w_{gt}(C) = \sum_i w_{gt}(e_i) = 2 \times \sum_i \oint_{e_i} x\, dy = 2 \times \oint_C x\, dy = 2 \times \text{Area}(R_C)$$

The third equality follows from the linearity of integrals and the last equality follows from Corollary 7. If C is oriented clockwise, we get that $w_{gt}(C) = -2 \times \text{Area}(R_C)$. Hence the lemma. □

The following lemma establishes Theorem 6.

Lemma 8 *Let G be a directed planar graph. Then with respect to the weight function w_{gt}, for every pair of nodes u and v, if there is a directed path from u to v, then there is a* unique *path from u to v of minimum weight.*

Proof Suppose there are u, v so that there are two u to v paths P_1 and P_2 of minimum weight. We assume that the paths do not intersect on vertices other than the end points (otherwise we can find two vertices u' and v' along these paths that satisfies this property using a standard cut-and-paste argument and use these vertices instead). We have $w_{gt}(P_1) = w_{gt}(P_2)$. Now consider the graph G' that is same as G except that the path P_2 is reversed so that the set of edges $(P_1, -P_2)$ becomes a simple cycle in G' ($-P_2$ denotes the reversed path). Let C denote this cycle. Then $w_{gt}(C) = w_{gt}(P_1) + w_{gt}(-P_2) = w_{gt}(P_1) - w_{gt}(P_2) = 0$. The second equality because of the skew-symmetry of the weight function. This contradicts Lemma 7. □

It is clear that we can use Green's Theorem to design a class of min-unique weight functions. In fact any "nice" solution to the differential equation $\left(\frac{\partial Q}{\partial x} - \frac{\partial P}{\partial y}\right) = 1$ will yield such a weight function. For example, setting $P(x, y) = \frac{-y}{2}$ and $Q(x, y) = \frac{x}{2}$ to the left-hand side of Green's theorem yields the weight function $w(e) = (x_u y_v - x_v y_u)$ which is also min-unique.

Can we use such geometric techniques to design min-unique weight functions for larger classes of graphs? In [BTV09] it is observed that reachability in layered grid graphs over three dimensions is complete for NL. It might be possible to use generalizations of Green's theorem (such as Stokes' theorem) to design a min-unique weight function for three-dimensional layered grid graphs.

3.4 The BBRS Bound

We present the algorithm due to Barnes, Buss, Ruzzo, and Schieber [BBRS98] that solves the directed graph reachability problem in sub-linear space and polynomial time.

Theorem 9 [BBRS98] *For any k, there is a polynomial-time algorithm that given a directed graph G and two nodes s and t, decides whether there is a path from s to t in space $O(\frac{n}{2^{k\sqrt{\log n}}})$, where n is the number of vertices of G.*

Proof The algorithm uses a combination of BFS and Savitch's algorithm. For a parameter λ (this will be set to $2^{k\sqrt{\log n}}$ to get the desired bound), it constructs the levels of BFS tree that are at λ distance apart. Divide the vertex set into levels according to distance from s. That is, the level i vertex set is defined as:

$$V_i = \{v \mid d(s, v) = i\}, \text{ where } d \text{ is the distance function.}$$

Partition the set of vertices into λ equivalence classes $C_0, C_1, \ldots, C_{\lambda-1}$ where $C_j = \bigcup_{i=0}^{\lfloor n/\lambda \rfloor} V_{j+i\lambda}$. Since the C_is partition the vertex set, we have the following fact.

Fact 10 $\exists j^*$ so that $|C_{j^*}| \leq \lceil \frac{n}{\lambda} \rceil$

The PARTIAL-BFS algorithm (described below) constructs C_{j^*} level by level. Since we do not explicitly know which C_j has $\leq \frac{n}{\lambda}$ nodes, the algorithm will keep a counter to count the number of vertices and try from $j = 0$. At any point of the construction, if $|C_j| > \frac{n}{\lambda}$, it will abandon that j and try the next value for j. The algorithm will succeed for the first such j. This will only increase the space by an additive $O(\log n)$ factor and the time by a multiplicative factor of λ. Hence we assume that the algorithm knows j^*. Following is the description of the PARTIAL-BFS algorithm.

PARTIAL-BFS(G, s) /* Outputs C_{j^*}*/
  $V_0 = \{s\}$
  $V_{j^*} =$ CONSTRUCT(G, V_0, j^*)
  For $i = 1$ to $\lfloor \frac{n}{\lambda} \rfloor$
    $V_{i\lambda+j^*} =$ CONSTRUCT$(G, V_{(i-1)\lambda+j^*}, \lambda)$
    Add $V_{i\lambda+j^*}$ to C_{j^*}
  End-For
  Output C_{j^*}

In general, the procedure CONSTRUCT takes G and a set of nodes S and a parameter λ and returns the set of nodes that are at distance λ from some node in S. CONSTRUCT will use the bounded version of the reachability problem (Barnes et al. calls it *short path problem*) as subroutine.

$$\text{SPATH}(u, v, \lambda) = \text{true} \Leftrightarrow \text{there is a path of length} \leq \lambda \text{ from } u \text{ to } v \text{ in } G.$$

We can use an algorithm for SPATH as subroutine to solve CONSTRUCT as follows. Given (G, S, λ), to check whether $v \in V$ is at distance λ from some vertex in S, first check whether SPATH(u, v, λ) is true for some $u \in S$ and check for all $u \in S$, SPATH$(u, v, \lambda - 1)$ is false.

For a given algorithm for SPATH, let $T(n, \lambda)$ be its time complexity and $S(n, \lambda)$ be its space complexity. Then the time complexity of CONSTRUCT is $O(n^3)T(n, \lambda)$ and its space complexity is $O(\frac{n}{\lambda}) + S(n, \lambda)$. Moreover, once C_{j^*} is constructed, reachability can be solved by making $\lceil \frac{n}{\lambda} \rceil$ calls to SPATH(u, t, λ) (for all $u \in C_{j^*}$). Thus the total running time for the reachability algorithm will be $O(n^4)T(n, \lambda)$ and the space bound will be $O(\frac{n}{\lambda}) + S(n, \lambda)$.

We will now focus on SPATH. We will use a divide and conquer approach as in Savitch's algorithm to design an algorithm for SPATH. The problem with a direct application of Savitch's algorithm is its running time: at each level of recursion it cycles through all n nodes as a candidate for the middle node. This results in $O(n^{\log n})$ time. Since we are interested in keeping the time polynomial, we can not afford to cycle through all n nodes. Instead, we will divide the set of nodes into μ equivalence classes and use a Savitch-like divide and conquer on these equivalence classes (instead of the vertices). For $\mu = 2^{O(\sqrt{\log n})}$ the depth of recursion will be $O(\sqrt{\log n})$ and this approach will result in polynomial time.

For a parameter μ, partition the vertex set into μ equivalence classes $[1], [2], \ldots [\mu]$ where vertex $x \in [a] \Leftrightarrow x \equiv a \pmod{\mu}$. Each equivalence class has $\lceil \frac{n}{\mu} \rceil$ elements (except for the last one whose cardinality may be smaller). We will use $[a]$, $[b]$, $[c]$ etc to denote these equivalence classes of vertices. Although this is not a very standard notation, the ith vertex of the equivalence class $[a]$ (according to some fixed ordering) will be denoted by $[a](i)$.

Consider the procedure MODIFIED-SAVITCH$(G, [a], [b], X, l)$ where $[a]$ and $[b]$ are equivalence classes of vertices, X is an $\lceil \frac{n}{\mu} \rceil$ binary array, and l is a length parameter. This procedure returns a binary vector Y of size $\lceil \frac{n}{\mu} \rceil$, where

$$\begin{aligned} Y[j] = 1 \Leftrightarrow \exists i \text{ so that } X[i] = 1 \text{and there is a path} \\ \text{of length} \leq 2^l \text{from} [a](i)\, to\, [b](j) \end{aligned}$$

SPATH(u, v, λ) can be solved by one call to MODIFIED-SAVITCH with parameter $([a], [b], X_u, \lceil \log_2 \lambda \rceil)$ where $[a]$ = the equivalence class containing u, $[b]$ = the equivalence class containing v, and X_u is the vector with 1 in the index corresponding to u and 0 otherwise. There is a path from u to v if and only if there is a 1 in the index corresponding to v in the output vector Y. Below is a recursive version of the algorithm MODIFIED-SAVITCH.

MODIFIED-SAVITCH$(G, [a], [b], X, l)$
  If $l = 0$ then
    If $[a] = [b]$ then $Y \leftarrow X$
    Else $Y[j] = 1$ iff $\exists i$ such that there is an edge from $[a](i)$ to $[b](j)$
  Else
    $Y \leftarrow \vec{0}$
    For $c = 1$ to μ
      $Z \leftarrow$ MODIFIED-SAVITCH$(G, [a], [c], X, l - 1)$
      $Y_c \leftarrow$ MODIFIED-SAVITCH$(G, [c], [b], Z, l - 1)$
      $Y \leftarrow Y \vee Y_c$
  Return Y

Correctness of MODIFIED-SAVITCH is easy to prove. Its time and space bounds can be estimated using the following recurrences:

$$\begin{aligned} S(l) &= O(\frac{n}{\mu}) + S(l-1) \\ &= O(\frac{n}{\mu}) \times l \\ T(l) &= \mu \times 2 \times T(l-1) + O(n) \\ &= (2\mu)^{l+1} \times O(n) \end{aligned}$$

Setting $\mu = 2^{(k+1)\sqrt{\log n}}$ and $l = \lceil \log_2 \lambda \rceil$, we get an algorithm for SPATH with time complexity $T(n, \lambda) = O(2^{\log \lambda} \times 2^{(k+1)\sqrt{\log n}(\log \lambda + 1)} \times n)$ and space complexity $S(n, \lambda) = O(\frac{n}{2^{(k+1)\sqrt{\log n}}} \times \log \lambda)$.

For $\lambda = 2^{k\sqrt{\log n}}$, this results in polynomial time and space $O(\frac{n}{2^{k\sqrt{\log n}}})$ giving an algorithm for the reachability problem with polynomial running time and $O(\frac{n}{2^{k\sqrt{\log n}}})$ space bound. □

Acknowledgments I would like to thank Pavan Aduri for extended collaboration and discussions on the reachability problem, and for his valuable comments on an earlier draft of this article. I would like to thank the organizers, V. Arvind and Manindra Agrawal, of the Complexity and Logic Workshop at IIT Kanpur (in celebration of the 60th birthday of Somenath Biswas) for inviting me to give a talk at the workshop.

References

[ABC+09] E. Allender, D.A. Barrington, T. Chakraborty, S. Datta, S. Roy, Planar and grid graph reachability problems. Theor. Comput. Syst. **45**(4), 675–723 (2009)

[AD11] T. Asano, B. Doerr, in *Memory-constrained algorithms for shortest path problem.* (CCCG, 2011), pp. 135–138

[ADR05] E. Allender, S. Datta, S. Roy, in *The Directed Planar Reachability Problem* (FSTTCS, 2005), pp. 238–249

[ÀJ93] C. Àlvarez, B. Jenner, A very hard log-space counting class. Theor. Comput. Sci. **107**, 3–30 (1993)

[AM08] V. Arvind, P. Mukhopadhyay. Derandomizing the isolation lemma and lower bounds for circuit size. In APPROX-RANDOM (2008), pp. 276–289

[ARZ99] E. Allender, K. Reinhardt, S. Zhou, Isolation, matching, and counting uniform and nonuniform upper bounds. J. Comput. Syst. Sci. **59**(2), 164–181 (1999)

[BBRS98] G. Barnes, J.F. Buss, W.L. Ruzzo, B. Schieber, A sublinear space, polynomial time algorithm for directed s- t connectivity. SIAM J. Comput. **27**(5), 1273–1282 (1998)

[BJLR91] G. Buntrock, B. Jenner, K. Lange, P. Rossmanith, Unambiguity and fewness for logarithmic space, in *8th International Symposium on Fundamentals of Computation Theory* (1991), pp. 168–179

[BTV09] C. Bourke, R. Tewari, N.V. Vinodchandran, Directed planar reachability is in unambiguous logarithmic space. ACM Trans. Comput. Theor. **1**(1), 1–17 (2009)

[CR80] S.A. Cook, C. Rackoff, Space lower bounds for maze threadability on restricted machines. SIAM J. Comput. **9**(3), 636–652 (1980)

[EPA99] J. Edmonds, C.K. Poon, D. Achlioptas, Tight lower bounds for st-connectivity on the NNJAG model. SIAM J. Comput. **28**(6), 2257–2284 (1999)

[GM87] H. Gazit, G.L. Miller, *A parallel algorithm for finding a separator in planar graphs* (FOCS, 1987), pp. 238–248

[Imm88] N. Immerman, Nondeterministic space is closed under complementation. SIAM J. Comput. **17**(5), 935–938 (1988)

[INP+13] T. Imai, K Nakagawa, A. Pavan, N. V. Vinodchandran, O. Watanabe, A $n^{1/2+\epsilon}$-space and polynomial-time algorithm for directed planar reachability, in *Conference on Computational Complexity* (2013), pp. 277–286

[KV10] J. Kynčl, T. Vyskočil, Logspace reduction of directed reachability for bounded genus graphs to the planar case. ACM Trans. Comput. Theor. **1**(3), 1–11 (2010)

[MVV87] K. Mulmuley, U. Vazirani, V. Vazirani, Matching is as easy as matrix inversion. Cominatorica **7**(1), 105–113 (1987)

[Poo93] C.K. Poon, in *Space Bounds for Graph Connectivity Problems on Node-Named Jags and Node-Ordered Jags* (FOCS, 1993), pp. 218–227

[RA00] K. Reinhardt, E. Allender, Making nondeterminism unambiguous. SIAM J. Comput. **29**(4), 1118–1131 (2000)

[Rei08] O. Reingold, Undirected connectivity in log-space. J. ACM, **55**(4), 1–24 (2008)

[RTV06] O. Reingold, L. Trevisan, S. Vadhan, Pseudorandom walks on regular digraphs and the RL versus L problem, in *STOC '06: Proceedings of the thirty-eighth annual ACM Symposium on Theory of Computing*, (ACM, New York, NY, USA, 2006), pp. 457–466

[Sav70] W.J. Savitch, Relationships between nondeterministic and deterministic tape complexities. J. Comput. Syst. Sci. **4**(2), 177–192 (1970)

[SBV10] D. Stolee, C. Bourke, N.V. Vinodchandran, A log-space algorithm for reachability in planar dags with few sources, in *Proceedings of 25th IEEE Conference on Computational Complexity* (2010), pp. 131–138

[SV12] D. Stolee, N.V. Vinodchandran. Space-efficient algorithms for reachabilityin surface-embedded graphs, in *IEEE Conference on Computational Complexity* (2012), pp. 326–333

[Sze88] R. Szelepcsényi, The method of forced enumeration for nondeterministic automata. Acta Informatica **26**, 279–284 (1988)

[TV12] R. Tewari, N.V. Vinodchandran, Green's theorem and isolation in planar graphs. Inf. Comput. **215**, 1–7 (2012)

[TW09] T. Thierauf, F. Wagner, in *Reachability in $K_{3,3}$-free graphs and K_5-free graphs is in unambiguous log-space* (FCT, 2009), pp. 323–334

[Wig92] A. Wigderson, The complexity of graph connectivity. Math. Found. Comput. Sci. **1992**, 112–132 (1992)

[Wig94] A. Wigderson. NL/poly $\subseteq$ $\oplus$L/poly. In Proceedings of the 9th Structures in Complexity conference, pages 59–62, 1994

Chapter 4
Algebraic Complexity Classes

Meena Mahajan

Abstract This survey describes, at an introductory level, the algebraic complexity framework originally proposed by Leslie Valiant in 1979, and some of the insights that have been obtained more recently.

Keywords Algebraic complexity · Circuits · Formulas · Branching programs · Determinant · Permanent

Mathematics Subject Classification (2010) Primary 68Q15 · Secondary 68Q05

4.1 Introduction

In this survey, I am going to try and describe the algebraic complexity framework originally proposed by Leslie and Valiant [Val79, Val82], and the insights that have been obtained more recently. This entire article has an "as it appeals to me" flavour, but I hope this flavour will also be interesting to many readers. The article is not particularly in-depth, but it is an invitation to read [BCS97, Bür00a] and many recent papers on the topic, and to start attacking the open problems in the area.

Valiant started out with the mission of understanding the core essence of reductions and completeness, as witnessed in both recursive function theory and in computational complexity theory. He provided an algebraic framework in which to interpret the clustering of natural problems into completeness classes, even for problems of an algebraic rather than combinatorial nature. He had a remarkable hypothesis:

The idea for writing this survey came while the author was working on the Indo-French CEFIPRA-supported project 4702-1.

M. Mahajan (✉)
The Institute of Mathematical Sciences, CIT Campus, Chennai 600113, India
e-mail: meena@imsc.res.in

M. Agrawal and V. Arvind (eds.), *Perspectives in Computational Complexity*, Progress in Computer Science and Applied Logic 26, DOI: 10.1007/978-3-319-05446-9_4,

> Linear algebra offers essentially the only fast technique for computing multivariate polynomials of moderate degree.

Clearly, then, we are going to talk about polynomials, not languages or functions.

4.2 Valiant's Original Framework

Let $\mathbb{F}$ be any field, and let $\mathbb{F}[x_1, \ldots, x_n]$ be the ring of polynomials over indeterminates $x_1, \ldots, x_n$ with coefficients from $\mathbb{F}$. Consider a family (f) of polynomials $(f_n)_{n \geq 1}$, where each f_n is in $\mathbb{F}[x_1, \ldots, x_{s(n)}]$ for some function $s : \mathbb{N} \longrightarrow \mathbb{N}$. When should we say that (f) is tractable? Clearly, if there are too many variables to keep track of, there cannot be tractability. So we will henceforth demand that s is a polynomially bounded function ($\exists c, \forall n, s(n) \leq c + n^c$); then the nth polynomial f_n has $O(n^c)$ variables. But that is of course not enough.

There are many ways in which we can set the bar for tractability. Here's a first attempt. Can (f) be computed by a *formula* of reasonable size? To elaborate further, a formula is an expression defined recursively:

1. for each $c \in \mathbb{F}$, "c" is a formula of size 0 computing the polynomial c,
2. for each indeterminate x_i, "x_i" is a formula of size 0 computing the polynomial x_i, and
3. if F_1, F_2 are formulas computing polynomials f_1 and f_2, then "$(F_1 + F_2)$" and "$(F_1 \times F_2)$" are formulas of size $\text{size}(F_1) + \text{size}(F_2) + 1$ each, computing the polynomials $f_1 + f_2$ and $f_1 \times f_2$, respectively.

Notice that size(F) is just the number of ring/field operations used to construct F.

Instead of such a recursive definition, we could have a more intuitive picture: a formula is a rooted binary tree where internal nodes are labelled $+$ or $\times$ and leaf nodes are labelled from the set $\mathbb{F} \cup X$, where X is the set of indeterminates. The size is just the number of non-leaf nodes.

Now, for tractability, we could require that there is a polynomially bounded function $t : \mathbb{N} \longrightarrow \mathbb{N}$ and a family of formulas $(F_n)_{n \geq 1}$ such that each F_n computes f_n and has size at most $t(n)$. Let us use the notation VF to denote families of polynomials tractable in this sense. (VF: Valiant's Formulas—of course, Valiant didn't use this name! This class is also referred to as VP_e: Valiant's Polynomial-sized Expressions. Personally, I prefer VF. Also note, in formal logic, the formulas/expressions referred to above are called terms, hence VF means families with polynomial "termic complexity".)

Here's a second attempt: Can (f) be computed by a *straight-line program* of reasonable size? As before, we will declare polynomial size to be reasonable. Straight-line programs are programs where instructions involve adding or multiplying previously computed polynomials, no divisions and no conditionals (no if-then-else). In the more intuitive picture, they correspond to directed *acyclic* graphs where each node is a source node (indegree 0) labelled from the set $\mathbb{F} \cup X$, or has indegree 2 and is labelled + or ×. A designated sink node (outdegree 0) is the output node. Each node computes a polynomial in the obvious way, and the graph computes the polynomial at the output node. (The acyclicity constraint ensures that there is a linear ordering of the nodes such that each node, or instruction, only uses previously computed polynomials. This gives the straight-line program.) The size is the number of non-source nodes; again, this corresponds to the number of ring/field operations required. Such graphs are in fact exactly *algebraic circuits*, and we now look for polynomial size circuit families.

Clearly, this model generalises formulas. The catch is that it generalises it too much! To see why, consider the following circuit family: C_n has $n + 1$ nodes $v_0, v_1, \ldots, v_n$, and the labeling is $v_0 = x_1$, $v_{i+1} = v_i \times v_i$ for $i \in [n]$. The family of polynomials (f_n) computed by (C_n) is $f_n = x_1^{2^n}$. Even for small integer values of x_1, writing down the value of $f_n(x_1)$ is going to require exponentially many bits. How can we say that such a family (f_n) is tractable?

So we need to impose some additional restrictions. The obvious parameter to restrict is the degree of the polynomial. Say that the family (f_n) has moderate degree if for some polynomially bounded function $d : \mathbb{N} \longrightarrow \mathbb{N}$, the degree of each polynomial f_n is at most $d(n)$. If degree$(f_n) = D$ is polynomially bounded, then on integer arguments with b-bit representations, the value of f_n requires no more than poly(n, b) bits. (In general, it needs no more than poly(n, D, b) bits.) Henceforth, to qualify for the label tractable, a family (f_n) must have polynomially bounded degree.

(Why didn't we face this problem when considering *VF*? Simply because a formula of size t cannot compute a polynomial of degree more than $t + 1$. Don't just believe me; check this by induction on formula size.)

Now we have our second possible definition of tractability: (f_n) is tractable if the sequence degree(f_n) is polynomially bounded, and there is a polynomially bounded function $t : \mathbb{N} \longrightarrow \mathbb{N}$ and a family of straight-line programs, or algebraic circuits, $(C_n)_{n \geq 1}$, such that each C_n computes f_n and has size at most $t(n)$. Let us use the notation VP to denote families of polynomials tractable in this sense. (VP: Valiant's analogue of the Boolean complexity class P. Valiant called these families p-computable [Val82]).

The well-studied polynomial family from linear algebra, the determinant of a matrix of indeterminates, is known to be tractable in this second sense. (To define the family (Det$_n$), imagine an $n \times n$ matrix A_n with a new indeterminate x_{ij} at each position (i, j), and let Det$_n$ be the polynomial that represents the determinant of A_n. Thus Det$_1 = x_{11}$, Det$_2 = x_{11}x_{22} - x_{12}x_{21}$, and so on. Clearly, this family satisfies the mandatory conditions: Det$_n$ has n^2 variables and is of degree n.) This is not surprising; we know that the determinant can be computed efficiently (in polynomial time) over instantiated matrices using, say, Gaussian elimination. But to compute the symbolic determinant via a straight-line program, Gaussian elimination is apparently not directly of use because we can't search for nonzero pivots and eliminate them (remember, no divisions and no conditionals). However, Strassen [Str73] gave a generic method of converting any straight-line program with divisions to a division-free straight-line program; the resulting program's size is polynomially bounded in the original size, the number of variables and the degree. Thus, we can conclude that there are polynomial-sized straight-line programs for the symbolic determinant. There are more direct algorithms as well; Samuelson, Berkowitz, Csanky, ... See [MV97] for an explicit description of circuits of size $O(n^4)$ (my favourite one—no surprise!).

Whether the determinant can be computed efficiently by formulas (is Det_n in VF?) is still famously open. We know that it needs formula size at least $\Omega(n^3)$, see [Kal85]. But we do know that it can be computed by formulas of subexponential size $2^{O(\log^2 n)}$. This can be shown in many different ways, one of which we will look at a bit later, but the earliest demonstration of this follows from Csanky's algorithm [Csa76], which uses binary arithmetic operations and $O(\log^2 n)$ parallel time. Thus, if we use quasi-polynomial ($2^{\log^c n}$ for some constant c) formula-size as the defining property for tractability (giving a class that we can call VQF), then again the family (Det_n) has long been known to be tractable. We could also use quasi-polynomial circuit size as the defining property for tractability, giving a class that we can call VQP. But VQP obviously contains VP and VQF, so (Det_n) is in VQP; nothing new there. (Note: in defining VQF and VQP, the quasi-polynomial limit on formula or circuit size is over and above the requirement that the degree and number of variables are polynomially bounded.)

Does VP include essentially all interesting and natural polynomial families? We do not know. In fact, there is a large list of such polynomial families not known to be in VP. The most natural one is the permanent family (Perm_n) where Perm_n is the polynomial representing the permanent of an $n \times n$ matrix A_n of indeterminates. It is tantalisingly similar to the determinant; just the sign term is missing.

$$\mathrm{Det}_n = \sum_{\sigma \in S_n} \mathrm{sign}(\sigma) \prod_{i=1}^{n} x_{i\sigma(i)} \qquad \mathrm{Perm}_n = \sum_{\sigma \in S_n} \prod_{i=1}^{n} x_{i\sigma(i)}$$

Yet, it does not seem to be tractable. How "intractable" is it?

Mirroring the definitions of the Boolean complexity classes P and NP, Valiant proposed a notion of p-definability in [Val79]. A polynomial family (f_n) is p-definable if it can be written as an exponential sum, over partial Boolean instantiations, of another tractable family. Formally, a family (f_n) over $s(n)$ variables and of degree $d(n)$ is p-definable if $s(n)$ and $d(n)$ are polynomially bounded, as always, and further, there exist a polynomially bounded function m, and a family of polynomials (g_n) in VF, such that g_n has $s(n) + m(n)$ variables denoted $\{x_1, \ldots, x_{s(n)}, y_1, \ldots, y_{m(n)}\}$, and

$$f_n(\tilde{x}) = \sum_{y_1=0}^{1} \sum_{y_2=0}^{1} \cdots \sum_{y_{m(n)}=0}^{1} g_n(\tilde{x}, \tilde{y}).$$

This looks like an algebraic analogue $\sum \cdot \mathrm{VF}$ of the boolean class $\exists \cdot F$, where F is the class of languages decided by polynomial-size formulas. But it is well-known that $\exists \cdot F = \mathrm{NP}$, so this should be algebraic NP. Later, Valiant redefined p-definability (no, that is not a circular definition!) as exponential sums of families in VP, rather than VF; that is, VNP $= \sum \cdot$VP. For clarity, let us agree to temporarily refer to these two definitions as VNF (or VNP$_e$) and VNP. However, Valiant [Val82] showed that these two classes are in fact the same, so just VNP will do. The proof involves showing that VP is contained in $\sum \cdot$VF. And it is of course easier to show upper bounds with the definition of VNP rather than VNF.

Now Valiant observed that not only (Det$_n$), even (Perm$_n$) is p-definable. This should be similar to showing that the 0–1 permanent is in #P, right? Almost. We are dealing with symbolic polynomials, so we do not have the liberty of looking at an input value and deciding what to do next. Still, the basic idea is the same. For a statement S, let $[S]$ denote the 0–1 valued Boolean predicate that takes value 1 exactly when S is true. Then

$$\begin{aligned}
\mathrm{Perm}_n = \sum_{\sigma\in S_n}\prod_{i=1}^{n} x_{i\sigma(i)} &= \sum_{Y\in\{0,1\}^{n\times n}} \begin{bmatrix} Y \text{ is a } 0-1 \\ \text{permutation} \\ \text{matrix}\end{bmatrix} \cdot \prod_{i=1}^{n}\left(\sum_{j=1}^{n} Y_{ij}x_{ij}\right) \\
\begin{bmatrix} Y \text{ is a } 0-1 \\ \text{permutation} \\ \text{matrix}\end{bmatrix} &= [Y \text{ has at least one 1 in each row}] \\
&\quad \times\ [Y \text{ has at most one 1 in each line} \\
&\quad (\text{line} = \text{row or column})] \\
&= \left(\prod_{i=1}^{n}\sum_{j=1}^{n} Y_{ij}\right)\left(\prod_{\substack{(i,j)\neq(k.m);\\ i=k \text{ or } j=m}} (1 - Y_{ij}Y_{km})\right)
\end{aligned}$$

Clearly, the polynomial family

$$g_n = \left(\prod_{i=1}^{n}\sum_{j=1}^{n} Y_{ij}\right)\left(\prod_{\substack{(i,j)\neq(k.m);\\ i=k \text{ or } j=m}} (1 - Y_{ij}Y_{km})\right)\prod_{i=1}^{n}\left(\sum_{j=1}^{n} Y_{ij}x_{ij}\right)$$

has formulas of size $O(n^3)$, and $\mathrm{Perm}_n(\tilde{x}) = \sum_{Y\in\{0,1\}^{n\times n}} g_n(\tilde{x}, Y)$, so (Perm$_n$) is in VNP.

So we have some families in VP (even VF), and some in VNP but maybe not in VP. How do we compare families? For comparing languages, we have many-one reductions and Turing reductions—what is the algebraic analogue? Valiant proposed projections, a most restrictive kind of reduction when dealing with Boolean classes, but completely natural in the algebraic context. We say that $g \in \mathbb{F}[y_1, \ldots, y_m]$ is a projection of $f \in \mathbb{F}[x_1, \ldots, x_n]$ if g can be obtained from f by substituting a value in $\mathbb{F} \cup Y$ for each variable in X. (For instance, if $f = x_1x_2 + x_3x_4$, then the following are all projections of f: $y_1 + y_2$, $y_1y_2 + 5$, $y_1y_2 + y_2y_3$, $2y^2$. But $y_1^2y_2$, $y_1 + y_2 + y_3$ are not, because a projection cannot increase the degree or number of monomials.) Further, we say that a family (g_n) is a p-projection of a family (f_n) if each g_n is a projection of some f_m for an m not too far from n. That is, there is a polynomially bounded function t, and each g_n is a projection of $f_{t(n)}$. If we allow t to be quasi-polynomially bounded, we obtain qp-projections.

Using these notions of reductions, we have the usual notions of hardness and completeness for algebraic classes. Here's what Valiant showed:

1. (Det_n) is hard for VF under p-projections (and is known to be in VP).
2. (Det_n) is complete for the class of quasi-polynomial size formulas VQF under qp-projections.
3. Over fields with characteristic other than 2, (Perm_n) is complete for VNP under p-projections. Over fields of characteristic 2, Perm_n equals Det_n and hence is in VP and VQF.
4. Polynomial families associated with a number of NP-complete languages are complete for VNP under p-projections.

The first two follow from a proof that a polynomial computed by a size s formula is a projection of Det_{s+2}. (It uses the combinatorial definition of determinant. as the signed weighted sum of cycle covers in an associated graph.) The hardness of (Perm_n) for VNP mirrors the hardness of the Boolean permanent for the counting class $\#P$. As in the case of the upper bound, additional care is needed to take into account non-access to an input instance and fully symbolic computations; in particular, the proof requires a multiplicative inverse of two and hence fails over fields of characteristic 2. See [Val79, BCS97, Bür00a] for various versions of these proofs. See [Blä13] for a simplified gadget construction.

4.3 The Current Status

We now know much more about the classes VF, VP, VQP, VNP defined above, and about other similarly defined classes. Let's review these results one by one.

Say that a family of polynomials (f_n) is a p-family if the number of variables in f_n and the degree of F are polynomially bounded functions of n. We only consider p-families.

Recall that VP consists of p-families with polynomial-sized circuits. Also note that algorithmically, circuit size roughly corresponds to number of processors needed in a parallel algorithm (associate one processor per gate), while circuit depth—the length of a longest path from the output node to an input node—corresponds to parallel time.

A clever construction due to Hyafil [Hya79] shows that any polynomial of degree D in M variables, computable by a circuit of size t, can be computed in parallel time $O(\log D \times \log(D^2t + M))$. This is a depth-reduction of the circuit, and generalises

Csanky's result which was specifically tailored for the determinant. Further, this algorithm has parallel multiplicative depth only $O(\log D)$; that is, any root-to-leaf path goes through at most $O(\log D)$ multiplication nodes. This is worth noting since multiplication seems a more costly operation than addition or subtraction. Unfortunately, the resulting circuit, while shallow and depth-reduced, is rather large, roughly $t^{\log D}$. Applying this construction would take us from polynomial size circuits to shallow quasi-polynomial size circuits. Soon after this, an improved construction was presented by Valiant et al. [VSBR83]; they achieved the same depth-reduction (and also $O(\log D)$ multiplicative depth) with size polynomial in tD. In particular, applying this construction to a circuit family (C_n) witnessing that a polynomial family (f_n) is in VP, we see that (f_n) is in $\mathrm{VSAC}^1 \subseteq \mathrm{VNC}^2$.

Wait, what exactly are these new classes? Again, we can think of them as analogues of Boolean classes. The Boolean circuit class NC^i has circuits of polynomial size and $O(\log^i n)$ depth. The class SAC^i is similarly defined, polynomial size, $O(\log^i n)$ $\wedge$-depth, and negations only at inputs. That is, if $\vee$ nodes are allowed to have unbounded in-degree, but $\wedge$ nodes must have in-degree 2, then these circuits have depth $O(\log^i n)$. (Hence the name SAC, for semi-unbounded alternation.) Clearly, $\mathrm{NC}^i \subseteq \mathrm{SAC}^i \subseteq \mathrm{NC}^{i+1}$. Now define the classes VNC^i and VSAC^i as algebraic analogues of these, with $\times$ and $+$ playing the roles of $\wedge$ and $\vee$, respectively. In the Boolean world, we know that $\mathrm{NC}^1 \subseteq \mathrm{SAC}^1 \subseteq \mathrm{NC}^2 \subseteq \cdots \subseteq \mathrm{NC} \subseteq \mathrm{P}$. In the algebraic world, however, $\mathrm{VNC}^1 \subseteq \mathrm{VSAC}^1 = \mathrm{VNC}^2 = \cdots = \mathrm{VNC} = \mathrm{VP}$.

An important consequence of the depth reduction result of [VSBR83] is that the $(\mathrm{Det}_n) \in \mathrm{VQF}$ result generalises to all of VP; $\mathrm{VP} \subseteq \mathrm{VQF}$. Another important consequence is that at quasi-polynomial size, formulas are as powerful as circuits; VQF equals VQP. Such an equivalence is not known for p-families at polynomial size. (It holds at exponential size, because polynomials in any p-family have only exponentially many monomials. An explicit sum-of-monomials expression gives an exponential-sized formula.)

Even before the results of [Hya79, VSBR83], Brent [Bre74] had shown that depth-reduction is possible for VF. Any formula F can be rebalanced by identifying in it a suitably chosen node N and rewriting F as a linear form in N, say $AN + B$. If N is properly chosen, then the polynomials A and B are computed by small subformulas (size at most half of F) of F, and can be recursively rebalanced. The appropriate N is identified by using the tree separator lemma. This process yields a $O(\log \mathrm{size}(F))$ depth formula. Thus, we conclude $\mathrm{VF} = \mathrm{VNC}^1$.

The depth reduction for VP from [VSBR83] proceeds similarly, but works on "proof-trees" or *parse trees*. Unfolding a circuit into a formula by systematically duplicating reused nodes may yield an exponential-sized formula (recall the example X^{2^n}.) Let us nonetheless do so. Now, a minimal subformula that includes the output node, both children of an included $\times$ node, and exactly one child of an included $+$ node, computes a potential monomial whose degree is the number of leaf nodes in the subformula. Call such a subformula a proof tree. For a circuit computing a p-family of polynomials, we can ignore proof trees of super-polynomial size. For each polynomial-sized proof tree, the balancing technique described above should work. The catch is, there can be too many proof trees (there can be exponentially many monomials), and each proof tree could require cutting at a different node. The clever twist is the following: in the formula depth-reduction, A can be computed recursively because it is the partial derivative of F with respect to N. If F is now a circuit rather than a formula, then F may not be linear in N, so computing the partial derivative will not help. But if N is chosen to have degree more than half the degree of F, then this is indeed the case. So, the algorithm of [VSBR83] computes, for each pair of nodes N, N', a new polynomial $F(N, N')$; these polynomial are recursively constructed, and whenever 2degree(N) > degree(N'), $F(N, N')$ equals the partial derivative of N' with respect to N. Putting this together carefully gives the depth-required circuit. For details, see [VSBR83] itself. Also see [AJMV98a, Vol99] for *uniform* versions, where the task of describing the depth-reduced circuit given the original circuit is achieved using limited computational resources.

A couple of things slipped by almost unnoticed. We know what is meant by the degree of a polynomial, but what do we mean by degree(N)? This should be the degree of the polynomial computed at the node N, and indeed [VSBR83] use degree in this sense. But the uniform versions cannot do so, because computing the degree of a specified node in a given circuit is a completely non-trivial task! See the discussion about DegreeSLP in [ABKPM09, Kay10]. Fortunately, we can equally easily work with an upper bound on the degree of each node. And an upper bound $u(N)$ on the degree at each node N is easy to obtain: $u(N) = 1$ if N is a leaf, $u(N_1 + N_2) = \max\{u(N_1), u(N_2)\}$, $u(N_1 \times N_2) = u(N_1) + u(N_2)$. This upper bound is referred to as the complete formal degree of the circuit (as opposed to the degree of the polynomial it computes). However, just because the output node of C computes a polynomial of degree d, this does not imply that each node computes a polynomial of degree at most d. Higher degree monomials may get computed along the way, and get cancelled finally. Is it necessary, in terms of efficiency, to compute them? No! If C is of size s and computes a polynomial f of degree d, then we can construct a circuit C' of size $O(sd^2)$ computing the same polynomial and with each node computing a polynomial of degree at most d: just compute the homogeneous parts of f separately in the obvious way. Now C' will have complete formal degree $O(d^3 s)$. (See [MP08] for details.) Thus, we could have defined VP in terms of circuits of polynomial size and polynomially bounded complete formal degree as well.

There is a much simpler proof of the fact that VP is contained in VNC. This proof yields a weaker upper bound of $VSAC^2$ rather than $VSAC^1$, but is still beautiful, and is still enough to conclude that VP $\subseteq$ VQF. I first saw this proof in a survey talk by Koiran at Dagstuhl [Koi10], and I wish I had come up with it myself! Let (f_n) be in VP, as witnessed by a circuit family (C_n) with complete formal degree bounded by (d_n). To depth-reduce C_n, partition the nodes into $1 + \lceil \log d_n \rceil$ parts; part k has nodes with formal degree in $[2^{k-1}, 2^k)$. Treating the polynomials from parts $i < k$ as variables, the nodes in part k form a *skew* circuit, where each $\times$ node has at most one child that is not an input node. (Multiplying two nodes both in part k would create high degree, giving rise to a node in part $k + 1$.) Now, skew circuits can be depth-reduced to $VSAC^1$ rather easily, using a divide-and-conquer argument dating back to Savitch [Sav70]. Doing this separately for each part gives a $VSAC^2$ circuit.

We just introduced a new kind of circuit there: skew circuits. Are they as powerful as general circuits? We do not know! Let's define $\mathrm{VP}_{\mathrm{skew}}$; p-families of polynomials computed by polynomial-sized skew circuits. It turns out this is a great class to study, because it *exactly* characterises the complexity of the determinant. Recall what we have already seen; (Det_n) is hard for $\mathrm{VF} = \mathrm{VNC}^1$ and is in VP. The upper bound proof from [MV97] actually gives a skew circuit of size $O(n^4)$, but skew circuit constructions were known much earlier: in [Ven92], Venkateswaran first defined Boolean skew circuits to capture nondeterministic circuits, and subsequently many authors independently extended that study to arithmetic rings, [Dam91, Tod92, Vin91, Val92]. And the lower bound proof from [Val79] shows that polynomials computed by skew circuits are p-projections of the determinant, though it is not stated this way. Valiant showed that a formula can be converted to a certain kind of graph that we nowadays call an algebraic branching program or ABP (more about this below), and that polynomials computed by ABPs are p-projections of (Det_n). And we now know that ABPs are essentially skew circuits.

Time to define ABPs. These are directed acyclic graphs, with a designated source node s and a designated target sink node t (sometimes there may be multiple target nodes), and with edges labelled from $\mathbb{F} \cup X$ (similar to input nodes in a circuit). For any directed path ρ, the weight of ρ is the product of the labels of the edges on ρ. The polynomial p_v computed at a node v is the sum of the weights of all directed sv paths. The polynomial computed by the ABP is just p_t. Families computed by polynomial-size ABPs form the class VBP. (In some parts of the literature, edge labels are allowed to be linear forms in X. This does not significantly change the properties of ABPs as we discuss here. We'll stick to the convention that labels are in $\mathbb{F} \cup X$.)

So why are ABPs and skew circuits essentially the same? ABPs to skew circuits: clearly, $p_s = 1$, and for any other source node (in-degree 0) s', $p_{s'} = 0$. Look at an edge $u \to v$ of the ABP with label ℓ. Then p_v has a contribution from $p_u \times \ell$. Summing this over all incoming edges at v gives a small circuit computing p_v from previously computed values, and this circuit is skew. For the reverse simulation, reverse this construction: (1) introduce a source node s, (2) for each input node u labelled ℓ, add an edge $s \to u$ labelled ℓ, (3) for each node $v = u + u'$, create edges $u \to v$ and $u \to v$ labelled 1, and (4) for each node $v = u \times \ell$, create an edge $u \to v$ labelled ℓ.

So now we can add to the list of results at the end of Sect. 4.2: (Det_n) is complete for $\mathrm{VBP} = \mathrm{VP}_{\mathrm{skew}}$ under p-projections.

In fact, we can add more. What makes the simulation from skew circuits to ABPs possible is the fact that at each $\times$ gate, one argument is *easy*. Toda [Tod92] took this argument further—it is enough if one argument is independent of the rest of the circuit. That is, for each $\times$ node $\alpha = \beta \times \gamma$, the entire sub-circuit rooted at either β or γ has no connection to the rest of the circuit except via this edge to α. (Equivalently, one of the edges into α is a bridge in the circuit.) Call such circuits *weakly skew* circuits. Toda showed that weakly skew circuits can be converted to skew circuits with linear size blow-up. See also [MP08], where Malod and Portier made the size bounds in the conversions even more precise. So now we can say $\mathrm{VBP} = \mathrm{VP}_{\mathrm{skew}} = \mathrm{VP}_{ws}$, where the subscript ws stands for weakly skew.

(Note: Neither [Tod92] not [MP08, Mal03] actually claimed linear size blow-up. However, their constructions from weakly skew circuits to ABPs, with the standard conversion from ABPs to skew circuits, does give linear blow-up. As far as I can see, linear blow-up for weakly skew to skew circuits was explicitly observed in [KK08, Jan08, Gre12a].)

Taking this idea further, Malod and Portier provide a brilliant characterization of the class VP. Say that a circuit is *disjoint* if at every node $\alpha = \beta \circ \gamma$, where $\circ$ could be $+$ or $\times$, the sub-circuits rooted at β and γ are disjoint. This is just a fancy (convoluted?) way of saying that the circuit is a formula. But now relax this constraint a bit. Say that a circuit is *multiplicatively disjoint* or MD if at every $\times$ node $\alpha = \beta \times \gamma$, the sub-circuits rooted at β and γ are disjoint. No restrictions apply to $+$ nodes. Like formulas, MD circuits of size s have complete formal degree bounded by s. But the MD restriction seems to allow more computation than formulas; for instance, weakly skew circuits are MD, and so MD circuits can compute (Det_n) in polynomial size. Malod and Portier showed that in fact polynomial size MD circuits can compute everything in VP, but nothing more. That is, $\mathrm{VP} = \mathrm{VP}_{\mathrm{MD}}$. While this fact can also be deduced once we have depth reduction to VSAC^1, Malod and Portier give a completely self-contained combinatorial proof which is very neat. Basically, imagine that each node in the VP circuit is labelled with its formal degree. Now make multiple copies of each node, inversely proportional to the formal degree. By carefully deciding which copies of its children to use to construct a copy of a node, multiplicative disjointness can be achieved with only polynomial blow-up in size.

A nice consequence of this characterisation of VP is a simpler proof of the fact that VP is contained in $\sum \cdot \mathrm{VF}$. The key observation used is that a circuit is multiplicatively disjoint exactly when every proof tree is already a subgraph of the circuit (even without any unfolding into a formula). See [MP08] for details.

Before we move on, we note another surprising relation between ABPs and formulas: VF equals the class of p-families computed by polynomial-size ABPs of constant width. What is this resource "width"? Recall that an ABP is a DAG with edges going "in the direction from s to t". Suppose we impose a layering constraint. The nodes of the DAG must be laid out at the vertices of a rectangular $w \times \ell$ grid, the node s must be at position $(S, 1)$ for some $S \in [w]$, the node t must be at position (T, ℓ) for some $T \in [w]$, and edges can only go across one layer, from (i, k) to $(j, k + 1)$ for some $i, j \in [w]$, $k \in [\ell - 1]$. Of course, any ABP can be converted to one of this form: just sub-divide edges when necessary and label the sub-division path so that its weight is the original edge's label (use lots of 1s). Now we say that w is the width of the layered ABP and ℓ is the length. A bounded-width branching program family (B_n) is one where for some absolute constant c, each B_n has width at most c. Seems quite a squeeze – if we view moving from s towards t as an incremental computation, then at each stage we can carry forward just c intermediate polynomials. We shouldn't be able to do much this way, right? Wrong! Ben-Or and Cleve [BOC92] showed, in a proof cleverly extending Barrington's famous characterisation [Bar89] of NC^1 by Boolean bounded-with branching programs, that every formula of depth D has an equivalent bounded-width branching program (that's quite a mouthful; let's agree to call it BWBP) of length 4^D and width just 3! Since we already know that formulas can be depth-reduced and VF equals VNC^1, we see that VF is contained in a class that we can name VBWBP: polynomial-sized constant-width ABPs. The converse inclusion is easily seen to hold, again using a Savitch-style divide-and-conquer. Thus, we have another characterisation of VF.

As a matter of curiosity, one may want to know: is the width-3 upper bound tight? Allender and Wang [AW11] recently settled this question affirmatively: they show that a very simple polynomial cannot be computed by any width-2 ABP, no matter what the length. On the other hand, width-3 ABPs are universal, since every polynomial family has some formula family computing it. The question is one of efficiency: which families have polynomial-size width 3 ABPs?

OK, so we've had a plethora of class definitions, but just a handful of distinct classes: $\mathrm{VF} = \mathrm{VP}_e = \mathrm{VNC}^1 = \mathrm{VBWBP}$, $\mathrm{VBP} = \mathrm{VP}_{\mathsf{skew}} = \mathrm{VP}_{\mathrm{ws}}$, $\mathrm{VP} = \mathrm{VP}_{\mathrm{MD}}$, $\mathrm{VQF} = \mathrm{VQP}$, $\mathrm{VNF} = \mathrm{VNP}$.

As stated in [Bür00a], *Valiant's hypothesis* says that $\mathrm{VNP} \not\subseteq \mathrm{VP}$, and *Valiant's extended hypothesis* says that $\mathrm{VNP} \not\subseteq \mathrm{VQP}$. Over fields of characteristic not equal to 2, these imply: Perm_n is not a p-projection of Det_n, and Perm_n is not a qp-projection of Det_n, respectively.

Some miscellaneous results, in no specific order:

1. Let SymDet_n be the polynomial that represents the determinant of a symmetric $n \times n$ matrix of indeterminates B_n. (For instance, $\mathrm{SymDet}_2 = x_{11}x_{22} - x_{12}^2$.) Clearly, (SymDet_n) is a p-projection of (Det_n). The converse is also almost true. As shown by Grenet et al. in [GKKP11], over any field of characteristic other than 2, Det_n is a projection of SymDet_{n^3}. Characteristic 2 is a problem: symmetric matrices correspond to undirected graphs, so each undirected cycle gives rise to two directed cycles, and so to get a projection we need division by 2. In characteristic 2, Det_n itself is provably not a projection of SymDet_m for any m; see [GMT13]. The best that we can currently say in characteristic 2 is that the squared determinant $(\mathrm{Det}_n)^2$ is a projection of SymDet_{2n^3+2}; this is also shown in [GKKP11].
2. VQP is also characterized by quasi-polynomial-size weakly skew circuits of polynomial degree. (From [VSBR83] it follows that $\mathrm{VQP} = \mathrm{VQF}$; hence the above characterization. A direct proof is presented in [MP08].) Several natural polynomials are complete for this class under qp-reductions: the (Det_n) family, of course, but also, the trace of iterated matrix product and the trace of a matrix power. These families are all complete for VBP under p-reductions.
3. While we do not know the exact relationship betweenVQP and VNP, (they both contain VP), we do know that VQP does not equal either VP or VNP. Bürgisser ([Bür00a], Sect. 8.2) has shown that there is an explicit family of polynomials (f_n) in VQP that is provably not in VNP, let alone in VP. This family is defined as follows: Consider numbers in base n. Let μ range over all such numbers with $m(n) = \lceil \log n \rceil$ digits. More precisely, let μ range over length-$m(n)$ sequences over the alphabet $\{0, 1, \ldots, n-1\}$, and let $k_n(\mu)$ denote the value of this sequence,

$k_n(\mu) = \sum_{j=1}^{m(n)} \mu_j n^{j-1}$. Define f_n as:

$$f_n(x_1, \ldots, x_{m(n)}) = \sum_{\mu \in \{0,\ldots,n-1\}^{m(n)}} 2^{2^{k_n(\mu)}} \prod_{j=1}^{m(n)} x_j^{\mu_j}$$

Exploiting the fact that the distinct double exponentials appear as coefficients in f_n, Bürgisser shows that f_n cannot be in VNP.

Furthermore, using $m(n) = \lceil \log^i n \rceil$ gives a family of polynomials f^i in VQP with size $O(n^{\log^i n})$ but provably not in size $O(n^{\log^{i-1} n})$, so within VQP there is a strict hierarchy.

4. From the qp-completeness of (Det_n) for VQP, and the p-completeness of (Perm_n) for VNP, it follows that VNP $\subseteq$ VQP if and only if (Perm_n) is a qp-projection of (Det_n). This is a very long-standing open question. Originally, the question of whether (Det_n) and (Perm_n) are p-equivalent was posed by Pólya [Pól13], who also showed that there is no way of expressing the permanent as the determinant by only changing the signs of selected entries (except for $n = 2$; flip the sign of a_{12} to get matrix B with $\mathrm{Det}(B) = \mathrm{Perm}(A)$). (I haven't myself seen Pólya's note, but have seen it referred to in various places.) Marcus and Minc [MM61] showed that there is no size-preserving transformation (Perm_n to Det_n), even if we relax the notion of projections to allow linear form substitions for each variable. For many years, a linear lower bound was the best known ($\Omega(\sqrt{2}n)$ due to [vzG87, Cai90, Mes89]), until Mignon and Ressayre [MR04] showed that over the fields of characteristic 0 (eg real or complex numbers), even if linear form substitutions are allowed in projections, to express Perm_n as a projection of Det_m, we need $m \geq n^2/2$. The same lower bound was obtained for fields of characteristic other than 2 by Cai et al. [CCL10]. From Ryser's work [Rys63] it follows that Perm_n is a projection of Det_m for some $m < n^2 2^n$. More recently, Grenet showed [Gre12b] via a very simple and neat construction that Perm_n is a projection of Det_m for $m = 2^n - 1$. This is the best known so far. Thus there is a huge gap between the lower and upper bounds on what is called the determinantal complexity of the permanent.
5. It is natural to believe that the complexity of a p-family (f_n) in this framework is closely related to the computational complexity of *evaluating* f_n for a given instantiation of its variables. In [Bür00b], Bürgisser gave this belief a firm footing. Consider a p-family (f_n) where f_n depends on n variables. Define its Boolean part BoolPart(f) as a string function mapping $x \in \{0, 1\}^n$ to the binary encoding of $f_n(x)$. Note that we have considered only Boolean values. Even so, evaluation may seem difficult, because the circuits for (f_n) can involve arbitrary constants from the field. Bürgisser showed that assuming the generalised Reimann hypothesis GRH, over fields of characteristic zero, BoolPart(VP) has non-uniform multi-output NC^3 circuits. Furthermore, assuming GRH, if Valiant's hypothesis is false over such a field, then the entire polynomial hierarchy has (non-uniform) NC circuits.

6. An *extreme* depth reduction result is given by the highly influential paper of Agrawal and Vinay [AV08]. To first see the context, note that any polynomial in n variables with degree d has an unbounded fan-in depth-2 circuit of size $2^{O(d+d\log\frac{n}{d})}$. (If $d \in \Omega(n)$, then $2^{O(d)}$ suffices, otherwise the second term in the exponent makes up.) This is because we can just explicitly compute all monomials of degree at most d, and add up the required ones with suitable weights. Now, can we find circuits substantially better than this, say even $2^{o(d+d\log\frac{n}{d})}$, if we allow depth to be increased a bit? Agrawal and Vinay showed that indeed this is possible, even with depth 4, provided there is some circuit (not necessarily depth-reduced) of that size to begin with. The idea is extremely simple. Peform the depth reduction from [VSBR83] or [AJMV98b], and ensure with some additional care that degree provably drops at $\times$ gates. (The price for this is small: a $\times$ gate may have fanin upto 6, instead of 2.) Now, choose a horizontal cut in the depth-reduced circuit so that for the sub-circuit above it, and for the sub-circuits below it rooted at gates on the cut, the "brute-force" construction described above is small. Obviously there is a trade-off: if the cut is too high up, the lower sub-circuits can have large explicit forms, but if it is too low down, the upper sub-circuit can have large explicit forms. Cut in the right place, and everything works out!

   Subsequently the extreme depth-reductions have been pushed further; see [Koi12, Tav13, GKKS13b]. The lower bound results of [GKKS13a, FLMS13] show that the depth reduction upper bound from [Tav13] is tight and cannot be pushed any further.

   This has significant implications for the quest for derandomizing algorithms for the well-studied problem ACIT (arithmetic circuit identity testing)—checking if a given circuit computes the identically zero polynomial. But that is not directly connected with this survey. One question it raises here is: what kind of extreme depth reduction can we achieve for VQP? Can we stay within quasi-polynomial size?

4.4 The Syntactic Multilinear World

Much of the study concerning VP and VNP involves the families (Det_n) amd (Perm_n). The polynomials in both families are *multilinear*. In principle, to compute a multilinear polynomial via a circuit, we need never compute intermediate polynomials that are not multilinear. Let us call such circuits, where the polynomial computed at each node is multilinear, *multilinear circuits*. However, often it is the case that allowing non-multilinear terms at intermediate stages, and eventually cancelling them out, allows more efficient computation (smaller circuits). This leads to the following quest: what kind of multilinear p-families have efficient multilinear formulas, or even multilinear circuits, where each intermediate polynomial is required to be multilinear? Even for the (Det_n) family, which we know is multilinear and in VP, we do not know of polynomial size multilinear circuits. That being the case, can we prove lower bounds?

This question is trickier than it seems at first glance, because given a circuit, even checking whether it is multilinear is non-trivial. Fournier, Malod and Mengel [FMM12] recently observed that checking multilinearity of a given circuit is computationally equivalent to the well-studied problem arithmetic circuit identity testing (ACIT)—checking if a given circuit computes the identically zero polynomial.

So we may want a notion of certifiably multilinear circuits. One such notion is that of syntactic multilinearity, SM. A circuit is said to be syntactically multilinear if at every $\times$ node $\alpha = \beta \times \gamma$, the sub-circuits rooted at nodes β and γ operate on disjoint sets of variables. Note that this is much more restrictive than multiplicative disjointness. But it certifies multilinearity, since no variable can ever get multiplied by itself. And syntactic multilinearity is easy to check computationally: it is violated if there is some node $\alpha = \beta \times \gamma$, some variable x, two input nodes I, I' labelled x, and paths from I to β and I' to γ.

If a family has efficient (polynomial-sized) SM circuits, then it has efficient multilinear circuits. The converse may not be true. But it is true if we look at formulas. Given a multilinear formula, identify an SM violation α, β, γ, x as above. Then we know by multilinearity of the polynomial $p(\alpha)$ that x does not appear in either $p(\beta)$ or $p(\gamma)$. In the appropriate subformula, set all instances of x to 0; the polynomials computed at and above α remain unchanged. Doing this systematically gives an SM formula of size no more than the original multilinear formula.

In the first major breakthrough, Raz [Raz09] showed that for computation by SM formulas, and hence by multilinear formulas, both (Det_n) and (Perm_n) need size $n^{\Omega(\log n)}$. Clearly, this also means that they are not in SM-VNC^1.

Since (Det_n) is in VP and even in VBP, SM-VF is strictly weaker than VBP. But this is hardly a fair comparison: we have restricted VF to be SM, but not VBP and VP. Can we say that SM-VF is strictly weaker than SM-VBP or SM-VP? We do not know whether (Det_n) is in multilinear VP, let alone SM-VP, so a different family is needed as a separating example. Such an example was provided soon thereafter, again by Raz [Raz06]. He constructed an explicit polynomial family that is in SM-VP and even in SM-VSAC^1, and showed that it needs SM-formula size $n^{\Omega(\log n)}$ and hence is not in SM-VNC^1. Improved lower bounds for constant-depth circuits and subclasses of formulas were subsequently obtained by Raz, Shpilka and Yehudayoff (see for instance [RY09, RSY08]).

Let's step back a bit. Why did we say "in SM-VP, and even in SM-VSAC^1"? Aren't VP and VSAC^1 the same? Well, we know that VP and VF can be depth-reduced. But can we assume that these depth reduction tehniques preserve syntactic multilinearity? Fortunately, they do; Raz and Yehudayoff [RY08] showed that the depth reduction of [VSBR83] preserves SM, so indeed SM-VP= SM-VSAC^1. Similarly, in [JMR12] it is observed that the formula depth reduction of [Bre74] also does preserves SM, so SM-VF= SM-VNC^1.

What about other relationhips between the algebraic classes? We had considered ABPs—what certifies multilinearity there? It is easy to see that a read-once restriction, where on each path in the ABP each variable appears as a label at most once, does so. Let us therefore use read-once as the definition of syntactic multilinearity in ABPs. Then, as observed in [JMR12], the Savitch-style divide-and conquer argument preserves SM. So does the conversion from formulas to ABPs, [Val79]. But the conversion from formulas to width-3 ABPs, [BOC92], does not. In fact, Rao [Rao10] showed that even a significant generalisation of Ben-Or and Cleve's technique, using polynomially many registers instead of just 3, cannot preserve syntactic multilinearity. Of course, there may be other ways of going from SM-VF to SM-VBWBP, but it could equally well be that the classes are distinct.

To get back perspective, in the SM world what we have seen so far is:

$$\text{SM-VBWBP} \subseteq \text{SM-VF} \subseteq \text{SM-VBP} \subseteq \text{SM-VP}$$

As mentioned earlier, Raz [Raz06] showed that the inclusion from SM-VF to SM-VP is proper. Very recently, this was improved by Dvir et al. [DMPY12]. They showed that in fact the inclusion SM-VF $\subseteq$ SM-VBP is strict. Whether the first and the last inclusion are strict is still open.

The proof of [DMPY12] is a clever adaptation of the original technique from [Raz06]. Let us briefly examine this.

The central ingredient in Raz's proof is randomly partitioning the variables and analysing the rank of the resulting partial derivatives matrix. Consider a polynomial f on $2n$ variables $X = \{x_1, \dots, x_{2n}\}$, and consider a partition of X into equi-sized sets Y, Z. Consider a $2^n \times 2^n$ matrix $M_f^{Y,Z}$ where rows and columns are indexed by subsets of Y and Z (equivalently, multilinear monomials over Y and Z, respectively). The entry (m_y, m_z) is the coefficient of the monomial $m_y \cdot m_z$ in f. Intuitively, if $M_f^{Y,Z}$ has high rank, then f should be hard. But high rank with respect to what partition? Raz showed that if multilinear f has small SM-formula size, then for at least one partition (Y, Z) of X, $M_f^{Y,Z}$ will have low rank. (The existence of the partition witnessing low rank is proved using the probabilistic method; choose a partition at random, and analyse the probability that the resulting matrix has rank exceeding some threshold.) He also constructed an explicit family g in SM-VSAC^1 and showed that for every partition (Y, Z) of X, $M_g^{Y,Z}$ has high rank; hence g is not in SM-VF.

The non-trivial adaptation done in [DMPY12] is to consider not all partitions, but a fairly small set of what they call arc-partitions. They showed that if f is in SM-VF, then for at least one arc-partition (Y, Z) of X, $M_f^{Y,Z}$ will have low rank. They consider an explicit family g in SM-VBP and show that for every arc-partition (Y, Z) of X, $M_g^{Y,Z}$ has high rank. Hence g is not in SM-VF. The low-rank proof is again probabilistic, but it has a very appealing combinatorial flavour. So does the very definition of an arc-partition.

4.5 More on Completeness

Assume that completeness is defined with respect to p-projections. If a family (f_n) is complete for a class, then understanding (f_n) better allows us to understand the class better. If a natural family is complete for a class, then this is evidence that the class itself is natural.

Valiant started off with a proof that Perm is VNP-complete. He also showed that polynomial families associated with a number of NP-complete languages are complete for VNP under p-projections. So let us agree that VNP is a natural class.

What about VP? The family that naturally contrasts with Perm is Det, but Det is not yet known to be complete for VP (unless we allow qp-projections; that is not quite satisfactory). If this turns out to be the case, it will solve a major open problem, showing that polynomial-degree polynomial size circuits are no more powerful than polynomial-size branching programs VBP. VBP seems a natural enough class, and Det and many other families are complete for it.

So what problems are complete for VP? One can construct a canonical family complete for VP. By canonical, I mean something similar to saying that

$$\{\langle M, x, 1^t\rangle \mid M \text{ is an NDTM that accepts } x \text{ in } t \text{ or fewer steps}\}$$

is NP-complete. Undoubtedly true, but it doesn't give any new intution about what NP is about. In the case of VP, the canonical family is not so trivial to construct (but not very difficult either).

The first description, with a very general completenes result, appears in [Bür00a] (see Sect. 5.6, Cor 5.32(b)). Bürgisser shows that for every p-family h, the *relativized* classes VP^h and VNP^h have complete families with respect to p-projections. Since $\mathrm{VP}^h = \mathrm{VP}$ and $\mathrm{VNP}^h = \mathrm{VNP}$ whenever h itself is in VP, this gives families complete for VP and VNP as well. (In fact, it shows the existence of VNP-complete families, independent of Valiant's original proof.) These complete families compute homogeneous components separately, to keep the degree small, and then add up the required parts. They are constructed by first defining *generic polynomials*, and then defining the appropriate projection/substitution. The generic polynomials capture the canonical notion referred to above.

Later, a more direct construction tailored for VP (as opposed to VP^h and VNP^h for all h) was described by Raz [Raz10], and also appears in [SY10]. Here, the proof of hardness exploits the fact that we can perform depth reduction on VP circuits. (This was not needed in Bürgisser's proof.) Roughly, here's how it goes: For each natural number N, consider a circuit C_N with nodes arranged in $2\log N + 1$ layers numberd $0, 1, \ldots, 2\log N$. All even layers have exactly N nodes, and compute polynomials $g_{i,j}$ where i is the layer number, $j \in [N]$. Odd layers are used to build these polynomials. At layer 0, the polynomials are just distinct variables, $g_{0.j} = x_j$. At higher layers, we have an inductive definition: $g_{i+1,j} = \sum_{k,\ell \in [N]} g_{i,k} \cdot g_{i,\ell} \cdot y_{i,j,k,\ell}$, where the $y_{i,j,k,\ell}$ are new variables. Thus the nodes at the odd layers are the fanin-3 $\times$ nodes, and nodes at even layers (other than the 0 layer) are $+$ nodes with large fanin. (We can reduce the fanins to 2 later; it won't change the polynomial computed.) The polynomial computed by this circuit at $g_{2\log N,1}$ is p_N. The total number of variables is $O(N^3 \log N)$, and the circuit is also of size $O(N^3 \log N)$. The degree of p_N is $2N - 1$. So (p_N) is in VP. Why is it VP-hard? Take any family (f_n) in VP. By the depth reduction of [VSBR83], it can be computed in VSAC^1. The

VSAC^1 circuit D_n can be normalised to have alternating $+$ and $\times$ nodes, with all $\times$ nodes having fanin 2, and all leaves at the same depth. Choose N at least as large as $\min\{\text{size}(D_n), 2^{\text{depth}(D_n)}\}$, and also at least as large as the number of variables in C_n. Now, the computation of D_n can be embedded into C_N: Choose the right number of $+$ nodes at each even layer, and by carefully assigning 0,1 values to the y variables, ensure that they compute the required combinations of polynomials from the previous even layer.

The circuits described above are called *universal circuits* in [SY10], because every circuit is a projection of the universal circuit of appropriate size. And if we start with VP circuits, the projections are p-projections.

So now we know that VP has complete families under p-projections as well. But generic polynomials, universal circuits, and the polynomials they compute, are rather artifical. Are there other families that are defined independent of circuits and are VP-complete? Actually, we know very few. Recently, Stefan Mengel [Men11] made further progress here, considering polynomial families associated with constraint satisfaction problems CSPs. (This builds on earlier work by Briquel, Koiran, Meer [BK09, BKM11], though they did not explicitly look for VP-completeness.) Let's first review what CSPs are. Think of them as generalising CNF–SAT. In CNF–SAT, each clause forbids one assignment to the variables in it. (e.g the clause $x_1 \vee \overline{x_3}$ forbids $x_1 = 0, x_3 = 1$.) In a CSP, variables can take values from a larger domain, not necessarily 0,1. Each constraint is like a clause; it has a set of variables, and it forbids certain combinations of assignments to these variables. (e.g on domain $\{a, b, c\}$ a constraint on x_1, x_2 could say that $x_1 \neq x_2$. That is, assignments aa, bb, cc are forbidden, the other 6 assignments satisfy this constraint.) As in SAT, we look for assignments satisfying all constraints. If the domain has size 2, the CSP is Boolean. If each constraint involves 2 (or less) variables, the CSP is binary. As usual, consider not just a CSP but a family of CSPs (Φ_n), where Φ_n has domain D_n. For tractability, we will require that the CSP is p-bounded; that is, the CSP has bounded arity (for some fixed constant c, each constraint in every Φ_n looks at no more than c variables), and it has polynomial-sized domains (in Φ_n, the variables take values from a set D_n, where the size of D_n is p-bounded). Now associate with each such CSP (Φ_n) a polynomial family $(Q_n = Q(\Phi_n))$, where Q_n is on the variable set $\{X_d \mid d \in D_n\}$ and is defined as follows:

$$\begin{aligned} Q(\Phi_n) &= \sum_{a:\text{var}(\Phi_n)\to D_n} [a \text{ satisfies } \Phi_n] \prod_{x\in\text{var}(\Phi_n)} X_{a(x)} \\ &= \sum_{a:\text{var}(\Phi_n)\to D_n} [a \text{ satisfies } \Phi_n] \prod_{d\in D_n} X_d^{|a^{-1}(d)|} \end{aligned}$$

(Recall, $[S]$ is Boolean, 1 if and only if statement S is true.) Mengel has this wonderful characterisation of the complexity of the family (Q_n). The characterisation involves associating with the CSP a graph G; this graph has a vertex for each variable and an edge between two variables if they occur simultaneously in some constraint. Now the treewidth and pathwidth of the graph (these parameters describe roughly how tree-like or path-like the graph is, if we can consider blobs of vertices. The smaller the blobs, the better the similarity. See [Bod98] for definitions and an overview.) relate to the complexity. It also involves an assignment bound: a CSP is c-assignment-bounded if for each constraint φ and each variable x in the constraint, the number of distinct values possible for x in assignments satisfying φ is bounded by c, even though the domain may be much larger. This seems like a strong condition, but recall that Boolean CSPs are by definition 2-assignment-bounded.

Enough of definitions! Here's what Mengel shows:

1. For each p-bounded CSP (Φ_n), $(Q(\Phi_n))$ is in VNP. Every family (f_n) in VNP is a p-projection of $(Q(\Phi_n))$ for some p-bounded (Φ_n).
2. For each p-bounded CSP (Φ_n) where G_n has bounded treewidth, $(Q(\Phi_n))$ is in VP. Every family (f_n) in VP is a p-projection of $(Q(\Phi_n))$ for some p-bounded binary (Φ_n) where G is a tree (treewidth 1).
3. For each p-bounded CSP (Φ_n) where G_n has bounded pathwidth, $(Q(\Phi_n))$ is in VBP. Every family (f_n) in VBP is a p-projection of $(Q(\Phi_n))$ for some p-bounded binary (Φ_n) where G is a path (pathwidth 1).
4. For each p-bounded c-assignment-bounded CSP (Φ_n) where G_n has bounded treewidth, $(Q(\Phi_n))$ is in VF. Every family (f_n) in VF is a p-projection of $(Q(\Phi_n))$ for some p-bounded 2-assignment-bounded binary (Φ_n) where G has pathwidth at most 26.

The hardness proofs involve looking at the structure of parse trees for VP, witnessing paths for VBP.

Note that as stated, this falls slightly short of providing a single complete family for VP. However, applying the hardness reduction from universal circuits will yield a single CSP family that is VP-complete. To the best of my knowledge, this is the first instance of a VP-hardness result for a family defined (almost) independent of circuits.

All the above results require that the CSP has bounded arity. Unbounded arity seems to immediately give rise to intractability. If arity is unconstrained, can other types of restrictions still result in families in VP? For further progress in this direction, see [DM11, CDM13].

4.6 Computing Integers

The questions concerning algebraic complexity classes are closely connected to another very intriguing question. Let $N > 1$ be any natural number. Suppose we want to build up N from 1, using only $+$, $-$ and $\times$. The most naive way of doing this would be $N = 1 + 1 + \cdots + 1$. But depending on N there can be many other ways. Which is the *most efficient* way? That is, which way uses the least number of $+$ or $\times$ operations? To do anything non-trivial, we must use $+$ at least once, and the first time we use it we will generate 2. So let us not even count this mandatory $+$. How many more operations are needed?

We can state this as a question about circuits. Each way of building up N is an arithmetic circuit, or a straight-line program (SLP), that uses no constants other than 1 and 2. Let us denote by $\tau(N)$ the size of the smallest such circuit computing N. (This is the τ complexity of N). By definition, $\tau(1) = \tau(2) = 0$, and for all $N > 2$, $\tau(N) > 0$. Algorithms for computing N give upper bounds on $\tau(N)$. For instance, to compute $N = 2^k$, here's an SLP: $g_0 = 2$, $g_{i+1} = 2 \times g_i$ for $0 \leq i \leq k-2$. Clearly, g_i computes 2^{i+1}, so $\tau(2^k) \leq k-1$. But I'm sure you can already see better ways of doing this. From the circuit viewpoint, an explanation of why this is not the best is that the circuit corresponding to this SLP is skew. Surely, we should be able to use non-skew gates and compute large numbers faster. Here's another SLP that computes big numbers fast: $f_0 = 2$, $f_{i+1} = f_i \times f_i$ for $0 \leq i \leq \ell - 1$. Clearly, f_i computes 2^{2^i}, so $\tau(2^{2^\ell}) \leq \ell$, a much better bound than the earlier $2^\ell - 1$ at least for numbers of this form. Note that the way we used non-skewness, we produced a circuit with exponential formal degree (the degree at f_i is 2^i), but we're not worried about that for now. Now, using these compact circuits for 2^{2^ℓ}, we can build a better circuit for 2^k by just using the binary expansion of k: $k = \sum_{i=0}^{t} b_i 2^i$, where $t = \lfloor \log k \rfloor$ and $b_t = 1$. So $2^k = 2^{\sum_{i=0}^{t} b_i 2^i} = \prod_{i=0}^{t} 2^{b_i \times 2^i} = \prod_{i:b_i=1} 2^{2^i}$. Compute all the double powers using t operations, and then multiply the required ones using at most t operations. Overall, $\tau(2^k) \leq 2t = 2\lfloor \log k \rfloor$.

We can use the same binary expansion idea to compute any N, not just a power of 2. Compute all powers of 2 upto $\log N$, and add the required ones. This shows that for all N, $\tau(N) \leq 2\lfloor \log N \rfloor - 1$.

So far we have not used any subtractions. But they can be very useful too. For instance, $\tau(2^{2^\ell} - 1) \leq \ell + 1$; compute 2^{2^ℓ} and subtract 1.

What about a lower bound? We can actually formalise the intuition that the exponential degree circuits we saw above for 2^{2^ℓ} produce the largest possible number in that size. Hence, for any N, $\tau(N) \geq \log \log N$.

In particular, $\tau(2^{2^\ell}) = \ell$. That sounds impressive – we know the exact value of τ for 2^{2^ℓ}. But essentially just for that; for all other numbers, we still seem to have a pretty large gap. If $N = 2^k$, then $\log \log N \leq \tau(N) \leq 2\lfloor \log k \rfloor = 2\lfloor \log \log N \rfloor$, so we know $\tau(N)$ within a factor of 2. But for general N, all we know is $\log \log N \leq \tau(N) \leq 2\lfloor \log N \rfloor - 1$. How can we reduce this gap? An obvious search for an efficient way where the last operation is $+$ or $-$ is to express N as $M \pm k$, compute M, compute $k = \pm(N - M)$, and combine, and to choose M that minimizes $\tau(M) + \tau(k) + 1$. (A similar approach can be used for factors of N and a $\times$ as the last operation.) But in computing M and $\pm(N - M)$ (or N/M), the complexity may be *subadditive* since we can reuse intermediate numbers from the program for M while computing $\pm(N - M)$ or N/M. (We are looking for circuits, not formulas.) It is identifying the extent of this reuse that is a challenge.

Similar to Shannon's bound for functions and circuits (most functions require exponential-sized circuits), de Melo and Svaiter [dMS96] showed that most numbers N have $\tau(N)$ closer to the upper bound. They showed that for every $\epsilon > 0$, most N

satisfy $\tau(N) \geq \frac{\log N}{(\log\log N)^{1+\epsilon}}$. Moreira [Mor97] improved this by showing that this holds even for $\epsilon = 0$. (He also showed that for all $\epsilon > 0$, there is an N_ϵ such that for all $N \geq N_\epsilon$, $\tau(N) \leq \frac{(1+\epsilon)\log N}{(\log\log N)}$). And yet, showing such lower bounds for specific numbers seems quite hard – the classic "searching for hay in a haystack" paradox.

Let's move over from individual numbers to sequences of numbers. Let $(a_n)_{n\geq 1}$ be some sequence of natural numbers. When can we say that the sequence is easy to compute? Each number in the sequence should be "easy" relative to its position in the sequence. That is, the sequence (b_n), where $b_n = \tau(a_n)$, should not grow very fast. One possible definition is that b_n should be polynomially bounded in n. For instance, for $a_n = 2^{2^n}$, we know that $b_n = n$. Is that not moderate growth? Not really. Consider a function that maps a position n to not just the number $\tau(a_n) = b_n$ but to an SLP of size b_n computing a_n. For the sequence (2^{2^n}), this function takes an input n represented in $\Theta(\log n)$ bits, and outputs a circuit of size n, that is, exponential in the size of the input. That's not moderate growth!

OK, so let's say that a sequence (a_n) is easy to compute if for some polynomial $p(.)$, for each n, $\tau(a_n) \leq p(\log n)$, and otherwise it is hard to compute. We've set up this definition so that (2^{2^n}) is hard to compute, while the sequences (n), (2^n) are easy to compute. Makes sense? Now let's ask, what other sequences are easy? And what sequences are hard?

A sequence with famously open status is $(n!)$. The completely naive SLP that constructs the first n numbers with $n-2$ increments and then multiplies them shows that $\tau(n!) \leq 2n-4$. But can this be improved significantly? Or is this sequence hard? The best we know is that $\tau(n!) \in O(\sqrt{n}\log^2 n)$; see [BCS97]. Here is the interesting connection to algebraic circuit complexity. Building on a sequence of constructions by Cheng [Che04] and Koiran [Koi05], Bürgisser [Bür09] showed that if $(n!)$ is hard to compute, then any algebraic circuit for the (Perm_n) family that uses only the constants $-1, 0, 1$ must be of superpolynomial size. If we can't even compute the numbers $n!$ easily, then we cannot compute the polynomials Perm_n efficiently, unless we allow the use of constants that cannot themselves built up efficiently.

Analogous to the τ complexity of natural numbers, we can define the τ complexity of polynomial families. Let $\tau(f)$ denote the size of the smallest algebraic circuit using only the constants $-1, 0, 1$—call such a circuit *constant-free*—and computing f. We say that the family (f_n) has polynomially bounded τ complexity if for some polynomial $p(n)$, and for each n, $\tau(f_n) \leq p(n)$. Bürgisser's result can now be stated as: if $\tau(\mathrm{Perm}_n)$ is polynomially bounded, then $(n!)$ is easy to compute.

Let's examine this a bit closely. Why do we state the hypothesis as "$\tau(\mathrm{Perm}_n)$ is polynomial"? Is this not equivalent to saying (Perm_n) is in VP, and hence VNP = VP? Actually, it may not be equivalent. It is possible that (Perm_n) has polynomial-sized circuits but no polynomial-sized constant-free circuits. Conceivably, using other constants in intermediate computation and then cancelling them out could help. Recall that the proof of VNP-hardness of (Perm_n) uses constants other than $-1, 0, 1$; $1/2$ is needed. (As another example, recall how in showing that Det_n is a projection of SymDet_n, we needed the constant $1/2$, even though all coefficients in Det_n are

$-1, 0, 1$.) So we can define a subclass of VP: families with constant-free circuits of polynomial size.

What can we say about such a subclass? As described above, Bürgisser has shown that if this subclass contains (Perm_n), then $(n!)$ is easy to compute. Under the same hypothesis, he also shows that the sequences $\lfloor 2^n e \rfloor$, $\lfloor (3/2)^n \rfloor$ and $\lfloor 2^n \sqrt{2} \rfloor$ are easy to compute.

Malod [Mal03] observed that unlike in the case of VP, for constant-free circuits we may not be able to bound complete formal degree. For VP, if the polynomial computed by a circuit of size s had degree d, we could find an equivalent circuit with formal degree d, and another with complete formal degree $O(d^3 s)$, with only polynomial blow-up in size. Not so if constants aren't freely available! Consider the polynomial family $f_n = 2^{2^n}(x_1 + \cdots + x_n)$. With arbitrary constants, we have a circuit of size n. With only $-1, 0, 1$, we have a circuit of size $2n+1$: build 2, build 2^{2^n}, build the linear form, multiply. But this circuit has exponential formal degree, and in fact, using only the constants $-1, 0, 1$, any circuit must have exponential formal degree to build up 2^{2^n}. So this polynomial is in VP, it has constant-free circuits of polynomial size, but it does not have constant-free polynomial size circuits with polynomially bounded complete formal degree.

This leads to a definition of a further subclass VP^0, first defined in [Mal03]: polynomial families computed by constant-free circuits with polynomially bounded complete formal degree. Define VNP^0 analogous to VNP as $\sum \cdot \mathrm{VP}^0$. Check back; our proof that (Perm_n) is in VNP also shows that (Perm_n) is in VNP^0.

The hypothesis $(\mathrm{Perm}_n) \in \mathrm{VP}^0$ is stronger than saying that $\tau(\mathrm{Perm}_n)$ is polynomially bounded. What does it imply? Can it lead to more sequences being easier to compute? First, note that $(\mathrm{Perm}_n) \in \mathrm{VP}^0$ does not immediately imply $\mathrm{VP}^0 = \mathrm{VNP}^0$. All we can say is the following, shown by Koiran [Koi05]: If (Perm_n) is in VP^0, then for every family $(f_n) \in \mathrm{VNP}^0$, there is some polynomially bounded function $p(n)$ such that the family $(2^{p(n)} f_n)$ is in VP^0. That is, a "shifted" version of f_n is in VP^0. The precise shift can be described as follows—we know that f_n is a projection of $\mathrm{Perm}_{q(n)}$ for some polynomially bounded $q(n)$, we assumed that $\mathrm{Perm}_{q(n)}$ can be computed by a circuit C_n of size and formal degree bounded by a polynomial function of n, we take $p(n)$ to be the formal degree of C_n. Now C_n can be massaged to compute $2^{p(n)} f_n$ instead of $\mathrm{Perm}_{q(n)}$.

This motivates another variant of easy-to-compute. Let's say that a sequence (a_n) of natural numbers is ultimately easy to compute if at least some shifted version of it is easy to compute. That is, there is some other integer sequence A_n such that the sequence $a_n A_n$ is easy to compute. Note that if (a_n) is not ultimately easy, then for infinitely many n, all nonzero multiples of a_n have large τ complexity. Using this property, under the hypothesis that $n!$ is not even ultimately easy to compute, we can obtain a non-trivial derandomization of the Arithmetic-Circuit-Identity-Testing problem; see the last section of [ABKPM09]. Earlier, Koiran showed in [Koi05] that if $n!$ is not even ultimately easy to compute, then we have some separation: either $\mathrm{VP}^0 \neq \mathrm{VNP}^0$, or $\mathrm{P} \neq \mathrm{PSPACE}$. This is curious: we have a consequence involving Boolean classes as well. But it should not be so surprising. VP^0 and VNP^0 are computed by (sums of) constant-free poly-formal-degree algebraic circuits, and these are the arithmetic circuits that arise when we consider counting classes like #P that count accepting paths of Turing machines. This does not mean that $\mathrm{VNP}^0 = \#\mathrm{P}$; the former is a collection of polynomial families whereas the latter is a collection of functions from strings to whole numbers. But the complexity of evaluating polynomial families in the former collection, at Boolean arguments, is closely related to what the latter collection refers to. Koiran's proof actually shows the contrapositive: he first shows that if $\mathrm{VP}^0 = \mathrm{VNP}^0$ and P = PSPACE, then the sequence $\tau((2^\ell)!)$ is polynomially bounded in ℓ. So consider instead of each $n!$ the possibly larger factorial $(2^{\ell(n)})!$, where $2^{\ell(n)-1} < n \leq 2^{\ell(n)}$. Then the sequence $(b_n) = ((2^{\ell(n)})!)$ is easy to compute, and each b_n is a multiple of $n!$, so $(n!)$ is ultimately easy to compute.

Since Perm_n is not known to be complete for VNP^0, what is? It turns out that for several other VNP-complete families, the hardness proofs use no constants other than $-1, 0, 1$ and the membership proofs use circuits with small formal degree; hence these families are complete for VNP^0 as well. As a concrete example, consider the Hamilton cycle polynomial family HC_n defined as follows: Let distinct variables $x_{i,j}$ label the edges of the complete directed graph D_n. Let C_n denote the set of all directed Hamiltonian cycles in D_n; elements of C_n can be described by cyclic permutations $\sigma \in S_n$. Then

$$\mathrm{HC}_n(x_{11}, \ldots, x_{nn}) = \sum_{\sigma \in C_n} \prod x_{i,\sigma(i)}$$

This family is complete for VNP^0; [Mal03].

Returning to the question "What does $(\mathrm{Perm}_n) \in \mathrm{VP}^0$ imply?"; Koiran [Koi05] showed that it implies the sequence $\lfloor 2^n \ln 2 \rfloor$ is easy to compute. He also improved the earlier-mentioned result in two ways, from "$[(\mathrm{VP}^0 = \mathrm{VNP}^0) \wedge (\mathrm{P} = \mathrm{PSPACE})] \Rightarrow (n!)$ is ultimately easy to compute" to "$[(\mathrm{Perm}_n \in \mathrm{VP}^0) \wedge (\mathrm{P} = \mathrm{PSPACE})] \Rightarrow (n!)$ is easy to compute".

Under the stronger hypothesis that $\mathrm{VP}^0 = \mathrm{VNP}^0$, we can show more (again due to [Koi05]). If $\mathrm{VP}^0 = \mathrm{VNP}^0$, then the sequences $(\sum_{i=1}^{2^n} 2^{i^2-1})$, $\lfloor 2^{2^n} \ln 2 \rfloor$, $\lfloor 2^{2^n} \ln 3 \rfloor$, $\lfloor 2^{2^n} \pi \rfloor$, all have polynomially bounded complexity, something that is not yet known unconditionally.

Acknowledgments I thank Arvind and Manindra for inviting me to contribute to this volume in honour of Somenath Biswas, a wonderful professional colleague and friend. I thank CEFIPRA for supporting an Indo-French collaboration (project 4702-1); many of my ideas for how to present this survey were crystallised during my visit to University of Paris-Diderot during May–June 2012. I have picked material I found interesting, and have not really attempted an exhaustive coverage. I apologise in advance to those whose favourite results I have omitted. I gratefully acknowledge many insightful discussions with Eric Allender, V. Arvind, Hervé Fournier, Bruno Grenet, Nutan Limaye, Guillaume Malod, Stefan Mengel, Sylvain Perifel, B. V. Raghavendra Rao, Nitin Saurabh, Karteek Sreenivasaiah, Srikanth Srinivasan, V. Vinay. I thank the organisers of the Dagstuhl Seminars on Circuits, Logic and Games (Feb 2010) and Computational Counting (Dec 2010) for inviting me and giving me the opportunity to discuss these topics. The survey by Pascal Koiran at the Dagstuhl seminar on Computational Counting in Dec 2010 was particularly helpful.

References

[ABKPM09] E. Allender, P. Bürgisser, J. Kjeldgaard-Pedersen, P.B. Miltersen, On the complexity of numerical analysis. SIAM J. Comput. **38**(5) 1987–2006 (2009)

[AJMV98a] E. Allender, J. Jiao, M. Mahajan, V. Vinay, Non-commutative arithmetic circuits: depth reduction and size lower bounds. Theoret. Comput. Sci. **209**, 47–86 (1998)

[AJMV98b] E. Allender, J. Jiao, M. Mahajan, V. Vinay, Non-commutative arithmetic circuits: Depth reduction and size lower bounds. Theor. Comput. Sci. **209**(1–2) 47–86 (1998)

[AV08] M. Agrawal, V. Vinay, Arithmetic circuits: a chasm at depth four, in *FOCS*, pp. 67–75 (2008). See also ECCC TR15-062, 2008

[AW11] E. Allender, F. Wang, On the power of algebraic branching programs of width two. ICALP **1**, 736–747 (2011)

[Bar89] D.A. Barrington, Bounded-width polynomial size branching programs recognize exactly those languages in NC^1. J. Comput. Syst. Sci. **38**, 150–164 (1989)

[BCS97] P. Bürgisser, M. Clausen, M.A. Shokrollahi, *Algebraic Complexity Theory* (Springer, Berlin, 1997)

[BK09] I. Briquel, P. Koiran, A dichotomy theorem for polynomial evaluation, in *MFCS*, pp. 187–198 (2009)

[BKM11] I. Briquel, P. Koiran, K. Meer, On the expressive power of cnf formulas of bounded tree- and clique-width. Discrete Appl. Math. **159**(1), 1–14 (2011)

[Blä13] M. Bläser. Noncommutativity makes determinants hard, in *Proceedings of ICALP*, vol. 7965 of Lecture Notes in Computer Science, pp. 172–183, Springer, ECCC TR 2012–142 (2013)

[BOC92] M. Ben-Or, R. Cleve, Computing algebraic formulas using a constant number of registers. SIAM J. Comput. **21**, 54–58 (1992)

[Bod98] H.L. Bodlaender, A partial k-arboretum of graphs with bounded treewidth. Theor. Comput. Sci. **209**(12) 1–45 (1998)

[Bre74] R.P. Brent, The parallel evaluation of general arithmetic expressions. J. ACM **21**, 201–206 (1974)

[Bür00a] P. Bürgisser, *Completeness and Reduction in Algebraic Complexity Theory*, vol. 7 of Algorithms and Computation in Mathematics (Springer, Berlin, 2000)

[Bür00b] P. Bürgisser, Cook's versus Valiant's hypothesis. Theor. Comput. Sci. **235**(1), 71–88 (2000)

[Bür09] P. Bürgisser, On defining integers and proving arithmetic circuit lower bounds. Comput. Complex. **18**(1), 81–103 (2009)

[Cai90] J.-Y. Cai, A note on the determinant and permanent problem. Inf. Comput. **84**(1), 119–127 (1990)

[CCL10] J.-Y. Cai, X. Chen, D. Li, Quadratic lower bound for permanent versus determinant in any characteristic. Comput. Complex. **19**(1), 37–56 (2010)

[CDM13] F. Capelli, A. Durand, S. Mengel, The arithmetic complexity of tensor contractions, in *STACS*, vol. 20 of LIPIcs, pp. 365–376 (2013)

[Che04] Q. Cheng, On the ultimate complexity of factorials. Theor. Comput. Sci. **326**(1–3), 419–429 (2004)

[Csa76] L. Csanky, Fast parallel inversion algorithm. SIAM J. Comput. **5**, 818–823 (1976)

[Dam91] C. Damm, DET= $L^{\#L}$. Technical Report Informatik-Preprint 8, Fachbereich Informatik der Humboldt-Universität zu Berlin (1991)

[DM11] A. Durand, S. Mengel, On polynomials defined by acyclic conjunctive queries and weighted counting problems. CoRR abs/1110.4201 (2011)

[DMPY12] Z. Dvir, G. Malod, S. Perifel, A. Yehudayoff, Separating multilinear branching programs and formulas, in *STOC*, pp. 615–624 (2012)

[dMS96] W. de Melo, B.F. Svaiter, The cost of computing integers. Proc. Am. Math. Soc. **124**(5), 1377–1378 (1996)

[FLMS13] H. Fournier, N. Limaye, G. Malod, S. Srinivasan, Lower bounds for depth 4 formulas computing iterated matrix multiplication. Electron. Colloquium Comput. Complex. (ECCC) **20** 100 (2013) to appear in STOC 2014

[FMM12] H. Fournier, G. Malod, S. Mengel, Monomials in arithmetic circuits: complete problems in the counting hierarchy, in *STACS*, pp. 362–373 (2012)

[GKKP11] B. Grenet, E. Kaltofen, P. Koiran, N. Portier, Symmetric determinantal representation of weakly-skew circuits, in *STACS*, pp. 543–554 (2011)

[GKKS13a] A. Gupta, P. Kamath, N. Kayal, R. Saptharishi, Approaching the chasm at depth four, in *IEEE Conference on Computational Complexity*, (2013)

[GKKS13b] A. Gupta, P. Kamath, N. Kayal, R. Saptharishi, Arithmetic circuits: a chasm at depth three, in *IEEE Foundations of Computer Science* (FOCS), ECCC 2013–026 (2013)

[GMT13] B. Grenet, T. Monteil, S. Thomassé, Symmetric determinantal representations in characteristic 2. Linear Alg. Appl. **439**(5), 1364–1381 (2013)

[Gre12a] B. Grenet, Représentation des polynômes, algorithmes et bornes infÃrieures. Ph.D. thesis, École Normale SupÃrieure de Lyon, (2012)

[Gre12b] B. Grenet, *An Upper Bound for the Permanent Versus Determinant Problem* manuscript, (2012)

[Hya79] L. Hyafil, On the parallel evaluation of multivariate polynomials. SIAM J. Comput. **8**(2), 120–123 (1979)

[Jan08] M.J. Jansen, Lower bounds for syntactically multilinear algebraic branching programs, in *MFCS*, vol. 5162 of Lecture Notes in Computer Science, pp. 407–418 (Springer, Berlin, 2008)

[JMR12] M. Jansen, M. Mahajan, B.V. Raghavendra Rao, Resource trade-offs in syntactic multilinear arithmetic circuits. Computational Complexity **22**(3), 517–564 (2013)

[Kal85] K. Kalorkoti, A lower bound for the formula size of rational functions. SIAM J. Comput. **14**(3), 678–687 (1985)

[Kay10] N. Kayal, Algorithms for arithmetic circuits. Electron. Colloquium Comput. Complex. (ECCC) **17** 73 (2010)

[KK08] E. Kaltofen, P.Koiran, Expressing a fraction of two determinants as a determinant, in *ISSAC*, pp. 141–146, ACM (2008)

[Koi05] P. Koiran, Valiant's model and the cost of computing integers. Comput. Complex. **13**(3–4), 131–146 (2005)

[Koi10] P. Koiran, Complexity of arithmetic circuits (a skewed perspective), in Slides from Dagstuhl seminar 10481. DROPS, 2010. http://www.dagstuhl.de/Materials/Files/10/10481/10481.KoiranPascal.Slides.pdf

[Koi12] P. Koiran, Arithmetic circuits: the chasm at depth four gets wider. Theor. Comput. Sci. **448**, 56–65 (2012)

[Mal03] G. Malod, PolynoĹmes et coefficients, Ph.D. thesis, University Claude Bernard Ü Lyon 1, (2003)

[Men11] S. Mengel, Characterizing arithmetic circuit classes by constraint satisfaction problems—(extended abstract). ICALP **1**, 700–711 (2011)

[Mes89] R. Meshulam, On two extremal matrix problems. Linear Algebra Appl. 114(115), 261–271 (1989). Special Issue Dedicated to A.J. Hoffman

[MM61] M. Marcus, H. Minc, On the relation between the determinant and the permanent. Ill. J. Math. **5**, 376–381 (1961)

[Mor97] C.G.T. de A. Moreira, On asymptotic estimates for arithmetic cost functions. Proc. Am. Math. Soc. **125**(2) 347–353 (1997)

[MP08] G. Malod, N. Portier, Characterizing valiant's algebraic complexity classes. J. Complex. **24**(1), 16–38 (2008)

[MR04] T. Mignon, N. Ressayre, A quadratic bound for the determinant and permanent problem, in *International Mathematics Research Notices*, pp. 2004–4241, (2004)

[MV97] M. Mahajan, V. Vinay, Determinant: combinatorics, algorithms, complexity. Chicago J. Theor. Comput. Sci. http://www.cs.uchicago.edu/publications/cjtcs, 1997:5, Dec 1997. Preliminary version in *Proceedings of the Eighth Annual ACM-SIAM Symposium on Discrete Algorithms SODA*, pp. 730–738 (1997)

[Pól13] G. Pólya. Aufgabe 424. Archiv der Mathematik und Physik **3**(20) 271 (1913)

[Rao10] B.V. Raghavendra Rao, A Study of Width Bounded Arithmetic Circuits and the Complexity of Matroid Isomorphism, Ph.D. thesis. The Institute of Mathematical Sciences, Chennai, India., 2010. http://www.imsc.res.in/xmlui/handle/123456789/177

[Raz06] R. Raz, Separation of multilinear circuit and formula size. Theory Comput. **2**(1) 121–135 (2006). preliminary version in FOCS 2004

[Raz09] R. Raz. Multi-linear formulas for permanent and determinant are of super-polynomial size. J. ACM, **56**(2), (2009). preliminary version in STOC 2004

[Raz10] R. Raz, Elusive functions and lower bounds for arithmetic circuits. Theory Comput. **6**(1), 135–177 (2010)

[RSY08] R. Raz, A. Shpilka, A. Yehudayoff, A lower bound for the size of syntactically multilinear arithmetic circuits. SIAM J. Comput. **38**(4), 1624–1647 (2008)

[RY08] R. Raz, A. Yehudayoff, Balancing syntactically multilinear arithmetic circuits. Comput. Complex. **17**(4), 515–535 (2008)

[RY09] R. Raz, A. Yehudayoff, Lower bounds and separations for constant depth multilinear circuits. Comput. Complex. **18**(2), 171–207 (2009)

[Rys63] H.J. Ryser, *Combinatorial Mathematics* (Carus mathematical monographs, Mathematical Association of America, 1963)

[Sav70] J. Walter, Savitch, relationships between nondeterministic and deterministic tape complexities. J. Comput. Syst. Sci. **4**(2), 177–192 (1970)

[Str73] V. Strassen, Vermeidung von divisionen. J. Reine U. Angew Math **264**, 182–202 (1973)

[SY10] A. Shpilka, A. Yehudayoff, Arithmetic circuits: a survey of recent results and open questions. Found. Trends Theor. Comput. Sci. **5**(3–4), 207–388 (2010)

[Tav13] S. Tavenas, Improved bounds for reduction to depth 4 and depth 3, in *MFCS*, vol. 8087 of Lecture Notes in Computer Science, pp. 813–824 (Springer, Berlin, 2013)

[Tod92] S. Toda, Classes of arithmetic circuits capturing the complexity of computing the determinant. IEICE Trans. Inf. Syst. **E75-D**, 116–124 (1992)

[Val79] L.G. Valiant, Completeness classes in algebra, in *STOC*, pp. 249–261 (1979)

[Val82] L.G. Valiant, Reducibility by algebraic projections, in *Logic and Algorithmic: International Symposium in honour of Ernst Specker*, vol. 30, pp. 365–380. Monograph. de l'Enseign. Math. (1982)

[Val92] L.G. Valiant, Why is boolean complexity theory difficult? in *Boolean Function Complexity*, ed. by M.S. Paterson (Cambridge University Press, London Mathematical Society Lecture Notes Series 169, 1992)

[Ven92] H. Venkateswaran, Circuit definitions of nondeterministic complexity classes. SIAM J. Comput. **21**, 655–670 (1992)

[Vin91] V. Vinay, Counting auxiliary pushdown automata and semi-unbounded arithmetic circuits, in *Proceedings of 6th Structure in Complexity Theory Conference*, pp. 270–284 (1991)

[Vol99] H. Vollmer, *Introduction to Circuit Complexity: A Uniform Approach* (Springer, New York, 1999)

[VSBR83] L.G. Valiant, S. Skyum, S. Berkowitz, C. Rackoff, Fast parallel computation of polynomials using few processors. SIAM J. Comput. **12**(4), 641–644 (1983)

[vzG87] J. von zur Gathen, Permanent and determinant. Linear Algebra Appl. **96**, 87–100 (1987)

Chapter 5
A Selection of Lower Bounds for Arithmetic Circuits

Neeraj Kayal and Ramprasad Saptharishi

It is convenient to have a measure of the amount of work involved in a computing process, even though it may be a very crude one ... We might, for instance, count the number of additions, subtractions, multiplications, divisions, recordings of numbers,...

from *Rounding-off errors in matrix processes,*
Alan M. Turing, 1948

Abstract This article is a survey of techniques used in arithmetic circuit lower bounds.

Keywords Arithmetic circuits · Lower bounds · Determinant · Permanent

Mathematics Subject Classification (2010) Primary 68Q25, 68W30, Secondary 12E05

5.1 Introduction

Polynomials originated in classical mathematical studies concerning geometry and solutions to systems of equations. They feature in many classical results in algebra, number theory, and geometry—e.g. in Galois and Abel's resolution of the solvability

To Somenath Biswas, on his 60th Birthday.

N. Kayal (✉) · R. Saptharishi
Microsoft Research, Bangalore 560001, India
e-mail: neeraka@microsoft.com

R. Saptharishi
e-mail: ramprasad@cmi.ac.in

M. Agrawal and V. Arvind (eds.), *Perspectives in Computational Complexity*,
Progress in Computer Science and Applied Logic 26, DOI: 10.1007/978-3-319-05446-9_5,

via radicals of a quintic, Lagrange's theorem on expressing every natural number as a sum of four squares and the impossibility of trisecting an angle (using ruler and compass). In modern times, computer scientists have begun to investigate as to what functions can be (efficiently) computed. Polynomials being a natural class of functions, one is naturally led to the following question:

What is the optimum way to compute a given (family of) polynomial(s)?

Now the most natural way to compute a polynomial $f(x_1, x_2, \ldots, x_n)$ over a field $\mathbb{F}$ is to start with the input variables $x_1, x_2, \ldots, x_n$ and then apply a sequence of basic operations such as additions, subtractions, and multiplications[1] in order to obtain the desired polynomial f. Such a computation is called a straight-line program. We often represent such a straight-line program graphically as an arithmetic circuit—wherein the overall computation corresponds to a directed acylic graph whose source nodes are labelled with the input variables $\{x_1, x_2, \ldots, x_n\}$ and the internal nodes are labelled with either $+$ or $\times$ (each internal node corresponds to one computational step in the straight-line program). We typically allow arbitrary constants from the underlying field on the incoming edges of a $+$ gate so that a $+$ gate can in fact compute an arbitrary $\mathbb{F}$-linear combination of its inputs. The complexity of the computation corresponds to the number of operations, also called the size of the corresponding arithmetic circuit. With arithmetic circuits being the relevant model, the informal question posed above can be formalized by defining the optimal way to compute a given polynomial as the smallest arithmetic circuit in terms of the size that computes it. While different aspects of polynomials have been studied extensively in various areas of mathematics, what is unique to computer science is the endeavor to prove upper and lower bounds on the size of arithmetic circuits computing a given (family of) polynomials. Here we give a biased survey of this area, focusing mostly on lower bounds. Note that there are already two excellent surveys of this area—one by Avi Wigderson [Wig02] and the other by Amir Shpilka and Amir Yehudayoff [SY10].[2] Our intention in writing the survey is the underlying hope that revisiting and assimilating the known results pertaining to circuit lower bounds will in turn help us make progress on this beautiful problem. Consequently, we mostly present here those results which we for some reason felt we did not understand comprehensively enough. We conclude with some recent lower bound results for homogeneous bounded depth formulas. Some notable lower bound results that we are unable to present here due to space and time constraints are as follows. A quadratic lower bound for depth three circuits by Shpilka and Wigderson [SW01], for bounded occur bounded depth formulas by Agrawal, Saha, Saptharishi, and Saxena [ASSS12] and the $n^{1+\Omega(1/r)}$ lower bound for circuits of depth r by Raz [Raz10].

[1] One can also allow more arithmetic operations such as division and square roots. It turns out, however, that one can efficiently simulate any circuit with divisions and square roots by another circuit without these operations while incurring only a polynomial factor increase in size.

[2] A more specialized survey by Chen, Kayal, and Wigderson [CKW11] focuses on the applications of partial derivatives in understanding the structure and complexity of polynomials.

Overview. The state of affairs in arithmetic complexity is such that despite a lot of attention we still have only modest lower bounds for general circuits and formulas. In order to make progress, recent work has focused on restricted subclasses. We first present the best-known lower bound for general circuits due to Baur and Strassen [BS83], and a lower bound for formulas due to Kalorkoti [Kal85]. The subsequent lower bounds that we present follow a common roadmap and we articulate this in Sect. 5.4, and present some simple lower bounds to help the reader gain familiarity. We then present (a slight generalization of) an exponential lower bound for monotone circuits due to Jerrum and Snir [JS82]. Moving on to some other restricted (but still nontrivial and interesting) models, we first present an exponential lower bound for depth three circuits over finite fields due to Grigoriev and Karpinski [GK98] and multilinear formulas. We conclude with some recent progress on lower bounds for homogeneous depth four circuits.

Remark Throughout the article, we shall use Det_n and Perm_n to refer to the determinant and permanent, respectively, of a symbolic $n \times n$ matrix $\left(\left(x_{ij}\right)\right)_{1 \le i,j \le n}$.

5.2 Existential Lower Bounds

Before we embark on our quest to prove lower bounds for interesting families of polynomials, it is natural to ask as to what bounds one can hope to achieve. For a multivariate polynomial $f(\mathbf{x}) \in \mathbb{F}[\mathbf{x}]$, denote by $S(f)$ the size of the smallest arithmetic circuit computing f.

Theorem 1 (Folklore) *For "most" polynomials $f(\mathbf{x}) \in \mathbb{F}[\mathbf{x}]$ of degree d on n variables we have*

$$S(f) \ge \Omega\left(\sqrt{\binom{n+d}{d}}\right).$$

Sketch of Proof We prove this here only in the situation where the underlying field $\mathbb{F}$ is a finite field and refer the reader to another survey ([CKW11], Chap. 4) for a proof in the general case. So let $\mathbb{F} = \mathbb{F}_q$ be a finite field. Any line of a straight-line program computing f can be expressed as taking the product of two $\mathbb{F}_q$-linear combinations of previously computed values. Hence the total number of straight-line programs of length s is at most $q^{O(s^2)}$. On the other hand, there are $q^{\binom{n+d}{d}}$ polynomials of degree d on n variables. Hence most n-variate polynomials of degree d require straight-line programs of length at least (equivalently arithmetic circuits of size at least) $s = \Omega\left(\sqrt{\binom{n+d}{d}}\right)$. □

Hrubes and Yehudayoff [HY11] showed that in fact most n-variate polynomials of degree d *with zero-one coefficients* have complexity at least $\Omega\left(\sqrt{\binom{n+d}{d}}\right)$. Now it turns out that this is in fact a lower bound on the number of multiplications in any circuit computing a random polynomial. Lovett [Lov11] complements this nicely by giving a matching upper bound. Specifically, it was shown in [Lov11] that for any polynomial f of degree d on n variables there exists a circuit computing f having at most $\left(\sqrt{\binom{n+d}{d}}\right) \cdot (nd)^{O(1)}$ multiplications.

5.3 Weak Lower Bounds for General Circuits and Formulas

Despite several attempts by various researchers to prove lower bounds for arithmetic circuits or formulas, we only have very mild lower bounds for general circuits or formulas thus far. In this section, we shall look at the two modest lower bounds for general circuits and formulas.

5.3.1 Lower Bounds for General Circuits

The only super-linear lower bound we currently know for general arithmetic circuits is the following result of Baur and Strassen [BS83].

Theorem 2 [BS83] *Any fan-in two circuit that computes the polynomial* $f = x_1^{d+1} + \cdots + x_n^{d+1}$ *has size* $\Omega(n \log d)$.

5.3.1.1 An Exploitable Weakness

Each gate of the circuit Φ computes a local operation on the two children. To formalize this, define a new variable y_g for every gate $g \in \Phi$. Further, for every gate g define a quadratic equation Q_g as

$$Q_g = \begin{cases} y_g - (y_{g_1} + y_{g_2}) & \text{if } g = g_1 + g_2 \\ y_g - (y_{g_1} \cdot y_{g_2}) & \text{if } g = g_1 \cdot g_2. \end{cases}$$

Further if y_o corresponds to the output gate, then the system of equations

$$\left\{Q_g = 0 \ : \ g \in \Phi\right\} \cup \{y_o = 1\}$$

completely characterize the computations of Φ that results in an output of 1.

The same can also be extended for *multi-output* circuits that compute several polynomials simultaneously. In such cases, the set of equations

$$\left\{Q_g = 0 \ : \ g \in \Phi\right\} \cup \left\{y_{o_i} = 1 \ : \ i = 1, \ldots, n\right\}$$

completely characterize computations that result in an output of all ones. The following classical theorem allows us to bound the number of common roots to a system of polynomial equations.

Theorem 3 (Bézout's theorem) *Let $g_1, \ldots, g_r \in \mathbb{F}[X]$ such that* $\deg(g_i) = d_i$ *such that the number of common roots of $g_1 = \cdots = g_r = 0$ is finite. Then, the number of common roots (counted with multiplicities) is bounded by $\prod d_i$.*

Thus in particular, if we have a circuit Φ of size s that *simultaneously* computes $\{x_1^d, \ldots, x_n^d\}$, then we have d^n inputs that evaluate to all ones (where each x_i must be a d-th root of unity). Hence, Bézout's theorem implies that

$$2^s \geq d^n \implies s = \Omega(d \log n).$$

Observe that $\{x_1^d, \ldots, x_n^d\}$ are all first-order derivatives of $f = x_1^{d+1} + \cdots + x_n^{d+1}$ (with suitable scaling). A natural question here is the following—if f can be computed an arithmetic circuit of size s, what is the size required to compute all first-order partial derivatives of f simultaneously? The naïve approach of computing each derivative separately results in a circuit of size $O(s \cdot n)$. Baur and Strassen [BS83] show that we can save a factor of n.

Lemma 4 [BS83] *Let Φ be an arithmetic circuit of size s and fan-in 2 that computes a polynomial $f \in \mathbb{F}[X]$. Then, there is a multi-output circuit of size $O(s)$ computing all first-order derivatives of f.*

Note that this immediately implies that any circuit computing $f = x_1^{d+1} + \cdots + x_n^{d+1}$ requires size $\Omega(d \log n)$ as claimed by Theorem 2.

5.3.1.2 Computing All First-Order Derivatives Simultaneously

Since we are working with fan-in two circuits, the number of edges is at most twice the size. Hence let s denote the number of edges in the circuit Φ, and we shall prove by induction that all first-order derivatives of Φ can be computed by a circuit of size at most $5s$. Pick a non-leaf node v in the circuit Φ closest to the leaves with both its children being variables, and say x_1 and x_2 are the variables feeding into v. In other words, $v = x_1 \odot x_2$ where $\odot$ is either $+$ or $\times$.

Let Φ' be the circuit obtained by deleting the two edges feeding into v, and replacing v by a new variable. Hence, Φ' computes a polynomial $f' \in \mathbb{F}[X \cup \{v\}]$ and has at most $(s-1)$ edges. By induction on the size, we can assume that there is a circuit $\mathbb{D}(\Phi')$ consisting of at most $5(s-1)$ edges that computes all the first-order derivatives of f'.

Observe that since $f' \mid_{(v=x_1 \odot x_2)} = f(\mathbf{x})$, we have that

$$\frac{\partial f}{\partial x_i} = \left(\frac{\partial f'}{\partial x_i}\right)_{v=x_1 \odot x_2} + \left(\frac{\partial f'}{\partial v}\right)_{v=x_1 \odot x_2} \left(\frac{\partial(x_1 \odot x_2)}{\partial x_i}\right).$$

Hence, if $v = x_1 + x_2$ then

$$\frac{\partial f}{\partial x_1} = \left(\frac{\partial f'}{\partial x_1}\right)_{v=x_1+x_2} + \left(\frac{\partial f'}{\partial v}\right)_{v=x_1+x_2}$$
$$\frac{\partial f}{\partial x_2} = \left(\frac{\partial f'}{\partial x_2}\right)_{v=x_1+x_2} + \left(\frac{\partial f'}{\partial v}\right)_{v=x_1+x_2}$$
$$\frac{\partial f}{\partial x_i} = \left(\frac{\partial f'}{\partial x_i}\right)_{v=x_1+x_2} \qquad \text{for } i > 2.$$

If $v = x_1 \cdot x_2$, then

$$\frac{\partial f}{\partial x_1} = \left(\frac{\partial f'}{\partial x_1}\right)_{v=x_1\cdot x_2} + \left(\frac{\partial f'}{\partial v}\right)_{v=x_1\cdot x_2} \cdot x_2$$
$$\frac{\partial f}{\partial x_2} = \left(\frac{\partial f'}{\partial x_2}\right)_{v=x_1\cdot x_2} + \left(\frac{\partial f'}{\partial v}\right)_{v=x_1\cdot x_2} \cdot x_1$$
$$\frac{\partial f}{\partial x_i} = \left(\frac{\partial f'}{\partial x_i}\right)_{v=x_1\cdot x_2} \qquad \text{for } i > 2.$$

Hence, by adding at most 5 additional edges to $\mathbb{D}(\Phi')$, we can construct $\mathbb{D}(\Phi)$ and hence size of $\mathbb{D}(\Phi)$ is at most $5s$. □ (Lemma 4)

5.3.2 Lower Bounds for Formulas

This section is devoted to the proof of Kalorkoti's lower bound [Kal85] for formulas computing Det_n, Perm_n.

Theorem 5 [Kal85] *Any arithmetic formula computing* Perm_n *(or* Det_n*) requires* $\Omega(n^3)$ *size.*

The exploitable weakness in this setting is again to use the fact that the polynomials computed at intermediate gates share many polynomial dependencies.

Definition 6 (*Algebraic independence*) A set of polynomials $\{f_1, \ldots, f_m\}$ is said to be *algebraically independent* if there is no nontrivial polynomial $H(z_1, \ldots, z_m)$ such that $H(f_1, \ldots, f_m) = 0$.

The size of the largest algebraically independent subset of $\mathbf{f} = \{f_1, \ldots, f_m\}$ is called the *transcendence degree* (denoted by $\mathrm{trdeg}(f)$).

The proof of Kalorkoti's theorem proceeds by defining a *complexity measure* using the above notion of algebraic independence.

The Measure: For any subset of variables $Y \subseteq X$, we can write a polynomial $f \in \mathbb{F}[X]$ of the form $f = \sum_{i=1}^{s} f_i \cdot m_i$ where m_i's are distinct monomials in the variables in Y, and $f_i \in F[X \setminus Y]$. We shall denote by $\mathrm{td}_Y(f)$ the transcendence degree of $\{f_1, \ldots, f_s\}$

Fix a partition of variables $X = X_1 \sqcup \cdots \sqcup X_r$. For any polynomial $f \in \mathbb{F}[X]$, define the map $\Gamma^{[\mathrm{Kal}]} : \mathbb{F}[X] \to \mathbb{Z}_{\geq 0}$ as

$$\Gamma^{[\mathrm{Kal}]}(f) = \sum_{i=1}^{r} \mathrm{td}_{X_i}(f).$$

The lower bound proceeds in two natural steps:

1. Show that $\Gamma^{[\mathrm{Kal}]}(f)$ is *small* whenever f is computable by a *small* formula.
2. Show that $\Gamma^{[\mathrm{Kal}]}(\mathsf{Det}_n)$ is *large*.

5.3.2.1 Upper Bounding $\Gamma^{[\mathrm{Kal}]}$ for a Formula

Lemma 7 *Let f be computed by a fan-in two formula Φ of size s. Then for any partition of variables $X = X_1 \sqcup \cdots \sqcup X_r$, we have $\Gamma^{[\mathrm{Kal}]}(f) = O(s)$.*

Proof For any node $v \in \Phi$, let $\textsc{Leaf}(v)$ denote the leaves of the subtree rooted at v and let $\textsc{Leaf}_{X_i}(v)$ denote the leaves of the subtree rooted at v that are in the part X_i. Since the underlying graph of Φ is a tree, it follows that the size of Φ is bounded by twice the number of leaves. For each part X_i, we shall show that $\mathrm{td}_{X_i}(f) = O(\left|\textsc{Leaf}_{X_i}(\Phi)\right|)$, which would prove the required bound.

Fix an arbitrary part $Y = X_i$. Define the following three sets of nodes:

$$\begin{aligned}
V_0 &= \left\{v \in \Phi \;:\; \left|\textsc{Leaf}_Y(v)\right| = 0 \;\text{ and }\; \left|\textsc{Leaf}_Y(\textsc{Parent}(v))\right| \geq 2\right\}\\
V_1 &= \left\{v \in \Phi \;:\; \left|\textsc{Leaf}_Y(v)\right| = 1 \;\text{ and }\; \left|\textsc{Leaf}_Y(\textsc{Parent}(v))\right| \geq 2\right\}\\
V_2 &= \left\{v \in \Phi \;:\; \left|\textsc{Leaf}_Y(v)\right| \geq 2\right\}.
\end{aligned}$$

Each node $v \in V_0$ computes a polynomial in $f_v \in \mathbb{F}[X \setminus Y]$, and we shall replace the subtree at v by a node computing the polynomial f_v. Similarly, any node $v \in V_1$ computes a polynomial of the form $f_v^{(0)} + f_v^{(1)} y_v$ for some $y_v \in Y$ and $f_v^{(0)}, f_v^{(1)} \in \mathbb{F}[X \setminus Y]$. We shall again replace the subtree rooted at v by a node computing $f_v^{(0)} + f_v^{(1)} y_v$.

Hence, the formula Φ now reduces to a smaller formula Φ_Y with leaves being the nodes in V_0 and V_1 (and nodes in V_2 are unaffected). We would like to show that the size of the reduced formula, which is at most twice the number of its leaves, is $O(\left|\textsc{Leaf}_Y(\Phi)\right|)$.

Observation 8 $|V_1| \leq \left|\textsc{Leaf}_Y(\Phi)\right|$.

Proof Each node in V_1 has a distinct leaf labelled with a variable in Y. Hence, $|V_1|$ is bounded by the number of leaves labelled with a variable in Y. □ (Obs)

This shows that the V_1 leaves are not too many. Unfortunately, we cannot immediately bound the number of V_0 leaves, since we could have a long chain of V_2 nodes each with one sibling being a V_0 leaf. The following observation would show how we can eliminate such long chains.

Observation 9 *Let u be an arbitrary node, and v be another node in the subtree rooted at u with $\textsc{Leaf}_Y(u) = \textsc{Leaf}_Y(v)$. Then the polynomial g_u computed at u and the polynomial g_v computed at v are related as $g_u = f_1 g_v + f_2$ for some $f_1, f_2 \in \mathbb{F}[X \setminus Y]$.*

Proof If $\textsc{Leaf}_Y(u) = \textsc{Leaf}_Y(v)$, then every node on the path from u to v must have a V_0 leaf as the other child. The observation follows as all these nodes are $+$ or $\times$ gates. □ (Obs)

Using the above observation, we shall remove the need for V_0 nodes completely by augmenting each node u (computing a polynomial g_u) in Φ_Y with polynomials $f_u^{(0)}, f_u^{(1)} \in \mathbb{F}[X \setminus Y]$ to enable them to compute $f_u^{(1)} g_u + f_u^{(0)}$. Let this augmented formula be called $\hat{\Phi}_Y$. Using Observation 9, we can now contract any two nodes u and v with $\textsc{Leaf}_Y(u) = \textsc{Leaf}_Y(v)$, and eliminate all V_0 nodes completely. Since all V_2 nodes are internal nodes, the only leaves of the augmented formula $\hat{\Phi}_Y$ are in V_1. Hence, the size of the augmented formula $\hat{\Phi}_Y$ is bounded by $2\,|V_1|$, which is bounded by $2\left|\textsc{Leaf}_Y(\Phi)\right|$ by Observation 8.

Suppose Φ computes a polynomial f, which can be written as $f = \sum_{i=1}^t f_i \cdot m_i$ with $f_i \in \mathbb{F}[X \setminus Y]$ and m_i's being distinct monomials in Y. Since $\hat{\Phi}_Y$ also computes f, each f_i is a polynomial combination of the polynomials $S_Y = \left\{f_u^{(0)}, f_u^{(1)} \;:\; u \in \hat{\Phi}_Y\right\}$. Since $\hat{\Phi}_Y$ consists of at most $2\left|\textsc{Leaf}_Y(\Phi)\right|$ augmented nodes, we have that $\mathrm{td}_Y(f) \le |S_Y| \le 4\left|\textsc{Leaf}_Y(\Phi)\right|$. Therefore,

$$\mathrm{td}_Y(f) = \mathrm{trdeg}\,\{f_i \;:\; i \in [t]\} \le 4\left|\textsc{Leaf}_Y(\Phi)\right|$$

Hence,

$$\Gamma^{[\mathrm{Kal}]}(\Phi) = \sum_{i=1}^{r} \mathrm{td}_{X_i}(f_i) \le 4\left(\sum_{i=1}^{r} \left|\textsc{Leaf}_{X_i}\right|\right) = O(s).$$

□

5.3.2.2 Lower Bounding $\Gamma^{[\mathrm{Kal}]}(\mathsf{Det}_n)$

Lemma 10 *Let $X = X_1 \sqcup \cdots \sqcup X_n$ be the partition as defined by $X_t = \left\{x_{ij} \;:\; i - j \equiv t \bmod n\right\}$. Then, $\Gamma^{[\mathrm{Kal}]}(\mathsf{Det}_n) = \Omega(n^3)$.*

Proof By symmetry, it is easy to see that $\mathrm{td}_{X_i}(\mathsf{Det}_n)$ is the same for all i. Hence, it suffices to show that $\mathrm{td}_Y(\mathsf{Det}_n) = \Omega(n^2)$ for $Y = X_n = \{x_{11}, \ldots, x_{nn}\}$.

To see this, observe that the determinant consists of the monomials $\left(\frac{x_{11} \ldots x_{nn}}{x_{ii} x_{jj}}\right) \cdot x_{ij} x_{ji}$ for every $i \neq j$. Hence, $\mathrm{td}_Y(\mathsf{Det}_n) \ge \mathrm{trdeg}\left\{x_{ij} x_{ji} \;:\; i \neq j\right\} = \Omega(n^2)$. Therefore, $\Gamma^{[\mathrm{Kal}]}(\mathsf{Det}_n) = \Omega(n^3)$. □

The proof of Theorem 5 follows from Lemma 7 and Lemma 10.

5.4 "Natural" Proof Strategies

The lower bounds presented in Sect. 5.3 proceeded by first identifying a *weakness* of the model, and exploiting it in an explicit manner. More concretely, Sect. 5.3.2 presents a promising strategy that could be adopted to prove lower bounds for various models of arithmetic circuits. The crux of the lower bound was the construction of a good map Γ that assigned a number to every polynomial. The map $\Gamma^{[\text{Kal}]}$ was useful to show a lower bound in the sense that any f computable by a *small* formula had *small* $\Gamma^{[\text{Kal}]}(f)$. In fact, all subsequent lower bounds in arithmetic circuit complexity have more or less followed a similar template of a "natural proof". More concretely, all the subsequent lower bounds we shall see would essentially follow the outlined plan.

Step 1 (normal forms) For every circuit in the circuit class $\mathcal{C}$ of interest, express the polynomial computed as a *small sum of simple building blocks*.

For example, every $\Sigma\Pi\Sigma$ circuit is a *small* sum of *products of linear polynomials* which are the building blocks here. In this case, the circuit model naturally admits such a representation but we shall see other examples with very different representations as sum of building blocks.

Step 2 (complexity measure) Construct a map $\Gamma : \mathbb{F}[x_1, \ldots, x_n] \to \mathbb{Z}_{\geq 0}$ that is *sub-additive* i.e., $\Gamma(f_1 + f_2) \leq \Gamma(f_1) + \Gamma(f_2)$.

In most cases, $\Gamma(f)$ is the rank of a large matrix whose entries are linear functions in the coefficients of f. In such cases, we immediately get that Γ is sub-additive.

The strength of the choice of Γ is determined by the next step.

Step 3 (potential usefulness) Show that if B is a *simple building block*, then $\Gamma(B)$ is *small*. Further, check if $\Gamma(f)$ for a *random polynomial* f is large (potentially).

This would suggest that if any f with large $\Gamma(f)$ is to be written as a sum of $B_1 + \cdots + B_s$, then sub-additivity and the fact that $\Gamma(B_i)$ is small for each i and $\Gamma(f)$ is large immediately imply that s must be large. This implies that the complexity measure Γ does indeed have a potential to prove a lower bound for the class. The next step is just to replace the *random polynomial* by an explicit polynomial.

Step 4 (explicit lower bound) Find an explicit polynomial f for which $\Gamma(f)$ is large.

These are usually the steps taken in almost all the known arithmetic circuit lower bound proofs. The main ingenuity lies in constructing a useful complexity measure, which is really to design Γ so that it is small on the *building blocks*.

Of course, there could potentially be lower bound proofs that do not follow the roadmap outlined. For instance, it could be possible that Γ is not small for a random polynomial, but specifically tailored in a way to make Γ large for the Perm_n. Or perhaps Γ need not even be sub-additive and maybe there is a very different way to argue that all polynomial in the circuit class have small Γ. However, this has been the roadmap for almost all lower bounds so far (barring very few exceptions).

As a warmup, we first present some very simple applications of the above plan to prove lower bounds for some very simple subclasses of arithmetic circuits in the next section. We then move on to more sophisticated proofs of lower bounds for less restricted subclasses of circuits.

5.5 Some Simple Lower Bounds

Let us start with the simplest complete[3] class of arithmetic circuits—depth-2 circuits or $\Sigma\Pi$ circuits.

5.5.1 Lower Bounds for $\Sigma\Pi$ circuits

Any $\Sigma\Pi$ circuit of size s computes a polynomial $f = m_1 + \cdots + m_s$ where each m_i is a monomial multiplied by a field constant. Therefore, any polynomial computed by a *small* $\Sigma\Pi$ circuit must have a *small* number of monomials. Hence, it is obvious that any polynomial that has many monomials requires large $\Sigma\Pi$ circuits.

This can be readily rephrased in the language of the outline described in the last section by defining $\Gamma(f)$ to simply be the number of monomials present in f. Hence, $\Gamma(f) \leq s$ for any f computed by a $\Sigma\Pi$ circuit of size s. Of course, even a polynomial like $f = (x_1 + x_2 + \cdots + x_n)^n$ has $\Gamma(f) = n^{\Omega(n)}$ giving the lower bound.

5.5.2 Lower Bounds for $\Sigma\wedge\Sigma$ Circuits

A $\Sigma\wedge\Sigma$ circuit of size s computes a polynomial of the form $f = \ell_1^{d_1} + \cdots + \ell_s^{d_s}$ where each ℓ_i is a linear polynomial over the n variables.[4]

Clearly, as even a single ℓ^d could have exponentially many monomials, the Γ defined above cannot work in this setting. Nevertheless, we shall try to design a similar map to ensure that $\Gamma(f)$ is *small* whenever f is computable by a *small* $\Sigma\wedge\Sigma$ circuit.

In this setting, the *building blocks* are terms of the form ℓ^d. The goal would be to construct a *sub-additive* measure Γ such that $\Gamma(\ell^d)$ is *small*. Here is the key observation to guide us towards a good choice of Γ.

Observation 11 *Any k-th order partial derivative of ℓ^d is a constant multiple of ℓ^{d-k}.*

---

[3] in the sense that any polynomial can be computed in this model albeit of large size.

[4] such circuits are also called *diagonal depth-3 circuits* in the literature.

Hence, if $\partial^{=k}(f)$ denotes the set of k-th order partial derivatives of f, then the space spanned by $\partial^{=k}(\ell^d)$ has dimension 1. This naturally leads us to define Γ exploiting this weakness.

$$\Gamma_k(f) \stackrel{\text{def}}{=} \dim\left(\partial^{=k}(f)\right)$$

It is straightforward to check that Γ_k is indeed sub-additive and hence $\Gamma_k(f) \leq s$ whenever f is computable by a $\Sigma \wedge \Sigma$ circuit of size s. For a random polynomial f, we should be expecting $\Gamma_k(f)$ to be $\binom{n+k}{k}$ as there is unlikely to be any linear dependencies among the partial derivatives. Hence, all that needs to be done is to find an explicit polynomial with large Γ_k.

If we consider Det_n or Perm_n, then any partial derivative of order k is just an $(n-k) \times (n-k)$ minor. Also, these minors consist of disjoint sets of monomials and hence are linearly independent. Hence, $\Gamma_k(\mathsf{Det}_n) = \binom{n}{k}^2$. Choosing $k = n/2$, we immediately get that any $\Sigma \wedge \Sigma$ circuit computing Det_n or Perm_n must be of size $2^{\Omega(n)}$.

5.5.3 Low-Rank $\Sigma\Pi\Sigma$

A slight generalization of $\Sigma \wedge \Sigma$ circuits is a *rank-r* $\Sigma\Pi\Sigma$ *circuit* that computes a polynomial of the form

$$f = T_1 + \cdots + T_s$$

where each $T_i = \ell_{i1} \ldots \ell_{id}$ is a product of linear polynomials such that $\dim\{\ell_{i1}, \ldots, \ell_{id}\} \leq r$.

Thus, $\Sigma \wedge \Sigma$ is a rank-1 $\Sigma\Pi\Sigma$ circuit, and a similar partial-derivative technique for lower bounds works here as well.

In the setting where r is much smaller than the number of variables n, each T_i is essentially an r-variate polynomial masquerading as an n-variate polynomial using an affine transformation. In particular, the set of n first-order derivatives of T have rank at most r. This yields the following observation.

Observation 12 *Let $T = \ell_1 \ldots \ell_d$ with $\dim\{\ell_1, \ldots, \ell_d\} \leq r$. Then for any k, we have*

$$\Gamma_k(T) \stackrel{def}{=} \dim\left(\partial^{=k}(T)\right) \leq \binom{r+k}{k}.$$

Thus once again by sub-additivity, for any polynomial f computable by a rank-r $\Sigma\Pi\Sigma$ circuit of size s, we have $\Gamma_k(f) \leq s \cdot \binom{r+k}{r}$. Note that a random polynomial is expected to have $\Gamma_k(f)$ close to $\binom{n+k}{k}$, which could be much larger for $r \ll n$. We already saw that $\Gamma_k(\mathsf{Det}_n) = \binom{n}{k}^2$. This immediately gives the following lower bound, the proof of which we leave as an exercise to the interested reader.

Theorem 13 *Let $r \leq n^{2-\delta}$ for some constant $\delta > 0$. For $k = \epsilon n^{\delta}$, where $\epsilon > 0$ is sufficiently small, we have*

$$\frac{\binom{n}{k}^2}{\binom{r+k}{k}} = \exp\left(\Omega(n^{\delta})\right).$$

Hence, any rank-r $\Sigma\Pi\Sigma$ *circuit computing* Det_n *or* Perm_n *must have size* $\exp\left(\Omega(n^{\delta})\right)$. □

This technique of using the rank of partial derivatives was introduced by Nisan and Wigderson [NW97] to prove lower bounds for *homogeneous depth*-3 *circuits* (which also follows as a corollary of Theorem 13). The survey of Chen, Kayal and Wigderson [CKW11] give a comprehensive exposition of the power of the *partial derivative method.*

With these simple examples, we can move on to other lower bounds for various other more interesting models.

5.6 Lower Bounds for Monotone Circuits

This section presents a slight generalization of a lower bound by Jerrum and Snir [JS82]. To motivate our presentation here, let us first assume that the underlying field is $\mathbb{R}$, the field of real numbers. A monotone circuit over $\mathbb{R}$ is a circuit having $+, \times$ gates in which all the field constants are *nonnegative* real numbers. Such a circuit can compute any polynomial f over $\mathbb{R}$ all of whose coefficients are nonnegative real numbers, such as for example the permanent. It is then natural to ask whether there are small monotone circuits over $\mathbb{R}$ computing the permanent. Jerrum and Snir [JS82] obtained an exponential lower bound on the size of monotone circuits over $\mathbb{R}$ computing the permanent. Note that this definition of monotone circuits is valid only over $\mathbb{R}$ (actually more generally over ordered fields but not over say finite fields) and such circuits can only compute polynomials with nonnegative coefficients. Here we will present Jerrum and Snir's argument in a slightly more generalized form such that the circuit model makes sense over any field $\mathbb{F}$ and is complete, i.e., can compute any polynomial over $\mathbb{F}$. Let us first explain the motivation behind the generalized circuit model that we present here. Observe that in any monotone circuit over $\mathbb{R}$, there is no cancellation as there are no negative coefficients. Formally, for a node v in our circuits let us denote by f_v the polynomial computed at that node. For a polynomial f let us denote by $\mathrm{Mon}(f)$ the set of monomials having a nonzero coefficient in the polynomial f.

1. If $w = u + v$ then
$$\mathrm{Mon}(f_w) = \mathrm{Mon}(f_u) \cup \mathrm{Mon}(f_v).$$

2. If $w = u \times v$ then
$$\mathrm{Mon}(f_w) = \mathrm{Mon}(f_u) \cdot \mathrm{Mon}(f_v) \stackrel{\text{def}}{=} \{m_1 \cdot m_2 \ : \ m_1 \in \mathrm{Mon}(f_u), m_2 \in \mathrm{Mon}(f_v)\}.$$

This means that for any node v in a monote circuit over $\mathbb{R}$ one can determine $\mathrm{Mon}(f_v)$ in a very syntactic manner starting from the leaf nodes. Let us make precise this syntactic computation that we have in mind.

Definition 14 (*Formal Monomials*) Let Φ be an arithmetic circuit. The *formal monomials* at any node $v \in \Phi$, which shall be denoted by $\mathrm{FM}(v)$, shall be inductively defined as follows:

If v is a leaf labelled by a variable x_i, then $\mathrm{FM}(v) = \{x_i\}$. If it is labelled by a constant, then $\mathrm{FM}(v) = \{1\}$.

If $v = v_1 + v_2$, then $\mathrm{FM}(v) = \mathrm{FM}(v_1) \cup \mathrm{FM}(v_2)$.

If $v = v_1 \times v_2$, then

$$\begin{aligned}\mathrm{FM}(v) &= \mathrm{FM}(v_1) \cdot \mathrm{FM}(v_2)\\ &\stackrel{\text{def}}{=} \{m_1 \cdot m_2 \ : \ m_1 \in \mathrm{FM}(v_1), m_2 \in \mathrm{FM}(v_2)\}\,.\end{aligned}$$

Note that for any node v in any circuit we have $\mathrm{Mon}(f_v) \subseteq \mathrm{FM}(v)$ but in a monotone circuit over $\mathbb{R}$ this containment is in fact an equality at every node. This motivates our definition of a slightly more general notion of a monotone circuit as follows:

Definition 15 (*Monotone circuits*) A circuit C is said to be *syntactically monotone* (simply monotone for short) if $\mathrm{Mon}(f_v) = \mathrm{FM}(v)$ for every node v in C.

The main theorem of this section is the following:

Theorem 16 [JS82] *Over any field $\mathbb{F}$, any syntactically monotone circuit C computing Det_n or Perm_n must have size at least $2^{\Omega(n)}$.*

The proof of this theorem is relatively short assuming the following structural result (which is present in standard depth-reduction proofs [VSBR83, AJMV98]).

Lemma 17 *Let f be a degree d polynomial computed by a monotone circuit of size s. Then, f can be written of the form $f = \sum_{i=1}^{s} f_i \cdot g_i$ where the f_i's and g_i's satisfy the following properties:*

1. *For each $i \in [s]$, we have $\frac{d}{3} < \deg g_i \le \frac{2d}{3}$.*
2. *For each i, we have $\mathrm{FM}(f_i) \cdot \mathrm{FM}(g_i) \subseteq \mathrm{FM}(f)$.*

We shall defer this lemma to the end of the section and first see how this would imply Theorem 16. The complexity measure $\Gamma(f)$ in this case is just the number of monomials in f, but it is the above *normal form* that is crucial in the lower bound.

Proof of Theorem 16 Suppose Φ is a circuit of size s that computes Det_n. Then by Lemma 17,

$$\mathsf{Det}_n = \sum_{i=1}^{s} f_i \cdot g_i$$

with $\text{FM}(f_i) \cdot \text{FM}(g_i) \subseteq \text{FM}(\text{Det}_n)$. The building blocks are terms of the form $T = f \cdot g$, where $\text{FM}(f) \cdot \text{FM}(g) \subseteq \text{FM}(\text{Det}_n)$.

Since all the monomials in Det_n are products of variables from distinct columns and rows, the rows (and columns) containing the variables f depends on is disjoint from the rows (and columns) containing variables that g depends on. Hence, there exists sets of indices $A, B \subseteq [n]$ such that f depends only on $\{x_{jk} \ : \ j \in A, k \in B\}$ and g depends only on $\{x_{jk} \ : \ j \in \overline{A}, k \in \overline{B}\}$.

Further, since Det_n is a homogeneous polynomial of degree n, we also have that both f and g must be homogeneous as well. Also, as all monomials of g using distinct row and column indices from $\overline{A}$ and $\overline{B}$ respectively, we see that $\deg g = |\overline{A}| = |\overline{B}|$ and $\deg f = |A| = |B|$. Using Lemma 17, let $|A| = \alpha n$ for some $\frac{1}{3} \leq \alpha \leq \frac{2}{3}$. This implies that $\Gamma(f) \leq (\alpha n)!$, and $\Gamma(g) \leq ((1-\alpha)n)!$ and hence

$$\Gamma(f \cdot g) \ \leq \ (\alpha n)!((1-\alpha)n)! \leq \frac{n!}{\binom{n}{n/3}}$$

as $\frac{1}{3} \leq \alpha \leq \frac{2}{3}$. Also, Γ is clearly sub-additive and we have

$$\Gamma(f_1 g_1 + \cdots + f_s g_s) \leq s \cdot \frac{n!}{\binom{n}{n/3}}.$$

Since $\Gamma(\text{Det}_n) = n!$, this forces $s \geq \binom{n}{n/3} = 2^{\Omega(n)}$. □

We only need to prove Lemma 17 now.

5.6.1 Proof of Lemma 17

Without loss of generality, assume that the circuit Φ is homogeneous,[5] and consists of alternating layers of $+$ and $\times$ gates. Also, assume that all $\times$ gates have fan-in two, and orient the two children such that the formal degree of the left child is at least as large as the formal degree of the right child. Such circuits are also called *left-heavy* circuits.

Definition 18 (*Proof tree*) A proof tree of an arithmetic circuit Φ is a sub-circuit Φ' such that

- The root of Φ is in Φ'
- If a multiplication gate with $v = v_1 \times v_2 \in \Phi'$, then v_1 and v_2 are in Φ' as well.
- If an addition gate $v = v_1 + \cdots + v_s \in \Phi'$, then exactly one v_i is in Φ'.

[5] It is a forklore result that any circuit can be *homogenized* with just a polynomial blowup in size. Further, this process also preserves monotonicity of the circuit. A proof of this may be seen in [SY10].

Such a sub-circuit Φ', represented as a tree (duplicating nodes if required), shall be called a proof tree of Φ.

Let PROOFTREES(Φ) denote the set of all proof trees of Φ. It is easy to see that any proof tree of Φ computes a monomial over the variables. Further, if Φ was a monotone circuit computing a polynomial f, then every proof tree computes a monomial in f. Therefore,

$$f = \sum_{\Phi' \in \text{PROOFTREES}(\Phi)} [\Phi']$$

where $[\Phi']$ denotes the monomial computed by Φ'. Of course, the number of proof trees is exponential and the above expression is huge. However, we could use a divide-and-conquer approach to the above equation using the following lemma.

Lemma 19 *Let Φ' be a left-heavy formula of formal degree d. Then there is a node v on the left-most path of Φ' such that $\frac{d}{3} \leq \deg(v) < \frac{2d}{3}$.*

Proof Pick the lowest node on the leftmost path that has degree at least $\frac{2d}{3}$. Then, its left child must be a node of degree less than $\frac{2d}{3}$, and also at least $\frac{d}{3}$ (because the formula is left-heavy). □

For any proof tree Φ' and a node v on its leftmost path, define $[\Phi' : v]$ to be the output polynomial of the proof tree obtained by replacing the node v on the leftmost path by 1. If v does not occur on the leftmost path of Φ', define $[\Phi' : v]$ to be 0. We will denote the polynomial computed at a node v by f_v. Then, the above equation can now be rewritten as:

$$\begin{aligned} f &= \sum_{\Phi' \in \text{PROOFTREES}(\Phi)} [\Phi'] \\ &= \sum_{\substack{v \in \Phi \\ \frac{d}{3} \leq \deg v < \frac{2d}{3}}} f_v \cdot \left(\sum_{\Phi' \in \text{PROOFTREES}(\Phi)} [\Phi' : v] \right) \\ &= \sum_{\substack{v \in \Phi \\ \frac{d}{3} \leq \deg v < \frac{2d}{3}}} f_v \cdot g_v \quad \text{where } g_v = \sum_{\Phi' \in \text{PROOFTREES}(\Phi)} [\Phi' : v]. \end{aligned}$$

Since $\frac{d}{3} \leq \deg v < \frac{2d}{3}$, we also have that $\frac{d}{3} < \deg g_v \leq \frac{2d}{3}$ and the last equation is what was required by Lemma 17. □

5.7 Lower Bounds for Depth-3 Circuits over Finite Fields

This section presents the lower bound of Grigoriev and Karpinski [GK98] for Det_n. A follow-up paper of Grigoriev and Razborov [GR00] extended the result over function fields, also including a weaker lower bound for the permanent, but we shall present a slightly different proof that works for the permanent as well.

Theorem 20 [GK98] *Any depth-3 circuit computing* Det_n *(or* Perm_n*) over a finite field* $\mathbb{F}_q$ *(*$q \neq 2$*) requires size* $2^{\Omega(n)}$.

Main idea: Let $q = |\mathbb{F}|$. Suppose $C = T_1 + \cdots + T_s$, where each T_i is a product of linear polynomials. Define $\text{rank}(T_i)$ as in Sect. 5.5.3 to be the dimension of the set of linear polynomials that T_i is a product of.

In Sect. 5.5.3, we saw that the dimension of partial derivatives would handle *low rank* T_i's. As for the high rank T_i's, since T_i is a product of at least r linearly independent linear polynomials, a random evaluation keeps T_i nonzero with probability at most $\left(1 - \frac{1}{q}\right)^r$. Since q is a constant, we have that a random evaluation of a high rank T_i is almost always zero. Hence, in a sense, C can be "approximated" by just the low-rank components.

Grigoriev and Karpinski [GK98] formalize the above idea as a measure by combining the partial derivative technique seen in Sect. 5.5.3 with evaluations to show that Det_n cannot be approximated by a low-rank $\Sigma\Pi\Sigma$ circuit.

5.7.1 The Complexity Measure

For any polynomial $f \in \mathbb{F}[x_{11}, \ldots, x_{nn}]$, define the matrix $M_k(f)$ as follows—the columns of $M_k(f)$ are indexed by k-th order partial derivatives of f, and rows by elements of $\mathbb{F}^{n^2}$, with the entry being the evaluation of the partial derivative (column index) at the point (row index).

The rank of $M_k(f)$ could be a possible choice of a complexity measure but Grigoriev and Karpinski make a small modification to handle the high rank T_is. Instead, they look at the matrix $M_k(f)$ and remove a few *erroneous* evaluation points and use the rank of the resulting matrix. For any $\mathcal{A} \subseteq \mathbb{F}^{n^2}$, let us define $M_k(f; \mathcal{A})$ to be the matrix obtained from $M_k(f)$ by only taking the rows whose indices are in $\mathcal{A}$. Also, let $\Gamma^{[\text{GK}]}_{k,\mathcal{A}}(f)$ denote $\text{rank}(M_k(f; \mathcal{A}))$.

5.7.2 Upper Bounding $\Gamma^{[\text{GK}]}_{k,\mathcal{A}}$ for a Depth-3 Circuit

Our task here is to give an upper bound on the complexity measure for a $\Sigma\Pi\Sigma$-circuit of size s. We first see that the task reduces to upper bounding the measure for a single term via subadditivity. It follows from the linearity of the entries of the matrix.

Observation 21 (Sub-additivity) $\Gamma^{[GK]}_{k,\mathcal{A}}(f+g) \leq \Gamma^{[GK]}_{k,\mathcal{A}}(f) + \Gamma^{[GK]}_{k,\mathcal{A}}(g)$.

Now fix a threshold $r_0 = \beta n$ for some constant $\beta > 0$ (to be chosen shortly), and let $k = \gamma n$ for some $\gamma > 0$ (to be chosen shortly). We shall call a term $T = \ell_1 \ldots \ell_d$ to be of *low rank* if $\text{rank}(T) \leq r_0$, and *large rank* otherwise. By the above observation, we need to upper bound the measure $\Gamma^{[GK]}_{k,\mathcal{A}}$ for each term T, for a suitable choice of $\mathcal{A}$.

Low rank terms ($\text{rank}(T) \leq r_0$)
Suppose $T = \ell_1 \ldots \ell_d$ with $\{\ell_1, \ldots, \ell_r\}$ being a maximal independent set of linear polynomials (with $r \leq r_0$). Then, T can be expressed as a linear combination of terms from the set $\left\{\ell_1^{e_1} \ldots \ell_r^{e_r} \;:\; e_i \leq d \quad \forall i \in [r]\right\}$. And since the matrix $M_k(f)$ depends only on evaluations in $\mathbb{F}^{n^2}$, we can use the relation that $x^q = x$ in $\mathbb{F}$ to express the function $T : \mathbb{F}^{n^2} \to \mathbb{F}$ as a linear combination of $\left\{\ell_1^{e_1} \ldots \ell_r^{e_r} \;:\; e_i < q \quad \forall i \in [r]\right\}$. Therefore, for any set $\mathcal{A} \subseteq \mathbb{F}^{n^2}$, we have that

$$\Gamma^{[GK]}_{k;\mathcal{A}}(T) \leq \text{rank}(M_k(f)) \leq q^r \leq q^{\beta n}.$$

High rank terms ($\text{rank}(T) > r_0$)
Suppose $T = \ell_1 \ldots \ell_d$ whose rank is greater than $r_0 = \beta n$, and let $\{\ell_1, \ldots, \ell_r\}$ be a maximal independent set. We want to use the fact that since T is a product of at least r independent linear polynomials, most evaluations would be zero. We shall be choosing our $\mathcal{A}$ to be the set where all k-th order partial derivatives evaluate to zero.

On applying the product rule of differentiation, any k-th order derivative of T can be written as a sum of terms each of which is a product of at least $r-k$ independent linear polynomials. Let us count the *erroneous points* $\mathcal{E}_T \subseteq \mathbb{F}^{n^2}$ that keep at least $r-k$ of $\{\ell_1, \ldots, \ell_r\}$ nonzero, or in other words makes at most k of $\{\ell_1, \ldots, \ell_r\}$ zero.

$$\Pr_{\mathbf{a} \in \mathbb{F}^{n^2}}[\text{at most } k \text{ of } \ell_1, \ldots, \ell_r \text{ evaluate to zero}] \leq \sum_{i=0}^{k} \binom{r}{i} \left(\frac{1}{q}\right)^i \left(1 - \frac{1}{q}\right)^{r-i}$$

Hence, we can upper bound $|\mathcal{E}_T|$ as

$$\begin{aligned} |\mathcal{E}_T| &\leq \sum_{i=0}^{k} \binom{r}{i} (q-1)^{r-i} q^{n^2 - r} \\ &= O\left(k \cdot \binom{r}{k} \left(1 - \frac{1}{q}\right)^{r-k} q^{n^2}\right) \quad \text{if } r > qk \\ &= q^{n^2} \cdot \alpha^n \quad \text{for some } 0 < \alpha < 1. \end{aligned}$$

By choosing $\mathcal{A} = \mathbb{F}^{n^2} \setminus \mathcal{E}$ where $\mathcal{E} = \bigcup_{T \text{ of large rank}} \mathcal{E}_T$, we have that $M_k(T; \mathcal{A})$ is just the zero matrix and hence $\Gamma^{[GK]}_{k,\mathcal{A}}(T) = 0$.

Putting it together, if $C = T_1 + \cdots + T_s$, then

$$\Gamma_{k,\mathcal{A}}^{[\mathrm{GK}]}(C) \leq s \cdot q^{\beta n}. \tag{5.1}$$

where $\mathcal{A} = \mathbb{F}^{n^2} \setminus \mathcal{E}$ for some set $\mathcal{E}$ of size at most $s \cdot \alpha^n \cdot q^{n^2}$ for some $0 < \alpha < 1$.

5.7.3 Lower Bounding $\Gamma_{k,\mathcal{A}}^{[\mathrm{GK}]}$ for Det_n and Perm_n

We now wish to show that $M_k(\mathsf{Det}_n; \mathcal{A})$ has large rank. The original proof of Grigoriev and Karpinski is tailored specifically for the determinant, and does not extend directly to the permanent. The following argument is a proof communicated by Srikanth Srinivasan [Sri13] that involves an elegant trick that he attributes to [Kou08]. The following proof is presented for the determinant, but immediately extends to the permanent as well.

Note that if we were to just consider $M_k(\mathsf{Det}_n)$, it would have been easy to show that the rank is full by looking at just those evaluation points that keep exactly one $(n-k) \times (n-k)$ minor nonzero (set the main diagonal of the minor to ones, and every other entry to zero). Hence, $M_k(\mathsf{Det}_n)$ has the identity matrix *embedded inside* and hence must be full rank. However, we are missing a few of the evaluations (since a small set $\mathcal{E}$ of evaluations is removed) and we would still like to show that the matrix continues to have full column-rank.

Lemma 22 *Let $p(X)$ be a nonzero linear combination of $r \times r$ minors of the matrix $X = ((x_{ij}))$. Then,*

$$\Pr_{A \in \mathbb{F}_q^{n^2}} [p(A) \neq 0] \geq \Omega(1).$$

This immediately implies that for every linear combinations of the columns of $M_k(\mathsf{Det}_n)$, a constant fraction of the coordinates have nonzero values. Since we are removing merely a set $\mathcal{E}$ of size $(1-o(1))q^{n^2}$, there must continue to exist coordinates that are nonzero. In other words, no linear combination of columns of $M_k(\mathsf{Det}_n; \mathcal{A})$ results in the zero vector.

The proof of the above lemma would be an induction on the number of minors contributing to the linear combination. As a base case, we shall use a well-known fact about Det_n and Perm_n of random matrices.

Proposition 23 *If A is a random $n \times n$ matrix with entries from a fixed finite field $\mathbb{F}_q$, then for $q \neq 2$ we have*

$$\Pr[\det(A) \neq 0] \geq \frac{q-2}{q-1} = \Omega(1).$$

We shall defer the proof of this proposition for later, and proceed with the proof of Lemma 22.

Proof of Lemma 22 If $p(X)$ is a scalar multiple of a single nonzero minor, then we already have the lemma from Proposition 23. Hence, let us assume that there are at least two distinct minors participating in the linear combination $p(X)$. Without loss of generality, assume that the first row occurs in some of the minors, and does not in others. That is,

$$\begin{aligned} p(X) &= \left(\sum_{i:\mathrm{Row}_1 \in M_i} c_i M_i \right) + \left(\sum_{j:\mathrm{Row}_1 \notin M_j} c_j M_j \right) \\ &= \left(x_{11} M_1' + \cdots + x_{1n} M_n' \right) + M'' \quad \text{(expanding along the first row).} \end{aligned}$$

To understand a random evaluation of $p(X)$, let us first set rows $2, \ldots, n$ to random values, and then setting row 1 to random values.

$$\begin{aligned} \Pr_A[p(A) \neq 0] \geq \Pr[& x_{11} M_1' + \cdots + x_{1n} M_n' + M'' \neq 0 \mid \text{some } M_i' \neq 0] \\ & \times \Pr[\text{some } M_i' \neq 0] \end{aligned}$$

Note that once we have set rows $2, \ldots, n$ to random values, $p(X)$ reduces to a linear polynomial in $\{x_{11}, \ldots, x_{1n}\}$. Further, a random evaluation of any nonconstant linear polynomial is zero with probability exactly $\left(1 - \frac{1}{q}\right)$. Hence,

$$\begin{aligned} \Pr_A[p(A) \neq 0] \geq \Pr[& x_{11} M_1' + \cdots + x_{1n} M_n' + M'' \neq 0 \mid \text{some } M_i' \neq 0] \\ & \times \Pr[\text{some } M_i' \neq 0] \\ &= \left(1 - \frac{1}{q}\right) \cdot \Pr[\text{some } M_i' \neq 0]. \end{aligned}$$

Now comes Koutis' Trick: the term $\left(1 - \frac{1}{q}\right) \cdot \Pr[\text{ some } M_i' \neq 0]$ is exactly the probability that $x_{11} M_1' + \cdots + x_{1n} M_n'$ is nonzero! Hence,

$$\begin{aligned} \Pr_A[p(A) \neq 0] &= \Pr[x_{11} M_1' + \cdots + x_{1n} M_n' + M'' \neq 0] \\ &\geq \Pr[x_{11} M_1' + \cdots + x_{1n} M_n' \neq 0] \\ &= \Pr\left[\left(\sum_{i:\mathrm{Row}_1 \in M_i} c_i M_i \right) \neq 0 \right]. \end{aligned}$$

which is just the linear combination obtained by only considering those minors that contain the first row. Repeating this process for other rows/columns until only one minor remains, we have

$$\Pr_A[p(A) \neq 0] \geq \Pr_A[\det(A) \neq 0] = \frac{q-2}{q-1} \quad \text{(by Proposition 23).}$$

□

We now give a proof of Proposition 23.

Proof of Proposition 23 We shall fix random values to the first row of A. Then,

$$\begin{aligned}\Pr_A[\mathsf{Det}_n(A) = 0] &\leq \Pr[a_{11}M_1 + \cdots + a_{1n}M_n = 0 \mid \text{some } a_{1i} \text{ nonzero}] \\ &\quad + \Pr[a_{11} = \cdots = a_{1n} = 0] \\ &= \Pr[a_{11}M_1 + \cdots + a_{1n}M_n = 0 \mid \text{some } a_{1i} \text{ nonzero}] \\ &\quad + \frac{1}{q^n}.\end{aligned}$$

Whenever there is some a_{1i} that is nonzero, then $a_{11}M_1 + \cdots + a_{1n}M_n$ is a nonzero linear combination of minors. By a similar argument as in the proof of Lemma 22, we have that

$$\Pr[a_{11}M_1 + \cdots + a_{1n}M_n = 0 \mid \text{not all } a_{1i} \text{ are zero}] \leq \Pr[\mathsf{Det}_{n-1}(A) = 0].$$

Unfolding this recursion, we have

$$\begin{aligned}\Pr[\mathsf{Det}_n(A) = 0] &\leq \frac{1}{q} + \frac{1}{q^2} + \cdots + \frac{1}{q^n} = \frac{1}{q-1} \\ \implies \Pr[\mathsf{Det}_n(A) \neq 0] &\geq \left(1 - \frac{1}{q-1}\right) = \frac{q-2}{q-1}.\end{aligned}$$

□

5.7.4 Putting It All Together

Hence, if Det_n is computed by a depth-3 circuit of top fan-in s over $\mathbb{F}$, then

$$\begin{aligned}s \cdot q^{\beta n} &= \Omega\left(\binom{n}{k}^2\right) \\ &= \Omega\left(2^{2H(\gamma)\cdot n}\right) \\ \implies \log s &= \Omega((2H(\gamma) - \beta \log q)n)\end{aligned}$$

where $H(\gamma)$ is the binary entropy function.[6] By choosing $\gamma < q^{-q/2}$, we can find a β such that $q\gamma < \beta$ (which was required in Sect. 5.7.2) and $2H(\gamma) > \beta \log q$, yielding the lower bound

---

[6] The binary entropy function is defined as $H(\gamma) \stackrel{\text{def}}{=} -\gamma \log_2(\gamma) - (1-\gamma)\log_2(1-\gamma)$. It is well known that $\binom{n}{k} \approx 2^{nH(k/n)}$.

$$\begin{aligned} s &= \exp\left(\Omega(q^{-q/2} \cdot q \log q \cdot n)\right) \\ &= 2^{\Omega(n)}. \end{aligned}$$

□ (Theorem 20)

5.8 Lower Bounds for Multilinear Models

Raz [Raz09] showed that multilinear formulas computing the Det_n or Perm_n must be of size $n^{\Omega(\log n)}$. The complexity measure used by Raz also led to exponential lower bounds for constant depth multilinear circuits [RY09] and superlinear lower bounds for syntactic multilinear circuits [RSY08]. We shall first give some intuition behind the complexity measure before actually seeing the lower bounds.

5.8.1 The Partial Derivative Matrix

Intuition A natural first step is to try the simpler task of proving lower bounds for depth-3 multilinear circuits.

$$f \quad = \ell_{11} \dots \ell_{1d} + \dots + \ell_{s1} \dots \ell_{sd}$$

The task is now to construct a measure Γ such that $\Gamma(\ell_1 \dots \ell_d)$ is small whenever each ℓ_i is a linear polynomial and different ℓ_i's are over disjoint sets of variables. Consider the simplest case of $f = (a_1 + b_1 x)(a_2 + b_2 y)$. An observation is that the coefficients of f are given by the 2×2 matrix obtained as $[a_1\ b_1]^T [a_2\ b_2] = \begin{bmatrix} a_1 a_2 & a_1 b_2 \\ a_2 b_1 & b_1 b_2 \end{bmatrix}$. In other words, a polynomial $f = a_0 + a_1 x + a_2 y + a_3 xy$ factorizes into two variable disjoint factors if and only if the matrix $\begin{bmatrix} a_0 & a_1 \\ a_2 & a_3 \end{bmatrix}$ has rank 1. A straightforward generalization of this to multiple variables yields the *partial derivative matrix* (which was first introduced by Nisan [Nis91] in the context of non-commutative ABPs)

Definition 24 For any given partition of variables $X = Y \sqcup Z$, define the partial derivative matrix $M_{Y,Z}(f)$ to be the matrix described as follows—the rows are indexed by monomials in Y, columns indexed by monomials in Z, and the (i, j)-th entry of the matrix is the coefficient of the monomial $m_i(Y) \cdot m_j(Z)$ *in* f. We shall use $\Gamma_{Y,Z}^{[\mathrm{Raz}]}(f)$ to denote $\mathrm{rank}(M_{Y,Z}(f))$. Further, we shall call a polynomial f to be full-rank if $M_{Y,Z}(f)$ is full-rank.

Here are some basic properties of the partial derivative matrix which would be extremely useful in later calculations.

Observation 25 (Sub-additivity) *For any partition $X = Y \sqcup Z$ and any pair of multilinear polynomials f and g in $\mathbb{F}[X]$ we have $\Gamma^{[\mathrm{Raz}]}_{Y,Z}(f+g) \leq \Gamma^{[\mathrm{Raz}]}_{Y,Z}(f) + \Gamma^{[\mathrm{Raz}]}_{Y,Z}(g)$.*

Proof Follows from the linearity of the matrix. □

Observation 26 (Multiplicativity) *If $f_1 \in \mathbb{F}[Y_1, Z_1]$ and $f_2 \in \mathbb{F}[Y_2, Z_2]$ with $Y = Y_1 \sqcup Y_2$ and $Z = Z_1 \sqcup Z_2$, then*

$$\Gamma^{[\mathrm{Raz}]}_{Y,Z}(f_1 \cdot f_2) = \Gamma^{[\mathrm{Raz}]}_{Y_1,Z_1}(f_1) \cdot \Gamma^{[\mathrm{Raz}]}_{Y_2,Z_2}(f_2).$$

Proof It is not hard to see that $M_{Y,Z}(f_1 \cdot f_2)$ is the tensor product $M_{Y_1,Z_1}(f_1) \otimes M_{Y_2,Z_2}(f_2)$, and the rank of a tensor product of two matrices is the product of the ranks. □

Observation 27 $\Gamma^{[\mathrm{Raz}]}_{Y,Z}(f) \leq 2^{\min(|Y|,|Z|)}$.

Proof The number of rows is $2^{|Y|}$ and number of columns is $2^{|Z|}$, and hence the rank is upper bounded by the minimum. □

Let us get back to lower bounds for multilinear models, and attempt to use $\Gamma^{[\mathrm{Raz}]}_{Y,Z}(f)$ defined above. Unfortunately, there are examples of simple polynomials like $f = (y_1 + z_1) \dots (y_n + z_n)$ with $\Gamma^{[\mathrm{Raz}]}_{Y,Z}(f) = 2^n$. Raz's idea here was to look at $\Gamma^{[\mathrm{Raz}]}_{Y,Z}(f)$ for a *random partition*, and show that with high probability the rank of the partial derivative matrix is far from full. As a toy example, we shall see why this has the potential to give lower bounds for depth-3 multilinear circuits.

Lemma 28 *Let $f(X) = \ell_1 \dots \ell_d$ be an n-variate multilinear polynomial. If $X = Y \sqcup Z$ is a random partition with $|Y| = |Z| = |X|/2$, then with high probability we have*

$$\Gamma^{[\mathrm{Raz}]}_{Y,Z}(f) \leq 2^{|X|/2} \cdot 2^{-|X|/16}.$$

It is to be noted that we should expect a random polynomial to be full-rank with respect to any partition, so the measure $\Gamma^{[\mathrm{Raz}]}_{Y,Z}(f)$ is expected to be $2^{|X|/2}$ which should yield a lower bound of $2^{\Omega(|X|)}$.

Sketch of Proof Without loss of generality we can assume that each ℓ_i depends on at least two variables as removing the ℓ_i's that depend on just one variable does not alter $\Gamma^{[\mathrm{Raz}]}_{Y,Z}(f)$ with respect to any partition. Let $|X| = n$.

Using Observation 26, $\Gamma^{[\mathrm{Raz}]}_{Y,Z}(f) \leq 2^d$ and hence if $d < n/3$ then we are done. Hence assume that $d \geq n/3$. By a simple averaging argument, there must hence be at least $d/4$ of the ℓ_i's that depend on at most 3 variables; we shall refer to these as the *small* ℓ_i's.

Since the partition is chosen at random, on expectation a quarter of the small ℓ_i's would have all its variables mapped to either Y or Z, hence not contributing to $\Gamma^{[\mathrm{Raz}]}_{Y,Z}(f)$. Therefore, with high probability,

$$\Gamma^{[\text{Raz}]}_{Y,Z}(f) \leq 2^d \cdot 2^{-d/16} \leq 2^{n/2} \cdot 2^{-n/16}. \qquad \square$$

More generally, if $f = g_1(X_1)\dots g_t(X_t)$ where the X_i's are mutually disjoint, then a random partition is very unlikely to partition all the X_i's into almost equal parts. This is formalized in the next section to prove the lower bound for multilinear formulas.

5.8.2 Lower Bound for Multilinear Formulas

We now present the lower bound for multilinear formulas due to [Raz09]. The first step of our roadmap is to find a suitable normal form for multilinear formulas. The normal form that we use is from the survey by Shpilka and Yehudayoff [SY10].

5.8.2.1 Formulas to Log-Product Sums

The following structural lemma shows that any multilinear formula can be converted in to a small sum of *log-product* polynomials. The techniques of the following lemma can also be used in other settings with minor modifications, and we encounter a different version of this lemma later as well.

Definition 29 A multilinear polynomial $f \in \mathbb{F}[X]$ is called a multilinear log-product polynomial if $f = g_1 \dots g_t$ and there exists a partition of variables $X = X_1 \sqcup \cdots \sqcup X_t$ such that

- $g_i \in \mathbb{F}[X_i]$ for all $i \in [t]$.
- $\frac{|X|}{3^i} \leq |X_i| \leq \frac{2|X|}{3^i}$ for all i, and $|X_t| = 1$.

Lemma 30 *Let Φ be a multilinear formula of size s computing a polynomial p. Then f can be written as a sum of $(s+1)$ log-product multivariate polynomials.*

Proof Similar to Lemma 19, let v be a node in Φ such that set of variables X_v that it depends on satisfies $\frac{|X|}{3} \leq |X_v| \leq \frac{2|X|}{3}$. If Φ_v is the polynomial computed at this node, then f can be written as

$$f = \Phi_v \cdot g_1 + \Phi_{v=0} \quad \text{for some } g_1 \in \mathbb{F}[X \setminus X_v].$$

where $\Phi_{v=0}$ is the formula obtained by replacing the node v by zero. Note that the subtree at the node v is completely disjoint from $\Phi_{v=0}$. Hence the sum of the sizes of Φ_v and $\Phi_{v=0}$ is at most s. Hence, $g_1 \in \mathbb{F}[X \setminus X_v]$ and $\frac{|X|}{3} \leq |X \setminus X_v| \leq \frac{2|X|}{3}$. Inducting on the formulas Φ_v and $\Phi_{v=0}$ gives the lemma. $\square$

5.8.2.2 Log-Products Are Far from Full-Rank on a Random Partition

The main technical part of the proof is to show that log-product multivariate polynomials are far from full-rank under a random partition of variables. This would let us show that a sum of log-product multivariate polynomials cannot be full rank unless it is a very large sum.

{**Main idea** Suppose $f = g_1 \ldots g_t$ where each $g_i \in \mathbb{F}[X_i]$. Let $X = Y \sqcup Z$ be a random partition with $|Y| = |Z| = |X|/2$, and $Y_i = Y \cap X_i$ and $Z_i = Z \cap X_i$. Let $d_i = \left|\frac{|Y_i|-|Z_i|}{2}\right|$ measure the imbalance between the sizes of Y_i and Z_i, and we shall say X_i is k-imbalanced if $d_i \geq k$. Let $b_i = \frac{|Y_i|+|Z_i|}{2} = \frac{|X_i|}{2}$.

By Observation 26, we know that

$$\begin{aligned}\Gamma_{Y,Z}^{[\mathrm{Raz}]}(f) &= \Gamma_{Y_i,Z_i}^{[\mathrm{Raz}]}(g_1)\ldots\Gamma_{Y_i,Z_i}^{[\mathrm{Raz}]}(g_t)\\ &\leq 2^{\min(|Y_1|,|Z_1|)}\ldots 2^{\min(|Y_t|,|Z_t|)}\\ &= 2^{b_1-d_1}\ldots 2^{b_t-d_t} = \frac{2^{|X|/2}}{2^{d_1+\cdots+d_t}}.\end{aligned}$$

Hence, even if one of the X_i's is a little imbalanced, the product is far from full-rank.

Lemma 30 shows that the size of X_i decreases slowly with i, and it is not hard to show that $|X_i| \geq \sqrt{|X|}$ for $i \leq t' \stackrel{\text{def}}{=} \frac{\log|X|}{100}$. We wish to show that the probability that none of g_i (for $i \leq t'$) is k-unbalanced for $k = |X|^{1/20}$ is very small. Let $\mathcal{E}_i$ be the event that X_i is not k-unbalanced. The goal is to upper bound the probability that all the events $\mathcal{E}_i$ hold. These probability calculations would follow from this lemma about the *hypergeometric distribution.*

Hypergeometric Distribution Fix parameters $n, g, r \geq 0$, and let $G \subseteq [n]$ with $|G| = g$. Informally, the hypergeometric distribution is the distribution obtained on the intersection sizes of a random set of size r with a fixed set of size g from a universe of size n. Formally, the random variable $\mathcal{H}(n, g, r)$ is defined as:

$$\Pr[\mathcal{H}(n,g,r) = k] = \Pr_{R\subseteq[n],|R|=r}[|R\cap G| = k] = \frac{\binom{g}{k}\binom{n-g}{r-k}}{\binom{n}{r}}.$$

The following lemma shows that for a fairly large range of parameters, the hypergeometric distribution does not put too much mass on any value.

Lemma 31 *Let n, g, r be parameters such that $\frac{n}{4} \leq r \leq \frac{3n}{4}$ and $0 \leq g \leq \frac{2n}{3}$. Then for any $t \leq g$,*

$$\Pr[\mathcal{H}(n,g,r) = t] \leq O\left(\frac{1}{\sqrt{g}}\right).$$

The proof of this lemma follows from standard binomial coefficient estimates on the probability.

Let us go back to estimating the probability that all the events $\mathcal{E}_i$ hold.

$$\Pr\left[\mathcal{E}_1 \wedge \cdots \wedge \mathcal{E}_{t'}\right] = \Pr[\mathcal{E}_1] \cdot \Pr[\mathcal{E}_2 \mid \mathcal{E}_1] \cdots \Pr[\mathcal{E}_{t'} \mid \mathcal{E}_1 \wedge \cdots \wedge \mathcal{E}_{t'-1}].$$

The event $\mathcal{E}_1$ is just the probability that a random set Y of size $|X|/2$ intersects X_1 in t places where $t \in \left[\frac{|X_i|}{2} - k, \frac{|X_i|}{2} - k\right]$. This is just a particular setting of the hypergeometric distribution and Lemma 31 asserts that

$$\Pr[\mathcal{E}_1] \leq O\left(\frac{2k}{\sqrt{|X|}}\right).$$

To apply a similar bound for the other terms, consider the event $\mathcal{E}_i$ given that $\mathcal{E}_1, \ldots, \mathcal{E}_{i-1}$ hold. Let $X' = X \setminus (X_1 \cup \ldots \cup X_{i-1})$ and $Y' = Y \cap X'$. The fact that $\mathcal{E}_1, \ldots, \mathcal{E}_{i-1}$ hold means that the partition has been fairly balanced in the first $(i-1)$ parts and hence $|Y'| \leq |X'| + ik$. Hence, we would still be in the range of parameters in Lemma 31 to also get that

$$\begin{aligned}
\forall i \leq t' \quad \Pr[\mathcal{E}_i \mid \mathcal{E}_1 \wedge \cdots \wedge \mathcal{E}_{i-1}] &\leq O\left(\frac{2k}{\sqrt{|X|}}\right) \\
\implies \Pr\left[\mathcal{E}_1 \wedge \cdots \wedge \mathcal{E}_{t'}\right] &\leq |X|^{-\epsilon \log |X|} \quad \text{for some } \epsilon > 0 \\
\implies \Pr\left[\Gamma_{Y,Z}^{[\mathrm{Raz}]}(g_1 \ldots g_t) \leq 2^{(|X|/2) - |X|^{1/20}}\right] &\leq |X|^{-\epsilon \log |X|}.
\end{aligned}$$

Hence, if $g_1 \ldots g_t$ is a log-product multilinear polynomial, then with probability at least $\left(1 - |X|^{-\epsilon \log |X|}\right)$ we have that $\Gamma_{Y,Z}^{[\mathrm{Raz}]}(g_1 \ldots g_t) \leq 2^{(|X|/2) - |X|^{1/20}}$. Further, if f is computable by a multilinear formula of size s then, by Lemma 30, f can be written as a sum of $(s+1)$ log-product multilinear polynomials. Hence, with probability at least $\left(1 - (s+1)|X|^{-\epsilon \log |X|}\right)$ we have that

$$\Gamma_{Y,Z}^{[\mathrm{Raz}]}(f) \leq (s+1) \cdot 2^{(|X|/2) - |X|^{1/20}}.$$

Hence, if $(s+1) < |X|^{(\epsilon/2) \log |X|}$, then with high probability a random partition would ensure $\Gamma_{Y,Z}^{[\mathrm{Raz}]}(f) \ll 2^{|X|/2}$. Let us record this as a lemma.

Lemma 32 *Let $f \in \mathbb{F}[X]$ be computed by a multilinear formula of size $s < |X|^{(\epsilon/2) \log |X|}$ for a small enough constant $\epsilon > 0$. Then with probability at least $(1 - |X|^{-(\epsilon/2) \log |X|})$ we have*

$$\Gamma_{Y,Z}^{[\mathrm{Raz}]}(f) \leq (s+1) \cdot 2^{|X|/2} \cdot 2^{-|X|^{1/20}}$$

for a random partition $X = Y \sqcup Z$ with $|Y| = |Z| = |X|/2$.

5.8.2.3 Det_n and Perm_n have Large Rank

The last step of the proof would be to find an explicit polynomial whose partial derivative matrix under a random partition has large rank. As earlier, our candidate polynomial would be Det_n or Perm_n. Unfortunately, both these polynomials are over n^2 variables and degree n. It is not hard to verify that the rank of the partial derivative matrix of Det_n or Perm_n can never be greater than 2^{2n}. Hence directly using Lemma 32, we would have $2^{O(n)}$ competing with $2^{n^2/2-n^{O(1)}}$ which is simply futile. A simple fix is to first randomly restrict ourselves to fewer variables and then apply Lemma 32.

Let $m = n^{1/3}$. Let σ be a random restriction that assigns random values to all but $2m$ randomly chosen variables. We shall call this set of $2m$ variables as X, and randomly partition this into two sets Y and Z of size m each. Hence, $\sigma(\mathsf{Det}_n)$ reduces to a multilinear polynomial over $2m$ variables. It is also worth noting that a multilinear formula remains a multilinear formula under this restriction. The following claim is easy to verify.

Claim 33 *With probability at least* $1/2$*, the variables in* X *belong to distinct rows and columns.*

We shall restrict ourselves to only these random restrictions, and without loss of generality let the sets be $Y = \{x_{1,1}, x_{3,3}, \ldots, x_{2m-1,2m-1}\}$ and $Z = \{x_{2,2}, x_{4,4}, \ldots, x_{2m,2m}\}$. For ease of notation, we shall refer to $x_{2i-1,2i-1}$ as y_i and $x_{2i,2i}$ as z_i for $i = 1, \ldots, m$.

Consider the following restriction:

$$f = \mathsf{Det}\begin{bmatrix} y_1 & 1 & & & & & & \\ 1 & z_1 & & & & & & \\ & & \ddots & & & & & \\ & & & y_m & 1 & & & \\ & & & 1 & z_m & & & \\ & & & & & 1 & & \\ & & & & & & \ddots & \\ & & & & & & & 1 \end{bmatrix}$$

$$= (y_1 z_1 - 1) \ldots (y_m z_m - 1).$$

It is easy to check that $\Gamma^{[\mathrm{Raz}]}_{Y,Z}(f) = 2^m$. Although this is a single restriction with large rank, the Schwartz-Zippel-DeMillo-Lipton lemma immediately gives that random restriction would also have rank 2^m with high probability.[7] We shall record this as a lemma.

[7] provided the underlying field is large, but this isn't really a concern as we can work with a large enough extension if necessary.

Lemma 34 *With probability at least* $1/100$*, we have that* $\Gamma^{[\text{Raz}]}_{Y,Z}(\sigma(\mathsf{Det}_n)) = 2^m$ *where* σ *is a random restriction to* $2m$ *variables for* $m = n^{1/3}$.

Combining Lemma 34 with Lemma 32, we have the following theorem.

Theorem 35 [Raz09] *Any multilinear formula computing* Det_n *or* Perm_n *must be of size* $n^{\Omega(\log n)}$. □

5.8.3 Stronger Lower Bounds for Constant Depth Multilinear Formulas

Looking back at Lemma 32, we see that whenever $f(X)$ is computable by a size s multilinear formula $\Gamma^{[\text{Raz}]}_{Y,Z}(f)$ is exponentially smaller than $2^{|X|/2}$ with probability $\left(1 - s \cdot |X|^{-\epsilon \log |X|}\right)$. Hence we had to settle for a $n^{\Omega(\log n)}$ lower bound not because of the rank deficit but rather because of the bounds in the probability estimate. Unfortunately, this lower bound technique cannot yield a better lower bound for multilinear formulas as there are explicit examples of polynomials computable by poly-sized multilinear circuits with $\Gamma^{[\text{Raz}]}_{Y,Z}(f) = 2^{|X|/2}$ under *every* partition [Raz06]. However, the probability bound can be improved in the case of constant depth multilinear circuits to give stronger lower bounds.

Note that Lemma 32 was proved by considering *multilinear log-products* (Definition 29) as the building blocks. To show that a multilinear log product $g_1(X_1) \dots g_\ell(X_\ell)$ has small rank under a random partition, we argued that the probability that all the X_i's are partitioned in a roughly balanced fashion is quite small. This was essentially done by thinking of this as $\ell = O(\log n)$ close-to-independent events, each with probability $1/\text{poly}(n)$.

If ℓ was much larger than $\log n$ (with other parameters being roughly the same), it should be intuitively natural to expect a much lower probability of all the X_i's being partitioned in a roughly balanced manner. This indeed is the case for constant depth multilinear circuits, and we briefly sketch the key points where they differ from the earlier proof. The first is an analog of Definition 29 in this setting.

Definition 36 A multilinear polynomial f is said to be a multilinear t-product if f can be written as $f = g_1 \dots g_t$ with the following properties:

- The variable sets of the g_i are mutually disjoint
- Each g_i non-trivially depends on at least t variables

Lemma 37 *Let* f *be a multilinear polynomial of degree* d *over* n *variables that is computed by a depth-*Δ *multilinear formula* Φ *of size* s*. Then,* f *can be written as a sum of at most* s *multilinear* t*-products for* $t = (n/100)^{1/2\Delta}$*, and a multilinear polynomial of degree at most* $n/100$.

Proof If $d < n/100$, then the lemma is vacuously true. Since Φ is a formula of depth Δ and computes a polynomial of degree $d > n/100$, there must be at least one product gate v of fan-in at least $\left(\frac{n}{100}\right)^{1/\Delta} = t^2$. Then similar to Lemma 30,

$$f = \Phi_v \cdot f' + \Phi_{v=0}$$

As Φ_v is a product of t^2 polynomials, by grouping the factors together we have that $\Phi_v \cdot f'$ is a multilinear t-product. Further, $\Phi_{v=0}$ is a multilinear polynomial that is computable by a depth-Δ formula of smaller size and we can induct on $\Phi_{v=0}$. □

Lemma 38 *Let $f(X)$ be an n-variate polynomial computed by a depth-Δ multilinear formula of size s. If $X = Y \sqcup Z$ is a randomly chosen partition with $|Y| = |Z| = n/2$, then with probability at least* $(1 - s \cdot \exp(-n^{\Omega(1/\Delta)}))$ *we have*

$$\Gamma^{[\mathrm{Raz}]}_{Y,Z}(f) \leq (s+1) \cdot 2^{n/2} \cdot \exp(-n^{\Omega(1/\Delta)}).$$

Sketch of Proof By Lemma 37, we have that f can be written as $g_0 + g_1 + \cdots + g_s$ where $\deg(g_0) \leq n/100$ and $g_1, \ldots, g_s$ are multilinear t-products. Note that since g_0 is a multilinear polynomial of degree at most $(n/100)$, the number of monomials in g_0 is at most $\binom{n}{n/100} \leq 2^{n/10}$. Hence, $\Gamma^{[\mathrm{Raz}]}_{Y,Z}(g_0) \leq 2^{n/10}$.

For the other g_i's, we can bound the probability that $\Gamma^{[\mathrm{Raz}]}_{Y,Z}(g_i)$ is large in a very similar fashion as in Lemma 32, as the probability that all the factors of g_i are partitioned in a balanced manner is roughly the intersection of t independent events. By very similar estimates, this probability can be bounded by $(1/\mathrm{poly}(n))^t$. Hence, with high probability

$$\Gamma^{[\mathrm{Raz}]}_{Y,Z}(f) \leq \Gamma^{[\mathrm{Raz}]}_{Y,Z}(g_0) + \cdots + \Gamma^{[\mathrm{Raz}]}_{Y,Z}(g_s) \leq (s+1) \cdot 2^{n/2} \cdot \exp(-n^{\Omega(1/\Delta)}).$$

□

Combining Lemma 38 with Lemma 34, we have the following theorem of Raz and Yehudayoff.

Theorem 39 [RY09] *Any multilinear formula of depth Δ computing* Det_n *or* Perm_n *must be of size* $\exp(n^{\Omega(1/\Delta)})$. □

5.9 Lower Bounds for Depth-4 Circuits

This section addresses a recent technique for proving lower bounds for some depth-4 circuits.

Definition 40 A depth-4 circuit, also referred to as a $\Sigma\Pi\Sigma\Pi$ circuit, computes a polynomial of the form

$$f = Q_{11} \ldots Q_{1d} + \cdots + Q_{s1} \ldots Q_{sd}.$$

The number of summands s is called the top fan-in of the circuit.

Further, a $\Sigma\Pi^{[a]}\Sigma\Pi^{[b]}$ circuit is a depth-4 circuit computing a polynomial of the form

$$f = Q_{11} \ldots Q_{1a} + \cdots + Q_{s1} \ldots Q_{sa} \quad \text{where } \deg Q_{ij} \leq b \text{ for all } i, j.$$

5.9.1 Significance of the Model

In a surprising series of results on depth reduction, Agrawal and Vinay [AV08] and subsequent strengthenings of Koiran [Koi12] and Tavenas [Tav13] showed that depth-4 circuits more or less capture the complexity of general circuits.

Theorem 41 [AV08, Koi12, Tav13] *If f is an n variate degree-d polynomial computed by a size s arithmetic circuit, then f can also be computed by a $\Sigma\Pi^{[O(\sqrt{d})]}\Sigma\Pi^{[\sqrt{d}]}$ circuit of size* $\exp\left(O(\sqrt{d}\log s)\right)$.

Conversely, if an n-variate degree-d polynomial requires $\Sigma\Pi^{[O(\sqrt{d})]}\Sigma\Pi^{[\sqrt{d}]}$ circuits of size $\exp\left(\Omega(\sqrt{d}\log s)\right)$, *then it requires arbitrary depth arithmetic circuits of size $n^{\Omega(\log s/\log n)}$ to compute it.*

Thus proving strong enough lower bounds for this special case of depth-4 circuits imply lower bounds for general circuits. The main results of the section is some recent lower bound [GKKS13, KSS13, FLMS13] that comes very close to the required threshold.

5.9.2 Building the Complexity Measure

As a simpler task, let us first attempt to prove lower bounds for expressions of the form

$$f = Q_1^d + \cdots + Q_s^d$$

where each of the Q_i's are quadratics. This is exactly the problem studied by Kayal [Kay12], which led to the complexity measure for proving depth-4 lower bounds.

The goal is to construct a measure Γ such that $\Gamma(f)$ is small whenever f is a power of a quadratic. As a first attempt, let us look at the space of k-th order partial derivatives of Q^d (for a suitable choice of k). Unlike the case of $\Sigma\wedge\Sigma$-circuits where the space of k-th order partial derivatives of ℓ^d had dimension 1, the space of partial derivatives of Q^d could be as large as it can be expected. Nevertheless, the following simple observation would provide the key intuition.

Observation 42 *Any k-th order partial derivative of Q^d is of the form $Q^{d-k}p$ where p is a polynomial of degree at most k. Hence, if $k \ll d$, then all k-th order partial derivatives of Q^d share large common factors.*

This suggests that instead of looking at linear combinations of the partial derivatives of Q^d, we should instead be analyzing *low-degree polynomial combinations* of them.

Definition 43 Let $\partial^{=k}(f)$ refer to the set of all *k-th* order partial derivatives of f, and $\mathbf{x}^{\leq \ell}$ refer to the set of all monomials of degree at most ℓ. The shifted partials of f, denoted by $\langle \partial^{=k}(f) \rangle_{\leq \ell}$, is the vector space spanned by $\{\mathbf{x}^{\leq \ell} \cdot \partial^{=k}(f)\}$. The dimension of this space shall be denoted by $\Gamma^{[\text{Kay}]}_{k,\ell}(f)$.

The above observation shows that any element of $\langle \partial^{=k}(Q^d) \rangle_{\leq \ell}$ is divisible by Q^{d-k} and we thereby have the following lemma.

Lemma 44 *If $f = Q^d$ where Q is a quadratic, then $\Gamma^{[\text{Kay}]}_{k,\ell}(f) \leq \binom{n+k+\ell}{n}$, the number of monomials of degree $(k+\ell)$.*

Note that if f was instead a random polynomial, we would expect the measure $\dim\left(\langle \partial^{=k}(f) \rangle_{\leq \ell}\right)$ to be about $\binom{n+k}{n} \cdot \binom{n+\ell}{n}$, which is *much* larger than $\binom{n+k+\ell}{n}$ for suitable choice of k, ℓ. Hence this measure $\Gamma^{[\text{Kay}]}_{k,\ell}$ is certainly potentially useful for this model. Very similar to the above lemma, one can also show the following upper bound for the *building blocks* of $\Sigma\Pi^{[a]}\Sigma\Pi^{[b]}$ circuits.

Lemma 45 *Let $f = Q_1 \dots Q_a$ with $\deg Q_i \leq b$ for all i. Then,*

$$\Gamma^{[\text{Kay}]}_{k,\ell}(f) = \dim\left(\left\langle \partial^{=k}(f) \right\rangle_{\leq \ell}\right) \leq \binom{a}{k}\binom{n+(b-1)k+\ell}{n}.$$

It is easy to check that $\Gamma^{[\text{Kay}]}_{k,\ell}$ is a sub-additive measure, and we immediately have this corollary.

Corollary 46 *Let f be an n-variate polynomial computed by a $\Sigma\Pi^{[a]}\Sigma\Pi^{[b]}$ circuit of top fan-in s. Then,*

$$\Gamma^{[\text{Kay}]}_{k,\ell}(f) \leq s \cdot \binom{a}{k}\binom{n+(b-1)k+\ell}{n}.$$

Or in other words for any choice of k, ℓ, we have that any $\Sigma\Pi^{[a]}\Sigma\Pi^{[b]}$ circuit computing a polynomial f must have top fan-in s at least

$$\frac{\Gamma^{[\text{Kay}]}_{k,\ell}(f)}{\binom{a}{k}\binom{n+(b-1)k+\ell}{n}}.$$

Intuition from Algebraic Geometry Another perspective for the shifted partial derivatives comes from algebraic geometry. Any zero $a \in \mathbb{F}^n$ of Q is a zero of *multiplicity* d of Q^d. This implies that the set of common zeroes of all k-th order partial derivatives of Q^d (for $k \approx \sqrt{d}$) is *large*. On the other hand if f is a random polynomial, then with high probability there are no roots of large multiplicity.

In algebraic geometry terminology, the common zeroes of a set of polynomials is called the *variety* of the ideal generated by them. Further, there is also a well-defined notion of a *dimension of a variety* which measures how large a variety is. Let $\mathbb{F}[\mathbf{x}]_{\leq r}$ refer to the set of polynomials of degree at most r, and let $\gamma_I(r) = \dim\left(I \cap \mathbb{F}[\mathbf{x}]_{\leq r}\right)$. Intuitively, if $\gamma_I(r)$ is large, then there are *many constraints* and hence the variety is *small*. In other words the growth of $\gamma_I(r)$ is inversely related to the dimension of the variety of I, and this is precisely captured by what is known as the *Affine Hilbert function of* I. More about the precise definitions of the Affine Hilbert function, etc., can be found in any standard text in algebraic geometry such as [CLO07].

In our setting, the ideal we are interested in is $I = \left\langle \partial^{=k} f \right\rangle$. If f is a homogeneous polynomial, then $I \cap \mathbb{F}[\mathbf{x}]_{\leq r} = \left\langle \partial^{=k}(f) \right\rangle_{\leq \ell}$ where $\ell = r - (\deg(f) - k)$. Hence studying the dimension of shifted partial derivatives is exactly studying $\gamma_I(r)$ which holds all information about the dimension of the variety.

5.9.3 Lower Bounding Shifted Partials of Explicit Polynomials

For a random polynomial $R(\mathbf{x})$, we would expect that

$$\Gamma_{k,\ell}^{[\mathrm{Kay}]}(R) \approx \min\left\{ \binom{n+\ell+d-k}{n}, \binom{n+k}{n}\binom{n+\ell}{n} \right\}.$$

The terms on the RHS correspond to trivial upper bounds, where the first term is the total number of monomials of degree $(\ell + d - k)$ and the second term is the total number shifted partials.

Claim 47 *For $k = \epsilon\sqrt{d}$ for a small enough $\epsilon > 0$, and $\ell = \frac{cn\sqrt{d}}{\log n}$ for a large enough constant c, we have*

$$\frac{\min\left\{ \binom{n+\ell+d-k}{n}, \binom{n+k}{n}\binom{n+\ell}{n} \right\}}{\binom{O(\sqrt{d})}{k}\binom{n+(\sqrt{d}-1)k+\ell}{n}} = 2^{\Omega(\sqrt{d}\log n)}.$$

The proof of this claim is easily obtained by using standard asymptotic estimates of binomial coefficients. Note that using Corollary 46, the above claim shows that if we can find an explicit polynomial whose dimension of shifted partials are as large as above, then we would have an $\exp(\Omega(\sqrt{d}\log n))$ lower bound for the top fan-in of $\Sigma\Pi^{[\sqrt{d}]}\Sigma\Pi^{[\sqrt{d}]}$ circuits computing this polynomial.

If we have a set of polynomials with distinct leading monomials, then they are clearly linearly independent. Hence, one way of lower bounding the dimension of a space of polynomials is to find a sufficiently large set of polynomials with distinct monomials in the space. The vector space of polynomials we are interested is $\langle \partial^{=k}(f)\rangle_{\leq \ell}$, and if we choose a structured polynomial f we can hope to be able to estimate the number of distinct leading monomials in this vector space.

5.9.3.1 Shifted Partials of the Determinant and Permanent

The first lower bound for $\Sigma\Pi^{[\sqrt{d}]}\Sigma\Pi^{[\sqrt{d}]}$ circuits was by Gupta, Kamath, Kayal, and Saptharishi [GKKS13] for the determinant and the permanent polynomials. We shall describe the lower bound for Det_n, although it would carry over immediately to Perm_n as well. As mentioned earlier, we wish to estimate the number of distinct leading monomials in $\langle \partial^{=k}(\mathsf{Det}_n)\rangle_{\leq \ell} = \operatorname{span}\{\mathbf{x}^{\leq \ell}\partial^{=k}\mathsf{Det}_n\}$. [GKKS13] made a relaxation to merely count the number of distinct leading monomials among the generators $\{\mathbf{x}^{\leq \ell}\partial^{=k}\mathsf{Det}_n\}$ instead of their span.

The first observation is that any k-th order partial derivative of Det_n is just an $(n-k)\times(n-k)$ minor. Let us fix a monomial ordering induced by the lexicographic ordering on the variables:

$$x_{11} \succ x_{12} \cdots \succ x_{1n} \succ x_{21} \succ \cdots \succ x_{nn}.$$

Under this ordering, the leading monomial of any minor is just the product of variables on the main diagonal of the submatrix corresponding to the minor, and hence is a term of the form $x_{i_1 j_1}\cdots x_{i_{(n-k)},j_{(n-k)}}$ where $i_1 < \cdots < i_{n-k}$ and $j_1 < \cdots < j_{n-k}$; let us call such a sequence of indices as an $(n-k)$-increasing sequence in $[n]\times[n]$. Further, for any $(n-k)$-increasing sequence, there is a unique minor M whose leading monomial is precisely the product of the variables indexed by the increasing sequence. Therefore, the task of lower bounding distinct leading monomials in $\{\mathbf{x}^{\leq \ell}\partial^{=k}\mathsf{Det}_n\}$ reduces to the following combinatorial problem:

Claim 48 *For any $k, \ell > 0$, we have*

$$\Gamma^{[\mathrm{Kay}]}_{k,\ell}(\mathsf{Det}_n) \geq \#\left\{\begin{array}{c}\textit{monomials of degree } (\ell+n-k) \textit{ that}\\ \textit{contain an } (n-k)\textit{-increasing sequence}\end{array}\right\}.$$

We could start with an $(n-k)$-increasing sequence, and multiply by a monomial of degree ℓ to obtain a monomial containing an increasing sequence. Of course, the issue is that this process is not invertible and hence we might overcount. To fix this issue, [GKKS13] assign a *canonical increasing sequence* to every monomial that contains an increasing sequence and multiply by monomials of degree ℓ that do not change the canonical increasing sequence.

Definition 49 Let $D_2 = \{x_{1,1}, \ldots, x_{n,n}, x_{1,2}, x_{2,3}, \ldots, x_{n-1,n}\}$, the main diagonal and the diagonal just above it. For any monomial m define the canonical increasing sequence of m, denoted by $\chi(m)$, as $(n-k)$-increasing sequence of m that is entirely contained in D_2 and is ordered highest according to the ordering '$\succ$'. If m contains no $(n-k)$-increasing sequence entirely in D_2, then we shall say the canonical increasing sequence is empty.

The reason we restrict ourselves to D_2 is because it is easier to understand which monomials change the canonical increasing sequence and which monomials do not.

Lemma 50 *Let S be an $(n-k)$-increasing sequence completely contained in D_2, and let m_S be the monomial obtained by multiplying the variables indexed by S. There are at least $(2(n-k)-1)$ variables in D_2 such that if m is any monomial over these variables, then $\chi(m_S) = \chi(m \cdot m_S)$.*

Proof Note that for any $x_{i,j} \in D_2$ other than $x_{n,n}$, exactly one of $x_{i+1,j}$ or $x_{i,j+1}$ is in D_2 as well; let us refer to this element in D_2 as the *companion* of $x_{i,j}$. It is straightforward to check that for any $(n-k)$-increasing sequence S, the elements of S and their companions do not alter the canonical increasing sequence. □

It is a simple exercise to check that the number of $(n-k)$-increasing sequences contained in D_2 is $\binom{n+k}{2k}$. Further, as we are free to use the $n^2 - 2n + 1$ variables outside D_2, and the $2(n-k)-1$ variables that don't alter the canonical increasing sequence, we have the following lemma.

Lemma 51 *For any $k, \ell \geq 0$,*

$$\dim \left(\left\langle \partial^{=k} (\mathsf{Det}_n) \right\rangle_{\leq \ell} \right) \geq \binom{n+k}{2k} \binom{(n^2 - 2n + 1) + 2(n-k) - 1 + \ell}{\ell}.$$

Although this lower bound is not as large as expected for a random polynomial, this is still sufficient to give strong lower bounds for depth-4 circuits. By choosing $k = \epsilon\sqrt{n}$ for a small enough $\epsilon > 0$, and $\ell = n^2\sqrt{n}$, Lemma 51 with Corollary 46 yields the lower bound of Gupta, Kamath, Kayal and Saptharishi [GKKS13]

Theorem 52 *Any $\Sigma\Pi^{[O(\sqrt{n})]}\Sigma\Pi^{[\sqrt{n}]}$ circuit computing Det_n or Perm_n has top fanin $2^{\Omega(\sqrt{n})}$.* □

It is worth noting that although Claim 47 suggests that we should be able to obtain a lower bound of $\exp(\Omega(\sqrt{n}\log n))$ for Det_n, [GKKS13] also showed that the above estimate for the dimension of shifted partial derivatives for the determinant is fairly tight. Hence the dimension of shifted partials cannot give a stronger lower bound for the determinant polynomial. However, it is possible that the estimate is *not* tight for the permanent and the dimension of shifted partial derivatives of the permanent is provably strictly larger than that of the determinant! It is conceivable that one should be able to prove an $\exp(\Omega(\sqrt{n}\log n))$ lower bound for the permanent using this measure.

Indeed, subsequently an $\exp(\Omega(\sqrt{d}\log n))$ was proved [KSS13, FLMS13] for other explicit polynomials which we now outline.

5.9.3.2 Shifted Partials of the Nisan-Wigderson Polynomial

Very shortly after [GKKS13]'s $2^{\Omega(\sqrt{n})}$ lower bound, Kayal, Saha, and Saptharishi [KSS13] gave a stronger lower bound for a different polynomial. Their approach was to engineer an explicit polynomial F for which the dimension of shifted partial derivatives is easier to estimate. The main idea was that, if any k-th order partial derivative of the engineered polynomial is a monomial, then once again estimating $\dim\left(\langle\partial^{=k}(F)\rangle_{\leq \ell}\right)$ reduces to a monomial counting problem. If we could ensure that no two monomials of F have a gcd of degree k or more, then we would immediately get that all k-th order partial derivatives of F are just monomials (albeit possibly zero). If we were to interpret the set of nonzero monomials of F as just subsets over the variables, then the above constraint can be rephrased as a set system with *small pairwise intersection*. Such systems are well studied and are known as Nisan-Wigderson designs [NW94]. With this in mind, [KSS13] studied the following polynomial family inspired by an explicit construction of a Nisan-Wigderson design.

Definition 53 (*Nisan-Wigderson Polynomial*). Let n be a power of 2 and let $\mathbb{F}_n$ be the finite field with n elements that are identified with the set $\{1, \ldots, n\}$. For any $0 \leq k \leq n$, the polynomial NW_k is a n^2-variate polynomial of degree n defined as follows:

$$\mathrm{NW}_k(x_{1,1}, \ldots, x_{n,n}) = \sum_{\substack{p(t) \in \mathbb{F}_n[t] \\ \deg(p) < k}} x_{1,p(1)} \cdots x_{n,p(n)}.$$

It is easy to show that the above family of polynomials is in VNP. Further, since any two distinct univariate polynomials of degree less than k intersects in less than k places, we have the following observation.

Observation 54 *Any two monomials of* NW_k *intersect in less than k variables. Hence, any k-th order partial derivative of* $\mathrm{NW}_k(\mathbf{x})$ *is a monomial (which could possibly be zero).* □

Hence, the problem of lower bounding the shifted partials of NW_k reduces to the problem of counting distinct monomials of degree $\ell + d - k$ that are divisible by one of these k-th order derivatives. [KSS13] additionally used the observation that two random k-th order partial derivatives of NW_k are monomials that are *far* from each other. Using this, they estimate the number of distinct shifts of these monomials and showed that the dimension of shifted partial derivatives of NW_k is very close to the trivial upper bound as in Claim 47. We sketch the argument by Chillara and Mukhopadhyay [CM14]. Formally, for any two multilinear monomials m_1 and m_2, let the $\Delta(m_1, m_2)$ denote $\min\{|m_1| - |m_1 \cap m_2|, m_2 - |m_1 \cap m_2|\}$ (abusing notation by identifying the multilinear monomials with the set of variables that divide it).

Lemma 55 [CM14] *Let* $m_1, \ldots, m_s$ *be monomials over* N *variables such that* $\Delta(m_i, m_j) \geq d$ *for all* $i \neq j$*. Then the number of distinct monomials that may* *be obtained by multiplying some m_i by arbitrary monomials of degree ℓ is at least* $s\binom{N+\ell}{N} - \binom{s}{2}\binom{N+\ell-d}{N}$.

Proof For $i = 1, \ldots, s$, let A_i be the set of monomials that can be obtained by multiplying m_i with a degree ℓ monomial. By inclusion-exclusion,

$$\left|\bigcup_{i=1}^{s} A_i\right| \geq \sum_{i=1}^{s} |A_i| - \sum_{i<j} \left|A_i \cap A_j\right|.$$

Note that each A_i is of size exactly $\binom{N+\ell}{N}$. Further, since $\Delta(m_i, m_j) \geq d$, any monomial that is divisible by m_i and m_j must necessarily be divisible by m_i and the variables in m_j not in m_i. Hence, $\left|A_i \cap A_j\right| \leq \binom{N+\ell-d}{N}$. The lemma follows by substituting these above. □

Note that any two distinct monomials of NW_k intersect in at most k places. For each monomial m_i of NW_k, let m_i' be any nonzero k-th order partial derivative of m_i. Therefore, $\Delta(m_i', m_j') \geq n - 2k \geq \frac{n}{2}$ for $k = \epsilon\sqrt{n}$. Since we have n^k monomials of pairwise distance at least $n/2$, the above lemma immediately yields a lower bound for the shifted partials of NW_k.

Theorem 56 [KSS13] *Let $k = \epsilon\sqrt{d}$ for some constant $\epsilon > 0$. Then for any $\ell = \Theta\left(\frac{n^2\sqrt{n}}{\log n}\right)$,*

$$\dim\left(\left\langle \partial^{=k}\left(\mathrm{NW}_k\right)\right\rangle_{\leq \ell}\right) \geq \frac{n^k}{2} \cdot \binom{n^2+\ell}{n^2}$$

Sketch of Proof As mentioned earlier, we have n^k monomials $\{m_i'\}$ with pairwise distance at least $\frac{n}{2}$. Using Lemma 55, it suffices to show that

$$n^k \cdot \binom{n^2+\ell}{n^2} \geq 2 \cdot \binom{n^k}{2} \cdot \binom{n^2+\ell-\frac{n}{2}}{n^2}$$

and this follows easily from standard binomial coefficient estimates. □

Combining with Corollary 46, we have the lower bound of [KSS13] using standard estimates.

Theorem 57 [KSS13] *Any $\Sigma\Pi^{[O(\sqrt{n})]}\Sigma\Pi^{[\sqrt{n}]}$ computing the NW_k polynomial, where $k = \epsilon\sqrt{n}$ for a sufficiently small $\epsilon > 0$, must have top fan-in* $\exp(\Omega(\sqrt{n}\log n))$. □

[KSS13] used the above lower bound to give an $n^{\Omega(\log n)}$ lower bound for a subclass of formulas called *regular formulas*. The interested reader can refer to [KSS13] for more details.

5.9.3.3 Shifted Partials of the Iterated-Matrix-Multiplication Polynomial

Fourier, Limaye, Malod and Srinivasan [FLMS13] showed the same lower bound as [KSS13] but for the *iterated matrix multiplication* polynomial which is known to have polynomial sized circuits computing it.

Definition 58 (*Iterated matrix multiplication polynomial*) Let $M_1, \ldots, M_d$ be $n \times n$ matrices with distinct variables as entries, i.e., $M_k = \left(\!\left(x_{ij}^{(k)}\right)\!\right)_{i,j \leq n}$ for $k = 1, \ldots, d$. The polynomial $\mathrm{IMM}_{n,d}$ is a $(n^2 d)$-variate degree-d polynomial defined as the $(1, 1)$-th entry of the matrix product $M_1 \ldots M_d$:

$$\mathrm{IMM}_{n,d}(\mathbf{x}) = (M_1 \ldots M_d)_{1,1}\,.$$

A more useful perspective is to interpret this as a *canonical algebraic branching program.*

Definition 59 (*Algebraic branching program*) An algebraic branching program (ABP) comprises a layered directed graph G with $(d+1)$ layers of vertices, where the first and last layer consists of a single node (called source and sink respectively), all other layers consist of n vertices, and edges are only between successive layers and have linear polynomials as edge-weights. The ABP is set to compute the polynomial f defined as

$$f(\mathbf{x}) = \sum_{\textit{source-sink path } \rho} \mathrm{weight}(\rho)$$

where the weight of any path is just the product of the edge weights on the path.

The canonical ABP comprises a graph where the i-th vertex of layer $(\ell - 1)$ is connected to the j-th vertex of layer ℓ with edge-weight $x_{ij}^{(\ell)}$ for every choice of i, j and ℓ. It is easy to see that the polynomial computed by the canonical ABP is in fact $\mathrm{IMM}_{n,d}$.

To lower bound the dimension of shifted partial derivatives of $\mathrm{IMM}_{n,d}$, first note that a derivative with respect to any variable (or edge) simply results in the sum of all source-sink paths that *pass* through this edge. [FLMS13] uses the following simple but crucial observation to assist in bounding the dimension of shifted partials.

Observation 60 *Assume that d is even. Let $e_1, e_3, \ldots, e_{d-1}$ be an arbitrary set of edges such that e_i is between layer i and $i + 1$. Then, there is a unique path from source to sink that passes through all these edges.*

Proof Since these are edges in alternate layers, their starting and ending points uniquely determine the edges that are picked up from the even-numbered layers to complete the source-sink path. □

Since we are interested in k-th order derivatives for $k \approx \epsilon\sqrt{d}$, [FLMS13] consider the following restriction by removing some edges from the underlying graph:

- Select $(2k-1)$ layers $\ell_1, \ldots, \ell_{2k-1}$ that are roughly equally spaced between the first and the last layer. These layers, and the first and the last layers, shall be untouched and shall be called *pristine layers*.
- In all the other layers, retain only those edges connecting vertex i of this layer to vertex i of the next.

This restriction effectively makes the graph similar to an ABP with $2k+1$ layers. Let the polynomial computed by the restricted ABP be $\mathrm{IMM}'_{n,d}(\mathbf{x})$. Since $\mathrm{IMM}'_{n,d}$ was obtained by just setting some variables of $\mathrm{IMM}_{n,d}$ to zero, the dimension of shifted partial derivatives of $\mathrm{IMM}'_{n,d}$ can only be smaller than that of $\mathrm{IMM}_{n,d}$. Similar to Observation 60, we have the following observation.

Observation 61 *For every choice of k edges from odd-numbered pristine layers, there is a unique source-sink path that passes through them.*

In other words, for any choice of k variables chosen by picking one from each odd-numbered pristine layer, then the k-th order partial derivative of $\mathrm{IMM}'_{n,d}$ *with respect to these k variables is a nonzero monomial.*

Once again, we can lower bound the dimension of shifted partial derivatives of $\mathrm{IMM}'_{n,d}$ by a monomial counting problem. Similar to the earlier case, [FLMS13] show that the monomials thus obtained are *far* from one another. We state their main lemma below without proof.

Lemma 62 [FLMS13] *There are at least $n^{k/2}$ monomials of* $\mathrm{IMM}'_{n,d}$ *of pairwise distance at least $\frac{n}{4}$.*

Again, using Lemma 55 and standard binomial coefficient estimates, this implies that the shifted partial derivatives of $\mathrm{IMM}'_{n,d}$ is almost as large as the trivial upper bound.

Theorem 63 [FLMS13] *Let $k = \epsilon\sqrt{d}$ for a sufficiently small $\epsilon > 0$ and ℓ be an integer such that $n^{1/16} \leq \frac{N+\ell}{\ell} \leq n^{1/4}$ where N is the number of variables* $\mathrm{IMM}'_{n,d}$ *depends on. Then,*

$$\dim\left(\left\langle \partial^{=k}\left(\mathrm{IMM}_{n,d}\right)\right\rangle_{\leq \ell}\right) \geq \dim\left(\left\langle \partial^{=k}\left(\mathrm{IMM}'_{n,d}\right)\right\rangle_{\leq \ell}\right)$$
$$= \Omega\left(n^{k/2}\cdot\binom{N+\ell}{\ell}\right).$$

□

Combining with Corollary 46, we get the lower bound of [FLMS13].

Theorem 64 [FLMS13] *Any $\Sigma\Pi^{[O(\sqrt{d})]}\Sigma\Pi^{[\sqrt{d}]}$ circuit computing* $\mathrm{IMM}_{n,d}$, *with $d \leq n^{\delta}$ for a sufficiently small $\delta > 0$, has top fan-in* $\exp(\Omega(\sqrt{d}\log n))$. □

Similar to [KSS13], the above result also implies $n^{\Omega(\log n)}$ lower bounds for regular formulas computing $\mathrm{IMM}_{n,d}$.

5.10 Conclusion

The field of arithmetic complexity, like Boolean complexity, abounds with open problems and proving lower bounds for almost any natural subclass of arithmetic circuits is interesting especially if the currently known techniques/ complexity measures do not apply to that subclass.[8] The surveys [Wig02, SY10, CKW11] mark out the frontiers of this area in the form of many open problems and we invite the reader to try some of them.

References

[AJMV98] E. Allender, J. Jiao, M. Mahajan, V. Vinay, Non-commutative arithmetic circuits: depth reduction and size lower bounds. Theor. Comput. Sci. **209**(1–2), 47–86 (1998)

[ASSS12] M. Agrawal, C. Saha, R. Saptharishi, N. Saxena, Jacobian hits circuits: hitting-sets, lower bounds for depth-d occur-k formulas and depth-3 transcendence degree-k circuits. in *Symposium on Theory of Computing (STOC)* (2012), pp. 599–614

[AV08] M. Agrawal, V. Vinay, Arithmetic circuits: a chasm at depth four. in *Foundations of Computer Science (FOCS)* (2008), pp. 67–75

[BS83] W. Baur, V. Strassen, The complexity of partial derivatives. Theor. Comput. Sci. **22**, 317–330 (1983)

[CKW11] X. Chen, N. Kayal, A. Wigderson, Partial derivatives in arithmetic complexity (and beyond). Found. Trends Theor. Comput. Sci. **6**, 1–138 (2011)

[CLO07] D.A. Cox, J.B. Little, D. O'Shea, *Ideals* (Springer, Varieties and Algorithms. Undergraduate texts in mathematics, 2007)

[CM14] S. Chillara, P. Mukhopadhyay, Depth-4 lower bounds, determinantal complexity: a unified approach. in *Symposium on Theoretical Aspects of Computing (STACS)* (2014)

[FLMS13] H. Fournier, N. Limaye, G. Malod, S. Srinivasan, Lower bounds for depth 4 formulas computing iterated matrix multiplication. Electron. Colloquium Comput. Complex. **20**, 100 (2013)

[GK98] D. Grigoriev, M. Karpinski, An exponential lower bound for depth 3 arithmetic circuits. in *Symposium on Theory of Computing (STOC)* (1998), pp. 577–582

[GKKS13] A. Gupta, P. Kamath, N. Kayal, R. Saptharishi, Approaching the chasm at depth four. in *Conference on Computational Complexity (CCC)* (2013)

[GR00] D. Grigoriev, A.A. Razborov, Exponential lower bounds for depth 3 arithmetic circuits in algebras of functions over finite fields. Appl. Algebra Eng. Commun. Comput. **10**(6), 465–487 (2000)

[HY11] P. Hrubeš, A. Yehudayoff, Arithmetic complexity in ring extensions. Theor. Comput. **7**(8), 119–129 (2011)

[JS82] M. Jerrum, M. Snir, Some exact complexity results for straight-line computations over semirings. J. ACM **29**(3), 874–897 (1982)

[Kal85] K. Kalorkoti, A lower bound for the formula size of rational functions. SIAM J. Comput. **14**(3), 678–687 (1985)

[Kay12] N. Kayal, An exponential lower bound for the sum of powers of bounded degree polynomials. Technical report, Electronic Colloquium on Computational Complexity (ECCC) (2012)

[8] Some of the complexity measures that we describe here yield lower bounds for slightly more general subclasses of circuits.

[Koi12] P. Koiran, Arithmetic circuits: the chasm at depth four gets wider. Theor. Comput. Sci. **448**, 56–65 (2012)

[Kou08] I. Koutis, Faster algebraic algorithms for path and packing problems. in *ICALP* (2008), pp. 575–586

[KSS13] N. Kayal, C. Saha, R. Saptharishi, A super-polynomial lower bound for regular arithmetic formulas. Electron. Colloquium Comput. Complex. **20**, 91 (2013)

[Lov11] S. Lovett, Computing polynomials with few multiplications. Theor. Comput. **7**(13), 185–188 (2011)

[Nis91] N. Nisan, Lower bounds for non-commutative computation. in *Symposium on Theory of Computing (STOC)* (1991), pp. 410–418

[NW94] N. Nisan, A. Wigderson, Hardness versus randomness. J. Comput. Syst. Sci. **49**(2), 149–167 (1994)

[NW97] N. Nisan, A. Wigderson, Lower bounds on arithmetic circuits via partial derivatives. Comput. Complex. **6**(3), 217–234 (1997)

[Raz06] R. Raz, Separation of multilinear circuit and formula size. Theor. Comput. **2**(1), 121–135 (2006)

[Raz09] R. Raz, Multi-linear formulas for permanent and determinant are of super-polynomial size. J. ACM **56**(2), 1–17 (2009)

[Raz10] R. Raz, Tensor-rank and lower bounds for arithmetic formulas. in *Symposium on Theory of Computing (STOC)* (2010), pp. 659–666

[RSY08] R. Raz, A. Shpilka, A. Yehudayoff, A lower bound for the size of syntactically multilinear arithmetic circuits. SIAM J. Comput. **38**(4), 1624–1647 (2008)

[RY09] R. Raz, A. Yehudayoff, Lower bounds and separations for constant depth multilinear circuits. Comput. Complex. **18**(2), 171–207 (2009)

[Sri13] S. Srinivasan, personal communication (2013)

[SW01] A. Shpilka, A. Wigderson, Depth-3 arithmetic circuits over fields of characteristic zero. Comput. Complex. **10**(1), 1–27 (2001)

[SY10] A. Shpilka, A. Yehudayoff, Arithmetic circuits: a survey of recent results and open questions. Found. Trends Theor. Comput. Sci. **5**, 207–388 (2010)

[Tav13] S. Tavenas, Improved bounds for reduction to depth 4 and depth 3. in *Mathematical Foundations of Computer Science (MFCS)* (2013)

[VSBR83] L.G. Valiant, S. Skyum, S. Berkowitz, C. Rackoff, Fast parallel computation of polynomials using few processors. SIAM J. Comput. **12**(4), 641–644 (1983)

[Wig02] A. Wigderson, Arithmetic complexity—a survey. Lecture Notes (2002)

Chapter 6
Explicit Tensors

Markus Bläser

Abstract This is an expository article the aim of which is to introduce interested students and researchers to the topic of tensor rank, in particular to the construction of explicit tensors of high rank. We try to keep the mathematical concepts and language used as simple as possible to address a broad audience. This article is thought to be an appetizer and does not provide by any means a complete coverage of this topic.

Keywords Algebraic complexity theory · Tensor rank · Lower bounds

Mathematics Subject Classification (2010) 68Q17 · 15A69

6.1 Tensors and Rank

Let U and V be vector spaces over some field k. It is a well-known fact that every linear map $\alpha : U \to V$ is represented by a matrix $A = (a_{i,j}) \in k^{\ell \times m}$, where $\ell = \dim U$ and $m = \dim V$. The rank of the matrix A is the maximum number of rows that are linearly independent. There are a lot of equivalent definitions of the rank of a matrix, for instance,

- the maximum number of columns that are linearly independent,
- $\ell - \dim \ker \alpha$,
- $\dim \operatorname{im} \alpha$,

This work was supported through funds provided by the Deutsche Forschungsgemeinschaft (BL 511/10-1) and by the Indo-German Max-Planck Center for Computer Science (IMPECS).

M. Bläser (✉)
Computer Science, Saarland University, Saarbrücken, Germany
e-mail: mblaeser@cs.uni-saarland.de

M. Agrawal and V. Arvind (eds.), *Perspectives in Computational Complexity*, Progress in Computer Science and Applied Logic 26, DOI: 10.1007/978-3-319-05446-9_6,

- the minimum number of matrices of the form $x^T \cdot y$ with $x \in k^\ell$ and $y \in k^m$, so-called rank-one-matrices, such that A can be written as the sum of these matrices,

just to mention a few. The rank of linear maps and matrices is well understood. There are efficient algorithms to compute the rank, the most famous method is Gaussian elimination.

Let W be another vector space over k, $\dim W = n$. Let $\beta : U \times V \to W$ be a bilinear map. By choosing bases $u_1, \dots, u_\ell$ of U, $v_1, \dots, v_m$ of V and $w_1, \dots, w_n$ of W, we can associate structural constants $b_{h,i,j}$ with β:

$$\beta(u_h, v_i) = \sum_{j=1}^{n} b_{h,i,j} w_j, \qquad 1 \le h \le \ell, \quad 1 \le i \le m. \tag{6.1}$$

We can view $B = (b_{h,i,j}) \in k^{\ell \times m \times n}$ as a three-dimensional matrix, a so-called *tensor*. As we have seen, there are many ways to define the rank of a matrix, which is nothing but a tensor in $k^{\ell \times m \times 1}$. From the many equivalent definitions of rank of a matrix given above, it turns out that the appropriate one for tensors is the last one. We call a tensor $S = (s_{h,i,j})$ a rank-one tensor or triad, if there are vectors $a = (a_1, \dots, a_\ell)^T \in k^\ell$, $b = (b_1, \dots, b_m)^T \in k^m$, and $c = (c_1, \dots, c_n)^T \in k^n$ such that

$$s_{h,i,j} = a_h b_i c_j, \qquad 1 \le h \le \ell, \quad 1 \le i \le m, \quad 1 \le j \le n.$$

We will write $S = a \otimes b \otimes c$. Then the *rank of a tensor* T is the minimum number r such that there are rank-one tensors $S_1, \dots, S_r$ with

$$T = S_1 + \cdots + S_r.$$

We denote the rank of T by $R(T)$. The question of the rank of tensors of order three or equivalently, of the rank of the corresponding bilinear mapping is a central question in algebraic complexity theory. The flagship problem is of course matrix multiplication, which is a bilinear mapping $k^{n \times n} \times k^{n \times n} \to k^{n \times n}$. The current best upper bounds are bounds are $O(n^{2.38})$, see [CW90, Sto10, Wil12], whereas the best lower bound is the recent $3n^2 - o(n^2)$ by Landsberg [Lan12].

6.1.1 What is So Special About Matrices?

The rank of a matrix is a well-understood quantity. The maximum rank of a matrix in $k^{\ell \times m}$ is $\min\{\ell, m\}$ and it is easy to come up with a matrix that achieves this maximum. For instance, the identity matrix padded with rows or columns of zeroes does the job. Or Vandermonde matrices. The fact that we have a lot of equivalent characterizations of the rank of the matrix seems to be crucial for the fact that it is so easy to come up with explicit matrices of high rank. We can compute the rank of a matrix in polynomial time.

For tensors the situation is more complicated. Per se, it is even not clear what the maximum rank of tensors in $k^{\ell \times m \times n}$ is.[1] We will discuss this briefly in the next section. How to construct explicit tensors of high rank is a widely open problem. And finally, computing the rank of a tensor is an NP-hard problem.

Theorem 1 (Håstad [Hås90]) *Let k be a field that can be represented over* $\{0, 1\}^*$. *Let* `Tensor-Rank` *(over k) be the following problem: Given a tensor* $T \in k^{\ell \times m \times n}$ *and a bound b, decide whether* $R(T) \leq b$.

1. *Over finite fields,* `Tensor-Rank` *is* NP*-complete.*
2. *Over* $\mathbb{Q}$, `Tensor-Rank` *is* NP*-hard.*

What does it mean that "a field can be represented over $\{0, 1\}^*$"? Traditional complexity classes like P or NP are defined over some fixed alphabet, so we need to be able to encode the field elements by $\{0, 1\}$-strings. For instance, elements from finite fields can be represented in binary and rational numbers by tuples of integers represented in binary. The actual encoding does not matter as long as it is "reasonably nice", that is, all operations like addition, multiplication, etc., can be performed in polynomial time.

The hardness proof is the same over finite fields and over $\mathbb{Q}$. Over finite fields, the problems is also NP-easy; we just have to guess b rank-one tensors and check whether their sum is T. Over $\mathbb{Q}$, it is not clear whether this is possible, since we do not know an upper bound on the number of bits of the representation of the entries of the rank-one tensors. It could be the case that the rank of T is b, but all sums of b rank-one tensors involve rational numbers with a huge number of bits. ("Huge" means superpolynomial in the size of the representation of the input tensor.) To the best of my knowledge, it is even not known whether `Tensor-Rank` over $\mathbb{Q}$ is decidable.

Over $\mathbb{R}$, the situation is somewhat better: $\mathbb{R}$ itself is not representable over $\{0, 1\}$, but since $\mathbb{Q} \subseteq \mathbb{R}$, we can look at tensors T over $\mathbb{Q}$ and ask what is the minimum number of rank-one tensor with entries from $\mathbb{R}$ such that T is the sum of these rank-one tensors. This problem is decidable and even in PSPACE, since it can be reduced to the existential theory over the reals [Can88, Ren92].

Open Problem 1 What can you say about the approximability of `Tensor-Rank`? Is there a constant factor approximation algorithm? A PTAS? As far as I know, nothing in this direction is known.

Open Problem 2 What is the complexity of `Tensor-Rank` over $\mathbb{Q}$?

Another important property of matrix rank is that it is semicontinuous. If $(M_i) \in k^{n \times n}$ is a sequence of matrices that converges to a matrix M (componentwisely), then

$$R(M_i) \leq r \quad \text{for all } i \qquad \Rightarrow \qquad R(M) \leq r.$$

[1] Of course, ℓmn is an upper bound. However, it is not clear—and not true—that this is necessary.

Why does this hold? The fact that the rank of $R(M_i) \le r$ is equivalent to the fact that all $(r+1) \times (r+1)$ minors vanish. These minors are polynomials and hence continuous functions. Therefore all $(r+1) \times (r+1)$ minors of M vanish, too.

For tensors, this is not the necessarily true. Consider the following tensor t given by the following two slices:

$$(t_{1,i,j}) = \begin{pmatrix} 1 & 0 \\ 0 & 0 \end{pmatrix} \quad \text{and} \quad (t_{2,i,j}) = \begin{pmatrix} 0 & 1 \\ 1 & 0 \end{pmatrix} \tag{6.2}$$

Define

$$t_\epsilon = \frac{1}{\epsilon} \cdot (\epsilon, 1) \otimes (1, \epsilon) \otimes (1, \epsilon) - \frac{1}{\epsilon} \cdot (0, 1) \otimes (1, 0) \otimes (1, 0)$$

The two slices of t_ϵ are

$$\begin{pmatrix} 1 & \epsilon \\ \epsilon & \epsilon^2 \end{pmatrix} \quad \text{and} \quad \begin{pmatrix} 0 & 1 \\ 1 & \epsilon \end{pmatrix}.$$

So $t_\epsilon \to t$ when $\epsilon \to 0$. And $R(t_\epsilon) \le 2$ for all $\epsilon > 0$. On the other hand, $R(t) = 3$. We can prove this using the so-called *substitution method*. This method was first introduced by Pan [Pan66] to prove the optimality of the Horner scheme. See [BCS97] for more applications of this method and more references.

Let

$$t = \sum_{i=1}^{r} u_i \otimes v_i \otimes w_i \tag{6.3}$$

with

$$u_i = (u_{i,1}, u_{i,2}), \quad v_i = (v_{i,1}, v_{i,2}), \quad w_i = (w_{i,1}, w_{i,2}), \quad 1 \le i \le r,$$

be an optimal decomposition of t into rank-one tensors. Since $t_{1,1,1} = 1$, there is an i_0 such that $u_{i_0,1} \neq 0$, w.l.o.g. $i_0 = r$. Think of t as consisting of two slices as in (6.2). From the decomposition (6.3), we will construct a new tensor in $k^{1\times 2\times 2}$, which is a linear combination of the two slices, in such a way that the rank drops by one. Specifically,

$$\sum_{i=1}^{r} (\alpha u_{i,1} + u_{i,2}) \otimes v_i \otimes w_i = \begin{pmatrix} \alpha & 1 \\ 1 & 0 \end{pmatrix} := t'.$$

If we set $\alpha = -u_{r,2}/u_{r,1}$, this kills the rth rank-one tensor. Therefore,

$$R(t) \ge R(t') + 1.$$

But t' is a matrix whose rank is obviously two. Therefore, $R(t) \ge 3$. Since t has only three entries that are nonzero, there is a trivial decomposition of t of length 3.

6.2 Explicit Tensors of High Rank Imply Circuit Lower Bounds

6.2.1 Higher Order Tensors

We can generalize the concept of tensors of order three to higher orders. Let $V_1, \ldots, V_n$ be vector spaces, $\dim V_i = d_i$, $1 \le i \le n$. The tensor product $V_1 \otimes \cdots \otimes V_n$ can be built as follows: Choose bases $v_{i,j}$, $1 \le j \le d_i$, for each V_i. Then we formally built the elements $v_{1,j_1} \otimes \cdots \otimes v_{n,j_n}$, $1 \le j_1 \le d_1$, ..., $1 \le j_n \le d_n$. They form a basis of the vector space $V_1 \otimes \cdots \otimes V_n$. If we have arbitrary vectors $x_i = \alpha_{i,1} v_{i,1} + \cdots \alpha_{i,d_i} v_{i,d_i} \in V_i$, $1 \le i \le d_i$, then

$$x_1 \otimes \cdots \otimes x_n = \sum_{j_1=1}^{d_1} \cdots \sum_{j_n=1}^{d_n} \alpha_{1,j_1} \cdots \alpha_{n,j_n} v_{1,j_1} \otimes \cdots \otimes v_{n,j_n}.$$

An element $v_1 \otimes \cdots \otimes v_n$ with $v_i \in V_i$ is a rank-one tensor. As before, the rank of a tensor $t \in V_1 \otimes \cdots \otimes V_n$ is minimum number of rank-one tensors $s_1, \ldots, s_r$ such that

$$t = s_1 + \cdots + s_r.$$

The definition above is a coordinate-free definition of tensor rank. You can also think in coordinates, if you prefer that: V_i is isomorphic to k^{d_i}, simply choose bases. These isomorphims naturally extend to an isomorphism between $V_1 \otimes \cdots \otimes V_n$ and $k^{d_1} \otimes \cdots \otimes k^{d_n}$. There is also a way of defining tensor products without choosing bases at all by the universal property of turning multilinear mappings into linear ones.

6.2.2 Basic Properties

Let $\pi \in S_n$ be a permutation of $\{1, \ldots, n\}$. If $v_i \in V_i$, then $v_{\pi(1)} \otimes \cdots \otimes v_{\pi(n)} \in V_{\pi(1)} \otimes \cdots \otimes V_{\pi(n)}$. So π identifies the rank-one tensors of $V_1 \otimes \cdots \otimes V_n$ with the rank-one tensors of $V_{\pi(1)} \otimes \cdots \otimes V_{\pi(n)}$. We can extend this mapping to a linear mapping $V_1 \otimes \cdots \otimes V_n \to V_{\pi(1)} \otimes \cdots \otimes V_{\pi(n)}$. This mapping clearly is surjective and by comparing dimensions, we see that is is in fact an isomorphism. The image of any tensor t under this mapping is denoted by t^π.

If you think in coordinates, then the entries $t'_{i_1,\ldots,i_n}$ of t^π are defined by $t'_{i_1,\ldots,i_n} = t_{i_{\pi^{-1}(1)},\ldots,i_{\pi^{-1}(n)}}$, where $t = (t_{j_1,\ldots,j_n})$.

Fact 1 $R(t) = R(t^\pi)$.

Let $U_1, \ldots, U_n$ be vector spaces. Let $h_i : V_i \to U_i$ be homomorphism of vector spaces, $1 \le i \le n$. We get a mapping that maps the rank-one tensors of $V_1 \otimes \cdots \otimes V_n$ to the rank-one tensors of $U_1 \otimes \cdots \otimes U_n$ by

$$v_1 \otimes \cdots \otimes v_n \mapsto h_1(v_1) \otimes \cdots \otimes h_n(v_n).$$

Again, we can extend this to a linear mapping $V_1 \otimes \cdots \otimes V_n \to U_1 \otimes \cdots \otimes U_n$. We denote this mapping by $h_1 \otimes \cdots \otimes h_n$.

Fact 2 *$R(t) \geq R(h_1 \otimes \cdots \otimes h_n(t))$ for any tensor t. If all h_i are isomorphisms, then equality holds.*

Let $t \in V_1 \otimes \cdots \otimes V_n$ and $s \in U_1 \otimes \cdots \otimes U_n$. We can embed both tensors into the larger space $(V_1 \oplus U_1) \otimes \cdots \otimes (V_n \oplus U_n)$ as follows: Since each V_i is a subspace of $V_i \oplus U_i$, each rank-one tensor of $V_1 \otimes \cdots \otimes V_n$ is also a rank-one tensor of $(V_1 \oplus U_1) \otimes \cdots \otimes (V_n \oplus U_n)$. Every tensor t in $V_1 \otimes \cdots \otimes V_n$ is a sum of rank-one tensors, so t embeds into $(V_1 \oplus U_1) \otimes \cdots \otimes (V_n \oplus U_n)$ as well. The same works for s. $t \oplus s$ denotes the tensor that we get by viewing t and s as tensors in $(V_1 \oplus U_1) \otimes \cdots \otimes (V_n \oplus U_n)$ and forming their sum. The following fact follows immediately.

Fact 3 *$R(t \oplus s) \leq R(t) + R(s)$.*

Open Problem 3 Does $R(t \oplus s) = R(t) + R(s)$ hold for all tensors t and s? This is known as *Strassen's additivity conjecture*.

Finally, we define the tensor product of t and s, which is an element of $(V_1 \otimes U_1) \otimes \cdots \otimes (V_n \otimes U_n)$. For two rank-one tensors $x = v_1 \otimes \cdots \otimes v_n$ and $y = u_1 \otimes \cdots \otimes u_n$, their tensor product is defined as

$$x \otimes y = (v_1 \otimes u_1) \otimes \cdots \otimes (v_n \otimes u_n).$$

If we write $t = x_1 + \cdots + x_r$ as a sum of rank-one tensors and $s = y_1 + \cdots + y_p$, then we define their tensor product as

$$t \otimes s = \sum_{i=1}^{r} \sum_{j=1}^{p} x_i \otimes y_j. \tag{6.4}$$

It is easy to verify that this is well defined.

If you think in coordinates, then the tensor product of $t = (t_{i_1,\dots,i_n}) \in k^{d_1 \times \cdots \times d_n}$ and $s = (s_{j_1,\dots,j_n}) \in k^{e_1 \times \cdots \times e_n}$ is given by

$$t \otimes s = (t_{i_1,\dots,i_n} s_{j_1,\dots,j_n}) \in k^{d_1 e_1 \times \cdots \times d_n e_n}.$$

The pair (i_1, j_1) is interpreted as a number from $\{1, \dots, d_1 e_1\}$ and is used to index the first coordinate, (i_2, j_2) for the second, and so on.

From (6.4), the next fact follows easily.

Fact 4 *$R(t \otimes s) \leq R(t) R(s)$.*

Note that in this case, the inequality may be strict. For instance, the rank of 2×2-matrix multiplication is 7, however, the rank of $2^m \times 2^m$-matrix matrix multiplication is strictly less than 7^m for large enough m, since there are algorithms for matrix multiplication that are asymptotically faster than Strassen's algorithm.

6.2.3 From Tensor Rank Bounds to Formula Size Bounds

With a tensor $t = (t_{i_1,\ldots,i_d}) \in k^n \otimes \cdots \otimes k^n$, we associate the following polynomial in the nd variables $X_{i,j}$, $1 \leq i \leq d$, $1 \leq j \leq n$:

$$f_t = \sum_{i_1=1}^{n} \cdots \sum_{i_d=1}^{n} t_{i_1,\ldots,i_d} X_{1,i_1} \cdots X_{n,i_n}.$$

Raz [Raz10] proved the following result:

Theorem 2 *For any family of tensors t_n of order $d(n)$ such that $\omega(1) \leq d(n) \leq o(\log n / \log\log n)$ and $R(t_n) \geq n^{(1-o(1))d(n)}$, the polynomials f_{t_n} have superpolynomial formula size.*

Note that $R(t_n) \leq n^{d(n)}$ by the trivial decomposition. Therefore, the family t_n has "almost" highest rank possible. It is a major open problem to find a family of polynomials with superpolynomial formula size. So finding high rank tensors might be a way of doing so. The best lower bounds we have are due to Kalorkoti [Kal85] and are quadratic.

Several decades earlier, Strassen [St73] proved the following result.

Theorem 3 *For any family of tensors t_n of order 3, the circuit complexity of the trilinear forms f_{t_n} are bounded by $\Omega(R(t_n))$.*

This means that a family of tensors of superlinear rank yields a family of polynomials with superlinear circuit complexity, something which we do not know for general circuit models.

But there is a catch, as we will see in the next section. Finding some family of tensors/polynomials with the desired properties is easy, a random choice does the job. So what we really want is an *explicit* tensor. We call a family of tensors $t_n = (t_{n;i_1,\ldots,i_d})$ explicit if the mapping $(n; i_1, \ldots, i_d) \mapsto t_{n;i_1,\ldots,i_d}$ can be computed by an arithmetic circuit of size polynomial in d and $\log n$. One can think also of other notion of explicitness. For the purpose of this appetizer, any notion that prevents random tensors is fine. If the entries of the tensors are rational, we could also require that the mapping is computable in P/poly. Then, by using Valiant's criterion, we can use high rank tensors to separate classes in Valiants model, in particular, we could show that the permanent does not have polynomial size formulas, see [Bür00].

6.2.4 Random Tensors

Let V be a vector space of dimension n. A generic rank-one tensor in $V^{\otimes d}$ is described by dn variables,

$$(x_{1,1}, \ldots, x_{n,1}) \otimes \cdots \otimes (x_{1,d}, \ldots, x_{n,d}).$$

The sum of r generic rank-one tensors is described by rdn varibles. Its entries are multilinear polynomials in these variables. A generic tensor in $V^{\otimes d}$ is described by n^d variables, all of them begin algebraically independent. Therefore, $rnd \geq n^d$ is required to write every tensor as a sum of r rank-one tensor, that is,

$$r \geq \frac{n^{d-1}}{d}.$$

This is a very simple argument, but sufficient for our needs and almost optimal. With more sophisticated ones, we can get tighter bounds, see the work of Lickteig and Strassen for three-dimensional tensors [Lic85, Str83], see Landsberg's book for the general case [Lan11].

From the argument above, it follows that there is a tensor with rank at least n^{d-1}/d. But even random tensors have at least this rank with high probability. The entries of tensors that can be written as the sum of fewer rank-one tensors are algebraically dependent, since they can we written as polynomials in less than n^d variables. Therefore, these entries fulfill some polynomial relation. It is well known that random points do not fufill polynomial relations with high probability. In theoretical computer science, this fact is known as the Schwartz-Zippel lemma.[2]

Lemma 1 (Schwartz–Zippel) *Let F be a field. Let p be a nonzero polynomial in $F[X_1, \dots, X_n]$ of total degree d. Let $S \subseteq F$. Then*

$$\Pr_{r_1,\dots,r_n \in S}[p(r_1, \dots, r_n) = 0] \leq |S|/d.$$

Even if we do not know a bound on the polynomial describing the algebraic dependence, if the underlying field is large enough or even infinite, a random tensor will have rank $\geq n^{d-1}/d$ with high probability.

Although it sounds very simple, it is a major open problem to find a tensor of high rank that is explicit, i.e., whose entries can be constructed by a *deterministic* polynomial time algorithm.

6.3 Explicit Tensors from Bilinear Mappings

This section shows the present (poor) knowledge of how to construct explicit tensors of high rank.

6.3.1 The Rank of Bilinear Mappings and Algebras

Let $\phi : U \times V \to W$ be a bilinear mapping. Every bilinear map corresponds in a unique way to a tensor t_ϕ in $U^* \otimes V^* \otimes W$, see (6.1). Since a vector space and its

---

[2] The name of the lemma is justified because Schwartz and independently Zippel were the *last* to prove this lemma.

dual are isomorphic, we can also think if t_ϕ living in $U \otimes V \otimes W$. We define the rank of a bilinear map ϕ to be the rank of the corresponding tensor t_ϕ. If A is an finite dimensional associative algebra with unity, that is, A is a ring which is also a finite dimensional vector space over some field k, then the multiplication map in A is a bilinear mapping $A \times A \to A$. The rank $R(A)$ of A is the rank of its multiplication map.

If we think in coordinates, we get the tensor that corresponds to A as follows. Choose a basis $x_1, \dots, x_n$ of A. The product of any two elements of A is again an element of A and can be written as a linear combination of $x_1, \dots, x_n$. In particular

$$x_i \cdot x_j = \sum_{k=1}^{n} \alpha_{i,j,k} x_k.$$

The so-called *structural constants* $\alpha_{i,j,k}$ are the entries of the tensor (with respect to the chosen basis). Since a change of basis is an isomorphim of vector spaces, we get that the rank of this tensor is independent of the chosen basis.

The best lower bounds for the rank of an algebra and for any other tensor of order three are of the form $3 \dim A - o(\dim A)$. Very recently, Landsberg proved this for the algebra $k^{n \times n}$ of $n \times n$-matrices. An earlier example with an easier proof is the algebra $k[X_1, \dots, X_n]/I_d$ where I_d is the ideal generated by all monomials of degree d, see [Blä01]. Because the families of algebras have a "regular" structure, it is clear that the corresponding tensors are explicit. We just have to compute the structural constants. For instance, in the case of the algebra $k^{n \times n}$ with the standard basis, we have

$$e_{i,i'} e_{j,j'} = \begin{cases} e_{i,j'} & \text{if } i' = j, \\ 0 & \text{otherwise.} \end{cases}$$

(Note that we use double indices since $\dim k^{n \times n} = n^2$.) In the second case, if we take all monomials of degree $< d$ as a basis, we get a similar expression.

It is a major open problem to find explicit tensors or explicit families of algebras with a larger rank.

Open Problem 4 1. Is there an explicit family of tensors $t_n \in k^n \otimes k^n \otimes k^n$ with $R(t_n) \geq (3 + \epsilon)n$ for some $\epsilon > 0$.
2. Can we even achieve this for tensors corresponding to the multiplication in an algebra, i.e., is there an explicit family of algebras A_n with $R(A_n) \geq (3 + \epsilon) \dim A_n$ for some $\epsilon > 0$. Of course, $\dim A_n$ should go to infinity.

6.3.2 From Tensors of Order Three to Higher Order Tensors

We can use the lower bounds of the rank of tensors of order three to obtains bounds for the rank of higher order tensors. Up to lower order terms, they match the current best lower bounds (see the next section).

Let $t \in V_1 \otimes \cdots \otimes V_n$. Let $I_1, \ldots, I_m$ be a partition of $\{1, \ldots, n\}$, that is, the I_j are pairwise disjoint and their union is $\{1, \ldots, n\}$. Let $U_j = \bigotimes_{i \in I_j} V_i$ for $1 \le j \le m$. We can view t as an element of $U_1 \otimes \cdots \otimes U_m$. Note that the rank of t as an element of $U_1 \otimes \cdots \otimes U_m$ is a lower bound for the rank of t as an element of $V_1 \otimes \cdots \otimes V_n$. Why? Any rank-one tensor $v_1 \otimes \cdots \otimes v_n \in V_1 \otimes \cdots \otimes V_n$ induces a rank-one tensor $u_1 \otimes \cdots \otimes u_m \in U_1 \otimes \cdots \otimes U_m$ by setting $u_j = \bigotimes_{k \in I_j} v_k$. When it is not clear from context, whether we think of t being a tensor in $U_1 \otimes \cdots \otimes U_m$ or $V_1 \otimes \cdots \otimes V_n$, we add it as a subscript.

Lemma 2 $R_{U_1 \otimes \cdots \otimes U_m}(t) \le R_{V_1 \otimes \cdots \otimes V_n}(t)$. □

The rank can indeed become smaller. Consider $\langle n, n, n \rangle \in k^{n \times n} \otimes k^{n \times n} \otimes k^{n \times n}$, the tensor of matrix multiplication. If we consider it as a tensor in $(k^{n \times n} \otimes k^{n \times n}) \otimes k^{n \times n}$, then it is a matrix of size $n^4 \times n^2$. Its rank is at most n^2. However, we know a lower bound of $3n^2 - o(n^2)$ for the rank of $\langle n, n, n \rangle$ as a tensor in $k^{n \times n} \otimes k^{n \times n} \otimes k^{n \times n}$ [Lan12]. In fact, it is an old open problem, whether the so-called exponent of matrix multiplication is two.

6.3.3 Explicit Tensors of Higher Order

Let d be even and let $N = n^{d/2}$ Take any full rank matrix $M \in k^{N \times N}$, for instance the identity matrix. It has rank $n^{d/2}$. By Lemma 6.2,

$$R_{\bigotimes_{i=1}^{d} k^n}(M) \ge n^{d/2}. \tag{6.5}$$

The tensor M is obviously explicit, an entry $m_{i_1,\ldots,i_d} = 1$ if $(i_1, \ldots, i_{d/2}) = (i_{d/2+1}, \ldots, i_d)$ and 0 otherwise. Note that if we could achieve $n^{(1-o(1))d}$, then this will lead to formula lower bounds.

It is a sad state of affairs that (6.5) is the asymptotically best lower bound for an explicit tensor that we currently know; further improvements just concern the constant factor.

Here is one such improvement which uses a lower bound by Hartmann [Har85]: Let k be a field and K be an extension field of dimension n. Consider the multiplication of the K-left module $K^{1 \times m}$ as a bilinear map over k. (We take $x \in K$ and $(y_1, \ldots, y_m) \in K^{1 \times m}$ and map them to $(xy_1, \ldots, xy_m) \in K^{1 \times m}$). However, we view this as a k-bilinear map and not as a K-bilinear map. Let $\hat{s}$ be the corresponding tensor. Hartmann showed that

$$R(K^{1 \times m}) \ge (2n - 1)m = 2nm - m. \tag{6.6}$$

If we now set $m = n^{e-1}$ and let $s \in K^{\otimes(2e+1)}$ be the tensor corresponding to $\hat{s}$, we get

$$R(s) \ge 2n^e - n^{e-1}$$

for a tensor of order $d = 2e + 1$. Note that if d is odd, then the approach above using an invertible matrix only gives the lower bound n^e.

If we take $K = k[X]/(X^n)$ instead of an extension field, then we can show the same bound as (6.6) and get another example of an explicit tensor. As a basis of K, we choose the basis $1, X, \dots, X^{n-1}$. This induces a basis of $K^{1,m}$ in the natural way. How does the tensor of the multiplication of $K^{1\times m}$ look like? First consider the case $m = 1$. The tensor looks like

$$\begin{pmatrix} 1 & 2 & 3 & \dots & n \\ 2 & 3 & & \iddots & 0 \\ 3 & & \iddots & 0 & 0 \\ \vdots & \iddots & \iddots & \vdots & \vdots \\ n & 0 & \dots & 0 & 0 \end{pmatrix} \tag{6.7}$$

How to interpret this? It is a $\{0, 1\}$-valued tensor of size $n \times n \times n$. An entry $k > 0$ in position (i, j) means that the entry in position (i, j, k) is 1. All other entries are 0 whether it is explicitly indicated or not. The tensor for arbitrary m looks like follows:

$$T = \begin{pmatrix} T_1 \\ T_2 \\ \vdots \\ T_m \end{pmatrix}.$$

Each T_j is a copy of the tensor in (6.7). However, these tensors T_j live in different slices. T is a tensor in $k^{n\times nm\times nm}$, each T_j lives in the slices $(j-1)n+1, \dots, jn$ in the third component. Now, as in the beginning, we want to apply substitution method. We will only work with the copy T_1, kill $2n-1$ products, and then we can simply apply induction. In an optimal decomposition of T into rank-one tensors

$$T = \sum_{i=1}^{r} u_i \otimes v_i \otimes w_i,$$

we can assume that $w_1, \dots, w_{n-1}$ restricted to the first $n-1$ coordinates, are linearly independent. Let h be the projection along the linear span of $w_1, \dots, w_{n-1}$ onto $\langle e_n, e_{n+1}, \dots, e_{mn}\rangle$. Here, $e_i \in k^{mn}$ is the ith unit vector and $\langle \dots \rangle$ the linear span. Applying h, we kill $n-1$ products in the decomposition of T. What happens to T under this homomorphism? Note that only the first n rows of T are affected, which just contain the copy T_1. In (6.7), h maps multiples of the slices $1, \dots, n-1$ onto the nth one. The result is a lower triangular matrix with all 1s on the diagonal. Therefore, the matrix has full rank. Like before, we can now kill another n products and the tensor that is still computed is

$$T' = \begin{pmatrix} T_2 \\ T_3 \\ \vdots \\ T_m \end{pmatrix}.$$

Therefore, we can proceed by induction and get a lower bound of $(2n-1)m$.

6.4 Explicit Tensors by Combinatorial Constructions

The currently best lower bound for an explicit tensor is due to [AF11]. It improves on the lower order term of the construction in the last section.

Let $\ell = \lfloor \log_2 n \rfloor$. For $i \in \{0, \ldots, \ell\}$, we recursively define the following matrices:

1. $S_{1,0} = (1)$.
2. For even $n = 2m > 1$,

$$S_{n,i} = \begin{cases} \begin{pmatrix} 0 & 0 \\ S_{m,i} & 0 \end{pmatrix} & \text{if } i < \ell \\ \begin{pmatrix} I_m & 0 \\ 0 & I_m \end{pmatrix} & \text{otherwise} \end{cases}.$$

3. For odd $n = 2m + 1 > 1$,

$$S_{n,i} = \begin{cases} \begin{pmatrix} 0 & 0 & 0 \\ 0 & 0 & 0 \\ S_{m,i} & 0 & 0 \end{pmatrix} & \text{if} i < \ell \\ \begin{pmatrix} 0 & 0 & 0 \\ I_m & 0 & 0 \\ 0 & I_m & 0 \end{pmatrix} & \text{otherwise} \end{cases}.$$

Finally, let T_n be the tensor consisting of slices $S_{n,0}, \ldots, S_{n,\ell}$. The format of T_n is $n \times n \times (\ell + 1)$. T_n is certainly explicit, we can determine the entries by following the recursive structure.

Theorem 4 *$R(T_n) \geq 2n - 2h(n) + 1$ where $h(n)$ is the number of* 1*s in the binary expansion of n.*

Note that $h(n) \leq \log n$. We can use the substitution method to prove the theorem. Certainly $R(T_1) = 1$ holds. If $n = 2m$ is even, then we can "substitute away" the two identity matrices, killing n products. The remaining tensor is $T_{n/2}$. Therefore, we get the recurrence

$$R(T_{2m}) \geq R(T_m) + 2m.$$

In the same way, we get

$$R(T_{2m+1}) \geq R(T_m) + 2m.$$

It is easy to verify that the bound stated in the theorem is the solution of this recurrence. We can now extend this to a tensor of size $n \times (n+1) \times n$ by extending the slices above by one column and then add $n - (\ell + 1)$ linearly independent slices, which just have a one in the extra column. These slices can be substituted away and we get a tensor of size $n \times (n+1) \times n$ with rank bounded by $3n - \Theta(\log n)$. If we set $n = m^e$ and add just m extra slices, we get a tensor of size $m^e \times (m^e + 1) \times m^e$ the rank of which can be lower bounded by $2n + m - \Theta(\log n) = 2m^e + m - \Theta(d \log m)$. As before, we can interpret this tensor as a tensor of order $2e + 1$. (The $+1$ in $m^e + 1$ the second component disturbs this construction a little bit, to remedy this, we can start with a tensor of size $(n-1) \times (n-1) \times (n-1)$ and then extend this to a tensor of size $n \times n \times n$ by adding zeros.)

If you look at the construction closely, we can view this again as a tensor related to an algebra, as pointed by Landsberg [Lan13]. For simplicity, assume that n is a power of 2. Otherwise, the tensor will just have some additional zeros. It is quite easy to see, that up to permutations, the tensor T_n are just the slices $1, 2, \ldots, 2^i, \ldots, 2^\ell$ of the tensor of the algebra $k[X]/(X^n)$. For instance, T_8 has the form

$$T_8 = \begin{pmatrix} 1 & 2 & & 3 & & & & 4 \\ 2 & & 3 & & & & 4 & \\ & 3 & & & & 4 & & \\ 3 & & & & 4 & & & \\ & & & 4 & & & & \\ & & 4 & & & & & \\ & 4 & & & & & & \\ 4 & & & & & & & \end{pmatrix}$$

Looking at this, it is easy to see that we can project away the lower four rows and four columns to the righthand side. This reduces the rank by 8 (in general by n) and affects only the fourth slice. We can remove this slice and get T_4. Using induction, we get a lower bound of $1 + 2 + 4 + 8 = 15$ (in general $2n - 1$, note that n is a power of 2).

6.5 Conclusions

In the examples we have seen, the bounds on the rank are proven via the substitution method. The bounds that are achievable with this method are usually limited by the sum of the dimension of the vector spaces. Up to lower order terms, our constructions reach this limit. To get better bounds, new lower bound techniques are needed. One promising approach for this is the geometric complexity approach by Bürgisser and Ikenmeyer [BI11, BI12].

References

[AF11] B. Alexeev, M.A. Forbes, J. Tsimerman, Tensor rank: some lower and upper bounds, in *IEEE Conference on Computational Complexity* (2011), pp. 283–291

[Blä01] M. Bläser, *Improvements of the Alder-Strassen Bound: Algebras with Nonzero Radical* (ICALP, 2001), pp. 79–91

[BCS97] P. Bürgisser, M. Clausen, M.A. Shokrollahi, *Algebraic Complexity Theory* (Springer, Berlin, 1997)

[Bür00] P. Bürgisser, *Completeness and Reduction in Algebraic Complexity Theory* (Springer, Berlin, 2000)

[BI11] P. Bürgisser, C. Ikenmeyer, *Geometric Complexity Theory and Tensor Rank* (STOC 2011), pp. 509–518

[BI12] P. Brgisser, C. Ikenmeyer, Explicit lower bounds via geometric complexity theory. CoRR abs/1210.8368 (2012)

[Can88] J.F. Canny, in *Some Algebraic and Geometric Computations in PSPACE* (STOC, 1988), pp. 460–467

[CW90] D. Coppersmith, S. Winograd, Matrix multiplication via arithmetic progressions. J. Symb. Comput. **9**(3), 251–280 (1990)

[Har85] W. Hartmann, On the multiplicative complexity of modules over associative algebras. SIAM J. Comput. **14**(2), 383–395 (1985)

[Hås90] J. Håstad, Tensor rank is NP-complete. J. Algorithms **11**(4), 644–654 (1990)

[Lan12] J.M. Landsberg, New lower bounds for the rank of matrix multiplication (2012), arXiv:1206.1530v1 [cs.CC]

[Lan11] J.M. Landsberg, Tensors: geometry and applications. Graduate Studies in Mathematics, vol.128 (American Mathematical Society, Providence, 2012)

[Lan13] J.M. Landsberg, Nontriviality of equations and eplicit tensors in $\mathbb{C}^m \otimes \mathbb{C}^m \otimes \mathbb{C}^m$ of border rank at least $2m - 1$ (2013), arXiv:1209.1664v2

[Lic85] T. Lickteig, Typical tensor rank. Lin. Alg. Appl. **69**, 95–120 (1985)

[Kal85] K. Kalorkoti, A lower bound for the formula size of rational functions. SIAM J. Comput. **14**(3), 678–687 (1985)

[Pan66] V.Y. Pan, Methods for computing values of polynomials. Russ. Math. Surv. **21**, 105–136 (1966)

[Raz10] R. Raz, in *Tensor-Rank and Lower Bounds for Arithmetic Formulas* (STOC 2010), pp. 659–666

[Ren92] J. Renegar, On the computational complexity and geometry of the first-order theory of the reals, part i: introduction. Preliminaries. the geometry of semi-algebraic sets. The decision problem for the existential theory of the reals. J. Symb. Comput. **13**(3), 255–300 (1992)

[Sto10] A. Stothers, On the Complexity of Matrix Multiplication, Ph.D. thesis, University of Edinburgh, 2010

[St73] V. Strassen, V. von Divisionen, Crelle's. J. Reine Angew. Math. **264**, 184–202 (1973)

[Str83] V. Strassen, Rank and optimal computation of generic tensors. Lin. Alg. Appl. **52**, 645–685 (1983)

[Wil12] V.V. Williams, in *Multiplying matrices faster than Coppersmith-Winograd* (STOC, 2012), pp. 887–898

Chapter 7
Progress on Polynomial Identity Testing-II

Nitin Saxena

To my grand-advisor Professor Somenath Biswas

Abstract We survey the area of algebraic complexity theory; with the focus being on the problem of polynomial identity testing (PIT). We discuss the key ideas that have gone into the results of the last few years.

Keywords Arithmetic circuit · Identity testing · Hitting-set · Rank · Lower bound · Jacobian · Concentration · Shift · Morphism

Mathematics Subject Classification (2010) Primary 68Q25, 68W30, Secondary 12Y05, 13P25

7.1 Introduction

Algebraic complexity theory is the study of computation via *algebraic* models, hence, algebraic techniques. In this article we work with only one model—*arithmetic circuit* (in short, *circuit*). A circuit $C(x_1, \ldots, x_n)$, over a ring R, computes a polynomial f in $R[x_1, \ldots, x_n]$. Its description is in the form of a rooted tree; with the *leaves* having the variables or constants as input, the internal *nodes* computing addition or multiplication, and the *root* having the f as output. The edges in C, called *wires*, carry the intermediate polynomials and could also be used to multiply by a constant (from R). By the *size*, respectively the *depth*, of C we mean the natural notions (sometimes to avoid "trivialities" we might want to take into account the bit-size needed to represent an element in R).

N. Saxena (✉)
Department of CSE, IIT Kanpur, Kanpur 208016, India
e-mail: nitin@cse.iitk.ac.in

M. Agrawal and V. Arvind (eds.), *Perspectives in Computational Complexity*, Progress in Computer Science and Applied Logic 26, DOI: 10.1007/978-3-319-05446-9_7,

A moment's thought would suggest that a circuit is a rather compact way of representing polynomials. Example a circuit of size s could produce a polynomial of degree 2^s (hint: repeated squaring). In fact, a single product gate could multiply s linear polynomials and produce $n^{\Omega(s)}$ many monomials. Thus, a circuit is an 'exponentially' compact representation of some polynomial families (as opposed to simply writing it as a sum of monomials). Conversely, are there 'explicit' polynomial families (say n-variate n-degree) that require exponential (i.e. $2^{\Omega(n)}$) sized circuits? We "expect" almost every polynomial to be this hard, but, the question of finding an *explicit* family is open and is the main goal motivating the development of algebraic complexity.

One can try to directly give a good *lower bound* against circuits by designing an explicit polynomial family $\{f_n\}$ and prove that it requires a 'large'-sized circuit family $\{C_n\}$. The other, indirect, way is to design an efficient *hitting-set* $\mathcal{H}$ for the circuit family, i.e. if $C_n \neq 0$ then $\exists a \in \mathcal{H}, C_n(a) \neq 0$. This 'flip' from lower bounds to algorithms was first remarked by [HS80] and now it has several improved versions [KI04, Agr05, Agr06]. This is a remarkable phenomenon and is one of the primary motivations to study the question of PIT: Given a circuit C test it for zeroness, in time polynomial in size(C). The hitting-set version of PIT is also called *blackbox* PIT (contrasted with *whitebox* PIT).

The last 10 years have seen a decent growth of algebraic tools and techniques to understand the properties of polynomials that a circuit computes. The feeling is that these polynomials are special, different from general polynomials, but a strong enough algebraic 'invariant' or a combinatorial 'concept' is still lacking. There have been several articles surveying the known techniques and the history of PIT [Sax09, AS09, SY10, CKW11, Sap13]. In this survey we will attempt not to repeat what those surveys have already covered. So, we will focus only on the new ideas and assume that the reader has given at least a cursory glance at the older ones. We directly move on to the Leitfaden.

7.1.1 Survey Overview

This article deals mainly with three broad topics—the 'universality' of depth-3 circuits, the design of hitting-sets via 'faithful' morphisms and that via rank 'concentration'. A major emerging area that we skip in this article is that of PIT vis à vis GCT (geometric complexity theory) program [Mul11, Mul12a, Mul12b]; the algebraic-geometry interpretations there are interesting though any concrete PIT algorithm, or application, is yet to emerge.

Shallow circuits A depth-2 circuit (top + gate) of size s, over a field, essentially computes a sum of s monomials. Such polynomials are called *sparse* polynomials; blackbox PIT for them was solved few decades ago. So, our next stop is depth-3: Polynomials of the form

$$C = \sum_{i=1}^{k} \prod_{j=1}^{d} L_{i,j},$$

where $L_{i,j}$ are linear polynomials in $\mathbb{F}[x_1, \ldots, x_n]$. Significant research has been done with this model, but both subexponential PIT and exponential lower bounds are open here. Recently, a remarkable universality result was shown for depth-3 [GKKS13]: If an n-variate poly(n)-degree polynomial can be nontrivially computed by a circuit, then it can be nontrivially computed in depth-3. This 'squashing' of depth means that it suffices to focus on depth-3 for PIT purposes.

If we consider a depth-2 circuit (top $\times$ gate), over a *ring* R, then again we get some remarkable connections. Fix R to be the 2×2 matrix algebra $M_2(\mathbb{F})$, and consider the circuit

$$D = \prod_{i=1}^{d} L_i,$$

where L_i are linear polynomials in $R[x_1, \ldots, x_n]$. Traditionally, D is called a *width-2 algebraic branching program* (ABP). It was shown by [SSS09] that depth-3 PIT efficiently reduces to width-2 ABP PIT.

Faithful morphisms It was observed in the last few years that in all the known hitting-sets, the key idea in the proof is to work with a *homomorphism* φ and an algebraic *property* that the image of φ should preserve. [SS12] used a (Vandermonde-based) map $\varphi : \mathbb{F}[x_1, \ldots, x_n] \to \mathbb{F}[y_1, \ldots, y_k]$ that preserves the 'linear' rank of any k linear polynomials. This gave the first blackbox PIT for bounded top fanin depth-3, over any field.

Beecken et al. [BMS13] and Agrawal et al. [ASSS12] used a (Vandermonde and Kronecker-based) map $\varphi : \mathbb{F}[x_1, \ldots, x_n] \to \mathbb{F}[y_1, \ldots, y_k]$ that preserves the 'algebraic' rank (formally, *transcendence degree*) of certain k polynomials. This gave the first blackbox PIT (and lower bounds) for several well-studied classes of constant-depth circuits. One drawback of the technique is that it requires zero/large characteristic fields.

Rank concentration Inspired from the tensors, a restricted circuit model called multilinear read-once ABP (ROABP) has been intensively studied. Let R be the $w \times w$ matrix algebra $M_w(\mathbb{F})$ and let $\{S_i\}$ be a partition of $[n]$. Consider the circuit $D = \prod_{i=1}^{d} L_i$, where L_i are linear polynomials in $R[x_{S_i}]$ (i.e. the linear factors have disjoint variables). For D [FSS13] gave a hitting-set in time poly$(wn)^{\log w \cdot \log n}$, i.e. quasi-poly-time. The proof is based on the idea, following [ASS13], that after applying a small (Kronecker-based) 'shift', D gets the following property: The rank of its coefficients (viewed as $\mathbb{F}$-vectors) is concentrated in the 'low' support monomials. Thus, checking the zeroness of these low monomials is enough!

We conjecture that rank concentration, after a 'small' shift, should be attainable in any ABP D. But currently the proof techniques are not that strong. Recently, [AGKS13] have achieved rank concentration in multilinear depth-3 circuits where the partitions (corresponding to each product gate) are 'close' to each other in the sense of 'refinement'.

7.2 Shallow Circuits, Deep Interconnections

In this section, we exhibit the key ideas behind the universality of two shallow circuits.

7.2.1 The Depth-3 Chasm

In the study of circuits one feels that low-depth should already hold the key. This feeling was confirmed in a series of work [VSBR83, AV08, Koi12, Tav13]: Any poly(n)-degree n-variate polynomial computed by a poly(n)-sized circuit C can also be computed by a $n^{O(\sqrt{n})}$-sized depth-4 circuit!

The idea for this is, in retrospect, simple—since the degree is only poly(n), first, squash the depth of C to $O(\log n)$ by only a polynomial blowup in the size. This is done in a way so as to make the product gates quite *balanced*, i.e. their two inputs are roughly of the same degree. Next, identify a subcircuit C_2 by picking those gates whose output polynomial has degree at least $\sqrt{n}$, and call the remaining subcircuit C_1. We view C_2 as our circuit of interest that takes gates of C_1 as input. It can be shown that C_2 computes a polynomial of degree $\approx\sqrt{n}$ of its input variables (which are poly(n) many). Obviously, each gate of C_1 also computes a polynomial of degree $\approx\sqrt{n}$ of its input variables (which are $x_1, \ldots, x_n$). Thus, C_2 finally computes a sum of $\approx\binom{\text{poly}(n)+\sqrt{n}}{\sqrt{n}}$ products, each product has $\sqrt{n}$ factors, and each factor is itself a sum of $\approx\binom{n+\sqrt{n}}{\sqrt{n}}$ degree-$\sqrt{n}$ monomials. To put it simply (and ignoring the constant factors), C can be expressed as a $\sum\prod^{\sqrt{n}}\sum\prod^{\sqrt{n}}$ circuit of size $n^{O(\sqrt{n})}$. The details of this proof can be seen in [Tav13].

The strength of depth-4 is surprising. Recently, an even more surprising reduction has been shown [GKKS13]—that to depth-3 (again, $n^{O(\sqrt{n})}$ sized). We will now sketch the proof. It ties together the known results in an unexpected way.

Essentially, the idea is to modify a $\sum\prod^a\sum\prod^a$ circuit C of size $s := n^a$ (where $a := \sqrt{n}$) by using two polynomial identities that are in a way "inverse" to each other, and are to do with powers-of-linear-forms. First, replace the product gates using Fischer's identity:

Lemma 2.1 [Fis94] *Any degree a monomial can be expressed as a linear combination of 2^{a-1} ath powers of linear polynomials, as:*

$$y_1 \cdots y_a \;=\; (2^{a-1}\cdot a!)^{-1}\cdot \sum_{r_2,\ldots,r_a\in\{\pm 1\}} \left(y_1 + \sum_{i=2}^{a} r_i y_i\right)^a \cdot (-1)^{\#\{i|r_i=-1\}}.$$

We denote this type of a circuit by the notation $\sum \bigwedge^a \sum$, where the wedge signifies the powering by a. The above identity transforms the $\sum \prod^a \sum \prod^a$ circuit C to a $\sum \bigwedge^a \sum \bigwedge^a \sum$ circuit, of size poly(s). We reuse s for this size estimate.

Next, the two power gates are 'opened' up using an identity introduced by the author:

Lemma 2.2 [Sax08] *For any a, m, there exist degree-a univariate polynomials $f_{i,j}$ such that*

$$(y_1 + \cdots + y_m)^a = \sum_{i=1}^{ma+1} \prod_{j=1}^{m} f_{i,j}(y_j).$$

Let us carefully see the jugglery on C. The $\sum \bigwedge^a \sum \bigwedge^a \sum$ circuit C has the expression $C = \sum_i T_i$, where each T_i has the form $(\sum_{j=1}^{s} \ell_{i,j}^{e_{i,j}})^a$ with linear $\ell_{i,j}$'s. We want to open up the top power gate of C. By Lemma 2.2 we get

$$T_i = \sum_{u=1}^{sa+1} \prod_{j=1}^{s} f_{u,j}(\ell_{i,j}^{e_{i,j}}).$$

Since $f_{u,j}$ is a univariate, it splits into linear polynomials when the base field $\mathbb{F}$ is *algebraically closed*. As $\ell_{i,j}$ is already a linear polynomial, we deduce that T_i, and hence C, is a $\sum \prod \sum$ circuit of size poly(s).

Finally, note that for the above arguments to work, we require $\mathbb{F}$ to be algebraically closed and char$(\mathbb{F}) > a$. Lemma 2.2 has been generalized to all characteristics by [FS13b], so it is likely that this depth-3 reduction can be extended to *all* algebraically closed fields.

The optimality of $n^{\sqrt{n}}$-size, in this reduction, is open. However, [KSS13] showed that any decent reduction in this size bound would imply $VNP \neq VP$.

7.2.2 The Width-2 Chasm

Here we look at $\prod \sum$ circuits over a matrix algebra. Though the model $D = \prod_i L_i$, with linear $L_i \in R[x_1, \ldots, x_n]$, seems innocuous at first sight, a closer look proves the opposite! It can be shown fairly easily that a polynomial computed by a constant-depth circuit (over a field) can as well be computed by a D over a 3×3 matrix algebra [BC88]. On the other extreme, by taking $R = M_n(\mathbb{F})$ we can compute the *determinant* of a matrix in $\mathbb{F}^{n \times n}$ [MV97], hence, arithmetic *formulas* (not general circuits!) can be simulated in this model [Val79].

Perhaps surprisingly, [SSS09] showed that: A polynomial C computed by a depth-3 circuit (over a field) can be "almost"[1] computed by a D over a 2×2 matrix algebra.

[1] We are able to compute only a *multiple* of C. However, the extra factor is simply a product of poly-many linear polynomials. So, it suffices for PIT purposes.

This, togetherwith the previous subsection, makes the $\prod\sum$ circuits over $M_2(\mathbb{F})$ quite strong.

Say, we want to express the depth-3 circuit $C = \sum_{i=1}^{k} T_i$ in a 2×2 matrix product. First, we express a product $T_i = \prod_{j=1}^{d} \ell_{i,j}$ as:

$$\begin{bmatrix} \ell_{i,1} & 0 \\ 0 & 1 \end{bmatrix} \cdots \begin{bmatrix} \ell_{i,d-1} & 0 \\ 0 & 1 \end{bmatrix} \cdot \begin{bmatrix} 1 & \ell_{i,d} \\ 0 & 1 \end{bmatrix} = \begin{bmatrix} T_i' & T_i \\ 0 & 1 \end{bmatrix}, \text{ where } T_i' := T_i/\ell_{i,d}.$$

Once we have such k 2×2 matrices, each containing T_i in the $(1, 2)$th place, we would like to sum the T_i's in a 'doubling' fashion (instead of one-by-one).

We describe one step of the iteration. Let $\begin{bmatrix} L_1 & L_2 f \\ 0 & L_3 \end{bmatrix}$ and $\begin{bmatrix} M_1 & M_2 g \\ 0 & M_3 \end{bmatrix}$ be encapsulating two intermediate summands f and g. With the goal of getting (a multiple of) $f + g$ we consider the following, carefully designed, product:

$$\begin{bmatrix} L_1 & L_2 f \\ 0 & L_3 \end{bmatrix} \cdot \begin{bmatrix} L_2 M_3 & 0 \\ 0 & L_1 M_2 \end{bmatrix} \cdot \begin{bmatrix} M_1 & M_2 g \\ 0 & M_3 \end{bmatrix}$$
$$= \begin{bmatrix} L_1 M_1 L_2 M_3 & L_2 M_3 L_1 M_2 (f+g) \\ 0 & L_3 M_3 L_1 M_2 \end{bmatrix}$$

After $\log k$ such iterations, we get a *multiple* of C in the $(1, 2)$-th entry of the final 2×2 matrix product. Note that the middle matrix, introduced in the LHS above, potentially doubles (in the degree of the entry polynomials) in each iteration. Thus, finally, D is a product of $\text{poly}(d2^{\log k})$ linear polynomials over $M_2(\mathbb{F})$. Thus, the size blowup is only polynomial in going from depth-3 to width-2.

7.3 Faithful Morphisms, Hitting-Sets

In algebraic complexity the study of certain maps has been fruitful—homomorphisms $\varphi : \mathcal{R} := \mathbb{F}[x_1, \ldots, x_n] \to \mathbb{F}[y_1, \ldots, y_k] =: \mathcal{R}'$ such that the algebraic 'relationship' of certain polynomials $\{f_1, \ldots, f_k\}$ does not change in the image of φ. When f_i's are linear this boils down to a linear algebra question and we can easily design φ in time $\text{poly}(n)$ (hint: employ Vandermonde matrix). This business becomes complicated when f_i's are nonlinear. Then we have to ask how are f_i's represented. If they are given via monomials then we invoke the Jacobian criterion to design φ, but the time complexity becomes exponential in k. Several variants of such faithful maps are discussed in the Ph.D thesis [Mit13]. We sketch the ideas behind two basic maps here.

7.3.1 Bounded Fanin Depth-3 Blackbox PIT

Let $C = \sum_{i\in[k]} T_i$ be a depth-3 circuit. When k is constant, C is naturally called *bounded fanin* depth-3. This case of PIT has, by now, a rich history [DS07, KS07, KS11, SS11, KS09, SS13, SS12]. Several new techniques have sprung up from this model— a locally decodable code structure, a rank-preserving map via extractors, Sylvester-Gallai configurations (higher dimensions and all fields) and rank bounds. We will sketch here the main idea behind the poly-time blackbox PIT of bounded fanin depth-3. The details are quite technical and could be seen in [SS13, SS12].

Vandermonde map We define a homomorphism Ψ_β, for a $\beta \in \mathbb{F}$, as:

$$\forall i \in [n], \quad \Psi_\beta : x_i \mapsto \sum_{j=1}^{k} \beta^{ij} y_j,$$

and $\Psi_\beta(\alpha) = \alpha$ for all $\alpha \in \mathbb{F}$. This (naturally) defines the action of Ψ_β, on *all* the elements of $\mathcal{R}$, that preserves the ring operations. We have the following nice property, as a consequence of [GR08, Lemma 6.1]:

Lemma 3.1 [Ψ_β preserves k-rank] *Let S be a subset of linear forms in $\mathcal{R}$ with* $\mathrm{rk}(S) \le k$, *and* $|\mathbb{F}| > nk^2$. *Then* $\exists \beta \in \mathbb{F}$, $\mathrm{rk}(\psi_\beta(S)) = \mathrm{rk}(S)$.

Intuitively, Ψ_β is *faithful* to any algebraic object involving the elements in span(S). The proof of this lemma is by studying the coefficient-matrix of the linear polynomials in S, and its change under Ψ_β. This map has a role to play in bounded fanin depth-3 owing to a certain structural theorem from [SS13]—*certificate for a non-identity*.

To discuss this certificate we need a definition, that of 'paths' of 'nodes' in C (assumed to be nonzero). A *path* $\mathbf{p}$ with respect to an ideal I is a sequence of terms $\{p_1, p_2, \ldots, p_b\}$ (these are products of linear forms) with the following property. Each p_i divides T_i, and each p_i is a 'node' of T_i with respect to the ideal $\langle I, p_1, p_2, \ldots, p_{i-1}\rangle$.[2] So p_1 is a node of T_1 wrt I, p_2 is a node of T_2 wrt $\langle I, p_1\rangle$, etc.

Let us see an example of a path $(\langle 0\rangle, p_1, p_2, p_3)$ in Fig. 7.1. The oval bubbles represent the list of forms in a product gate, and the rectangles enclose forms in a node. The arrows show a path. Starting with the zero ideal, nodes $p_1 := x_1^2$, $p_2 := x_2(x_2 + 2x_1)$, and $p_3 := (x_4 + x_2)(x_4 + 4x_2 - x_1)(x_4 + x_2 + x_1)(x_4 + x_2 - 2x_1)$ form a path. Initially, the path is just the zero ideal, so x_1^2 is a node. Note how p_2 is a power of x_2 modulo radsp$\langle p_1\rangle$, and p_3 is a power of x_4 modulo radsp$\langle p_1, p_2\rangle$.

The non-identity certificate theorem [SS13, Theorem 25] states that for any non-identity C, there exists a path $\mathbf{p}$ such that modulo $\langle \mathbf{p}\rangle$, C reduces to a single nonzero multiplication term.

---

[2] By a *node* p_i we mean that some nonzero constant multiple of p_i is identical to a power-of-a-linear-form modulo radsp$\langle I, p_1, p_2, \ldots, p_{i-1}\rangle$, where radsp is the ideal generated by the set of all the linear polynomials that divide p_j, $j \in [i-1]$ and the generators of I.

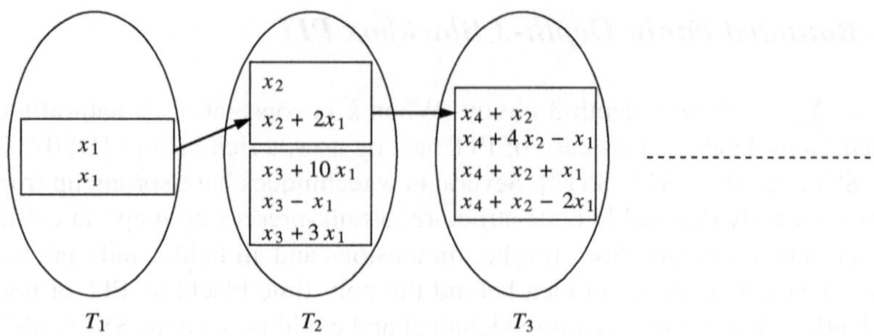

Fig. 7.1 Nodes and paths in $C = T_1 + T_2 + T_3 + \cdots$

Theorem 3.2 (Certificate for a non-identity) *Let I be an ideal generated by some multiplication terms. Let $C = \sum_{i\in[k]} T_i$ be a depth-3 circuit that is nonzero modulo I. Then $\exists i \in \{0, \ldots, k-1\}$ such that $C_{[i]}$*[3] *mod I has a path $\mathbf{p}$ satisfying: $C \equiv \alpha \cdot T_{i+1} \not\equiv 0 \pmod{I + \langle \mathbf{p} \rangle}$ for some $\alpha \in \mathbb{F}^*$.*

The proof of this theorem involves an extension of Chinese remaindering to ideals that are generated by multiplication terms. Once we have this structural result about depth-3, observe that we would be done if we could somehow ensure $T_{i+1} \notin \langle \mathbf{p} \rangle$ (in our application I is zero). How do we preserve this ideal non-membership under a cheap map?

Notice that the rank of the set S_0 of linear polynomials that divide the nodes in the path $\mathbf{p}$ is $<k$ (since path length is below k). Moreover, T_{i+1} factors into at most d linear polynomials, denote the set by S_1. So if we apply a map that preserves the rank of each of the d sets $S_0 \cup \{\ell\}$, $\ell \in S_1$, then, intuitively, the ideal non-membership should be preserved. As $\mathrm{rk}(S_0 \cup \{\ell\}) \le k$ we can employ the previously discussed map Ψ_β (over a field satisfying $|\mathbb{F}| > dnk^2$). This idea could be easily turned into a proof; details are in [SS12].

Finally, what we have achieved is the construction of a map Ψ_β, in time $\mathrm{poly}(dnk)$, that reduces the variables of C from n to k and preserves nonzeroness. Once this is done, the $\mathrm{poly}(nd^k)$ blackbox PIT follows from the brute-force hitting-set.

7.3.2 Depth $\geq$ 3 Results

Looking at the success of bounded fanin depth-3 one wonders about the analogous depth-4 model:

$$C = \sum_{i\in[k]} \prod_{j\in[d]} f_{i,j}, \quad \text{where } f_{i,j} \text{ are sparse polynomials.} \tag{7.1}$$

[3] We mean $C_{[i]} := \sum_{j\in[i]} T_j$.

Here we are thinking of a bounded k. But now even $k = 2$ seems non-trivial! In fact, a simpler PIT case than this is an old open question in a related area [vzG83].

This *bounded top fanin* depth-4 PIT is an important open question currently. What is doable are other restricted models of depth-4. Inspired from the last subsection we ask: Is there a notion of 'rank' for general polynomials, are there easy 'faithful' maps, and finally is all this useful in PIT?

There are several notions of rank in commutative algebra. The one we [BMS13] found useful is—*transcendence degree* (trdeg). We say that a set S of polynomials $\{f_1, \ldots, f_m\} \subset \mathbb{F}[x_1, \ldots, x_n]$ is *algebraically dependent* if there exists a nonzero annihilating polynomial $A(y_1, \ldots, y_m)$, over $\mathbb{F}$, such that $A(f_1, \ldots, f_m) = 0$. The largest number of algebraically *in*dependent polynomials in S is called trdeg(S). With this notion we call a homomorphism φ *faithful* if trdeg(S) = trdeg($\varphi(S)$). The usefulness of φ (assuming that one can come up with it efficiently) was first proved in [BMS13]:

Lemma 3.3 (Faithful is useful) *Let φ be a homomorphism faithful to $\mathbf{f} = \{f_1, \ldots, f_m\} \subset \mathbb{F}[\mathbf{x}]$. Then for any $C \in \mathbb{F}[\mathbf{y}]$, $C(\mathbf{f}) = 0 \Leftrightarrow C(\varphi(\mathbf{f})) = 0$.*

This implies that we can use a faithful map to 'reduce' the number of variables n without changing the nonzeroness of C. The strategy can be used in cases where trdeg($\mathbf{f}$) is small, say, smaller than a constant r.

The only missing piece is the efficiency of φ.[4] To do this we need three fundamental ingredients—an efficient criterion for algebraic independence (Jacobian), its behaviour under φ (chain rule), and standard maps (Vandermonde and Kronecker-based).

Lemma 3.4 (Jacobian criterion) *Let $\mathbf{f} \subset \mathbb{F}[\mathbf{x}]$ be a finite set of polynomials of degree at most d, and* trdeg($\mathbf{f}$) $\leq r$. *If* char($\mathbb{F}$) $= 0$ *or* char($\mathbb{F}$) $> d^r$, *then* trdeg($\mathbf{f}$) $=$ $\mathrm{rk}_{\mathbb{F}(\mathbf{x})}\, \mathcal{J}_{\mathbf{x}}(\mathbf{f})$, *where* $\mathcal{J}_{\mathbf{x}}(\mathbf{f}) := \left(\partial f_i / \partial x_j\right)_{m \times n}$ *is the* Jacobian matrix.

There are several proofs of this, see [Jac41, For91, BMS13, MSS12]. This gives us an efficient way to capture trdeg, when the characteristic is zero/large. Let us now see how the Jacobian matrix changes under φ.

Lemma 3.5 (Chain rule) $\mathcal{J}_{\mathbf{y}}(\varphi(\mathbf{f})) = \varphi\left(\mathcal{J}_{\mathbf{x}}(\mathbf{f})\right) \cdot \mathcal{J}_{\mathbf{y}}(\varphi(\mathbf{x}))$, *where φ applied to a matrix/set refers to the matrix/set obtained by applying φ to every entry.*

This is a simple consequence of the chain rule of 'derivatives'. It suggests that for φ to preserve the trdeg of the polynomials, we need to control—(1) the image of the original Jacobian under φ, and (2) the Jacobian of the image of $\mathbf{x}$. In our applications, the former is achieved by a Kronecker-based map (i.e. sparse PIT tricks, e.g. [BHLV09]) and the latter by Vandermonde map (as seen in the previous subsection).

This general 'recipe' has been successfully implemented to various circuit models. The case of the circuit $C'(\mathbf{x}) := C(\mathbf{f})$, where trdeg($\mathbf{f}$) $\leq r$ and f_i's are polynomials

[4] It can be shown, from first principles, that a faithful r-variate map always *exists* [BMS13].

of sparsity at most s, was worked out in [BMS13]. The proof follows exactly the above strategy. The time complexity is polynomial in size(C') and $(s \cdot \deg(C'))^r$, where the exponential dependence comes from the sparsity estimate of $\mathcal{J}_{\mathbf{x}}(\mathbf{f})$ (and of course the final brute-force hitting-set for the r-variate $\varphi(C')$).

Agrawal et al. [ASSS12] extended the recipe to depth-4 circuits (7.1) where the number of $f_{i,j}$'s where any variable appears is bounded by r.[5] This model is called *occur-r* depth-4; it generalizes the well-studied multilinear read-r depth-4. Interestingly, slightly modified techniques also provided *exponential* lower bounds against these special models. This required proving some combinatorial properties of the derivatives of immanant (e.g. permanent, determinant).

The faithful maps recipe has been able to unify all the assorted *poly-time* hitting-sets known. However, one drawback is that it needs the characteristic to be zero/large. Baby steps in resolving that issue have been taken by [MSS12].

7.4 Rank Concentration, Shift, Hitting-Sets

The hitting-sets that we saw till now were for models where some parameter was kept bounded. But we could also study models with a 'structural' restriction, e.g. multilinearity. This route has also been successful and enlightening. We call a depth-3 circuit $C = \sum_i T_i$ *multilinear* if the linear factors in T_i involve disjoint variables. Hence, each product gate T_i induces a partition $\mathcal{P}_i$ on the variables (or indices) $[n]$. Moreover, we call C *set-multilinear* if these partitions are the same across all T_i's.

There is a large body of work on the set-multilinear model [RS05, AMS10, FS12, FS13b, ASS13, FS13a, FSS13, AGKS13]. The motivation for this model is, on the one hand, the algebraic concept of *tensors*, and, on the other hand, the interest in read-once *boolean* branching programs [Nis92, IMZ12, Vad12]. Interestingly, [FSS13] has shown (extending the ideas of [ASS13]) that the *current* situation in the arithmetic world is exponentially better than that in the boolean one!

Here we will exhibit the key ideas of [ASS13] and [AGKS13] on two *toy* cases that are already quite instructive; this saves us from the gory technical machinery that drives the more general cases.

7.4.1 Multilinear ROABP

Agrawal et al. [ASS13] gave the first quasi-poly-time hitting-set for set-multilinear depth-3 (and extensions to constant-depth, non-multilinear versions). This was generalized by [FSS13] to *any* depth; in fact, they dealt directly with the *multilinear ROABP* $D = \prod_i L_i$ over $M_w(\mathbb{F})$, where L_i's are linear polynomials in disjoint variables. Both the papers proved 'low-support rank concentration' in their models.

[5] Note that this does not mean that $\text{trdeg}(f_{i,j}|i, j)$ is bounded.

For the following discussion we fix a base commutative ring $R = H_w(\mathbb{F})$ called the *Hadamard* algebra (instead of the $w \times w$ matrix algebra). This is basically $(\mathbb{F}^k, +, \star)$, where $+$ is the vector addition and $\star$ is the coordinate-wise vector product (called the Hadamard product).

ℓ-concentration. We say that a polynomial $f \in R[x_1, \ldots, x_n]$ is *ℓ-concentrated* if

$$\mathrm{rk}_{\mathbb{F}}\{\mathrm{coef}_f(x_S) \mid S \subseteq [n], |S| < \ell\} = \mathrm{rk}_{\mathbb{F}}\{\mathrm{coef}_f(x_S) \mid S \subseteq [n]\},$$

where coef_f extracts a coefficient in f.

I.e. the coefficient-vectors of 'lower' monomials already span every possible coefficient-vector in f. We are interested in studying whether circuits compute an ℓ-concentrated polynomial for small ℓ (say, $\log n$ instead of n). By itself this is not true, e.g. the trivial circuit $D = x_1 \cdots x_n$ is not even n-concentrated. But, maybe we can transform f a bit and then attain $(\log n)$-concentration? In this case, $D' := D(x_1 + 1, \ldots, x_n + 1)$ is suddenly 1-concentrated!

It was shown by [ASS13] that any D, above R, becomes $(\log k)$-concentrated after applying a 'small' shift; the price of which is $n^{\log k}$ time. Once we have this it directly applies to the set-multilinear depth-3 model. Since, a depth-3 $C = \sum_{i\in[k]} T_i$ can be rewritten as $C = [1, \ldots, 1] \cdot D$, where $D = \begin{bmatrix} T_1 \\ \vdots \\ T_k \end{bmatrix}$ is of the promised sort over $R = H_k(\mathbb{F})$ (since D completely factorizes into disjoint-variate linear polynomials). So, ℓ-concentration in D implies an easy way to check C for zeroness—test the coefficients of the monomials below ℓ-support in C.

Glimpse of a proof We now show how to achieve ℓ-concentration, $\ell = O(\log k)$, in the following toy model:

$$D = \prod_{i\in[n]} (1 + z_i x_i), \quad \text{where } z_i \in H_k(\mathbb{F}). \tag{7.2}$$

Because of the disjointness of the factors it can be seen, as a simple exercise, that: D is ℓ-concentrated iff $D_S := \prod_{i\in S}(1 + z_i x_i)$ is ℓ-concentrated, for all $S \in \binom{[n]}{\ell}$. Thus, from now on we assume, wlog, $n = \ell$.

Shift D by formal variables $\mathbf{t}$, and normalize, to get a new circuit:

$$D' = \prod_{i\in[\ell]} (1 + z_i' x_i), \quad \text{where } z_i' \in H_k(\mathbb{F}(\mathbf{t})).$$

We can express the new coefficients as:

$$z_i' = z_i/(1 + z_i t_i), \ \forall i \in [\ell].$$

Conversely, we write:

$$z_i = z_i'/(1 - z_i' t_i), \ \forall i \in [\ell]. \tag{7.3}$$

We write z_S for $\prod_{i\in S} z_i$. Now the goal is to 'lift' an $\mathbb{F}$-dependence of z_S's to the z'_S; which ultimately shows the condition on the shift that shall yield concentration.

Consider the 2^ℓ vectors $\{z_S \mid S \subseteq [\ell]\}$. If $\ell > \log k$ then there is a nontrivial linear dependence amongst these vectors, say,

$$\sum_{S\subseteq[\ell]} \alpha_S z_S = 0, \text{ where } \alpha_S \in \mathbb{F}.$$

Rewriting this in terms of z'_S we get:

$$\sum_{S\subseteq[\ell]} \alpha_S \cdot \prod_{i\in S} z'_i/(1 - z'_i t_i) = 0.$$

$$\text{Or, } \sum_{S\subseteq[\ell]} \alpha_S \cdot z'_S \cdot \prod_{i\in[\ell]\setminus S} (1 - z'_i t_i) = 0. \tag{7.4}$$

Let us collect the 'coefficient' of $z'_{[\ell]}$ in the above expression. It comes out to,

$$\sum_{S\subseteq[\ell]} \alpha_S \cdot (-1)^{|[\ell]\setminus S|} \cdot t_{[\ell]\setminus S}. \tag{7.5}$$

If we can ensure this expression to be nonzero then Eq. (7.4) tells us that $z'_{[\ell]}$ is in the $\mathbb{F}(\mathbf{t})$-span of the 'lower' z'_S. But, ensuring the nonzeroness of Eq. (7.5) is easy—use t_i's such that all the ($\leq\ell$)-support monomials t_S are *distinct*. We can use standard sparse PIT tricks [BHLV09] for this, in time poly(n^ℓ).

What we have shown is that, after applying a Kronecker-based shift, the circuit D becomes ℓ-concentrated; all this in time $n^{O(\log k)}$. This 'recipe' of studying the generic shift, via some combinatorial properties of the 'transfer' equations (7.3), is generalized in [ASS13] to other D; and further improved in [FSS13] to multilinear ROABP. The latter use a 'primal' interpretation of the 'transfer' matrix and show that the linear transformation– corresponding to a Kronecker shift together-with the truncation of the high-support monomials –behaves like a *rank-extractor*.

It is not known how to design such hitting-sets, even for the toy case, in *poly*-time.

7.4.2 Towards Multilinear Depth-3

It is tantalizing to achieve ℓ-concentration in multilinear depth-3 (before embarking on the general depth-3!). A partial result in that direction was obtained in [AGKS13]. We will sketch their ideas in a toy model.

Consider a multilinear depth-3 circuit C with only *two* partitions being induced by the product gates—$\mathcal{P}_1 = \{\{1\}, \cdots, \{n\}\}$ and an arbitrary partition $\mathcal{P}_2$. Say, the number of the corresponding product gates is k_1 respectively k_2 (summing to k).

We can say, naturally, that $\mathcal{P}_1$ is a *refinement* of $\mathcal{P}_2$ (denoted $\mathcal{P}_1 \leq \mathcal{P}_2$) because: For every color (or part) $S \in \mathcal{P}_2$ there exist colors in $\mathcal{P}_1$ whose union is *exactly* S. In this refinement situation [AGKS13] showed that, again, a suitable shift in the $\prod\sum$ circuit D (corresponding to C) achieves ℓ-concentration in time $\text{poly}(n^{\log k})$.

Glimpse of a proof We can assume $\mathcal{P}_2$ different from $\mathcal{P}_1$, otherwise this case is no different from the last subsection. We assume that the first k_1 product gates in $C = \sum_{i\in[k]} T_i$ respect $\mathcal{P}_1$ and the rest k_2 respect $\mathcal{P}_2$. The corresponding circuit D where we desire to achieve concentration is $D = \begin{bmatrix} T_1 \\ \vdots \\ T_k \end{bmatrix}$ over $R = H_k(\mathbb{F})$.

But now the linear factors of D are not necessarily in disjoint variables. Example $\begin{bmatrix} x_1x_2 \\ x_1 + x_2 \end{bmatrix} = \left(x_1 + \begin{bmatrix} 0 \\ 1 \end{bmatrix} \cdot x_2\right) \cdot \left(\begin{bmatrix} 0 \\ 1 \end{bmatrix} + \begin{bmatrix} 1 \\ 0 \end{bmatrix} \cdot x_2\right)$ over $H_2(\mathbb{F})$.

To get some kind of a reduction to the set-multilinear case, we prove rank concentration in parts. First, we consider those monomials (called $\mathcal{P}_1$-*type*) that could only be produced by the 'upper' part of D (i.e. the first k_1 product gates of C). Such a monomial, say indexed by $S \subseteq [n]$, is characterized by the presence of $i, j \in S$ that are in the same color of $\mathcal{P}_2$. For a fixed such i, j we can "access" all such monomials by the derivative $\partial^2 D/\partial x_i \partial x_j =: \partial_{i,j} D$. Notice that this differentiation kills the 'lower' part of D and only the $\mathcal{P}_1$-part remains. So, we can prove $(2 + \log k_1)$-concentration in the monomials containing i, j as in Sect. 7.4.1. This proves $O(\log k_1)$-concentration in the monomials of $\mathcal{P}_1$-type.

Next, we want to understand the remaining monomials (called $\mathcal{P}_2$-type); those that could be produced by the 'lower' part of D (i.e. the last k_2 product gates of C). These, obviously, could also be produced by the upper part of D. Let us fix such a monomial, say $x_1 \cdots x_\ell$. Assume that $S_1, \ldots, S_\ell \in \mathcal{P}_2$ are the colors that contain one of the indices $1, \ldots, \ell$. Consider the subcircuit D_ℓ that in its i-th coordinate, $\forall i \in [k]$, simply drops those factors of T_i that are free of the variables $S_1 \cup \cdots \cup S_\ell$. The problem here is that D_ℓ may be a 'high' degree circuit ($\approx n$ instead of ℓ) and so we cannot use a proof like in Sect. 7.4.1.

But, notice that all the degree-($\geq \ell$) monomials in D_ℓ are $\mathcal{P}_1$-type; where we know how to achieve ℓ-concentration. So, we only have to care about degree-($\leq\ell$) $\mathcal{P}_2$-type monomials in D_ℓ. There, again, $(\log k)$-concentration can be shown using Sect. 7.4.1 and the well-behaved transfer equations.

This sketch, handling two refined partitions, can be made to work for significantly generalized models [AGKS13]. But, multilinear depth-3 PIT is still open (nothing better than exponential time known).

Remark 4.1 Using a different technique [AGKS13] also proves *constant*-concentration, hence designs *poly*-time hitting-sets, for certain constant-width ROABP. These models are arithmetic analogs of the *boolean* ones—width-2 read-once branching programs [AGHP92, NN93] and constant-width read-once permutation branching programs [KNP11].

7.5 Open Ends

The search for a strong enough technique to study arithmetic circuits continues. We collect here some easy-to-state questions that interest us.

Top fanin-2 depth-4 Find a faithful map φ that preserves the algebraic independence of two products-of-sparse polynomials $\prod_i f_i$ and $\prod_j g_j$. If we look at the relevant 2×2 Jacobian determinant, say wrt variables $X := \{x_1, x_2\}$, then the question boils down to finding a hitting-set for the special *rational* function $\sum_{i,j} \frac{\det \mathcal{J}_X(f_i, g_j)}{f_i g_j}$. Can this version of *rational sparse* PIT be done in subexponential time?

Independence over $\mathbb{F}_p$ Currently, there is no subexponential time algorithm/heuristic known to test two given circuits for algebraic independence over a 'small' finite field $\mathbb{F}_p$. The reason is that something as efficient as the Jacobian criterion is not readily available, see [MSS12].

Model in Eqn (7.2) Find a *poly*-time hitting-set for this simple model. Note that a poly-time whitebox PIT is already known [RS05].

Multilinear depth-3 Achieve $o(n)$-concentration in multilinear depth-3 circuits, in $n^{o(n)}$ time. Here, the presence of an exponential lower bound against the model [RY09] is quite encouraging.

References

[AGHP92] N. Alon, O. Goldreich, J. Håstad, R. Peralta, Simple construction of almost k-wise independent random variables. Random Struct. Algorithms **3**(3), 289–304 (1992). (Conference version in FOCS 1990)

[AGKS13] M. Agrawal, R. Gurjar, A. Korwar, N. Saxena, Hitting-sets for low-distance multilinear depth-3. Electron. Colloquium Comput. Complex. **20**, 174 (2013)

[Agr05] M. Agrawal, in *Proving lower bounds via pseudo-random generators, Proceedings of the 25th Annual Foundations of Software Technology and Theoretical Computer Science* (FSTTCS, 2005), pp. 92–105

[Agr06] M. Agrawal, *Determinant versus permanent, Proceedings of the 25th International Congress of Mathematicians (ICM)*, vol. 3 (2006), pp. 985–997

[AMS10] V. Arvind, P. Mukhopadhyay, S. Srinivasan, New results on noncommutative and commutative polynomial identity testing. Comput. Complex. **19**(4), 521–558 (2010). (Conference version in, CCC 2008)

[AS09] M. Agrawal, R. Saptharishi, Classifying polynomials and identity testing. Indian Acad. Sci. **P1**, 1–14 (2009). Platinum Jubilee

[ASS13] M. Agrawal, C. Saha, N. Saxena, in *Quasi-polynomial hitting-set for set-depth-Δ formulas* (STOC, 2013), pp. 321–330

[ASSS12] M. Agrawal, C. Saha, R. Saptharishi, N. Saxena, in *Jacobian hits circuits: hitting-sets, lower bounds for depth-D occur-k formulas & depth-3 transcendence degree-k circuits* (STOC, 2012), pp. 599–614

[AV08] M. Agrawal, V. Vinay, in *Arithmetic circuits: a chasm at depth four*. (FOCS, 2008), pp. 67–75

[BC88] M. Ben-Or, R. Cleve, in *Computing Algebraic Formulas Using a Constant Number of Registers* (STOC, 1988), pp. 254–257

[BHLV09] M. Bläser, M. Hardt, R.J. Lipton, N.K. Vishnoi, Deterministically testing sparse polynomial identities of unbounded degree. Inf. Process. Lett. **109**(3), 187–192 (2009)

[BMS13] M. Beecken, J. Mittmann, N. Saxena, Algebraic independence and blackbox identity testing. Inf. Comput. **222**, 2–19, (2013). (Conference version in ICALP 2011)

[CKW11] X. Chen, N. Kayal, A. Wigderson, Partial Derivatives in Arithmetic Complexity (and beyond). Found. Trends Theor. Comput. Sci. **6**(1–2), 1–138 (2011)

[DS07] Z. Dvir, A. Shpilka, Locally decodable codes with two queries and polynomial identity testing for depth 3 circuits. SIAM J. Comput. **36**(5), 1404–1434 (2007). (Conference version in STOC 2005)

[Fis94] I. Fischer, Sums of like powers of multivariate linear forms. Math. Mag. **67**(1), 59–61 (1994)

[For91] K. Forsman, Constructive commutative algebra in nonlinear control theory, Ph.D. thesis, Dept. of Electrical Engg., Linköping University, Sweden, 1991

[FS12] M.A. Forbes, A. Shpilka, in *On identity testing of tensors, low-rank recovery and compressed sensing* (STOC, 2012), pp. 163–172

[FS13a] M.A. Forbes, A. Shpilka, Explicit Noether Normalization for Simultaneous Conjugation via Polynomial Identity Testing, APPROX-RANDOM, 2013, pp. 527–542

[FS13b] M.A. Forbes, A Shpilka, in *Quasipolynomial-time Identity Testing of Non-Commutative and Read-Once Oblivious Algebraic Branching Programs* (FOCS, 2013)

[FSS13] M.A. Forbes, R. Saptharishi, A. Shpilka, Pseudorandomness for multilinear read-once algebraic branching programs, in any order. Electron. Colloquium Comput. Complex. **20**, 132 (2013)

[GKKS13] A. Gupta, P. Kamath, N. Kayal, R. Saptharishi, in *Arithmetic circuits: a chasm at depth three* (FOCS, 2013)

[GR08] A. Gabizon, R. Raz, Deterministic extractors for affine sources over large fields. Combinatorica **28**(4), 415–440 (2008). (Conference version in FOCS 2005)

[HS80] J. Heintz, C.-P. Schnorr, in *Testing Polynomials which Are Easy to Compute (Extended Abstract)* (STOC, 1980), pp. 262–272

[IMZ12] R. Impagliazzo, R. Meka, D. Zuckerman, in *Pseudorandomness from shrinkage* (FOCS, 2012), pp. 111–119

[Jac41] C.G.J. Jacobi, De determinantibus functionalibus. J. Reine Angew. Math. **22**(4), 319–359 (1841)

[KI04] V. Kabanets, R. Impagliazzo, Derandomizing polynomial identity tests means proving circuit lower bounds. Comput. Complex. **13**(1–2), 1–46 (2004). (Conference version in STOC 2003)

[KNP11] V. Kabanets, R. Impagliazzo, Derandomizing polynomial identity tests means proving circuit lower bounds. Comput. Complex. **13**(1–2), 1–46 (2004). (Conference version in STOC 2003)

[Koi12] P. Koiran, Arithmetic circuits: The chasm at depth four gets wider. Theor. Comput. Sci. **448**, 56–65 (2012)

[KS07] N. Kayal, N. Saxena, Polynomial identity testing for depth 3 circuits. Comput. Complex. **16**(2), 115–138 (2007). (Conference version in, CCC 2006)

[KS09] N. Kayal, S. Saraf, in *Blackbox polynomial identity testing for depth-3 circuits* (FOCS, 2009), pp. 198–207

[KS11] Z.S. Karnin, A. Shpilka, Black box polynomial identity testing of generalized depth-3 arithmetic circuits with bounded top fan-in. Combinatorica **31**(3), 333–364 (2011). (Conference version in, CCC 2008)

[KSS13] N. Kayal, C. Saha, R. Saptharishi, A super-polynomial lower bound for regular arithmetic formulas. Electron. Colloquium Comput. Complex. **20**, 91 (2013)

[Mit13] J. Mittmann, Independence in Algebraic Complexity Theory, Ph.D. thesis, Mathematisch-Naturwissenschaftlichen Fakultät der Rheinischen Friedrich-Wilhelms-Universität Bonn, Germany, Dec 2013

[MSS12] J. Mittmann, N. Saxena, P. Scheiblechner, Algebraic independence in positive characteristic—A p-adic calculus. Electron. Colloquium Comput. Complex. **TR12-014** (2012). (accepted in Trans. Amer. Math. Soc. 2013)

[Mul11] K. Mulmuley, On P versus NP and geometric complexity theory: dedicated to Sri Ramakrishna. J. ACM **58**(2), 5 (2011)

[Mul12a] K. Mulmuley, *Geometric Complexity Theory V: Equivalence between Blackbox Derandomization of Polynomial Identity Testing and Derandomization of Noether's Normalization Lemma* (FOCS, 2012), pp. 629–638

[Mul12b] K. Mulmuley, The GCT program toward the P versus NP problem. Commun. ACM **55**(6), 98–107 (2012)

[MV97] M. Mahajan, V. Vinay, Determinant: combinatorics, algorithms, and complexity. Chicago J. Theor. Comput. Sci. **5**, 730–738 (1997). (Conference version in SODA 1997)

[Nis92] N. Nisan, Pseudorandom generators for space-bounded computation. Combinatorica **12**(4), 449–461 (1992). (Conference version in STOC 1990)

[NN93] J. Naor, M, Naor, Small-bias probability spaces: efficient constructions and applications. SIAM J. Comput. **22**(4), 838–856 (1993). (Conference version in STOC 1990)

[RS05] R. Raz, A Shpilka, Deterministic polynomial identity testing in non-commutative models. Comput. Complex. **14**(1), 1–19 (2005). (Conference version in, CCC 2004)

[RY09] R. Raz, A. Yehudayoff, Lower bounds and separations for constant depth multilinear circuits. Comput. Complex. **18**(2), 171–207 (2009). (Conference version in, CCC 2008)

[Sap13] R. Saptharishi, Unified Approaches to Polynomial Identity Testing and Lower Bounds, Ph.D. thesis, Department of CSE, IIT Kanpur, India, Apr 2013

[Sax08] N. Saxena, Diagonal circuit identity testing, lower bound. ICALP **1**, 60–71 (2008)

[Sax09] N. Saxena, Progress on polynomial identity testing. Bull. EATCS **90**, 49–79 (2009)

[SS11] N. Saxena, C. Seshadhri, An almost optimal rank bound for depth-3 Identities. SIAM J. Comput. **40**(1), 200–224 (2011). (Conference version in, CCC 2009)

[SS12] N. Saxena, C. Seshadhri, Blackbox identity testing for bounded top-fanin depth-3 circuits: the field doesn't matter. SIAM J. Comput. **41**(5), 1285–1298 (2012). (Conference version in STOC 2011)

[SS13] N. Saxena, C. Seshadhri, From Sylvester-Gallai configurations to rank bounds: Improved blackbox identity test for depth-3 circuits. J. ACM **60**(5), 33 (2013). (Conference version in STOC 2010)

[SSS09] C. Saha, R. Saptharishi, N. Saxena, in *The Power of Depth 2 Circuits over Algebras* (FSTTCS, 2009), pp. 371–382

[SY10] A. Shpilka, A. Yehudayoff, Arithmetic Circuits, A survey of recent results and open questions. Found. Trends Theor. Comput. Sci. **5**(3–4), 207–388 (2010)

[Tav13] S. Tavenas, in *Improved Bounds for Reduction to Depth 4 and Depth 3* (MFCS, 2013), pp. 813–824

[Vad12] S.P. Vadhan, Pseudorandomness. Found. Trends Theor. Comput. Sci. **7**(1–3), 1–336 (2012)

[Val79] L.G. Valiant, in *Completeness classes in algebra* (STOC, 1979), pp. 249–261

[VSBR83] L.G. Valiant, S. Skyum, S.J. Berkowitz, C. Rackoff, Fast parallel computation of polynomials using few processors. SIAM J. Comput. **12**(4), 641–644 (1983)

[vzG83] J. von zur Gathen, *Factoring Sparse Multivariate Polynomials* (FOCS, 1983), pp. 172–179

Chapter 8
Malod and the Pascaline

Bruno Poizat

Presented to Somenath Biswas on the occasion of his 60th birthday

Abstract We make explicit the central role played by the binomial coefficients in the description of the coefficient-function of a polynomial computed in polynomial time (no bound on the degree), following the works of Guillaume Malod. Our results are obtained with the help of a universal polynomial which is simpler than the one used by Malod. As a corollary, with the help of a result of Peter Bürgisser, we establish in characteristic zero a connection between Leslie Valiant's question $\mathsf{VNP} = ?\mathsf{VP}$ (bounded degree) and its unbounded degree version, generalizing what Malod had previously done in finite characteristic.

Keywords Complexity · Polynomials · Summations

8.1 Polynomials, Functions and Arithmetic Circuits

We compute polynomials, in several variables, with integer coefficients. For that, we consider *arithmetic circuits*, with input gates labeled either by a variable or by the constant -1; the other gates are binary (they receive two arrows) and perform either

---

Guillaume Malod defended his doctorate in 2003 in Lyon; he visited Kanpur as a participant to a cooperation program between IITK and Claude Bernard University. Blaise Pascal (1623–1662), philosopher, mathematician and physicist; he discovered the recurrence relation between the binomial coefficients ("Triangle de Pascal"), and invented the Pascaline, a mechanical device performing arithmetic operations.

B. Poizat (✉)
Institut Camille Jordan, Université Claude Bernard,
43, boulevard du 11 novembre 1918, 69622 Villeurbanne-cedex, France
e-mail: poizat@math.univ-lyon1.fr

M. Agrawal and V. Arvind (eds.), *Perspectives in Computational Complexity*, Progress in Computer Science and Applied Logic 26, DOI: 10.1007/978-3-319-05446-9_8,

an addition or a multiplication; there is only one output gate, where is obtained the polynomial computed by the circuit.

The size of the circuit is the number of its operation gates, and the *complexity* of a polynomial is the minimal size of a circuit computing it. We note that, contrarily to a widely admitted convention in Valiant's style calculus, we do not consider that the constants are given for free: under our convention, if a big integer N is needed for the computation of a polynomial f, the number of steps necessary to obtain N starting from the constant -1 should be included in the complexity of f.

A polynomial of complexity c has at most $c+1$ variables; its degree is bounded by 2^c; it is the sum of less than $2^{c(c+1)}$ monomials, whose coefficients are integers with absolute value bounded by 2^{2^c}. For more details on circuits you may consult [Poi95] or [Bür00], and the survey chapter of Meena Mahajan in this volume.

A special kind of circuits are the *terms* (some say *formulae*), of an arborescent nature, in which each gate is allowed to throw only one arrow. The degree of a polynomial computed by a term of size c is bounded by $c+1$, and its coefficients are bounded by $2c$. The *termic complexity* of a polynomial is the minimal size of a term which computes it.

Thank to the parallelization lemma of [MP76], terms correspond to computations in logarithmic depth: an (arithmetic) term of size c can be replaced by an equivalent circuit (or term!) of depth $O(\log c)$, computing the same polynomial.

We also consider functions, which by definition are applications from $\{0, 1\}^n$ into $\mathbb{Z}$; a function is *boolean* if it takes its values in $\{0, 1\}$. It is easily seen that any function is the restriction to $\{0, 1\}^n$ of some polynomial with integer coefficients, and we define the complexity of the function as the minimal complexity of such a polynomial, that is, as the minimal size of an arithmetic circuit computing the function when we restrict its entry variables to the ≪ boolean ≫ values 0 and 1. This gives a fair evaluation of the usual complexity of a boolean function since, on one hand, one can consider the arithmetic circuit modulo 2, and on the other addition and multiplication modulo 2 can be simulated in characteristic zero by $x + y - 2 \cdot x \cdot y$ and $x \cdot y$, respectively. This is also the case for termic complexity, thanks to parallelization.

We use letters like $x, y, z, \ldots$ to denote the variables in our polynomials; letters like $u, v, w, \ldots$ are used for *boolean variables,* that is, variables that are assumed to take only the values 0 and 1.

8.2 Two Functions, Pascaline and Factorial

To avoid a conflict between french and english notations, we denote by $C(m, n)$ the number of subsets with n elements of a set with m elements. The n'th instance of the Pascaline is the following function, depending of $2n + 2$ boolean variables: $\mathrm{Pasc}_n(u_0, \ldots, u_n; v_0, \ldots, v_n) = C(U, V)$, where U and V are the two numbers whose developments in binary figures are respectively $\underline{u}$ and $\underline{v}$; in other words, $U = u_0 + 2 \cdot u_1 + \cdots + 2^n \cdot u_n$ and $V = v_0 + 2 \cdot v_1 + \cdots + 2^n \cdot v_n$.

We consider also the Factorial, which is defined by: $\mathrm{Fact}_n(u_0, \ldots, u_n) = U!$.

These numbers are admittedly big, but their size (they are twice exponential) is not in itself an obstacle to a polynomial complexity. Nevertheless, it seems very unlikely, or at least highly unwishable, that they be computable in a polynomial number of steps, since this would provoke a cryptographic tsunami.

Let us first observe that, if the Pascaline has a polynomial complexity, the same is true of the Factorial. Indeed, from the identity $(2 \cdot U)! = C(2 \cdot U, U) \cdot (U!)^2$ we deduce the following induction formula: $\mathrm{Fact}_n(u_0, u_1, \dots, u_n) = (1 + u_0(u_1 \cdot 2 + \cdots + u_n \cdot 2^n)) \cdot C_n(0, u_1, \dots, u_n; u_1, \dots, u_n, 0) \cdot (\mathrm{Fact}_{n-1}(u_1, \dots, u_n)^2)$.

In the other direction, it is not clear that a fast computation for the Factorial would provide the same for the Pascaline, in the absence of division.

If the Factorial were easy to compute, then factorization would be possible using the following well-known algorithm: consider two positive integers $U < V$, given in figures; compute $U!$ modulo V, and $\gcd(U!, V)$: this decides whether or not V has a factor smaller than U; in a small number of steps we localize the smallest factor of V, and finally discompose it.

We also consider two boolean functions, depending on some more boolean variables, which are the Pascaline and the Factorial in figures, defined by:

$\mathrm{Pascfig}_n(u_0, \dots, u_n; v_0, \dots, v_n; w_0, \dots, w_n) =$ the $W°$ digit of the representation of $\mathrm{Pasc}_n(u_0, \dots, u_n; v_0, \dots, v_n)$ in base 2, where $W = w_0 + 2 \cdot w_1 + \cdots + 2^n \cdot w_n$

$\mathrm{Factfig}_n(u_0, \dots, u_n; w_0, \dots, w_{n+\log(n)}) =$ the $W°$ digit of $\mathrm{Fact}_n(u_0, \dots, u_n)$

These two (sequences of) functions belong to PSPACE. The best argument for that is that every sequence of boolean functions appearing in an algorithmic context belongs to PSPACE, unless it is manufactured to be a counter-example; for something more formal, you may consult this time [Poi08] or [Bür09].

8.3 Polynomials as Sums of Monomials

A Valiant summation is an expression of the form:

$$f(x_1, \dots, x_n) = \sum_{\underline{u} \text{ boolean}} \varphi(u_0, \dots, u_m; x_1, \dots, x_n)$$

where φ is a polynomial in which we have separated the variables into two blocks. This summation of exponentially many terms should have an explosive effect on the complexity of polynomials, but this is an open question, We define the Σ-*complexity* of f as the minimal complexity of φ. It seems unlikely that the Pascaline be of polynomial Σ-complexity, but this fact, if false, should be hard to refute since it is implied by P = PSPACE (the Pascaline is obtained from the Pascaline in figures by a summation; see below the details on the exponential polynomial).

If it were true, it is not clear that it would imply a low Σ-complexity for the Factorial, because the induction formula contains a square; the Fubini formula $\left(\sum_{\underline{u}} \varphi(\underline{u})\right) \cdot \left(\sum_{\underline{v}} \psi(\underline{v})\right) = \sum_{\underline{u},\underline{v}} \varphi(\underline{u}) \cdot \psi(\underline{v})$ is valid only if the tuples of variables $\underline{u}$ and $\underline{v}$ are disjoint; in case of a squaring, $\left(\sum_{\underline{u}} \varphi(\underline{u})\right)^2 = \sum_{\underline{u},\underline{v}} \varphi(\underline{u}) \cdot \varphi(\underline{v})$, so that the computation of φ must be done twice: when we repeat n times, we explode. Similarly, we can define the Π-complexity, replacing summations by productions; it is equally unclear if the Pascaline, as the Factorial, has a polynomial Π-complexity.

A polynomial is the sum of its monomials. This basic truth explains the importance of summations. Let us first define exponential as the following polynomial, where $\underline{y} = (y_0, \ldots, y_m)$:

$$x^y = (y_0 \cdot x + 1 - y_0) \cdot (y_1 \cdot x^2 + 1 - y_1) \cdots (y_i \cdot x^{2^i} + 1 - y_i) \cdots (y_m \cdot x^{2^m} + 1 - y_m)$$

Observe that if we give boolean values to $\underline{y}$, then $x^{\underline{u}} = x^U$, where U is the integer $U = u_0 + 2 \cdot u_1 + \cdots + 2^n \cdot u_n$. In these conditions, a polynomial $f(\underline{x})$ depending of n variables of degree at most 2^m can be written as:

$$f(\underline{x}) = \sum_{\underline{u}} \mathrm{Cf}(\underline{u}_1; \ldots; \underline{u}_n) \cdot x_1^{\underline{u}_1} \ldots x_n^{\underline{u}_n}$$

where $\mathrm{Cf}(\underline{u})$ is a function depending of $m \cdot n$ boolean variables, that we call the coefficient-function of $f(\underline{x})$.

Since polynomials correspond bijectively to their coefficient-functions, it is tempting to establish a correlation between the complexities of these two kinds of objects. It is easy to bound the complexity of $x^{\underline{y}}$ by 6m, so that a polynomial Σ-complexity for the coefficient-function implies a polynomial Σ-complexity for the polynomial. What about the other direction? We shall see that the Pascaline plays a crucial role in this question.

8.4 Operations on the Polynomials and their Effect on the Coefficient-Functions

Given a polynomial $f(y, \underline{x})$, what is the effect of a projection, i.e., of the substitution of some of the variables by constants, on the coefficient-function? When we replace the variable y by the constant a, how do we obtain the coefficient-function of $g(\underline{x}) = f(a, \underline{x})$ from the coefficient-function of $f(y, \underline{x})$? We must reconstruct partially the polynomial: if we note $\underline{v}$ the boolean variables describing the degree of the variable y, and leave the others in the dark, $\mathrm{Cg} = \sum_{\underline{v}} \mathrm{Cf}(\underline{v}) \cdot a^{\underline{v}}$. The substitution has therefore only a polynomial effect on the Σ-complexity of the coefficient-functions.

This is not the case of the termic Σ-complexity, since the exponential involves a^{2^n} which is obtained by a succession of n squarings. Fortunately enough, there are three integers whose powers remain at a reasonable distance:

- $a = 0$; Cg is the constant term, $\mathrm{Cg} = \mathrm{Cf}(\underline{0})$
- $a = 1$; all the powers are equal to 1, $\mathrm{Cg} = \sum_{\underline{v}} \mathrm{Cf}(\underline{v})$
- $a = -1$; we change the sign according to parity, $\mathrm{Cg} = \sum_{\underline{v}} (1 - 2v_0) \cdot \mathrm{Cf}(\underline{v})$

We have no increase of Σ-complexity, even termic, in the first two cases, and a minuscule one in the third. Similarly, when we reconstruct a polynomial with boolean variables, satisfying $u^n = u$ when $n > 0$, there is no need to rise them to powers, so that a summation will have only a benign effect on the Σ-complexity, and on the termic Σ-complexity, of the coefficient-functions.

We note in passing, although we shall not use this kind of operations, that replacements of variables also have a mild effect: for instance, when we substitute z to x and y in $f(x, y)$, to obtain $g(z) = f(z, z)$, $\mathrm{Cg}(\underline{w}) = \sum_{\underline{u},\underline{v}} S(\underline{u}, \underline{v}, \underline{w}) \cdot \mathrm{Cf}(\underline{u}, \underline{v})$, where S is an easily computable boolean function taking the value 1 if $W = U + V$, and 0 if not.

There is a similar formula for the product of two polynomials, so that the coefficient-function of a polynomial computed by a circuit of size c will be computed by a circuit of comparable size with summation gates deeply buried in it; but in general there is no known way to drag smoothly these summation gates at the output of the circuit (this is possible when the circuit is multiplicatively disjoint; see the definition in Sect. 8.6). Circuits with summation gates define $\mathsf{VPSPACE}$, the analog for polynomials of PSPACE for boolean functions: this is the only class above the class VNPmd^0 (that will be defined in Sect. 8.6) that we are certain to be closed by taking coefficient-functions; for the details, see [Poi08]; see also [KP07] for an approach of the class $\mathsf{VPSPACE}$ via coefficient-functions in figures.

Now the Pascaline again. Consider the polynomial $b(x, \underline{y}) = (1 + x)^{\underline{y}} = (y_0 \cdot (1+x)+1-y_0) \cdot \cdots \cdot (y_n \cdot (1+x)^{2^n} + 1 - y_n)$, whose complexity is less than $6(n+1)$. Giving to $\underline{y}$ a boolean value $\underline{w}$, by identification in the binomial formula $(1+x)^{\underline{w}} = \sum_{\underline{v} \leq \underline{w}} \mathrm{Pasc}_n(\underline{w}; \underline{u}) \cdot x^{\underline{u}}$ we obtain $\mathrm{Pasc}_n(\underline{w}, \underline{u}) = \sum_{\underline{v}_0,\ldots,\underline{v}_n} \mathrm{Cb}(\underline{u}, v_0, \ldots, v_n) \cdot w_0^{v_0} \cdot \cdots \cdot w_n^{v_n}$, so that we have an easy polynomial bound of the Σ-complexity of the Pascaline in function of the Σ-complexity of the coefficient-function of $b(x, \underline{y})$. Note in passing that such a simple operation as the substitution of x by $1 + x$ may have a drastic effect on the complexity of the coefficient-function.

In conclusion, if the Σ-complexity of the coefficient-function can be polynomially bounded in function of the Σ-complexity of the polynomial, then the Pascaline has a polynomial Σ-complexity; to establish the reciprocal, we shall construct in the next section a universal polynomial whose coefficient-function has a simple expression in function of the Pascaline.

Remark Since $b(x, \underline{y})$ is a product, its coefficient-function can be expressed as a summation from the coefficient-functions of the factors. This give an expression, with a summation, of the Pascaline in function of $\Gamma_n(\underline{v}) = C(2^n, V)$: the Pascaline has a polynomial Σ-complexity iff this is true also for this function. There is no mystery in this reduction: it is closely related to the formula $C(U_1 + \cdots + U_k, V) = \sum_{V_1+\cdots+V_k=V} C(U_1, V_1) \cdot \cdots \cdot C(U_k, V_k)$, which generalizes Pascal's Triangle. There is no apparent induction on V leading to a computation of the Pascaline with the help of the constants $C(2^i, 2^j)$.

8.5 A Universal Polynomial

We consider the polynomial P_n depending of $n(n^2 - 1)/6$ variables x_{ijk}, where $0 \leq k < j < i \leq n$, which is defined by the following induction:

$$P_0 = P_1 = 1$$
$$P_n(\ldots x_{ijk}, \ldots) = \sum_{\{(i,j)|0 \leq j < i < n\}} x_{nij} \cdot P_i \cdot P_j$$

We say that n is the *head* of the variable x_{nij}, and that i and j are its *tails*.

This polynomial, of complexity less than $n^3/2$, is able to simulate any circuit of size less than $n/4$ just by replacing some of its variables by 0, 1 or -1. To see that, we consider circuits as straight line programs, that is, we order their gates in such a way that each operation gate receive its two arrows from anterior gates. If we want that P_i simulate an input gate labeled by the variable x_{i10}, we equate all the other variables of head i to 0. If we want to simulate the product of P_j and P_k, where $1 < k < j < i$, we equate x_{ijk} to 1 and the other variables of head i to 0. If we want to simulate the addition of P_j and P_k, where $1 < j < i$ and $1 < k < i$, we equate x_{ij1} and x_{ik0} to 1, and the other variables of head i to 0. The only constraint in the simulation is that we cannot multiply directly a gate by itself: we must before duplicate it by a neutral operation, such that a multiplication by 1, tripling (yes) the size of the circuit; note that the entry gates count in the length of the straight line program, but not in the size of the circuit: all this explains the bound $n/4$.

There is no need to make substitutions of variables, since we can assume that, in a circuit, the variables are associated injectively to the input gates; but of course one of the variable has to be replaced by -1.

Let us now evaluate the coefficient-function of P_n. When we develop brutally P_n, just by distributing the product on the sum, we express it has the sum of a twice exponential family of products of variables that we call *arborescent monomials*; the actual expression of P_n as a sum of monomials is obtained by grouping together the arborescent monomials corresponding to a same monomial, those which have the same degree in each of the variables.

The arborescent monomials are obtained as follows: we start from a variable x_{nij} whose head is n, then we multiply it by a variable of head i and a variable of head j (if possible, that is if $i > j \geq 2$), and repeat the process till we reach at the extremity of every branch a terminal variable x_{k01}; when the small tail is 0 or 1, but not the big one, there is only one variable that follows.

In the following example, for $n = 6$, we do not write the x's, but only their indexes; the monomial associated to this arborescent monomial is:

$$654.540.(432)^2.321.310.(210)^3.$$

$$
\begin{array}{ccccccccc}
 & & & & & & 321 & \to & 210 \\
 & & & & & \nearrow & & & \\
 & & 540 & \to & 432 & & & & \\
 & & & & & \searrow & & & \\
 & \nearrow & & & & & 210 & & \\
654 & & & & & & & & \\
 & & & & 310 & & & & \\
 & \searrow & & \nearrow & & & & & \\
 & & 432 & & & & & & \\
 & & & \searrow & & & & & \\
 & & & & 210 & & & &
\end{array}
$$

The condition for a monomial, given by the expansion in figures of the degree of each of the variables, to be associated to an arborescent monomial is easily determined: the largest index n must appear only once, and in head position, and all the others i, $n > i \geq 2$, must appear as many times in head position than in tail position. This condition can be expressed by a boolean function of the degrees of low termic complexity, taking the value 1 when satisfied and 0 if not; let us call it the compatibility condition.

To check that this condition is sufficient, we observe that, before placing the variables of head i, we can place the variables of head bigger than i; this done, we can add to the tree the variables of head i, since the number of tail position corresponds, and then the variables of tail $i - 1$, etc.

For instance, in the example above there is only one second arborescent monomial giving the same monomial (therefore the coefficent of this monomial is 2), the only degree of freedom being the choice for the placement of the two variables with head 3. This second arborescent monomial is:

$$
\begin{array}{ccccccc}
 & & & & & & 310 \\
 & & & & & \nearrow & \\
 & & 540 & \to & 432 & & \\
 & & & & & \searrow & \\
 & \nearrow & & & & & 210 \\
654 & & & & & & \\
 & & & & 321 & \to & 310 \\
 & \searrow & & \nearrow & & & \\
 & & 432 & & & & \\
 & & & \searrow & & & \\
 & & & & 210 & &
\end{array}
$$

To compute the coefficient-function of P_n, we must count how many arborescent monomials are produced by a given monomial. If we note C_i the number of ways to place the variables of head i once the variables of bigger head are placed, number that is independent of the anterior (and posterior) placements, then the coefficient of the monomial will be the product of the C_i: this is only a matter of matching heads and tails.

To compute C_i, let us note k the number of occurencies of i as tail; k is less than 2^n. The variables with head i are splitted in s sorts of respective degrees $k_1, \ldots, k_s$, and $k = k_1 + \cdots + k_s$; s is smaller than $i(i-1)/2$. We must choose the k_1 i's in tail position that will be followed by a variable of the first kind, making $C(k, k_1)$ choices; then we must affect the variables of the second sort, making $C(k - k_1, k_2)$ choices, and so on. Taking into account the fact that $C(0, 0) = 1$, which permits the neutralization of certain binomials coefficients, we see that C_i is expressible as the product of $i(i-1)/2$ pascalines of size n, whose arguments are numbers in figures easily computable from the degrees: they have a polynomial term complexity.

In short, the coefficient-function of P_n is expressible as the product of the compatibility condition and of less than $n^3/6 \geq \sum i(i-1)/2$ pascalines of size n.

Since the coefficient-function of a polynomial of complexity n is obtained by the following operations from the coefficient-function of P_{4n}: equating to 0 the variables representing the degrees of the variables that we replace by 0, summing over the degrees of the variables that we replace by 1 or -1 (after a small multiplication concerning the unique variable that we replace by -1), it is now clear that if the Pascaline has a polynomial Σ-complexity, then the Σ-complexity of the coefficient-function of an arbitrary polynomial is polynomially bounded in function of the Σ-complexity of the polynomial.

8.6 A Short Review of Malod's Thesis

We shall further on follow a common (and vicious) tendency, to which we have resisted till now, to express complexity hypothesises with the help of sequences of polynomials, or functions, with asymptotic properties. We adopt the notations of [Mal03, Mal07].[1]

VPnb^0 denotes the class of sequences of polynomials whose complexity is polynomially bounded; V is for Valiant, P for polynomial time, nb indicates that there is no bound on the degree of the polynomials in the sequence, and ^{0} indicates that -1 is the only constant that is used in the computations. Note that the class is nonuniform: a sequence of boolean functions is in VPnb^0 if and only if it is in $\mathsf{P}/poly$.[2]

VNPnb^0 is the class of sequences of polynomials which are obtained by a summation from a sequence in VPnb^0; the letter N, which is not a fortunate choice, has been introduced in this context by Valiant as an analogy with the class NP. We have shown that VNPnb^0 is closed by taking the coefficient-functions if and only if the Pascaline belongs to it.

---

[1] We have resisted to a crave for changing them!.

[2] It is customary in Valiant's calculus to use nonuniform classes, based on circuit size; were we insist on uniformity for their Boolean counterparts, we could uniformise the Valiant's classes as well; but for what concerns us here, uniformity would play only a decorative role.

The classes VP and VNP in general use by Valiant and his followers differs from them by two aspects: on one hand, there is a polynomial bound on the degree of the polynomials in the sequence; on the other, the entries of the circuits may be labeled by arbitrary constants, living in an unprecised loka (integers are not always sufficient, since the inversion of 2 is essential in the celebrated proof of the universality of the Permanent).

Various devices have been invented preventing a circuit to produce a polynomial of exponential degree, but the truly adequate one has been defined by Malod: a circuit is multiplicatively disjoint if each of its multiplication gates receives its two arrows from disjoint subcircuits (note that a term is both additively and multiplicatively disjoint). A *multiplicatively disjoint* circuit of size c produces a polynomial of degree at most c, and in fact, when we allow arbitrary gratuitous constants, the class VP is equal to the class VPmd of sequences of polynomials with a polynomially bounded md-complexity (the first step of the reduction uses a computation by homogeneous parts, renouncing to any control on the constants; see the details in [MP06, MP08]).

We shall use the class VPmd^0 of sequences computed by multiplicatively disjoint circuits of polynomial size, using only the constant -1, and also its termic analogue VPt^0; it is not known if VPt^0 is a proper subclass of VPmd^0, but terms and md-circuits are equivalent in the presence of summations; in other words $\mathsf{VNPmd}^0 = \mathsf{VNPt}^0$; this last class is closed under taking coefficient-functions.

To study the unbounded case, Malod constructs a universal polynomial whose coefficient-function is computable provided that the Pascaline is so. He then observes that an old theorem of [Luc78] shows that the Pascaline is computable modulo p,[3] where p is a *fixed* prime number, to obtain the following results concerning computations of polynomials with coefficients in $\mathbb{F}_p = \mathbb{Z}/p\mathbb{Z}$:

(i) the class $\mathsf{VNPnb}^0 \pmod p$ is closed for taking coefficient-functions
(ii) any sequence in $\mathsf{VNPnb}^0 \pmod p$ is obtained from a sequence in $\mathsf{VNPmd}^0 \pmod p$ by replacing variables by (simply exponential) powers of variables
(iii) $\mathsf{VNPmd}^0 \pmod p = \mathsf{VPmd}^0 \pmod p \iff \mathsf{VNPnb}^0 \pmod p = \mathsf{VPnb}^0 \pmod p$.

The right to left implication of (iii) rests on the fact that the only constants appearing in a computation are the integers modulo p, which form a finite set.

With the help of our simpler universal polynomial, and a result of Peter Bürgisser, we shall say more on what happens in characteristic zero.

8.7 Characteristic Zero

Bürgisser [Bür09] shows that the hypothesis $\mathsf{VNPmd}^0 = \mathsf{VPmd}^0$ collapses a certain hierarchy within $\mathsf{PSPACE}/poly$, to which belongs the Pascaline in figures; in short, $\mathsf{VNPmd}^0 = \mathsf{VPmd}^0 \implies$ Pascfig $\in \mathsf{P/poly}$.

[3] If we develop the integers m and n in base p , $C(m,n)$ is equal modulo p to the product of the binomial coefficients $C(u_i, v_i)$ of their digits.

Let us remind that, by the Parsimonious Reduction Lemma, to any boolean function $f(\underline{u})$, computed by an arithmetic circuit modulo 2 of complexity c, is associated a boolean function $g(\underline{u}, \underline{v})$, computed by an arithmetic term modulo 2 of size $4c$, such that, if $f(\underline{u}) = 0$, then $g(\underline{u}, \underline{v}) = 0$ for every $\underline{v}$, and if $f(\underline{u}) = 1$, then $g(\underline{u}, \underline{v}) = 0$ for every $\underline{v}$, except for one for which it takes the value 1 (the length of $\underline{v}$ is equal to c, and the exceptional $\underline{v}$ corresponds to the values that are computed at the gates of the circuit). After parallelization, the computation of g can be simulated in characteristic zero by a term of small complexity, and, because of the uniqueness of the exceptional $\underline{v}$, $f(\underline{u})$ is obtained by a summation in front of this last term (we do not claim that there is a way to simulate in characteristic zero summations in characteristic 2; see [Poi]). As a consequence, any sequence of boolean functions which is in VPnb^0, i.e., in $\mathsf{P/poly}$, is in VNPmd^0.

Assume that $\mathsf{VNPmd}^0 = \mathsf{VPmd}^0$. If, in the expression of the Pascaline as a summation from the Pascaline in figure we replace the numbers 2^{2^i} by variables y_i, we obtain therefore a (sequence of) multilinear polynomial(s) in VNPmd^0; using a second time our hypothesis, we see that this polynomial is in VPmd^0, and, after substitution of the y_i by the adequate powers of 2, that the Pascaline is in VPnb^0 (we remind that this implies an easy factorization). This is more than enough for VNPnb^0 to be closed for the coefficient-function.

Moreover, in the expression of a polynomial as the summation of its monomials, we can also replace the power $x_i^{2^j}$ of a variable x_i by a new variable x_{ij}. From our description of the coefficient-functions of the projections of the universal polynomial, we conclude that any sequence of polynomials in VNPnb^0 is obtained by plugging in a sequence in VNPmd^0 simply exponential powers of 2 and of variables. We reach the same conclusion as Malod in characteristic p, this time not as a fact, but as a consequence of an hazardous hypothesis.

The conclusion is that $\mathsf{VNPmd}^0 = \mathsf{VPmd}^0$ implies $\mathsf{VNPnb}^0 = \mathsf{VPnb}^0$.

The reciprocal is problematic: if $\mathsf{VNPnb}^0 = \mathsf{VPnb}^0$, VNPmd^0 is included in VPnb^0; since a polynomial in VNPmd^0 has a small degree, a computation by homogeneous parts truncates its VPnb-computation into a small multiplicatively disjoint circuit, which unfortunately may use at its entry gates some twice exponential integers: we know no way to get rid of these big numbers.

Nevertheless we should keep in a corner of our mind that all these hypothesizes are probably equivalent, and false!

References

[Bür00] P. Bürgisser, *Completeness and Reduction in Algebraic Complexity Theory. Number 7 in Algorithms and Computation in Mathematics.* (Springer, Berlin, 2000)

[Bür09] P. Bürgisser, On defining integers and proving arithmetic circuit lower bounds. Comput. Complex. **18**(1), 81–103 (2009)

[KP07] P. Koiran, S. Perifel, VPSPACE and a transfer theorem over the complex field, Proc. MFCS, LNCS **4708**, 359–370 (2007)

[Luc78] E. Lucas, Théorie des fonctions numériques simplement périodiques. Am. J. Math. **1**(2), 184–196 (1878)
[Mal03] G. Malod, *Polynômes et coefficients*. Ph.D. thesis, Universitè Claude Bernard, 2003.
[Mal07] G. Malod, The complexity of polynomials and their coefficient functions, in *Computational Complexity, 2007. CCC '07. Twenty-Second Annual IEEE Conference* pp. 193–204, 2007
[MP76] D.E. Muller, F.P. Preparata, Restructuring of arithmetic expressions for parallel evaluation. J. Assoc. Comput. Mach. **23**(3), 534–543 (1976)
[MP06] G. Malod, N. Portier, Characterizing valiant's algebraic complexity classes. Lect. Notes Comput. Sci. **4162**, 267–279 (2006)
[MP08] G. Malod, N. Portier, Characterizing valiant's algebraic complexity classes. J. Complex. **24**(1), 16–38 (2008)
[Poi] B. Poizat. Changing the domain in numerical computations. Submitted
[Poi95] B. Poizat, Les petits cailloux, une introduction modèle-théorique à l'algorithmie. Nur al-mantiq wal-Ma'rifah n° 3, (1995)
[Poi08] B. Poizat, A la recherche de la définition de la complexité d'espace pour le calcul des polynômes à la manière de valiant. **73**, 1179–1201 (2008)

Chapter 9
A Tutorial on Time and Space Bounds in Tree-Like Resolution

Jacobo Torán

Abstract Tree-like resolution is a well-known method for proving the unsatisfiability of a given formula. Lower bounds for the size and space in tree-like resolution imply lower bounds for many of the algorithms used in practice to solve satisfiability problems. We review a combinatorial game that can be used to prove lower and upper bounds for size and space in tree-like resolution and show some of its applications.

Keywords Resolution · Combinatorial games

9.1 Introduction

Let us first introduce some notation needed to understand the tutorial. The Boolean formulas considered here are in conjunctive normal form (CNF), that is, they are a conjunction of clauses. A clause is a disjunction of literals (variables or negated variables). We will also talk about a set of clauses to refer to a formula. An assignment α for a formula F is (partial) mapping from the set of variables in the formula to $\{0, 1\}$. We denote by $F\alpha$ the result of substituting in F the variables assigned by α by the corresponding value and reducing the formula in the intuitive way. If $F\alpha$ is 1 then we say that α is a satisfying assignment for F. If some satisfying assignment for F exists, then we say that F is satisfiable, otherwise it is unsatisfiable.

Robinson introduced in [Rob65] the concept of *resolution*, a method for deciding if a given formula in conjunctive normal form is unsatisfiable. Due to its simplicity and to its importance in automatic theorem proving and logic programming systems, resolution is one of the best studied refutation systems. The only inference rule in this proof system is the resolution rule:

J. Torán (✉)
Department of Theoretical Computer Science, University of Ulm, Ulm, Germany
e-mail: jacobo.toran@uni-ulm.de

M. Agrawal and V. Arvind (eds.), *Perspectives in Computational Complexity*, Progress in Computer Science and Applied Logic 26, DOI: 10.1007/978-3-319-05446-9_9,

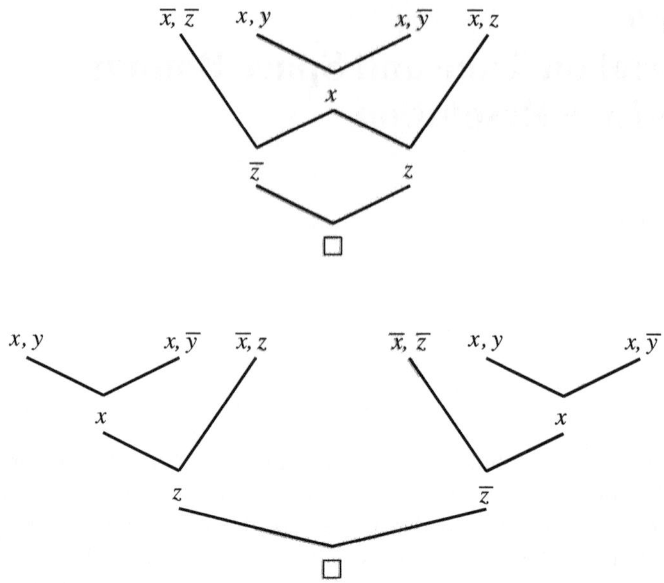

Fig. 9.1 A refutation and a tree-like refutation of the set of clauses $\{x \vee y, x \vee \overline{y}, \overline{x} \vee z, \overline{x} \vee \overline{z}\}$

$$\frac{C \vee x \qquad D \vee \bar{x}}{C \vee D}.$$

Cutting variable x from clauses $C \vee x$ and $D \vee \bar{x}$ we get the *resolvent* clause $C \vee D$. It is easy to see that if a set of clauses S is satisfiable and a clause C is the resolvent of two clauses in S, then $S \cup C$ is also satisfiable. A resolution refutation of a CNF formula F is a sequence of clauses $C_1, \ldots, C_s$ where each C_i is either a clause from F or is inferred from earlier clauses by the resolution rule, and C_s is the empty clause (denoted by $\square$). By the above observation, if $\square$ can be derived from a set S of clauses by resolution, then S is unsatisfiable. In other words, the resolution system is *correct*. We will soon observe that the system is also *complete*, that is, if a set of clauses in unsatisfiable then there is always a way to derive the empty clause from it by resolution.

A resolution refutation can be seen as a directed acyclic graph, a *dag*, in which the clauses are the vertices, and if two clauses are resolved then there is a directed edge going from each of the two clauses to the resolvent. If the underlying graph in a refutation happens to be a tree, we call it a *tree-like* resolution. See Fig. 9.1. Tree-like resolution is also a complete refutation system.

Theorem 1.1 *Let F be a set of clauses. If F is unsatisfiable, then there is a tree-like resolution refutation for it.*

Proof We argue by induction on the number n of variables in F. If $n = 1$, then there is only one variable x_1 and since F is unsatisfiable it must contain the clauses x_1

and $\overline{x_1}$. Resolving them we obtain $\Box$. For the induction step, by giving variable x_n the values 0 and 1 we obtain two new sets of clauses $F\{x_n = 0\}$ and $F\{x_n = 1\}$ containing at most $n - 1$ variables. Since F is unsatisfiable, so are the new sets of clauses and applying the induction hypothesis we know that there exist tree-like resolution refutations R_0 and R_1 for $F\{x_n = 0\}$ and $F\{x_n = 1\}$. The clauses of $F\{x_n = 0\}$ are either clauses of F (not containing variable x_n) or are clauses C for which $C \vee x_n$ is in F. By adding x_n to the initial clauses of this type in R_0 we either get a tree-like derivation of x_n from clauses in F (variable x_n is never resolved) or a derivation of $\Box$ (in case none of the clauses in R_0 contain variable x_n). In the later case we are done, otherwise we can do the same process on R_1 obtaining either a tree-like derivation of $\overline{x_n}$ from F or a refutation of F. Again, in the second case we are done, otherwise we put together the tree-like resolution derivations of x_n and $\overline{x_n}$ and the resolvent of this two clauses to obtain a tree-like refutation of F. $\Box$

From the proof of the above Theorem it follows that the number of clauses needed in a tree-like refutation of a formula F is at most exponential in the number of its variables.

It is known that for certain formulas general resolution can produce shorter refutations than tree-like resolution [BEGJ02, BIW04]. The reason for this is that, contrary to general resolution, in tree-like resolution if a clause is needed more than once it must be re-derived from the initial clauses each time. Even if tree-like refutations can be larger than in general resolution, the study of this proof system is well motivated by many methods used in practice for testing satisfiability (SAT solvers). These methods are in fact concrete implementations of tree-like resolution. Most SAT solvers are based on the basic backtracking algorithm:

proc Backtracking (F : set of clauses, α : assignment) : **bool**
// outputs 1, if Fα is satisfiable and 0 otherwise
if $\Box \in F\alpha$ **then return** 0
if $F\alpha = \emptyset$ **then return** 1
else (choose a variable $x \in \text{Var}(F\alpha)$)
  if Backtracking ($F, \alpha\{x = 0\}$) **then return** 1
    else return Backtracking ($F, \alpha\{x = 1\}$)

In order to test if a formula is satisfiable, the algorithms starts with the empty assignment. This backtracking algorithm is in principle more efficient than a straightforward test of all possible total assignments for the variables because when for a partial assignment α, the algorithms obtains $\Box \in F\alpha$, then it follows $F\beta = 0$ for all extensions β from α. The Algorithm does not need to explore the sub-tree with the possible extensions from α and this decreases the search space.

In the algorithm we have not specified the way to choose the next variable to be considered. There are many ways to do this. Modern SAT solvers use refined methods for choosing the next variable to be assigned as well as many other techniques for efficiently pruning the search tree (see e.g., [ST13]).

Exercise 1.2 *Show that in the backtracking algorithm there is a strategy for choosing the next variable to be assigned so that on input a formula F in CNF with n variables and at most k variables in each clause, the algorithm produces at most* $(2^k - 1)^{n/k}$ *recursive calls.*

A proof of the unsatisfiability of a formula with the backtracking algorithm is in fact a tree-like resolution refutation for the formula as can be seen in the next well-known result:

Theorem 1.3 *Let F be a nonsatisfiable formula in CNF and let r be the minimum number of recursive calls made by the backtracking algorithm on input* $(F, \emptyset)$ *(by any ordering of the variables). Then there is a tree-like resolution refutation of F of size at most r.*

Proof The backtracking algorithm running on the input $(F, \emptyset)$ defines a tree. The nodes of the tree indicate what variable is being assigned. From an interior node two edges come out, one for each of the two possible values (0 and 1) that the variable can take. We can associate a resolution refutation with this tree. Every node in the tree is reached from the root by a partial assignment and can be identified with an initial clause falsified by this assignment in the following way: Every leaf in the recursion tree is identified with a clause of F falsified by the assignment reaching the leaf. Such a clause must exist because otherwise this particular assignment would satisfy the formula. For an inner node corresponding to a variable x and whose children are identified with clauses K_i and K_j, we associate a clause K. If both clauses K_i and K_j contain variable x, then in one of the clauses the variable is positive and in the other clause it is negated. We define then the clause K as the resolvent of K_i and K_j. Otherwise we associate with the node the clause not containing variable x from one of its children. One can see that in both cases K is falsified by the assignment that goes from the root of the tree to the node. Moreover, every clause in the tree is either the resolvent of its children or one of the clauses from the children. The root in the recursion tree is associated to the empty clause □ since this is the only clause falsified by the empty assignment. The tree with its associated clauses defines a tree-resolution refutation for F with at most r clauses. □

In this result, we have already used the natural complexity measure of *size*. The size of a refutation is the number of clauses it contains. It is known that certain families of propositional formulas need (general) resolution refutations with a number of clauses that is exponential in the formula size [Hak85, Urq87, CS88, BP96].

Another natural complexity measure is the *space*. Intuitively, the resolution space of a CNF formula is the minimal number of clauses that must be kept simultaneously in order to refute a formula. The formal definition [ET01, ABRW02] is the following:

Definition 1.4 Let $k \in \mathbb{N}$, we say that an unsatisfiable CNF formula F has resolution refutation bounded by space k if there is a series of CNF formulas $F_1, \ldots, F_s$, such that $F_1 \subseteq F$, $\Box \in F_s$, in any F_i there are at most k clauses, and for each $i < s$, F_{i+1} is obtained from F_i by:

(1) Deleting a clause from F_i.
(2) Adding the resolvent of two clauses from F_i.
(3) Adding a clause from F (initial clause).

The space needed for the resolution refutation of an unsatisfiable formula is the minimum k for which the formula has a refutation bounded by space k. Note that initial clauses do not need much space because they can be added at any moment and at most two of them are needed simultaneously. The only clauses that consume space are the ones derived at intermediate stages. From the proof of Theorem 1.1, it can be observed that the space needed in a refutation of a formula with n variables is at most $n+1$. In [Tor99, ET01, ABRW02] it is shown that the refutations for certain families of formulas need linear space in the number of variables. It was observed in [ET01] that the space required for the refutation of a CNF formula F, corresponds to the minimum number of pebbles needed in the following game played on the graph of a refutation of F.

Definition 1.5 Given a connected directed acyclic graph G with one sink the aim of the pebble game is to put a pebble on the sink of the graph, the only node with no outgoing edges, following this set of rules:

(1) A pebble can be placed in any initial node, that is, a node with no predecessors.
(2) Any pebble can be removed from any node at any time.
(3) A node can be pebbled provided all its predecessors are pebbled.
(4) If all the predecessors of node are pebbled, instead of placing a new pebble on it, one can shift a pebble from a predecessor.

We denote by Peb(G) the minimum number of pebbles needed in order to put a pebble on the sink of G following the above rules.

Exercise 1.6 *Show that for a complete binary tree of depth d, T_d (with the directed edges pointing to the root), Peb(T_d) $= d+1$. Show that this is also true for any tree T with the property that d is the depth of the biggest complete binary tree embedded*[1] *in T.*

Lemma 1.7 [ET01] *Let F be an unsatisfiable CNF formula. The space needed in a resolution refutation of F coincides with the number of pebbles needed for the pebble game played on the graph of a resolution refutation of F.*

In this tutorial we will concentrate on tree-like refutations. The size of such a refutation is the number of nodes in the resolution tree, and the space is the number of pebbles needed to play the game in such a refutation tree. There is also a way to define the space measure for tree-like resolution without using the pebbling game [ET01].

There is a relation between the space and size in tree-like resolution. As the following result shows, a lower bound on the tree-like space of a formula implies an exponentially larger lower bound on the size of a tree-like resolution refutation.

---

[1] We say that a graph G is embedded in another graph H if H can be obtained from G by adding new vertices and edges and placing new vertices in the middle of an edge.

Theorem 1.8 *Let F be an unsatisfiable CNF formula with a tree-like refutation of size s, then F has a tree-like resolution refutation of space $\lceil \log s \rceil + 1$.*

Proof As mentioned in Exercise 1.6, the resolution tree in the refutation of F can be pebbled with $d + 1$ pebbles, where d is the depth of the biggest complete binary tree embedded in the refutation tree. As the biggest possible complete binary tree embedded in a tree of size s has depth $\lceil \log s \rceil$, the result follows. □

Exercise 1.9 *Show that if in an unsatisfiable formula F all its clauses have at most 2 literals, then there is a tree-like resolution refutation for F with constant space and polynomial size.*

9.2 A Combinatorial Game for Proving Lower Bounds in Tree-Like Resolution

Impagliazzo and Pudlák introduced in [PI00] a combinatorial game for proving lower bounds on the size of tree-like refutations. This game was also used in [BIW04]. We will show that this game exactly characterizes tree-like resolution space.

The Prover-Delayer game:
The game is played in rounds on an unsatisfiable set of clauses F by two players: Prover and Delayer.[2] Prover wants to falsify some initial clause and Delayer tries to retard this as much as possible. In each round Prover chooses a variable in F and asks Delayer for a value for this variable. Delayer can answer either 0,1 or $*$. In this last case Prover can choose the truth value (0 or 1) for the variable and Delayer scores one point. The variable is set to the selected value and the next round begins. The game ends when a clause in F is falsified (all its literals are set to 0) by the partial assignment constructed in this way. The goal of Delayer is to score as many points as possible and Prover tries to prevent this. The outcome of the game is the number of points scored by Delayer.

Definition 2.1 Let F be an unsatisfiable formula in CNF. We denote by $g(F)$ the maximum number of points that Delayer can score while playing the game on F with an optimal strategy of Prover.

As an example we show how this game applies to the family of formulas for the general Pigeonhole Principle PHP_n^m. These formulas express the fact that it is not possible to fit m pigeons in n pigeonholes (for $m > n$). For the case $m = n + 1$, $\neg\mathrm{PHP}_n^m$ was the first example of a family of formulas with an exponential resolution size lower bound [Hak85]. The contradiction $\neg\mathrm{PHP}_n^m$ can be written as a CNF formula in the following way: The variables of the formula are $x_{i,j}$, $i \in [m]$, $j \in [n]$. $x_{i,j}$ has the intuitive meaning that pigeon i is mapped to hole j. There are mn variables. The clauses of the formula are:

[2] For clarity in the exposition Delayer is a female player.

(1) $x_{i,1} \vee x_{i,2} \vee \cdots \vee x_{i,n}$ for $i \in [m]$, and
(2) $\overline{x}_{i,k} \vee \overline{x}_{j,k}$ for $i, j \in [m], k \in [n],\ i < j$.

Clauses of type (1) express the fact that every pigeon is mapped to some hole, while the clauses of type (2) indicate that at most one pigeon can be mapped to any hole.

The number of clauses in $\neg\mathrm{PHP}^m_n$ is $m + \binom{m}{2}n < m^2n$.

Lemma 2.2 $g(\neg PHP^m_n) \geq n$

Proof We give a strategy for Delayer scoring at least n points. When asked for the value of a variable $x_{i,j}$ Delayer answers 0 if for some $i' \neq i$, $x_{i',j}$ has been assigned value 1, otherwise she assigns value $*$ to $x_{i,j}$. With this strategy the game can only end when a clause of Type 1 has been falsified. Observe that since Delayer does not assign any 1's, for every 0 she assigns to a variable $x_{i,j}$, she must have assigned before a $*$ to another variable $x_{i',j}$ that is set to 1 by Prover. When a clause $\bigvee_{j=1}^n x_{i,j}$ is falsified because all its variables have value 0, Delayer has assigned a $*$ to at least one variable $x_{i',j}$ corresponding to position j for each position, and therefore she has scored at least n points. □

Exercise 2.3 *Show by designing a suitable strategy for Prover that it also holds* $g(\neg PHP^m_n) \leq n$.

We show now that for an unsatisfiable CNF formula F, the space needed in a tree-like resolution refutation of F is exactly $g(F) + 1$. In our example this implies that the tree-like resolution space for $\neg\mathrm{PHP}^m_n$ is exactly $n + 1$ independently of m.

We show first that $g(F) + 1$ is an upper bound for the tree-like resolution space.

Theorem 2.4 [ET03] *If a CNF formula F requires tree-like resolution space s, then Delayer has a strategy in which at least $s - 1$ points can be scored.*

Proof Let s be the minimum space needed in any tree-like resolution refutation of F. We give a strategy for Delayer for playing the combinatorial game on F that scores at least $s - 1$ points with any strategy of Prover. We prove the result by induction on n, the number of variables in F.

For the base case $n = 1$, F contains just one variable and therefore $s \leq 2$. Delayer just needs to answer $*$ to the only variable asked by Prover.

For $n > 1$, let x be the first variable asked by Prover and let $F\{x = 1\}$ and $F\{x = 0\}$ the CNF formulas obtained after given value 1 and 0 respectively to variable x in F. Any tree-like refutation of F requires s pebbles and therefore either

(i) the tree-like space for refuting each of $F\{x = 1\}$ and $F\{x = 0\}$ is at least $s - 1$ or
(ii) for one of the formulas (say $F\{x = 1\}$) the tree-like resolution space is at least s.

Any other possibility would imply that F could be refuted in space less than s. In the first case, Delayer can answer $*$ and she scores one point. By induction hypothesis Delayer can score $s-2$ more points playing the game in any of the formulas $F\{x = 1\}$ or $F\{x = 0\}$. In the second case, Delayer answers the value leading to the formula that requires tree-like resolution space s ($x = 1$ in this case) and the game is played on $F\{x = 1\}$ in the next round. □

On the other hand $g(F)$ is also a lower bound for the tree-like resolution space.

Theorem 2.5 [ET03] *The tree-like space needed for refuting a CNF F is at least $g(F)+1$.*

Proof Let s be the minimum space needed in a tree-like refutation of F. We describe a Strategy for Prover in which the number p of points scored by Delayer is at most $s-1$. Let R be a tree-like resolution refutation for F with $\text{Peb}(R) = s$. Prover chooses the variables in the order induced by the refutation in the following way: He starts at the empty clause in R and in general at the end of a round moves to a clause C that is falsified by the partial assignment constructed so far. In the next round, Prover chooses the resolved variable x from the two parent clauses of C. Let α_i be the partial assignment constructed after i rounds of the game and $R\alpha_i$ be the subtree of the refutation that has its root at the node reached from the root of R by the path specified by α_i and let p_i be the number of points scored by Delayer after round i. If Delayer assigns to x a value 0 or 1 (by setting the weights in the way indicated above) then Prover moves to the parent clause that is falsified by the constructed partial assignment and the new round starts. When Delayer assigns value $*$ to a variable x at step i then Prover gives x value 0 if $\text{Peb}(R\alpha_{i-1}\{x=0\}) \leq \text{Peb}(R\alpha_{i-1}\{x=1\})$ and assigns x value 1 otherwise. We show by induction on the number i of rounds that with this strategy

$$p_i \leq s - \text{Peb}(R\alpha_i).$$

The result follows from this fact because when the game has reached a contradiction of an initial clause after constructing an assignment α, then $\text{Peb}(R\alpha) = 1$ and the inequality shows $p \leq s-1$.

In the beginning of the game $R\alpha_0$ is the whole tree and Delayer has scored 0 points.

For the inductive step, if at round $i+1$ Prover chooses variable x and Delayer assigns values 0 or 1 to it then she does not score any points, and we get

$$p_{i+1} = p_i \leq s - \text{Peb}(R\alpha_i) \leq s - \text{Peb}(R\alpha_{i+1})$$

since $R\alpha_{i+1}$ is a subtree of $R\alpha_i$.

If Delayer assigns $*$ to x, since Prover select the value corresponding to the subtree from $R\alpha_i$ with the smallest pebbling number and $\min\{\text{Peb}(R\alpha_i\{x=0\}), \text{Peb}(R\alpha_i\{x=1\})\} < \text{Peb}(R\alpha_i)$ we get

$$p_{i+1} = p_i + 1 \leq s - \text{Peb}(R\alpha_i) - 1 \leq s - \text{Peb}(R\alpha_{i+1})$$

and the result follows. □

As mentioned before, the combinatorial game was defined in [PI00] as a tool for proving lower bound for the size of tree-like resolution refutation. This application can be seen considering the relationship between tree-like space and size from Theorem 1.8:

Corollary 2.6 *For any unsatisfiable CNF formula F, if Delayer has a strategy on F which scores r points then any tree-like resolution refutation of F has size at least $2^{r+1} - 1$.*

Buss and Pitassi [BP97] showed that for for any $m > n$, $\neg\mathrm{PHP}_n^m$ needs tree-like resolution refutation of size at least 2^n. The given estimation for the points obtained by Delayer on the formula $\neg\mathrm{PHP}_n^m$ with the space characterization and the above corollary provide an alternative proof for this result. The bound has been improved to $2^{\Omega(n \log n)}$ in [IM99, DR01].

Computing the number of points that can be scored by Delayer on a formula is in general not an easy task. Hertel and Urquhart [HU07] have shown that in fact this problem is PSPACE complete. We finish this section with some exercises to compute the points scored on certain formulas related to pebbling.

Exercise 2.7 *Consider the following pebbling formulas defined by Ben-Sasson and Wigderson in* [BW01]. *They express the principle that in a directed acyclic graph, pebbling the source nodes and following the rule that if all the predecessors of a node v contain a pebble then v also gets one, implies that a pebble will be placed on the sink.*

Let $G = (V, E)$ be a directed acyclic graph in which every vertex has fan-in 2 or 0 with a unique sink s. We call a graph with these properties a circuit graph. We associate 2 distinct Boolean variables v_1, v_2 with every vertex $v \in V$. Peb_G, the pebbling contradiction of G, is the conjunction of:

- **Source axioms:** $v^1 \vee v^2$ *for each source v.*
- **Pebbling axioms:** $\overline{u}^i \vee \overline{v}^j \vee w^1 \vee w^2$ *for u and v the two predecessors of w and $i, j \in \{1, 2\}$.*
- **Sink axioms:** $\overline{s}^i$ *for the sink s and $i \in \{1, 2\}$.*

Show that for a constant c, if G is a tree then $g(Peb_G) = Peb(G) + c$

Exercise 2.8 [Nor12] *Consider now a variation of the pebbling formulas, defined by Ben-Sasson and Nordström in* [BN08] *expressing the same principle but using the parity function instead of disjunction. For simplicity we just define these formulas for line graphs $G = (V, E)$ with n vertices $v_1 \dots v_n$ with unique source v_1, unique sink v_n and with a directed edge (v_1, v_{i+1}) for $1 \le i \le n-1$. We associate 2 distinct Boolean variables v_i^1, v_i^2 with every vertex $v_i \in V$. $Peb_G^{\oplus}$ is the conjunction of (the clauses expressing in conjunctive normal form the formulas):*

- **Source axioms:** $v_1^1 \oplus v_1^2$.
- **Pebbling axioms:** $v_i^1 \oplus v_i^2 \to v_{i+1}^1 \oplus v_{i+1}^2$ *for* $1 \le i \le n-1$.
- **Sink axioms:** $\neg(v_n^1 \oplus v_n^2)$.

Show that for a line graph G with n vertices, $g(Peb_G^{\oplus}) = \Theta(\log n)$.

9.3 The Asymmetric Prover-Delayer Game

We have seen in the previous section that the original Prover-Delayer game gives an exact characterization of tree-like resolution space and provides a good tool for proving size lower bounds in tree-like resolution size. However, this lower bound might not be tight in case the refutation trees are not well balanced. Beyersdorff, Galesi, and Lauria [BGL10, BGL11, BGL13] have developed a refinement of the game that improves in some cases the size lower bounds obtained by using the original game.

The asymmetric game is defined exactly as the original one, only the way of scoring points by Delayer is different. In each round, Prover selects an unassigned variable x as before, and Delayer assigns two weights p_0 and p_1 to the two possible variable values, satisfying:

$$p^0 \geq 0, \quad p^1 \geq 0, \quad \text{and } p^0 + p^1 = 1.$$

Prover selects one of the values in $\{0, 1\}$ for x and the number of points that Delayer gets is $-\log p^b$.

Definition 3.1 Let F be an unsatisfiable formula in CNF. We denote by $G(F)$ the maximum number of points that Delayer can score while playing the game on F with an optimal strategy of Prover.

By setting the weights to either $(0, 1)$ or $(1, 0)$, Delayer forces Prover to choose one of the two values, since by choosing the other one he would loose an infinite amount of points. Also, by setting the weights to be $(\frac{1}{2}, \frac{1}{2})$ Delayer scores one point no matter what the choice of Prover is. These two cases show that this game is an extension of the original one.

The new game can be used to obtain a characterization of tree-like resolution size. The next theorem shows that the size of a resolution tree implies an upper bound on the number of points scored by Delayer.

Theorem 3.2 [BGL11] *Let F be an unsatisfiable formula in CNF. If F has a tree-like refutation of size at most S then $G(F) \leq \log\lceil\frac{S}{2}\rceil$.*

Proof The proof is very similar to that of Theorem 2.5 but considering the new scoring rules. Let R be a tree-like resolution refutation of size S for F and let $L(R)$ be the number of leaves in R. Observe that $L(R) = \lceil\frac{S}{2}\rceil$. We describe a Strategy for Prover in which the number p of points scored by Delayer is at most $\log(L(R))$.

Prover chooses the variables in the order induced by the refutation starting at the empty clause in R and at the end of a round moves to the parent clause C that is falsified by the partial assignment constructed so far. Let α_i be the partial assignment constructed after i rounds of the game and $R\alpha_i$ be the subtree of the refutation that has its root at the node reached from the root of R by the path specified by α_i and let

p_i be the number of points scored by Delayer after round i. When Delayer assigns values (p^0, p^1) to a variable x at step i then Prover gives x value 0 if $L(R\alpha_{i-1}\{x = 0\}) \le p^0 L(R\alpha_{i-1})$ and assigns x value 1 otherwise. By the property of the weights p^0 and p^1 and by the fact that $L(R\alpha_{i-1}) = L(R\alpha_{i-1}\{x = 0\}) + L(R\alpha_{i-1}\{x = 1\})$ it follows that in this last case $L(R\alpha_{i-1}\{x = 1\}) \le p^1 L(R\alpha_{i-1})$.

We show by induction on the number i of rounds that with this strategy

$$2^{p_i} \le \frac{L(R)}{L(R\alpha_i)}.$$

Again the result follows from this because when the game has reached a contradiction of an initial clause after constructing an assignment α then $L(R\alpha) = 1$ and the inequality shows $2^p \le L(R)$.

In the beginning of the game $R\alpha_0$ is the whole tree and Delayer has scored 0 points. For the inductive step, if at round $i+1$ Prover chooses variable x and Delayer assigns weights p^0, p^1 to it, and Prover selects value $b \in \{0, 1\}$ for the variable, we get

$$2^{p_{i+1}} = 2^{p_i - \log p^b} = \frac{2^{p_i}}{p^b} \le \frac{L(R)}{p^b L(R\alpha_i)} \le \frac{L(R)}{L(R\alpha_{i+1})}$$

and the result follows. □

The extended game completely characterizes the size of a tree-like resolution proof. This is a consequence of Theorem 3.2 and the converse result:

Theorem 3.3 [BGL13] *Let F be an unsatisfiable formula in CNF. If the smallest tree-like refutation for F has size S then $G(F) \ge \log\lceil\frac{S}{2}\rceil$.*

Proof Let $L(F)$ be the number of leaves in the shortest tree-like refutation of F. When a new variable x is selected by Prover and α is the partial assignment computed so far, Delayer assigns weights according to the following rules: for $b \in \{0, 1\}$,

$$p^b = \frac{L(F\alpha\{x = b\})}{L(F\alpha\{x = 0\}) + L(F\alpha\{x = 1\})}.$$

We show by induction on n, the number of variables in F, that Delayer scores at least $\log L(F)$ points. This implies the result, since a tree-like refutation of size S has exactly $\lceil\frac{S}{2}\rceil$. leaves. The base case is trivial; if there is only one variable, the resolution tree has two leaves, and Delayer can always score 1 point. For $n > 1$, let x be the first variable chosen by Prover and let $b \in \{0, 1\}$ be the value assigned by him to it. The score of the game is $-\log p^b + X$ where X is the score achieved in the subsequent steps. By induction hypothesis we have $X \ge \log L(F\{x = b\})$. The total score is then at least

$$
\begin{aligned}
G(F) &\geq -\log p^b + \log L(F\{x=b\}) \\
&= \log\left(\frac{L(F\alpha\{x=0\}) + L(F\alpha\{x=1\})}{L(F\alpha\{x=b\})}\right) + \log L(F\{x=b\}) \\
&= \log(L(F\alpha\{x=0\}) + L(F\alpha\{x=1\})) \geq \log L(F).
\end{aligned}
$$

□

We can see that the asymmetric game can improve the size lower bounds for tree-like resolution for the pigeon hole principle achieved by using the original game [BGL10]. For this we analyze the asymmetric game on PHP formulas and then use Theorem 3.2.

Theorem 3.4 *For* $m > n$, $G(\neg PHP_n^m, c_0, c_1) = \Omega(n \log n)$

Proof We give a simplified proof from [Bey11] describing a strategy for Delayer for which the number of scored points is $\Omega(n \log n)$.

For $i \in [m]$ we define the function h_i that for a partial assignment α

$$h_i(\alpha) = |\{k \in [n] \mid \alpha(x_{i,k}) = 0 \text{ and } \alpha(x_{j,k}) \neq 1 \text{ for all } j \in [m]\}|.$$

$h_i(\alpha)$ indicates the number of holes that are still free but are excluded for pigeon i under α.

The strategy of Delayer for the variable $x_{i,j}$ when the partial assignment constructed so far is α, is the following:

If there is some other variable $x_{i',j}$ assigned to 1 (hole j is already used) or there is some $x_{i,j'}$ assigned to 1 (pigeon i already has a hole) then set $x_{i,j}$ to 0 (by setting the weights to $(1, 0)$). Otherwise if $h_i(\alpha) \geq \frac{n}{2}$ then set $x_{i,j}$ to 1. Otherwise set the weights of $x_{i,j}$ to (p^0, p^1). The values of p^0 and p^1 are the same for all the variables satisfying this property and will be specified later.

Intuitively, when the number of free holes excluded for pigeon i are at least $\frac{n}{2}$ then Delayer tries to put i in a free hole. As in Lemma 2.2 with this strategy only a clause of type 1 $\bigvee_{j=1}^n x_{i,j}$ can be falsified. We show that when this happens at the end of the game, at least $\frac{n}{2}$ variables have been assigned to 1. Consider the last round r after which $h_i(\alpha_r) < \frac{n}{2}$. If at the end of the game $h_i(\alpha_r) < \frac{n}{2}$, then there are less than $\frac{n}{2}$ holes that are free for pigeon i and therefore more than $\frac{n}{2}$ are occupied. The corresponding variables are set to 1. Otherwise let z be the number of variables in $\bigvee_{j=1}^n x_{i,j}$ set to 0 after round r. There are exactly $\frac{n}{2}$ of them that correspond to free holes excluded for pigeon i. For the other $z - \frac{n}{2}$ variables the corresponding hole is already occupied, that is, there is a variable for each of these holes set to 1. After round r every time one of the $n - z$ remaining variables $x_{i,j}$ is set to 0 in $\bigvee_{j=1}^n x_{i,j}$ this is done by Delayer and because there is some other variable $x_{i',j}$ already set to 1. The number of 1's at the end of the game is then at least $z - \frac{n}{2} + n - z = \frac{n}{2}$. W.l.o.g. let us call the first $\frac{n}{2}$ variables that are set to 1 at the end of the game by x_{i,j_i}, $i \in 1 \ldots \frac{n}{2}$. We analyze now how many points Delayer scores for each of them. If variable x_{i,j_i} has been assigned by Prover, then Delayer scores $-\log(p^1)$ points. If it is Delayer the one who gave the variable value 1, then at that point there were at least $\frac{n}{2}$ free holes that were excluded for pigeon i. All the 0's indicating the exclusion, had been set by Prover because Delayer only sets 0's when a hole is not free or a pigeon has already a hole. Therefore, Delayer scores at least $-\frac{n}{2}\log(p^0)$ points for this. Observe that since Delayer does not allow a pigeon to be in more than one hole, the 0's set by Prover are different for every pigeon i. Summarizing, Delayer receives either $-\log(p^1)$ or $-\frac{n}{2}\log(p^0)$ points for each of the $\frac{n}{2}$ variables x_{i,j_i} set to 1.

Let us now define the values for p^0 and p^1. Intuitively Delayer has to score more points when Prover sets a variable to 1 than we he sets it to 0 because the former brings the game quicker to an unsatisfying assignment. We define:

$$p^1 = \frac{\log n}{n} \text{ and } p^0 = 1 - \frac{\log n}{n} = \Omega\left(e^{-\frac{\log n}{n}}\right) = 2^{\Omega\left(\frac{-\log n}{n}\right)}$$

The number of points scored for each of the $\frac{n}{2}$ variables is then either $-\log(p^1) = \Omega(\log n)$ or $-\frac{n}{2}\log(p^0) = \Omega(\log n)$, and the total number of points $\Omega(n\log n)$. □

With more complicated arguments the authors of the original paper [BGL10] provide a lower bound of $\frac{n}{2}\log(\frac{n}{2}+1)$ for $G(\neg\text{PHP}^m_n)$. The best existing lower bounds for the tree-like resolution size for PHP can be seen as consequences of this result and Theorem 3.2:

Theorem 3.5 [IM99, DR01] *For $m > n$ the size of a tree-like resolution refutation of $\neg PHP^m_n$ is at least $2^{\frac{n}{2}\log(\frac{n}{2}+1)}$.*

9.4 Conclusions

We have shown in this tutorial that variations of a combinatorial game played on formulas, characterize exactly the concepts of space and size in tree-like resolution. It is interesting to observe that the outcome of these games depends only on the structure of the input formulas. This means that the concepts of tree-like resolution space and size are complexity measures intrinsic to the formulas and completely independent of the notion of resolution. It is an important open question whether such game characterizations also exist for the case of general resolution.

References

[ABRW02] M. Alekhnovich, E. Ben-Sasson, A.A. Razborov, A. Wigderson, Space complexity in propositional calculus. SIAM J. Comput. **31**(4), 1184–1211 (2002)

[BP96] P. Beame, T. Pitassi, Simplified and improved resolution lower bounds, in *37th Annual IEEE Symposium on Foundations of Computer Science*, pp. 274–282 (1996)

[BIW04] E. Ben-Sasson, R. Impagliazzo, A. Wigderson, Near-optimal separation of tree-like and general resolution. Combinatorica **24**(4), 585–603 (2004)

[BN08] E. Ben-Sasson, J. Nordström, Short proofs may be spacious: an optimal separation of space and length in resolution, in *Proceedings of 49th FOCS Conference*, pp. 709–718 (2008)

[BW01] E. Ben-Sasson, A. Wigderson, Short proofs are narrow—resolution made simple. J. ACM **48**(2), 149–169 (2001)

[Bey11] O. Beyersdorff, in *Proofs and Games*. Lecture notes 2011, http://www.thi.uni-hannover.de/fileadmin/mitarbeiter/beyersdorff/ESSLLI11-course-material.pdf

[BGL10] O. Beyersdorff, N. Galesi, M. Lauria, A lower bound for the pigeon hole principle in tree-like resolution by asymmetric prover-delayer games. Inf. Process. Lett. **110**, 1074–1077 (2010)

[BGL11] O. Beyersdorff, N. Galesi, M. Lauria, Parameterized complexity of DPLL search procedures, in *Proceedings of 14th Conference on Theory and Applications of Satisfiability Testing*, LNCS, vol. 6695, pp. 5–18, 110 (2011)

[BGL13] O. Beyersdorff, N. Galesi, M. Lauria, A characterization of tree-like resolution size. Inf. Process. Lett. **113**, 666–671 (2013)

[BEGJ02] M.L. Bonet, J.L. Esteban, N. Galesi, J. Johannsen, On the relative complexity of resolution refinements and cutting planes proof systems. SIAM J. Comput. **30**(5), 1462–1484 (2002)

[BP97] S. Buss, T. Pitassi, Resolution and the weak pigeonhole principle, in *Proceedings of Computer Science Logic 97*, Springer Verlag. LNCS, vol. 1414, pp. 149–156 (1997)

[CS88] V. Chvátal, E. Szemerédi, Many hard examples for resolution. J. ACM **35**, 759–768 (1988)

[DR01] S. Dantchev, S. Riis, Tree resolution proofs and the weak pigeon hole principle, in *Proceedings of 16th IEEE Conference on Computational Complexity*, pp. 69–75 (2001)

[ET01] J.L. Esteban, J. Torán, Space bounds for resolution. Inf. Comput. **171**(1), 84–97 (2001)

[ET03] J.L. Esteban, J. Torán, Combinatorial characterization of tree-like resolution space. Inf. Process. Lett. **87**(6), 295–300 (2003)

[Hak85] A. Haken, The intractability of resolution. Theoret. Comput. Sci. **39**(2–3), 297–308 (1985)

[HU07] A. Hertel, A. Urquhart, Game characterizations and the PSPACE-completeness of tree resolution space, in *Proceedings of CSL 2007*, pp. 527–541 (2007)

[IM99] K. Iwama, S. Miyazaki, Tree-like resolution is superpolynomially slower than dag-like resolution, in *Proceedings of 10th ISAAC*, LNCS, vol. 1741, pp. 133–142 (1999)

[Nor12] J. Nordström, Personal communication (2012)

[PI00] P. Pudlák, R. Impagliazzo, A lower bound for DLL algorithms for k-SAT, in *Proceedings of 11th Annual ACM-SIAM Symposium on Discrete Algorithms*, pp. 128–136 (2000)

[Rob65] J.A. Robinson, A machine oriented logic based on the resolution principle. J. ACM **12**(1), 23–41 (1965)

[ST13] U. Schöning, J. Torán, *The Satisfiability Problem SAT, Algorithms and Analyses* (Lehmanns media, Berlin, 2013)

[Tor99] J. Torán, Lower bounds for the space used in resolution, in *Proceedings of 13th Computer Science Logic Conference, Springer*. Lecture Notes in Computer Science, vol. 1683, pp. 362–373 (1999)

[Urq87] A. Urquhart, Hard examples for resolution. J. ACM **34**, 209–219 (1987)

Chapter 10
An Entropy-Based Proof for the Moore Bound for Irregular Graphs

S. Ajesh Babu and Jaikumar Radhakrishnan

Abstract We provide proofs of the following theorems by considering the entropy of random walks.

Theorem 1 (Alon, Hoory and Linial) *Let G be an undirected simple graph with n vertices, girth g, minimum degree at least 2 and average degree $\bar{d}$.*

Odd girth *If $g = 2r + 1$, then* $n \geq 1 + \bar{d} \sum_{i=0}^{r-1} (\bar{d} - 1)^i$.

Even girth *If $g = 2r$, then* $n \geq 2 \sum_{i=0}^{r-1} (\bar{d} - 1)^i$.

Theorem 2 (Hoory) *Let $G = (V_L, V_R, E)$ be a bipartite graph of girth $g = 2r$, with $n_L = |V_L|$ and $n_R = |V_R|$, minimum degree at least 2 and the left and right average degrees d_L and d_R. Then,*

$$n_L \geq \sum_{i=0}^{r-1} (d_R - 1)^{\lceil \frac{i}{2} \rceil} (d_L - 1)^{\lfloor \frac{i}{2} \rfloor},$$
$$n_R \geq \sum_{i=0}^{r-1} (d_L - 1)^{\lceil \frac{i}{2} \rceil} (d_R - 1)^{\lfloor \frac{i}{2} \rfloor}.$$

This work was done while S. Ajesh Babu was at School of Technology and Computer Science, Tata Institute of Fundamental Research, Homi Bhabha Road, Mumbai 400005, India.

J. Radhakrishnan (✉)
School of Technology and Computer Science, Tata Institute of Fundamental Research, Mumbai 400005, India
e-mail: jaikumar@tifr.res.in

S. A. Babu
Microsoft Research India, Bangalore 560001, India
e-mail: ajesh.babu@gmail.com

M. Agrawal and V. Arvind (eds.), *Perspectives in Computational Complexity*, Progress in Computer Science and Applied Logic 26, DOI: 10.1007/978-3-319-05446-9_10,

Keywords Girth · Moore bound · Graphs

Mathematics Subject Classification (2010) Primary 05C35 · Secondary 05C81

10.1 Introduction

The Moore bound (see Theorem 3.1) gives a lower bound on the order of any simple undirected graph, based on its minimum degree and girth. Alon et al. [AHL02] showed that the same bound holds with the minimum degree replaced by the average degree. Later, Hoory [Hoo02] obtained a better bound for simple bipartite graphs. We reprove the results of Alon et al. [AHL02] and Hoory [Hoo02] using information theoretic arguments based on nonreturning random walks on the graph.

The chapter has three sections: In Sect. 10.2 we introduce the relevant notation and terminology. In Sect. 10.3, we present the information theoretic proof of the result of Alon et al. [AHL02]; in Sect. 10.4, we present a similar proof of the result of Hoory [Hoo02] for bipartite graphs.

10.2 Preliminaries

For an undirected simple graph $G = (V, E)$, let $\vec{G} = (V, \vec{E})$, be the directed version of G, where for each undirected edge of the form $\{v, v\}$ in E, we place two directed edges in $\vec{E}$, one of the form (v, v) and another of the form (v, v). Similarly, for an undirected bipartite graph $G = (V_L, V_R, E)$, let $\vec{G} = (V_L, V_R, \vec{E}_{LR} \cup \vec{E}_{RL})$ be the directed version of G, where for each undirected edge of the form $\{v, v\}$ in E, with $v \in V_L$ and $v \in V_R$, we place one directed edge of the form (v, v) in $\vec{E}_{LR}$, and another of the form (v, v) in $\vec{E}_{RL}$.

We will consider *nonreturning* walks on $\vec{G}$, that is, walks where the edges corresponding to the same undirected edge of G do not appear in succession. For a vertex v, let $n_i(v)$ denote the number of nonreturning walks in $\vec{G}$ starting at v and consisting of i edges. For an edge $\vec{e}$, let $n_i(\vec{e})$ denote the number of nonreturning walks in $\vec{G}$ starting with $\vec{e}$ and consisting of exactly $i + 1$ edges (including $\vec{e}$).

Our proofs will make use of information theoretic ideas. Similar ideas have been employed in various combinatorial proofs to succinctly present arguments that involve averaging and convexity. More examples can be found in the references [CT91, Kah02, LL13, Rad99, Rad01].

Let X be a random variable taking values in a finite set. Let support(X) be the set of values that X takes with positive probability. The entropy of X is

$$H[X] = - \sum_{x \in \text{support}(X)} \Pr[X = x] \log_2 \Pr[X = x].$$

For random variables X and Y, taking values in finite sets according to some joint distribution, and $y \in \text{support}(Y)$, let X_y be the random variable taking values in support(X) such that $\Pr[X_y = x] = \Pr[X = x \mid Y = y]$. Then, the conditional entropy of X given Y is

$$H[X \mid Y] = \sum_{y \in \text{support}(Y)} \Pr[Y = y] H[X_y].$$

We will use of the following standard facts about entropy [CT91].

$$H[X] \leq \log_2 |\text{support}(X)|;$$

$$H[X_1 X_2 \ldots X_k \mid Y] = \sum_{i=1}^{k} H[X_i \mid X_1 X_2 \ldots X_{i-1} Y].$$

10.3 Moore Bound for Irregular Graphs

In Sect. 10.3.1, we recall the proof of the Moore bound; in Sect. 10.3.2, we review and reprove the theorem of Alon et al. [AHL02] assuming Lemma 3.4. In Sect. 10.3.3, we prove this lemma using an entropy- based argument.

10.3.1 Proof of the Moore Bound

The Moore bound provides a lower bound for the order of a graph in terms of its minimum degree and girth.

Theorem 3.1 (The Moore bound [Big93, p. 180]) *Let G be a simple undirected graph with n vertices, minimum degree δ and girth g.*

Odd girth *If $g = 2r + 1$, then $n \geq 1 + \delta \sum_{i=0}^{r-1} (\delta - 1)^i$.*

Even girth *If $g = 2r$, then $n \geq 2 \sum_{i=0}^{r-1} (\delta - 1)^i$.*

The key observation in the proof of the Moore bound is the following. If the girth is $2r + 1$, then two distinct nonreturning walks of length at most r starting at a vertex v lead to distinct vertices. Similarly, if the girth is $2r$, then nonreturning walks of length at most r starting with (some directed version of) an edge e lead to distinct vertices. We will need this observation again later, so we record it formally.

Observation 3.2 *Let G be an undirected simple graph with n vertices and girth g.*

Odd girth *Let $g = 2r + 1$. Then, for all vertices v,*

$$n \geq n_0(v) + n_1(v) + \cdots + n_r(v).$$

Even girth *Let $g = 2r$. Let e be an edge of G and suppose $\vec{e}_1$ and $\vec{e}_2$ are its directed versions in $\vec{G}$. Then,*

$$n \geq \sum_{i=0}^{r-1} [n_i(\vec{e}_1) + n_i(\vec{e}_2)].$$

Proof of Theorem 3.1 The claim follows immediately from Observation 3.2 by noting that for such a graph G, for all vertices $v \in V$ and edges $\vec{e} \in \vec{E}$,

$$n_i(v) \geq \delta(\delta - 1)^{i-1} \quad (\text{for } i \geq 1), \quad n_0(v) = 1; \tag{10.1}$$
$$n_i(\vec{e}) \geq (\delta - 1)^i \quad (\text{for } i \geq 0). \tag{10.2}$$

□

10.3.2 The Alon–Hoory–Linial Bound

Alon, Hoory, and Linial showed that the bound in Theorem 3.1 holds for any undirected graph even when the minimum degree δ is replaced by the average degree $\bar{d}$.

Theorem 3.3 (Alon et al. [AHL02]) *Let G be an undirected simple graph with n vertices, girth g, minimum degree at least 2 and average degree $\bar{d}$.*

Odd girth *If $g = 2r + 1$, then $n \geq 1 + \bar{d} \sum_{i=0}^{r-1} (\bar{d} - 1)^i$.*

Even girth *If $g = 2r$, then $n \geq 2 \sum_{i=0}^{r-1} (\bar{d} - 1)^i$.*

We will first prove this theorem assuming the following lemma, which is the main technical part of Alon et al. [AHL02]. This lemma shows that the bounds (10.1) and (10.2) holds with δ replaced by $\bar{d}$. In Sect. 10.3.3, we will present an information theoretic proof of this lemma.

Lemma 3.4 *Let G be an undirected simple graph with n vertices, girth g, minimum degree at least two and average degree $\bar{d}$.*

(a) *If $v \in V(G)$ is chosen with distribution π, where $\pi(v) = d_v/(2|E(G)|) = d_v/(\bar{d}n)$, then $\mathbb{E}[n_i(v)] \geq \bar{d}(\bar{d} - 1)^{i-1}$ $(i \geq 1)$.*
(b) *If $\vec{e}$ is a uniformly chosen random edge in $\vec{E}$, then $\mathbb{E}[n_i(\vec{e})] \geq (\bar{d} - 1)^i$ $(i \geq 0)$.*

Proof of Theorem 3.3 First, consider graphs with odd girth. From Observation 3.2, Lemma 3.4 (a) and linearity of expectation we obtain

$$n \geq \mathbb{E}[n_0(v) + n_1(v) + \cdots + n_r(v)] \geq 1 + \bar{d} \sum_{i=0}^{r-1} (\bar{d} - 1)^i,$$

where $v \in V(G)$ is chosen with distribution π (defined in Lemma 3.4 (a)).

Now, consider graphs with even girth. Let $\vec{e}_1$ be chosen uniformly at random from $\vec{E}$ and let $\vec{e}_2$ be its companion edge (going in the opposite direction). Note that $\vec{e}_2$ is also uniformly distributed in $\vec{E}$. Then, from Observation 3.2, Lemma 3.4 (b) and linearity of expectation we obtain

$$n \geq \mathbb{E}\left[\sum_{i=0}^{r} [n_i(\vec{e}_1) + n_i(\vec{e}_2)]\right] \geq 2 \sum_{i=0}^{r-1} (\bar{d} - 1)^i.$$

10.3.3 The Entropy-Based Proof of Lemma 3.4

The proof of Lemma 3.4 below is essentially the same as the one originally proposed by Alon, Hoory, and Linial but is stated more naturally using the language of entropy.

Proof of Lemma 3.4 (a) Consider the Markov process v, $\vec{e}_1$, $\vec{e}_2$, ..., $\vec{e}_i$, where v is a random vertex of G chosen with distribution π, $\vec{e}_1$ is a random edge of $\vec{G}$ leaving v (chosen uniformly from the d_v choices), and for $1 \leq j < i$, $\vec{e}_{j+1}$ is a random successor edge for $\vec{e}_j$ chosen uniformly from among the nonreturning possibilities. (If $\vec{e}_j$ has the form (x, y), then there are $d_y - 1$ possibilities for $\vec{e}_{j+1}$). Let $v_0 = v, v_1, v_2, \ldots, v_i$ be the vertices visited by this non-returning walk. We observe that each $\vec{e}_j$ is distributed uniformly in the set $E(\vec{G})$ and each v_j has distribution π. Then,

$$\begin{aligned}
\log_2 \mathbb{E}[n_i(v)] &\geq \mathbb{E}[\log_2 n_i(v)] \\
&\geq H[\vec{e}_1 \vec{e}_2 \ldots \vec{e}_i \mid v] \\
&= H[\vec{e}_1 | v] + \sum_{j=1}^{i-1} H[\vec{e}_{j+1} \mid \vec{e}_1 \vec{e}_2 \ldots \vec{e}_j v] \\
&= \mathbb{E}[\log_2 d_v] + \sum_{j=1}^{i-1} \mathbb{E}[\log_2 (d_{v_j} - 1)] \\
&= \mathbb{E}[\log_2 d_v (d_v - 1)^{i-1}] \\
&= \frac{1}{\bar{d}n} \sum_v d_v \log_2 d_v (d_v - 1)^{i-1} \\
&\geq \log_2 \bar{d} (\bar{d} - 1)^{i-1},
\end{aligned}$$

where to justify the first inequality we use Jensen's inequality for the concave function log, to justify the second we use the fact that the entropy of a random variable is at most the log of the size of its support, and to justify the last we use Jensen's inequality for the convex function $x \log_2 x(x-1)^{i-1}$ $(x \geq 2)$. The claim follows by exponentiating both sides.

(b) This time we consider the Markov process $\vec{e}_0 = \vec{e}, \vec{e}_1, \ldots, \vec{e}_i$, where $\vec{e}$ is chosen uniformly at random from $\vec{E}$, and for $0 \leq j < i$, $\vec{e}_{j+1}$ is a random successor edge for $\vec{e}_j$ chosen uniformly from among the nonreturning possibilities. Let $v_0, v_1, v_2, \ldots, v_{i+1}$ be the vertices visited by this nonreturning walk. As before observe that each v_j has distribution π. Then,

$$\begin{aligned}
\log_2 \mathbb{E}[n_i(e)] &\geq \mathbb{E}[\log_2 n_i(e)] \\
&\geq H[\vec{e}_1 \vec{e}_2 \ldots \vec{e}_i \mid \vec{e}_0] \\
&= \sum_{j=1}^{i} \mathbb{E}[\log_2(d_{v_j} - 1)] \\
&= \mathbb{E}[\log_2(d_{v_0} - 1)^i] \\
&= \frac{1}{\bar{d}n} \sum_{v} d_v \log_2(d_v - 1)^i \\
&\geq \log_2(\bar{d} - 1)^i,
\end{aligned}$$

where we justify the first two inequalities as before, and the last using Jensen's inequality applied to the convex function $x \log_2(x-1)^i$ $(x \geq 2)$. The claim follows by exponentiating both sides. □

Remark 3.5 We assumed above that the minimum degree is at least 2. It is possible to eliminate vertices of small degree and show that Theorem 3.3 holds for any graph with *average* degree at least 2. For details, see the proof of Theorem 1 in [AHL02].

10.4 Moore Bound for Bipartite Graphs

Following the proof technique of [AHL02], Hoory [Hoo02] obtained an improved Moore bound for bipartite graphs. In this section, we provide an information theoretic proof of Hoory's result.

10.4.1 The Hoory Bound

Theorem 4.1 (Hoory [Hoo02]) *Let $G = (V_L, V_R, E)$ be a bipartite graph of girth $g = 2r$, with $n_L = |V_L|$ and $n_R = |V_R|$, minimum degree at least 2 and the left and right average degrees d_L and d_R. Then,*

$$n_L \geq \sum_{i=0}^{r-1} (d_R - 1)^{\lceil \frac{i}{2} \rceil} (d_L - 1)^{\lfloor \frac{i}{2} \rfloor},$$

$$n_R \geq \sum_{i=0}^{r-1} (d_L - 1)^{\lceil \frac{i}{2} \rceil} (d_R - 1)^{\lfloor \frac{i}{2} \rfloor}.$$

For bipartite graphs the girth is always even. We then have the following variant of Observation 3.2.

Observation 4.2 *Let $G = (V_L, V_R, E)$ be an undirected bipartite graph with $|V_L| = n_L$ and $|V_R| = n_R$ and girth $g = 2r$. Let e be an edge of G and suppose $\vec{e}_1$ and $\vec{e}_2$ be its directed versions in $\vec{G}$, such that $\vec{e}_1 \in \vec{E}_{LR}$ and $\vec{e}_2 \in \vec{E}_{RL}$. Then,*

$$n_L \geq \sum_{i=0}^{\lfloor \frac{r}{2} \rfloor - 1} n_{2i+1}(\vec{e}_1) + \sum_{i=0}^{\lceil \frac{r}{2} \rceil - 1} n_{2i}(\vec{e}_2).$$

We will prove the Theorem 4.1, assuming the following lemma, which is the main technical part of Hoory [Hoo02]. In Sect. 10.4.2, we will present the proof of this lemma using the language of entropy.

Lemma 4.3 *Let $G = (V_L, V_R, E)$ be an undirected simple bipartite graph with n_L vertices on the left and n_R vertices on the right, girth g, minimum degree at least two and average left and right degrees, respectively d_L and d_R.*

(a) *If $\vec{e}$ is a uniformly chosen random edge in $\vec{E}_{LR}$, then $\mathbb{E}[n_{2i+1}(\vec{e})] \geq (d_R - 1)^{i+1}(d_L - 1)^i$ $(i \geq 1)$.*
(b) *If $\vec{e}$ is a uniformly chosen random edge in $\vec{E}_{RL}$, then $\mathbb{E}[n_{2i}(\vec{e})] \geq (d_R - 1)^i(d_L - 1)^i$ $(i \geq 1)$.*

Proof of Theorem 4.1 We will prove the bound for n_L. The proof for n_R case is similar. Let $\vec{e}_1$ be chosen uniformly at random from $\vec{E}_{LR}$ and let $\vec{e}_2$ be its companion edge (going in the opposite direction). Note that $\vec{e}_2$ is also uniformly distributed in $\vec{E}_{RL}$. Then, from Observation 4.3, Lemma 4.3 and linearity of expectation we obtain

$$n_L \geq \mathbb{E}\left[\sum_{i=0}^{\lfloor \frac{r}{2} \rfloor - 1} n_{2i+1}(\vec{e}_1) + \sum_{i=0}^{\lceil \frac{r}{2} \rceil - 1} n_{2i}(\vec{e}_2) \right] \geq \sum_{i=0}^{r-1} (d_R - 1)^{\lceil \frac{i}{2} \rceil} (d_L - 1)^{\lfloor \frac{i}{2} \rfloor}. \qquad \square$$

10.4.2 The Entropy-Based Proof of Lemma 4.3

The proof of Lemma 4.3 below is essentially the same as the one originally proposed by Hoory, but is stated in the language of entropy.

Proof of Lemma 4.3 (a) Consider a Markov process $\vec{e}_0, \vec{e}_1, \vec{e}_2, \ldots, \vec{e}_{2i+1}$, where $\vec{e}_0$ is a uniformly chosen random edge from $\vec{E}_{LR}$, and for $0 \leq j < 2i+1$, $\vec{e}_{j+1}$ is a random successor edge for $\vec{e}_j$ chosen uniformly from among the nonreturning possibilities. Let $v_0, v_1, v_2, \ldots, v_{2i+2}$ be the vertices visited by this non-returning walk. We observe that for $0 \leq j \leq i$ each $\vec{e}_{2j}$ and $\vec{e}_{2j+1}$ is respectively distributed uniformly in the set $\vec{E}_{LR}$ and $\vec{E}_{RL}$. Furthermore, for j even, $\Pr[v_j = v] = d_v/|E(G)|$ for all $v \in V_L$, and for j odd, $\Pr[v_j = v] = d_v/|E(G)|$ for all $v \in V_R$. Then,

$$\begin{aligned}
\log_2 \mathbb{E}[n_{2i+1}(e)] &\geq \mathbb{E}[\log_2 n_{2i+1}(e)] \\
&\geq H[\vec{e}_0\vec{e}_1 \ldots \vec{e}_{2i+1} \mid \vec{e}_0] \\
&= \sum_{j=0}^{i} H[\vec{e}_{2j+1}|\vec{e}_{2j}] + \sum_{j=1}^{i} H[\vec{e}_{2j}|\vec{e}_{2j-1}] \\
&= \sum_{j=0}^{i} \mathbb{E}[\log_2(d_{v_{2j+1}} - 1)] + \sum_{j=1}^{i} \mathbb{E}[\log_2(d_{v_{2j}} - 1)] \\
&\geq (i+1)\log_2(d_R - 1) + i\log_2(d_L - 1) \\
&= \log_2(d_R - 1)^{i+1}(d_L - 1)^i.
\end{aligned}$$

where to justify the first inequality we use Jensen's inequality for the concave function log, to justify the second we use the fact that the entropy of a random variable is at most the log of the size of its support, and to justify the last we use Jensen's inequality for the convex function $x\log_2(x-1)$ $(x \geq 2)$. The claim follows by exponentiating both sides.

(b) Similarly,

$$\log_2 \mathbb{E}[n_{2i}(e)] \geq \log_2(d_L - 1)^i(d_R - 1)^i. \qquad \square$$

References

[AHL02] N. Alon, S. Hoory, N. Linial, The Moore bound for irregular graphs. Graphs Comb. **18**(1), 53–57 (2002)

[Big93] N. Biggs, *Algebraic Graph Theory*, 2nd edn. (Cambridge University Press, Cambridge, 1993)

[CT91] T.M. Cover, J.A. Thomas, *Elements of Information Theory* (Wiley-Interscience, New York, 1991)

[Hoo02] S. Hoory, The size of bipartite graphs with a given girth. J. Comb. Theory, Ser. B, **86**(2):215–220 (2002)

[Kah02] J. Kahn, Entropy, independent sets and antichains: a new approach to Dedekind's problem. Proc. Amer. Math. Soc. **130**, 371–378 (2002)

[LL13] N. Linial, Z. Luria, Upper bounds on the number of Steiner triple systems and 1-factorizations. Random Struct. Algorithms **43**, 399–406 (2013)

[Rad99] J. Radhakrishnan, An entropy proof of Bregman's theorem. J. Comb. Theor. A **77**(1), 161–164 (1999)

[Rad01] J. Radhakrishnan, Entropy and counting, in *IIT Kharagpur Golden Jubilee Volume on Computational Mathematics, Modelling and Algorithms*, ed. by J.C. Mishra (Narosa Publishers, New Delhi, 2001)

Chapter 11
Permutation Groups and the Graph Isomorphism Problem

Sumanta Ghosh and Piyush P Kurur

Abstract In this article we discuss various algorithms for permutation group-theoretic problems and their connections to Graph Isomorphism. In the last part we examine the group representability problem on graphs, its connection to Graph Isomorphism, and discuss some open problems that arise in this context.

Keywords Graph isomorphism · Permutation groups

11.1 Introduction

One of the core ideas in mathematics is the notion of an isomorphism, i.e. structure preserving bijections between mathematical objects like groups, rings and fields. A natural computational question is to decide, given two such objects as input, whether they are isomorphic or not. In the context of undirected finite graphs, this problem is called the graph isomorphism problem and is the subject matter of this article. Informally, we say that two graphs are isomorphic if they are the same up to a renaming of their vertices, i.e. we have a bijection between the vertex sets that preserve the adjacency relation of edges. Many other isomorphism problems for explicitly presented finite mathematical structures like groups, for example, reduce to the graph isomorphism problem. Also, many problems that arise in practice, like studying the structure of chemical compounds, are essentially graph isomorphism in disguise. Hence, understanding this problem computationally is important.

S. Ghosh (✉) · P. P. Kurur
Department of Computer Science and Engineering, Indian Institute of Technology Kanpur, Kanpur 208016, Uttar Pradesh, India
e-mail: smghosh@cse.iitk.ac.in

P. P. Kurur
e-mail: ppk@cse.iitk.ac.in

M. Agrawal and V. Arvind (eds.), *Perspectives in Computational Complexity*, Progress in Computer Science and Applied Logic 26, DOI: 10.1007/978-3-319-05446-9_11,

There are efficient programmes and libraries (see NAUTY for instance) that can solve large instances of graph isomorphism that arise in practice. However, there are no known polynomial-time algorithm for the general case. In complexity theory, ever since the notion of NP-completeness has been formalised, the graph isomorphism problem has had an important status as it is believed to be a natural example of a problem of intermediate complexity [GJ79, Chap. 7], i.e. neither in P nor NP-complete: It is known that the graph isomorphism problem is in the complexity class co-AM [BHZ87], a randomised version of co-NP, and its NP hardness will result in the collapse of the polynomial hierarchy [BHZ87, Sch87]. Furthermore, Köbler et al. [KST92] showed that graph isomorphism is low for the counting class PP by showing its membership in LWPP. This was further improved by Arvind and Kurur [AK02] to SPP. As a result graph isomorphism is in and low for various counting complexity classes like classes like ⊕P etc. Thus, under reasonable complexity theoretic assumptions, graph isomorphism is *not* NP-hard. Ladner [Lad75] proved the existence of an infinite hierarchy of problems of intermediate complexity assuming that P is different from NP. The graph isomorphism problem, for reasons stated above, is believed to be a natural example.

In this article, we study graph isomorphism and related problems. There is now a vast literature on graph isomorphism and we really cannot do justice to the topic in such a short article. For a detailed study of graph isomorphism, mainly from a complexity theoretic view point, we refer the reader to the excellent book by Köbler et al. [KST93]. This paper concentrates on one of the aspects of the graph isomorphism problem, namely its intimate connection to permutation group algorithms. Permutation groups arise in the study of graph isomorphism problem because of its close relation to the graph automorphism problem. For a graph X, the automorphisms, i.e. isomorphisms from X to itself, forms a group under function composition. We can identify this group as a subgroup of the set of all permutations of the vertex set $V(X)$. Automorphisms, thus are symmetries of the graph. The computational problem of computing a generating set of the automorphism group is equivalent to the graph isomorphism problem [Mat79]. Most algorithms for graph isomorphism that make use of permutation group theory makes use of this connection. Understanding the automorphism group of a graph is also in tune with what is now a guiding principle in much of modern mathematics: understanding objects by understanding their symmetries.

11.2 Preliminaries

We briefly review the group theory required for this article mainly to fix notation and convention. For details please refer any standard book on group theory, for example, Marshall Hall [Hal59]. The *trivial group* that contains only the identity is denoted by 1. For a group G, we use the notation $H \leqslant G$ (or $G \geqslant H$) to denote that H is a subgroup of G. The *right coset* of the subgroup H of G associated with an element $g \in G$ is the set $Hg = \{hg | h \in H\}$. The set of all right cosets form a partition of G

and any subset T of G that has a unique coset representative from each right coset is called a *right transversal* of H in G. Analogously, we define left cosets and left transversals. In general, the right coset Hg and the left coset gH is different. We say that H is a *normal subgroup* if $gH = Hg$. We use the notation $H \trianglelefteq G$ (or $G \trianglerighteq H$) to denote that H is a normal subgroup of G.

A *simple group* is a group that has no non-trivial normal subgroups. A *composition series* of a group G is a tower of subgroups $G = G_0 \trianglerighteq G_1 \trianglerighteq \cdots \trianglerighteq G_t = 1$ such that each of the factor groups G_i/G_{i+1}, called the *composition factors*, are simple. The Jordan-Hölder theorem states that for any group G, its composition series is essentially unique, i.e. any two composition series are of equal length and the list of composition factors are equal up to a reordering. *Solvable groups* are those whose composition factors are abelian.

The set of all permutation of n elements forms a group called the *symmetric group* which we denote by S_n. In algorithmic settings, it is often useful to make the domain of n elements explicit: For a finite set Ω, the set Sym (Ω) denote the group of permutations on Ω. By a *permutation group* on Ω we mean a subgroup of the symmetric group Sym (Ω). As is customary, we use Wielandt's notation [Wie64]: Let α be any element of Ω and let g be a permutation in Sym (Ω), the image of α under g is denoted by α^g. The advantage of this notation is that it follows the familiar laws of exponentiation: $(\alpha^g)^h = \alpha^{gh}$. We can extend this notation to (i) subsets of permutations: $\alpha^A = \{\alpha^g | g \in A\}$, or to (ii) subsets of Ω: $\Sigma^g = \{\alpha^g | \alpha \in \Sigma\}$. In particular, for a permutation group on Ω, the set α^G is called the *G-orbit* of α. Given any two elements α and β of Ω the G-orbits α^G and β^G are either disjoint or are the same. Thus, orbits of G partition the underlying set Ω. A subset Σ of G is said to be *G-stable* if $\Sigma^g = \Sigma$. Clearly any G-orbit is G-stable. In general, a G-stable set is a union of G-orbits.

Let G be a permutation group acting on the set Ω and let Σ be a subset of Ω. The *point-wise stabiliser* of Σ is the subgroup of all g in G that is trivial on Σ, i.e. $\alpha^g = \alpha$ for all α in Σ. The *setwise stabiliser* is the subgroup that fixes the set Σ as a whole, i.e it is the subgroup of all g in G such that $\Sigma^g = \Sigma$.

A permutation group G is said to be *transitive* if the entire set Ω is a single orbit. Equivalently, G is transitive if for any two elements α and β in Ω there is a permutation g in G such that $\alpha^g = \beta$. For a transitive permutation G on Ω, a subset Σ is said to be a *G-block* if for any permutation g in G, the set Σ^g is either *identical to* or *disjoint from* the set Σ. Any singleton set is a block and so is the entire set Ω. These blocks are called the *trivial blocks* of G. For a transitive permutation group G on Ω and a permutation g in G, the set Σ^g is a G-block whenever Σ itself is one. Such a block Σ^g is called a *conjugate block* of Σ. The family of conjugate blocks $\{\Sigma^g | g \in G\}$ forms a partition of the set Ω which is called the *block system* associated with the block Σ. A permutation group that has no non-trivial block is called a *primitive permutation group*. An example of a primitive group is the group Sym (Ω). We have the following lemma about block systems which is more or less direct from the definition.

Lemma 2.1 *Let G be a transitive permutation group on Ω and let Σ be a block. Let N denote the subgroup of G that setwise stabilises all the elements in the Σ-block system $\mathcal{B}(\Sigma) = \{\Sigma^g | g \in G\}$. Then N is a normal subgroup of G and G/N acts as a permutation on Σ-block system $\mathcal{B}(\Sigma)$. In addition, if Σ is a maximal G-block then this action of the group G/N is primitive.*

In algorithms that deal with permutation groups, we need a succinct way to encode them which we now describe. Any permutation of Ω can be presented by an array of $\#\Omega$ elements and hence can be encoded as a string of size $O(n \lg n)$. A permutation group is presented via a list of permutations that generate the group. It is a well-known fact that any group G has a generating set of size less than $\lceil \lg \#G \rceil$ and hence this presentation of permutation group is reasonable. Thus, we assume that the input size, for an algorithm that takes a generating set S of a permutation group G on Ω, is $\#S + \#\Omega$. Similarly, an algorithm that is expected to produce a permutation group as output, should output a generating set of size polynomial in $\#\Omega$. For example, the strong generating set that we describe in the next section, is of size at most $\#\Omega^2$.

By a graph, we mean an undirected graph, i.e. a finite set of vertices and an edge set which is a subset of unordered pairs of vertices. We use $V(X)$ and $E(X)$ to denote the set of vertices and the set of edges of a graph X, respectively. A bijection f from $V(X)$ to $V(Y)$ is an *isomorphism* if for every two vertices u and v of X, the unordered pair $\{u, v\}$ is an edge of X if and only if $\{f(u), f(v)\}$ is an edge of Y. An *automorphism* of a graph X is an isomorphism from the graph to itself. The set of automorphism of a graph X, denoted by Aut (X), form a group under composition. In fact, Aut (X) is a permutation group on $V(X)$.

In the article, we assume that a graph of n vertices is encoded as an n^2-bit strings that represent its $n \times n$ adjacency matrix. We now define the *graph isomorphism problem*.

Problem 2.2 (*Graph isomorphsim problem*) The *graph isomorphism problem* (GI for short) is defined as follows: Given two undirected graphs X and Y via their adjacency matrix, decide whether they are isomorphic.

The counting version of the graph isomorphism problem, denoted by #GI, is the problem of computing the number of isomorphism between the two input graphs (0 when they are not isomorphic).

Graph isomorphism problem is closely related to the *automorphism problem* that we define next.

Problem 2.3 (*Automorphism problem*) The *automorphism problem* (AUT for short) is the problem of computing a strong generating set of the automorphism group Aut (X) of an input graph X.

Mathon [Mat79] proved that the problems GI, #GI and AUT are all polynomial-time Turing reducible to each other. Therefore, in the setting of permutation group algorithms, it is often the automorphism problem that is attacked for solving graph isomorphism.

A graph X is said to be *rigid* if it has no non-trivial automorphism, i.e. if Aut (X) is the trivial group. We now define the graph rigidity problem.

Problem 2.4 (*Graph rigidity problem*) Given an input graph X via its adjacency matrix, check whether the graph is rigid.

Clearly, an oracle for the automorphism problem, or by Mathon's result [Mat79], the graph isomorphism problem, is sufficient to decide the rigidity of a graph. However, the other direction is open.

Open problem 2.5 Is the graph rigidity problem polynomial-time equivalent to the graph isomorphism problem.

An important variant of graph isomorphism is the isomorphism of coloured graphs. For this article, a *c-colouring* of a graph X, where c a positive integer, is a map from the vertex set $V(X)$ to the set of integers $1, \ldots, c$. Given a c-colouring ψ, the ith *colour class* is subset $\psi^{-1}(i)$ of $V(X)$. A *coloured graph* is a tuple (X, ψ) of a graph X and colouring ψ. We often suppress the colouring ψ when it is understood from the context and just denote the coloured graph by X. Given two c-coloured graphs (X, ψ) and (Y, φ), an isomorphism f between the underlying graphs X and Y is a *coloured graph isomorphism* if it respects the vertex colours, i.e. for any vertex v of X, $\psi(v) = \varphi(f(v))$. An automorphism of a coloured graph is analogously defined. Clearly coloured graph isomorphism generalises graph isomorphism as we can assume an ordinary graph as 1-coloured graph. In the other direction, coloured graph isomorphism polynomial-time Turing reduces to the graph isomorphism problem. The key idea is the following *gadget construction*. For a coloured graph X, we construct a new graph $\tilde{X}$ by first adding, for each colour class i, a long path L_i (say of length $n + i + 1$). We then connect all the vertices of the colour class i to one of the end points of L_i. Given coloured graphs X and Y, any isomorphism between the modified graphs $\tilde{X}$ and $\tilde{Y}$ forces the vertices in a given colour class of X to be mapped to the vertices of the same colour class in Y due to the graph gadgets L_i. Therefore, the coloured graphs X and Y are isomorphic if and only if the modified graphs $\tilde{X}$ and $\tilde{Y}$ are isomorphic. The rigidity problem and the automorphism problem generalise naturally to coloured graphs as well.

The graph isomorphism problem and the automorphism problem can be defined for directed graphs as well. It turns out that these variants are polynomial-time Turing reducible to the undirected case. Therefore, in this article, we mostly concentrate on undirected graph isomorphism. Nonetheless, from the perspective of the isomorphism problem, there is an important subclass of directed graphs called *tournaments* that we define below.

Definition 2.6 A directed graph X is a *tournament* if for every two distinct vertices u and v, exactly one of the directed edge (u, v) or (v, u) exists in $E(X)$.

The automorphism group of a tournament cannot have a 2-cycle (why?), and hence has to be of odd order. This forces it to be *solvable* by Feit-Thompson theorem [FT63]. This property has been exploited by Babai and Luks [BL83] to give significantly efficient algorithms for tournament isomorphism.

11.3 Basic Polynomial-Time Algorithms

In this section, we mention some well-known polynomial-time algorithms for permutation group problems. The very first polynomial-time algorithm is the algorithm to compute the orbits of a permutation group. Let S be a generating set of the permutation group G then define a relation $\alpha \rightarrow_S \beta$ if there exists a g in S such that $\alpha^g = \beta$. It is easy to see that the symmetric, transitive closure of the relation $\rightarrow_s$ gives us all the G-orbits. We can thus compute the orbits efficiently by computing reachability.

Lemma 3.1 *There is a polynomial-time algorithm, which given a generating set S of a permutation group G on Ω and an $\alpha \in \Omega$, computes the orbit α^G.*

Many permutation group algorithms follows the general scheme of first reducing the problem to the transitive case by finding all the orbits of the group using the above lemma, and then restricting the group to the orbit. This is followed by a divide can conquer that is done on the blocks of the transitive action of the group. Thus, finding the blocks of a transitive permutation group is a crucial step in various algorithms. Let G be a transitive permutation group over Ω. Fix any two elements α and β in Ω and consider the graph $X_{\alpha,\beta}$ whose vertices are Ω and edges are $\{\alpha, \beta\}^G$. Let Σ be the smallest G-block containing both α and β then Sim's observed [Sim67] that vertices in any connected component C of the graph $X_{\alpha,\beta}$ is a G-block in the block system $\{\Sigma^g | g \in G\}$ associated with Σ. By running this algorithm on all pairs one can compute the set of minimal (as well as maximal) blocks of G-blocks.

Lemma 3.2 *There is a polynomial-time algorithm that takes as input the generating set of a transitive permutation group G on Ω and decides whether G is primitive or not. If the input group G is not primitive, then it computes a minimal (or maximal) G-block system.*

We already argued that a generating set of a group is a natural way to present a permutation group. A *strong generating set* is a special generating set of a permutation group that makes many computational tasks easy. Consider a permutation group G on the set Ω. Fix an ordering $\{\alpha_1, \cdots, \alpha_n\}$ on the set Ω and let $G^{(i)}$ denote the subgroup of G that fixes the first i elements of Ω, i.e. the subgroup of all elements g of G such that $\alpha_j^g = \alpha_j$ for all $1 \leq j \leq i$. Consider the *tower* $G = G^{(0)} \geqslant \cdots \geqslant G^{(n-1)} = 1$ of subgroups of G. Let C_i be any set of permutations that has exactly one element from each right coset of $G^{(i)}$ in $G^{(i-1)}$, i.e. C_i is a *right transversal* of $G^{(i)}$ in $G^{(i-1)}$. Given any permutation g in G, there is an unique element, say h_1, in C_1 which is in the same right coset of $G^{(1)}$ as that of g. It is easy to see that $g' = gh_1^{-1}$ is in $G^{(1)}$. Continuing this argument with g' and the group $G^{(1)}$, it is easy to see that any element g can be expressed as a product $h_1 \ldots h_{n-1}$, $h_i \in C_i$. In fact, if the transversals C_i's are fixed, the above product representation is unique. Thus, $\cup_i C_i$ forms a generating set of G which we call the *strong generating* set of G. Many computational tasks become trivial once the strong generating set is calculated. For example, the uniqueness of the product representation of g shows that the order of the group $\#G$ is the product of the sizes $\prod_i \#C_i$.

We now describe the algorithm to compute the strong generating set of a permutation group that was given in its complete form by Furst et al. [FHL80] based on ideas from Sims [Sim78]. It is based on the following lemma due to Schreier and hence it (and similar algorithms) are some times called *Schreier-Sims algorithm.*

Lemma 3.3 (Schreier's Lemma) *Let G be a group and H be a subgroup of G. Let T be any right transversal of H in G that contains 1 as a coset representative. For each g in G, let $\overline{g}$ denote the unique coset representative of Hg in T. Let S be a generating set for G then set*

$$S' = \{ts(\overline{ts})^{-1} | t, s \in S\}$$

generates the group H.

Let S be the generating set of a permutation group G. The main idea of the algorithm is that once we have a right transversal C_1 of $G^{(1)}$ in $G^{(0)} = G$, we can use Schreier's Lemma 11.3.3 to compute the Schreier generating set for $G^{(1)}$. We then recursively compute the strong generating set for $G^{(1)}$. At each stage of the algorithm, we compute the right transversal C_{i+1} and recurse on the Schreier generating set of $G^{(i+1)}$ obtained in that stage.

The right transversal C_1 is computed by starting with the set $T_0 = 1$ and inductively compute T_{i+1} as follows: T_{i+1} is the union of T_i and a subset of $T_i S$ such that T_{i+1} does not contain any redundant representative of same right coset of $G^{(1)}$, i.e. T_{i+1} does not contain two distinct elements g_1 and g_2 such that $\alpha_1^{g_1} = \alpha_1^{g_2}$. If at some point $T_{i+1} = T_i$, we stop the procedure. Since the set C_1 can at most have n elements this procedure has to terminate in polynomially many steps. The actual algorithm [FHL80] can be significantly more efficient by computing all the transversals C_i's simultaneously through a sifting procedure. We summarise all the polynomial-time solvable tasks that uses the Schreier-Sims procedure in the following lemma.

Lemma 3.4 (Furst, Hopegraft and Luks) *There are polynomial-time algorithms for the computational tasks:*

1. *computing a strong generating set,*
2. *computing the order of a permutation group,*
3. *checking the membership of a permutation $g \in Sym(\Omega)$ in a given permutation group G.*

The Schreier-Sims algorithm can be generalised to find the generating set of a subgroup H of a permutation group G, given indirectly by a membership oracle, provided the index $\frac{\#G}{\#H}$ is small. We state this in the next lemma.

Lemma 3.5 *There is algorithm that takes as input a permutation group G on Ω via a generating set S and computes the generating set of a subgroup H of G given via a membership oracle, i.e. a procedure to test whether a given element g of G is actually an element of H. The algorithm takes time polynomial in $\#S$, $\#\Omega$ and the index $\frac{\#G}{\#H}$.*

Consider a permutation group G on Ω and let Δ be any subset of Ω. The *point-wise stabiliser* of the set Δ, which we denote by $G(\Delta)$ is the subgroup of all elements of G that fix every element of G, i.e. $G(\Delta) = \{g|\delta^g = \delta, \ \forall\delta \in \Delta\}$. It is easy to see that finding the point-wise stabiliser of any subset of Ω can be done in polynomial-time by adapting the Schreier-Sims algorithm.

11.4 Divide and Conquer Algorithms for Permutation Groups

We now illustrate a general technique that is used in many permutation group algorithms by giving an algorithm to find the setwise stabiliser for special groups. Although superficially similar to point-wise stabiliser, computing the setwise stabiliser is a different ball game. It is at least as hard as graph isomorphism: For a graph X, consider the group $G = \mathrm{Sym}\,(V(X))$ acting on the set $\Omega = \binom{V(X)}{2}$. The automorphism group of the graph X is the set-wise stabiliser of the subset $E(X)$ of Ω. The setwise stabiliser problem is a variant of a more general problem which we define below.

Problem 4.1 (*Colour preserving subgroup*) Let G be a permutation group on Ω which is partitioned into k-colour classes $\mathcal{C} = \{C_i\}_{i=1}^k$. Compute the subgroup of G that stabilises each of the colour class C_i, i.e. compute the subgroup $\{g \in G|C_i^g = C_i\}$.

The setwise stabiliser problem is the special case when the number of colours is 2. While we cannot expect a polynomial-time algorithm for this problem in general without solving the graph isomorphism problem, for special groups, we can solve the above in polynomial-time. For example, if we know that the input group G is solvable then we have a polynomial-time algorithm. In fact, the polynomial-time algorithm of Luks [Luk82] for trivalent graphs uses such a subroutine as the group that occurs there is a 2-group and hence is solvable.

We now given a sketch of the algorithm, detail of which can be found in the paper by Luks [Luk82]. To avoid notation clutter we fix an input group G and the colouring $\mathcal{C}$ of Ω. We say that a permutation g preserves colours of all elements in the subset Σ if for all $\alpha \in \Sigma$, α and α^g are in the same colour class. Let H be a subgroup of G and Σ an H-stable subset of Ω. We denote $\mathrm{CP}\,(H, \Sigma)$ to be the subset of H that preserves the colours of elements of Σ. Our task is to compute $\mathrm{CP}\,(G, \Omega)$. For the divide and conquer algorithm to work, we need to generalise the problem to cosets of permutation groups: We need to compute $\mathrm{CP}\,(Hg, \Sigma)$ for the coset Hg of the subgroup H of G where the set Σ is H-stable. Note that Σ is not necessarily stabilised by elements of Hg.

The set $\mathrm{CP}\,(Hg, \Sigma)$ has the following crucial properties which follows more or less directly from the definitions.

Lemma 4.2 *1. The set* $\mathrm{CP}\,(H, \Sigma)$ *is a subgroup of* H.
 2. *The set* $\mathrm{CP}\,(Hg, \Sigma)$ *is either empty or is a coset of the group* $\mathrm{CP}\,(H, \Sigma)$.

3. *Suppose* Σ *is the disjoint union* $\Sigma_1 \uplus \Sigma_2$ *both of which are* H*-stable then* $\mathrm{CP}(Hg, \Sigma) = \mathrm{CP}(\mathrm{CP}(Hg, \Sigma_1), \Sigma_2)$.

It is crucial that $\mathrm{CP}(Hg, \Sigma)$ is a coset (item 2 in the above lemma) because we can then succinctly represent the set by giving a generating set of $\mathrm{CP}(H, \Sigma)$ and the coset representative.

We are now ready to give the outline of the divide and conquer algorithm for computing $\mathrm{CP}(Hg, \Sigma)$.

Reduction to transitive case Let $\Sigma' \subset \Sigma$ be any H-orbit. We first compute the group $\mathrm{CP}\left(Hg, \Sigma'\right)$ which is the transitive case of the above problem. Let $\Sigma = \Sigma' \uplus \Sigma''$. We use the fact that $\mathrm{CP}(Hg, \Sigma) = \mathrm{CP}\left(\mathrm{CP}\left(Hg, \Sigma'\right), \Sigma''\right)$.
Transitive case For this case H acts transitively on Σ. Let Δ be a *maximal* H-block of H and let $\mathcal{B}(\Delta) = \{\Delta_1, \ldots, \Delta_k\}$ be the associated block system. Let N be the normal subgroup of H that fixes all the blocks $\mathcal{B}(\Delta)$ setwise. Then we have $H = \uplus_x Nx$ as a disjoint union of cosets of N. We can then compute the set $\mathrm{CP}(Hg, \Sigma)$ by taking the union of all the cosets $\mathrm{CP}(Nxg, \Sigma)$ which are not empty.

If the number of cosets Nx are polynomially bounded then we can compute a generating set for N using Lemma 11.3.5. It then amounts to recursively computing the polynomially many cosets $\mathrm{CP}(Nxg, \Sigma)$ and combining the nonempty ones. The number of cosets Nx that is considered in the transitive case is the same as the order of the quotient group H/N. Since the block Δ that we choose is the maximal block, the quotient group H/N, as a permutation group on the set $\mathcal{B}(\Delta)$, is a primitive group (See Sect. 2.1). Thus, we need a bound on the size of a primitive permutation group. While the order of a primitive permutation group on n elements can be exponential in n, consider the case of the primitive group S_n for example, for solvable primitive permutation groups, a result by Pálfy [Pál82, Theorem 1] gives us the polynomial bound we are looking for.

Theorem 4.3 (pálfy) *There are absolute constants* C *and* c *such that any solvable primitive permutation group on* Ω *is of size less than* $C \cdot \#\Omega^c$.

The above bound has a generalisation to groups with bounded non-abelian composition factors: Let Γ_d denote the class of groups such that each composition factor is either abelian or is isomorphic to a subgroup of S_d. Babai et al. [BCP82] generalised the Pálfy's bound to the class Γ_d.

Theorem 4.4 (Babai, Cameron and pálfy) *There are absolute constants* C *and* c *such that for any positive integer* d*, any primitive permutation group on* Ω *in the class* Γ_d *is of size less than* $C\#\Omega^{cd}$.

As a result, the colour preserving subgroup problem is solvable in polynomial-time for groups that are in the class Γ_d.

Lemma 4.5 *Colour preserving subgroup problem is solvable in polynomial-time for the class of solvable groups and the class of groups in the family* Γ_d *for constant* d.

Fig. 11.1 Luks gadget for bounded valence graphs

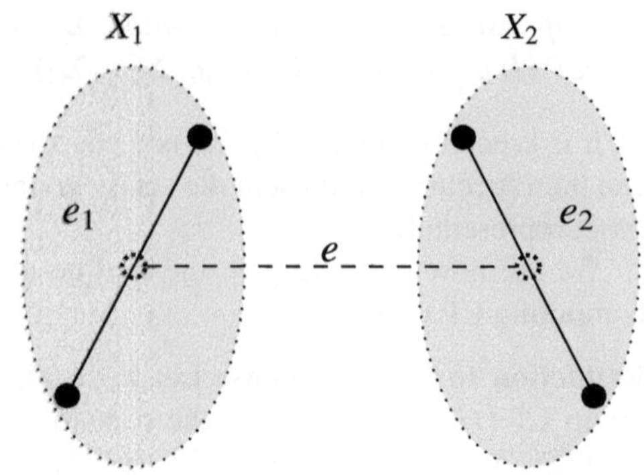

We now discuss a natural context where the colour stabiliser problem for groups in the class Γ_d occurs. Consider the graphs of valence d, i.e. all vertices are of degree less than or equal to d. Luks [Luk82] gives a polynomial-time algorithm for this class of graphs by reducing it to the colour preserving subgroup problem where the input group G is in the class Γ_{d-1}. We quickly give a sketch.

Fix the two input graphs X_1 and X_2. We assume that the graphs are connected, otherwise we run the algorithm for each pair of connected components. Furthermore, we restrict our attention to checking whether X_1 and X_2 are isomorphic via an isomorphism that maps a particular edge e_1 of X_1 to an edge e_2 of X_2: We just need to repeat the procedure for all such pairs of edges to know whether X_1 and X_2 are isomorphic.

Consider the new graph which we denote by Z which is essentially the disjoint union of the graph X_1 and X_2 with an additional edge e that connects the mid points of e_1 and e_2 (see Fig. 11.1). If $d > 2$ is the maximum degree of any vertex in the input graph then Z also has degree bounded by d. Luks algorithm for bounded valence computes the group subgroup $\text{Aut}\,(Z)_e$ of $\text{Aut}\,(Z)$ that maps the auxiliary edge e to itself. It is clear that the input graphs X_1 and X_2 are isomorphic if and only if there is at least one element in $\text{Aut}\,(Z)_e$ (and therefore in any generator set) that flips the edge e. The algorithm then proceeds to compute the group $\text{Aut}\,(Z)_e$. First the graph Z is layered as follows: Let the ith layer of Z, denoted by Z_i, be the subgraph which contains all the edges (as well as the end points) at a distance i from the auxiliary edge e. In particular, the graph Z_0 consists of just the edge e and its end points. All automorphisms of Z that stabilises the edge e has to preserve this layered structure. The group $\text{Aut}\,(Z)_e$ is then computed by inductively computing the groups $G_i = \text{Aut}\,(Z_i)_e$.

Inductively, Luks proves that G_i's are in the class Γ_{d-1} as follows: Let H_{i+1} denote the subgroup of G_i that is obtained by restricting elements of G_{i+1} on Z_i and let K_{i+1} be the associated kernel, i.e. the subgroup of G_{i+1} that is trivial when restricted to Z_i. For a fixed vertex u in layer i, i.e. in the graph $Z_i \setminus Z_{i-1}$, is connected to at most $d-1$ vertices in the layer $i+1$ and these vertices have to be mapped within themselves by elements of K_{i+1}. Therefore, K_{i+1} is a subgroup of a product of m many copies of the symmetric groups S_{d-1} for some positive integer m. The quotient group G_{i+1}/K_{i+1} is the group H_{i+1}, which itself is a subgroup G_i, a group in the class Γ_{d-1}. This is possible only if G_{i+1} is in Γ_{d-1}: Consider any composition series of G_{i+1} which passes through the normal subgroup K_{i+1}. The composition factors are either composition factors of H_i or that K_{i+1}.

Having computed G_i the algorithm computes G_{i+1} by computing (1) a generating set for the kernel K_{i+1} and (2) a set of elements of G_{i+1} whose restriction to G_i generates H_i. It is this inductive step that requires the solution of colour preserving subgroup problem and luckily the input group (G_i in our case) turns out to be in the class Γ_d and hence solvable by the algorithm in Lemma 11.4.5. Thus, we have the following theorem.

Theorem 4.6 (Luks) *Consider the family $\mathcal{G}_d$ of graphs whose vertices are of degree bounded by d. There is a $n^{O(d)}$ algorithm to decide isomorphism of graphs in $\mathcal{G}_d$.*

The current fastest algorithm for graph isomorphism [ZKT85] is based on a valence reduction step together with the application of the above theorem of Luks for bounded valence graphs. Therefore, any improvement on the bounded valence case will improve the state of the art for the general graph isomorphism problem.

11.5 Lexicographically Least Permutations

We now mention some results that make use of the ordering of permutations induced by the ordering on the domain Ω. First, note that a total ordering on the set Ω gives a total ordering on Sym (Ω): for distinct permutations g and h, $g < h$ if at the first (in the ordering on Ω) element α in Ω where they differ, we have $\alpha^g < \alpha^h$. We call this ordering the *lexicographic ordering* on the permutations. Under this ordering the lexicographically least permutation is the identity permutation. The first problem that we study is the problem of computing the lexicographically least element in a coset.

Problem 5.1 (*lexicographically least in a coset*) Given a permutation group G on Ω as a set of generators and an arbitrary permutation x on Ω, compute the lexicographically least element in the coset Gx.

Clearly if x is in G then the coset is the group G itself and the lexicographically least element of the coset is the identity element. We now sketch a polynomial-time algorithm for this problem.

Let α be the least element of Ω. The set of images of α under permutations in the coset Gx is given by the set $\alpha^{Gx} = (\alpha^G)^x$ which can be computed easily once the orbit α^G of α is computed. Clearly, the lexicographically least element of Gx should map α to the least element β of α^{Gx}. We can also compute, using the transitive closure algorithm for orbits, an element x_1 in the coset Gx such that $\alpha^{x_1} = \beta$. Therefore, the lexicographically least element of Gx is also the lexicographically least element in the coset $G_\alpha x_1$ as this coset is precisely the set of elements of Gx that maps α to β. A similar algorithm can be given for left cosets xG as well. We thus have the following lemma:

Lemma 5.2 *Computing the lexicographically least element in a coset can be done in polynomial-time.*

The above lemma is a key step in proving that the graph isomorphism problem is in the complexity class SPP.

Definition 5.3 (*SPP*) For a non-deterministic polynomial-time Turing machine let $\text{gap}(M, x)$ denote the *difference* in the number of accepting paths and rejecting paths of M on the input x. A language L is in the complexity class SPP if there is a polynomial time non-deterministic Turing machine M such that for all strings x in the language L, the $\text{gap}(M, x)$ is 1 and for all strings not in the language L, $\text{gap}(M, x)$ is 0.

Languages in SPP are believed to be of low complexity and are unlikely to be NP-hard. In particular, any *gap definable complexity* [FFK91] class not only contain SPP but also derive no extra computational power with a language in SPP as oracle, i.e. SPP is *low* for all these complexity classes. Gap definable complexity classes [FFK91] are counting classes defined using GapP functions, i.e. functions that are differences of accepting and rejecting paths of an NP machine, and contain many common counting complexity class like PP and ⊕P, etc.

The main idea involved in the proof is to design a polynomial-time algorithm A that makes queries to an NP language L with some restriction on the queries that A makes to L. We design a non-deterministic polynomial-time machine M for L such that for all queries the algorithm A makes, the machine M has at most one accepting path. Such an oracle machine can be converted to an SPP algorithm, i.e. an NP machine whose gap function is the characteristic function of GI, using standard techniques [KST92].

The base algorithm A is an inductive algorithm that builds the strong generating set of the automorphism group by computing the group G_i of all automorphisms that fix the first i vertices of the graph. To compute G_{i-1} from G_i, the algorithm has to query the NP-language L which essentially checks, given a $j > i$, whether there is an automorphism that maps i to j. The base polynomial-time machine can then find one such by doing a prefix search. However, we need to design an NP machine M for L such that for all queries asked by A, there is at most one accepting path. This we achieve as follows: The algorithm A also provides to L the generator set of G_i, i.e. queries to L are (encoding of) pairs $\langle S, j\rangle$ where S is a generating set of G_i at the i-th stage. We know that if there is an automorphism, say g in G_{i-1}, that maps i to j then the set of all such automorphisms form the coset $G_i g$. The machine M essentially guess the automorphism g that maps i to j if it exists and accepts only if g is the lexicographically least permutation in $G_i g$. Since there is only one such guess g, we know that for all queries that the algorithm A makes to L the machine M has at most one accepting path. The SPP result then follows as mentioned above.

Theorem 5.4 (Arvind and Kurur) *The graph isomorphism problem is in SPP.*

While computing the lexicographically least element in a coset has an efficient algorithm, consider the following generalisation to a double coset.

Problem 5.5 (*Lex-least in a double coset*) Given the generating sets of permutation groups G and H on a totally ordered set Ω and an arbitrary permutation x on Ω, compute the lexicographically least element in GxH.

The problem of computing the lex-least element in a double coset is intimately connected to the problem of graph canonisation which we define below.

Definition 5.6 (*Canonical forms for graphs*) Consider the class $\mathcal{G}(\Omega)$ of all graphs on the vertex set Ω. A function **CF** on $\mathcal{G}(\Omega)$ is a *canonical form* if it satisfies the following properties:

1. For every graph X in $\mathcal{G}(\Omega)$, $\mathbf{CF}(X)$ is isomorphic to X.
2. If X and Y are two isomorphic graphs then $\mathbf{CF}(X)$ is the same as $\mathbf{CF}(Y)$.

In other words, a canonical form **CF** picks a unique representative from each isomorphism class of graphs on Ω. Clearly graph isomorphism is solvable given a canonisation procedure. Therefore, one way of attacking the graph isomorphism problem is to give fast canonisation procedure. For many classes of graphs, Babai and Luks [BL83] gave an efficient canonisation procedure which is also currently the best general purpose algorithms. In particular, they were able to give an $O\left(n^{c \log n}\right)$ for tournaments. This canonisation procedure makes use of the fact that tournaments have a solvable automorphism group. They also show how Luks' polynomial-time algorithm for bounded valance [Luk82] can be modified and extended to a canonisation algorithm with essentially the same running time.

As opposed to computing the lexicographically least element in a coset, computing it in a double coset is known to be NP-hard [Luk93, Theorem 5.1] even when one of the group is solvable. However, in many contexts, particularly in relation with graph isomorphism and canonisation, we have some freedom to choose the underlying ordering of the set Ω. Can we reorder the set Ω so as to make it possible to apply the divide and conquer technique similar to that of the colour preserving subgroup problem that we saw in Sect. 11.4. Indeed this is the case provided the reordering is "compatible" with the divide and conquer structure of the group G. First, we need to generalise the lexicographically least element as follows: Consider an ordering $<$ on Ω. Let Δ be a G-stable set. We consider the restriction of the order $<$ on the set Δ. This gives a partial order on elements of permutations, we say that $g < h$ if for the least δ in Δ on which g and h differ, we have $\delta^g < \delta^h$. Under this restricted ordering there will be multiple lexicographically minimal elements. We now describe how to build a new ordering $\prec$ on Ω under which it is feasible to compute the lexicographically least element of the double coset GxH.

Ordering the orbit Fix an ordering between the G-orbits by picking say the least element in each of them. If Ω_1 and Ω_2 are two orbits such that $\Omega_1 < \Omega_2$ in the above chosen order, then we set every element of Ω_1 to be less than that of Ω_2 under the new ordering $\prec$. The motivation of this reordering is the following: Let $\Omega = \Omega_1 \uplus \Omega_2$ then computing the lexicographically least element with respect to the new ordering $\prec$ can be done by first computing the lexicographically minimal elements with respect to the restriction of $\prec$ on Ω_1 and then from them picking the lexicographically least element with respect to the restriction of $\prec$ on Ω_2.

Ordering within orbits When G is transitive, we do the reordering with respect to the blocks. We pick a maximal G-block Δ in a canonical way. The Δ block system partition the set Ω so reorder it pretty much the same way as in the previous case using the Δ block system instead. If N denotes the normal subgroup of G that fixes all the blocks in the Δ block system, we can recursively find the lex-least elements in double cosets $NgxH$ and find the minimal ones out of it. If G is in the class Γ_d, the number of subproblems are polynomially bounded by Theorem 11.4.4.

The above reordering can be formalised in terms of the *structure forest* of the group G. The structure forest is the collection of *structure trees* one for each G-orbit. For an orbit Σ, the structure tree is a tree where the leaves are elements of Σ. Each internal node v is associated with a G-block Δ_v with the following properties.

1. For any child u of v, the G-block Δ_u is a maximal block contained in Δ_v.
2. If $\{u_1, \ldots, u_k\}$ are the set of children of v then the blocks Δ_{u_i}'s are all conjugates of each other and partition the parent block Δ_v.

This structure forest captures the divide and conquer on G. The elements of Ω can be reordered once we compute the structure forest of G: The structure trees are ordered in the order of the least element in them. For each internal node v and children u_1 and u_2, $u_1 \prec u_2$ if the least element of the associated block in u_1 is smaller than that of u_2 according to the original ordering $<$. This will finally give an ordering $\prec$ on the entire set Ω. Thus, we have the following theorem (See the survey article by Luks [Luk93] for details).

Theorem 5.7 *Given a totally ordered set Ω, permutation groups G and H on Ω and a permutation x on Ω. Suppose G is in the class Γ_d then in time polynomial in n we can compute a new ordering $\leq_G$ such that computing the lex-least element of the double coset GxH can be done in $n^{O(d)}$.*

Notice that we do not have any restriction whatsoever on the group H.

11.6 Structure of Primitive Groups

In the last two sections, we saw how bounds on the order of primitive permutation group can be crucial in the runtime analysis of various divide and conquer algorithms for permutation groups. We now mention how knowing the actual structure of primitive groups are computationally useful. This section is mainly motivated by the study of *bounded colour class graph isomorphism problem*.

Problem 6.1 (*Bounded colour class graph isomorphism*) Fix a constant b. Given two coloured graph X and Y such that the number of vertices in any given colour class is bounded by b decide whether the graphs are isomorphic.

We abbreviate this problem as BCGI_b. This restricted graph isomorphism problem does have polynomial-time algorithm but what about fast parallel algorithms? Luks [Luk86] answered this question affirmatively by giving a reduction to a restricted point-wise stabiliser problem and solving it in NC. Further careful analysis by Arvind et al. [AKV05] showed that the problem lies in the Mod_kL hierarchy. Together with the hardness for this class [Tor04], we have a fairly tight classification of this variant of graph isomorphism.

We now explain the overall structure of the algorithm for BCGI_b. Both Luks [Luk86] and Arvind et al. [AKV05] reduce the bounded colour graph isomorphism problem to a restricted version of point-wise stabiliser problem which we now define.

Problem 6.2 (*bounded orbit point-wise stabiliser problem*) Given as input a set Ω, a subset Δ of Ω and a permutation group G on Ω such that the G-orbits are all of cardinality bounded by a constant c. Compute generating set of the point-wise stabiliser subgroup $G(\Delta)$.

We abbreviate this problem as PWS_c. As mentioned before, the above problem is solvable in polynomial-time. However, in this context, we are interested in providing a parallel algorithm. What needs to be exploited is that the G-orbits are bounded and thus G is actually a subgroup of a product of small symmetric groups, the symmetric groups on each of the G-orbits.

To see the connection of the PWS_c and BCGI_b isomorphism we consider the equivalent automorphism problem which we denote by AUT_b.

Lemma 6.3 (Luks) *The* AUT_b *problem logspace reduces to* $\text{PWS}_{2^{b^2}}$ *problem.*

Here is the sketch of this reduction. Let X be the coloured graph and let $C_1, \ldots, C_m$ denote the colour classes into which the vertex set $V(X)$ is partitioned. The automorphism group is a subgroup of the product group $G = \prod_i \text{Sym}(C_i)$. We expressed the automorphism group $\text{Aut}(X)$ as a point-wise stabiliser of G on its action on a different set Ω that we construct as follows: Define the set $C_{i,j}$ to be the set of unordered pairs $\{u, v\}$ where $u \in C_i$ and $v \in C_j$ and let $E_{i,j}$ be the subset of edges of X between the colour classes C_i and C_j. Define the set $\Omega_{i,j}$ to be the power set of $C_{i,j}$ then the edge sets $E_{i,j}$ are actually *points* or element of $\Omega_{i,j}$. Consider the natural action of G on the union $\Omega = \cup_{i,j} \Omega_{i,j}$ and let Δ be the subset of all the *points* $E_{i,j}$ of Ω. It is easy to see that the point-wise stabiliser of G with respect to the subset Δ is actually the automorphism group $\text{Aut}(X)$. Notice that each of the set $\Omega_{i,j}$ are G-stable and hence the orbits of G are at most of size $2^{\binom{b}{2}}$.

Both Luks [Luk86] and Arvind et al. [AKV05] then solve the PWS_c problem. While the actual algorithms of Luks [Luk86] and Arvind et al. [AKV05] are fairly technical, we attempt to explain the essence of the algorithm and the permutation group theory involved in those results.

The group G in question can be seen as a product of groups $G = \prod_i G_i$ where G_i is the action of G on the ith orbit. For each of the groups G_i, compute a special *normal series* $G_i = N_{i,0} \triangleright \cdots \triangleright N_{i,t}$ and let N_k denote the produce $\prod_i N_{i,k}$. The algorithm does a divide and conquer to compute $N_k(\Delta)$ going one level at a time. The base case of this divide and conquer is when the group $N_{i,s}$ hits a socle. The *socle* of a group G is the group generated by the set of all minimal normal subgroups of G. The main group theoretic result that is used is the O'Nan-Scott theorem (See the book by Dixon and Mortimer [DM91] for a proof of this result) on the structure of the socle of a primitive permutation group and its point-wise stabiliser.

For the details of the algorithm, we refer the reader to the conference paper of Arvind et al. [AKV05]. A detailed version is available in the Ph.D thesis of Kurur [Kur06, Chap. 5]

11.7 Representation of Groups on Graphs

In this section, we look at the group representability problem. This problem was defined and studied by Dutta and Kurur [DK09] to explore the connection between graph isomorphism and permutation group algorithms from a representation theoretic point of view. Representation theory is the study of homomorphisms from a group to the group $\mathrm{GL}(V)$, the automorphisms of a vector space V. In the context of graph isomorphism, we would like to understand homomorphisms between groups and automorphisms of graphs.

Definition 7.1 A *representation* of a group G *on a* graph X is a homomorphism from the group G to the automorphism group Aut (X) of the graph X.

There is always a trivial representation that sends all the elements of the group to the identity automorphism. What we are interested in is a non-trivial homomorphisms. The main problem of interest in this section is the following [DK09].

Problem 7.2 (*Group representability problem*) Given a group G and a graph X, decide whether G has a non-trivial representation on X.

The hardness of the problem depends on how the group G is presented. We assume, unless otherwise mentioned, that G is provided to the algorithm via a multiplication table. Therefore, one can assume that the input size is $\#G + \#V(X)$. In studying its connection to graph isomorphism, we can restrict the problem in two ways: (1) restrict the groups to come from a natural class like, for example, solvable or abelian or (2) restrict the class of graphs to say planar graphs or trees.

The very first result that we have in this context is the following [DK09].

Lemma 7.3 (Dutta and Kurur) *The graph isomorphism problem is log-space many-one reducible to the abelian group representability problem.*

The main idea behind the proof is the following: Consider an instances of graph isomorphism where we want to check whether the graphs X and Y are isomorphic. We assume they are connected and have n vertices. For a prime p, consider the graph Z which is the disjoint union of $p - 1$ copies of X and 1 copy of Y. Suppose that the group Aut (Z) has a p-cycle say g. Consider any vertex u of Z such that $u^g \neq u$. It is easy to see that the orbit of u under the cyclic group generated by g has to have p-elements. Furthermore, if any two of the elements in this orbit is in the same connected component of Z then the entire orbit is. If the prime p is chosen to be greater than n then such a p-cycle necessarily has to permutes the components as each of the connected components of Z are of cardinality at most $n < p$. This is only possible if some copy of X in Z is mapped to the copy of Y and hence X and Y have to be isomorphic. Conversely, for any prime p, if X and Y are isomorphic, then the group Aut (Z) has a p-cycle. Thus, to decide whether X and Y are isomorphic, we need to check the group representability of the additive group of $\mathbb{Z}/p\mathbb{Z}$, on the graph Z for some prime p greater than the number of vertices in X. By Bertrand's postulate (it is actually a theorem but the name seems to be stuck) there is always a prime p between n and $2n$ which we chose for this purpose.

Notice that for the previous lemma, all we needed is to pick a prime p such that Aut (X) does not have p-cycle. Recall that tournaments have odd order automorphism group and hence we have the following result.

Theorem 7.4 *Tournament isomorphism is reducible to $\mathbb{Z}/2\mathbb{Z}$ representability.*

In this context, we have the following open problem.

Open problem 7.5 Is graph isomorphism reducible to $\mathbb{Z}/2\mathbb{Z}$-representability (or for that matter any fixed group representability).

What about the other direction, i.e. reduction from representability to isomorphism? Dutta and Kurur [DK09] prove the following result for solvable group representability.

Lemma 7.6 (Dutta and Kurur) *The representability problem for solvable groups is polynomial-time Turing reducible to graph isomorphism.*

For a group G, let G' be the commutator subgroup. The main idea is the following group theoretic fact.

Lemma 7.7 *A solvable group G is representable on X if and only if there is a prime p that divides both the orders G/G' and* #Aut (X).

From the above lemma it follows that to check representability for solvable groups, all we need is a way to compute the orders $\#G/G'$ and that of #Aut (X). Clearly the former can be computed easily as the group is presented as a multiplication table and the latter using an oracle to the automorphism problem (or equivalently the graph isomorphism problem). We can even assume that the group is presented as a permutation group because there are efficient algorithms to compute the commutator subgroup of a permutation group [FHL80, Theorem 4].

Thus as far as group representability is concerned, as long as we restrict the problem to solvable groups, we are within the realm of graph isomorphism. However, even for the simplest of the non-solvable case we do not have a satisfactory answer:

Open problem 7.8 (A_5 representability problem) Given a graph X decide whether there is a subgroup of Aut (X) which is isomorphic to the alternating group of A_5.

The importance of A_5 here is that it is the smallest example of a non-solvable group. Since A_5 is a simple group, non-trivial homomorphisms from it to Aut (X) can only be injections.

Torán [Tor04] showed that graph isomorphism is hard for a lot of parallel complexity classes like ⊕L etc. An important open problem in the context of graph isomorphism is whether it is hard for the complexity class P (under suitable reductions). If this is the case, it would give evidence that it is unlikely to have efficient parallel algorithms for graph isomorphism. We would like to pose the same question for the group representability problem

Open problem 7.9 Is the group representability problem hard for the complexity class P.

Are there reasons to believe that the group representability problem is harder than graph isomorphism? We really do not know. However, for the restricted case of representability on trees, we already have a difficulty. Graph isomorphism on trees can be done in polynomial time. In fact, even for planar graphs isomorphism testing can be done in linear time [HW74] or, if one is interested in the space-bounded classes, in logarithmic space [DLNTW09]. In contrast, [DK09] proved the following result for representability on trees.

Theorem 7.10 (Dutta and Kurur) *The problem of group representability on trees is Turing equivalent to the problem of testing, given an integer n in unary and group G via multiplication table, whether there is a non-trivial homomorphism to the symmetric group S_n or not.*

We call the problem of checking whether a group G has a homomorphism to S_n as *permutation representability problem* and is motivated by Cayley's theorem that states that every finite group is a subgroup of a symmetric group. However, finding the smallest n for which G is a subgroup seems to be hard although we admit that there are no known hardness result for the above problem.

11.8 Conclusion

In this article, we discussed the complexity of some permutation group algorithms and its close connection to graph isomorphism. Most of these algorithms perform a divide and conquer and it is here the structure of permutation groups plays a crucial role. Of particular interest are permutation group theoretic structures like orbits and blocks whose computation allows us to often reduce the general case to the primitive case. Also in most of these cases the primitive case is solvable if the group is known to be in some special class like solvable or Γ_d. This makes use of bounds on the sizes of primitive groups or, in some cases, their explicit structure. We also saw how these classes naturally arose in study of restricted versions of graph isomorphism. We therefore believe that a better understanding of permutation groups and is relation to graph isomorphism is crucial in pinning down the computational complexity of this elusive problem.

References

[AK02] V. Arvind, P.P. Kurur, Graph Isomorphism is in SPP in 43rd Annual IEEE Symposium of Foundations of Computer Science, November 2002, pp. 743–750.

[AKV05] V. Arvind, P.P. Kurur, T.C. Vijayaraghavan, Bounded color multiplicity graph isomorphism is in the #L hierarchy in 20th IEEE Conference on Computational Complexity (CCC 2005), June 2005, pp. 13–27.

[BL83] L. Babai, E.M. Luks, Canonical labeling of graphs. in Proceedings of the Fifteenth Annual ACM Symposium on Theory of, Computing, 1983, pp. 171–183.

[BCP82] L. Babai, P.J. Cameron, P.P. Pálfy, On the order of primitive groups with restricted nonabelian composition factors. J. Algebra. **79**, 161–168 (1982)

[BHZ87] R. Boppana, J. Hastad, S. Zachos, Does co-NP have short interactive proofs. Inf. Process. Lett. **25**(2), 127–132 (1987)

[DM91] J.D. Dixon B. Mortimer, Permutation Groups. Graduate Texts in Mathematics, vol 163, (Springer, New York, 1991).

[DK09] S. Dutta, P.P. Kurur, Representating groups on graphs. CoRR, abs/0904.3941 (2009).

[DLNTW09] S. Dutta, N. Limaye, P. Nimbhorkar, T. Thierauf, F. Wagner, Planar graph isomorphism is in logspace. in Proceedings of the IEEE Conference on Computational Complexity, 2009, pp 124–203.

[FT63] W. Feit, J.G. Thompson, Solvability of groups of odd order. Pac. J. Math. 13(3), 775–1027, (1963). URL http://projecteuclid.org/euclid.pjm/1103053943

[FFK91] S.A. Fenner, L. Fortnow, S.A. Kurtz, Gap-definable counting classes. in Structure in Complexity Theory Conference, 1991, pp 30–42. URL http://citeseer.nj.nec.com/fenner92gapdefinable.html

[FHL80] M.L. Furst, J.E. Hopcroft, E.M. Luks, Polynomial-time algorithms for permutation groups. in IEEE Symposium on Foundations of Computer, Science, 1980, pp. 36–41.

[GJ79] M. Garey, D. Johnson, *Computers and intractability: a guide to the theory of NP-completeness* (W. H. Freeman, New York, 1979)

[Hal59] M. Hall Jr., *The Theory of Groups*, 1st edn. (The Macmillan Company, New York, 1959)

[HW74] J.E. Hopcroft, J.K. Wong, Linear time algorithm for isomorphism of planar graphs (preliminary report). in Proceedings of the Sixth Annual ACM Symposium on Theory of Computing, STOC '74 (ACM, New York, 1974), pp. 172–184.

[KST92] J. Köbler, U. Schöning, J. Torán, Graph isomorphism is low for pp. Comput. Complex. 2(4), 301–330 (1992) URL http://citeseer.nj.nec.com/obler92graph.html

[KST93] J. Köbler, U. Schöning, J. Torán, *The Graph Isomorphism Problem: Its Structural Complexity* (Birkhauser, Switzerland, 1993)

[Kur06] P.P. Kurur, Complexity upper bounds using permutation group theory. PhD thesis, Institute of Mathematical Sciences, Chennai, India, 2006.

[Lad75] R.E. Ladner, On the structure of polynomial time reducibility. J. ACM **22**(1), 155–171 (1975)

[Luk82] E.M. Luks, Isomorphism of graphs of bounded valence can be tested in polynomial time. J. Comput. Syst. Sci. **25**(1), 42–65 (1982)
[Luk86] E.M. Luks, Parallel algorithms for permutation groups and graph isomorphism, in Proceedings of the IEEE Foundations of Computer Science, IEEE Computer Society (1986), pp. 292–302
[Luk93] E.M. Luks, Permutation groups and polynomial time computations. DIMACS Ser. Discrete Math. Theoret. Comput. Sci. **11**, 139–175 (1993)
[Mat79] R. Mathon, A note on graph isomorphism counting problem. Inf. Process. Lett. **8**(3), 131–132 (1979)
[Pál82] P.P. Pálfy, A polynomial bound for the orders of primitive solvable groups. J. Algebra **77**, 127–137 (1982)
[Sch87] U. Schöning, Graph isomorphism is in the low hierarchy. in Symposium on Theoretical Aspects of Computer, Science (1987), pp. 114–124
[Sim67] C.C. Sims, Graphs and finite permutation groups. Mathematische Zeitschrift **95**, 76–86 (1967)
[Sim78] C.C. Sims, Some group theoretic algorithms. Top. Algebra 697, 108–124 (1978)
[Tor04] J. Torán, On the hardness of graph isomorphism. SIAM J. Comput. 33(5), 1093–1108 (2004)
[Wie64] H. Wielandt, *Finite Permutation Groups* (Academic Press, New York, 1964)
[ZKT85] V.N. Zemlyachenko, N.M. Korneenko, R.I. Tyshkevich, Graph isomorphism problem. J. Sov. Math. **29**, 1426–1481 (1985)

MIX
Papier aus verantwortungsvollen Quellen
Paper from responsible sources
FSC® C105338

If you have any concerns about our products,
you can contact us on
ProductSafety@springernature.com

In case Publisher is established outside the EU,
the EU authorized representative is:
Springer Nature Customer Service Center GmbH
Europaplatz 3, 69115 Heidelberg, Germany

Printed by Libri Plureos GmbH
in Hamburg, Germany

Stem Cell Biology and Regenerative Medicine

Series Editor
Kursad Turksen, Ph.D.
kursadturksen@gmail.com

For further volumes:
http://www.springer.com/series/7896

Tiziana A.L. Brevini
Editor

Stem Cells in Animal Species: From Pre-clinic to Biodiversity

Humana Press

Editor
Tiziana A.L. Brevini
Laboratory of Biomedical Embryology
UniStem
Centre for Stem Cell Research
Università degli Studi di Milano
Milan, Italy

ISSN 2196-8985 ISSN 2196-8993 (electronic)
ISBN 978-3-319-34557-4 ISBN 978-3-319-03572-7 (eBook)
DOI 10.1007/978-3-319-03572-7
Springer Cham Heidelberg New York Dordrecht London

Printed on acid-free paper

Humana Press is a brand of Springer
Springer is part of Springer Science+Business Media (www.springer.com)

Contents

Contributors

Iolanda Aquila Molecular and Cellular Cardiology, Department of Medical and Surgical Sciences, Magna Graecia University, Catanzaro, Italy

Ida Biunno Institute for Genetic and Biomedical Research of the National Research Council (IRGB-CNR), Milan, Italy

IRCCS Multimedica, Milan, Italy

Tiziana A.L. Brevini Laboratory of Biomedical Embryology, UniStem, Centre for Stem Cell Research, Università degli Studi di Milano, Milan, Italy

Christelle Correale Molecular and Cellular Cardiology, Department of Medical and Surgical Sciences, Magna Graecia University, Catanzaro, Italy

Fausto Cremonesi Reproduction Unit, Large Animal Hospital, Università degli Studi di Milano, Polo Veterinario di Lodi, Lodi, Italy

Pasquale DeBlasio Integrated Systems Engineering SRL, Milan, Italy

Silvia Erratico Ystem SRL, Milan, Italy

Andrea Farini Stem Cell Laboratory, Department of Pathophysiology and Transplantation, Università degli Studi di Milano, Fondazione IRCCS Ca' Granda Ospedale Maggiore Policlinico, Centro Dino Ferrari, Milan, Italy

Fulvio Gandolfi Laboratory of Biomedical Embryology, UniStem, Centre for Stem Cell Research, Università degli Studi di Milano, Milan, Italy

Martha C. Gómez Audubon Nature Institute Center for Research of Endangered Species, New Orleans, LA, USA

Ciro Indolfi Molecular and Cellular Cardiology, Department of Medical and Surgical Sciences, Magna Graecia University, Catanzaro, Italy

Anna Lange-Consiglio Reproduction Unit, Large Animal Hospital, Università degli Studi di Milano, Polo Veterinario di Lodi, Lodi, Italy

Sara Maffei Laboratory of Biomedical Embryology, UniStem, Centre for Stem Cell Research, Università degli Studi di Milano, Milan, Italy

Fabiola Marino Molecular and Cellular Cardiology, Department of Medical and Surgical Sciences, Magna Graecia University, Catanzaro, Italy

Mirella Meregalli Stem Cell Laboratory, Department of Pathophysiology and Transplantation, Università degli Studi di Milano, Fondazione IRCCS Ca' Granda Ospedale Maggiore Policlinico, Centro Dino Ferrari, Milan, Italy

Ystem SRL, Milan, Italy

Anjali Nandal Department of Animal and Avian Sciences, University of Maryland, College Park, MD, USA

Animal Bioscience and Biotechnology Laboratory, USDA Agricultural Research Service (USDA-ARS), Beltsville, MD, USA

Gianantonio Nappi Department of Cardio-Thoracic and Respiratory Sciences, Second University of Naples, Naples, Italy

C. Earle Pope Audubon Nature Institute Center for Research of Endangered Species, New Orleans, LA, USA

Mariangela Scalise Molecular and Cellular Cardiology, Department of Medical and Surgical Sciences, Magna Graecia University, Catanzaro, Italy

Clemetina Sitzia Stem Cell Laboratory, Department of Pathophysiology and Transplantation, Università degli Studi di Milano, Fondazione IRCCS Ca' Granda Ospedale Maggiore Policlinico, Centro Dino Ferrari, Milan, Italy

Bhanu Prakash V.L. Telugu Department of Animal and Avian Sciences, University of Maryland, College Park, MD, USA

Animal Bioscience and Biotechnology Laboratory, USDA Agricultural Research Service (USDA-ARS), Beltsville, MD, USA

Michele Torella Department of Cardio-Thoracic and Respiratory Sciences, Second University of Naples, Naples, Italy

Daniele Torella Molecular and Cellular Cardiology, Department of Medical and Surgical Sciences, Magna Graecia University, Catanzaro, Italy

Department of Cardio-Thoracic and Respiratory Sciences, Second University of Naples, Naples, Italy

Yvan Torrente Stem Cell Laboratory, Department of Pathophysiology and Transplantation, Università degli Studi di Milano, Fondazione IRCCS Ca' Granda Ospedale Maggiore Policlinico, Centro Dino Ferrari, Milan, Italy

Ystem SRL, Milan, Italy

Rajneesh Verma Monash Institute of Medical Research, Monash University, Melbourne, VIC, Australia

Institute of Molecular Bioscience, Mahidol University, Bangkok, Thailand

Paul John Verma Stem Cells and Reprogramming Group, Biological Engineering Laboratories, Faculty of Engineering, Monash University, Melbourne, VIC, Australia

South Australian Research and Development Institute (SARDI), Adelaide, Australia

Turretfield Research Centre, Rosedale, Australia

Carla Vicinanza Molecular and Cellular Cardiology, Department of Medical and Surgical Sciences, Magna Graecia University, Catanzaro, Italy

Stem Cell and Regenerative Biology Unit (BioStem), Research Institute for Sport & Exercise Sciences (RISES), Liverpool John Moores University, Liverpool, UK

About the Editor

Dr. Tiziana A.L. Brevini is an Associate Professor of Anatomy and Embryology at the University of Milan, Italy. She studied the interactions between the female genital tract and the maturing oocyte in the Department of Molecular Embryology at the University of Cambridge, UK. She completed her doctorate degree in Endocrinology and Metabolic Sciences at the University of Milan, Italy, and then carried out research at Monash University in Melbourne, Australia, and at the University of Adelaide in Adelaide, Australia. Her main areas of research focus on the understanding of cell differentiation and commitment, epigenetic control of cell fate, and pluripotency-related networks in mammalian cells and embryos.

Chapter 1
Stem Cells in Dystrophic Animal Models: From Preclinical to Clinical Studies

Clemetina Sitzia, Silvia Erratico, Andrea Farini, Yvan Torrente, and Mirella Meregalli

1 Introduction

To set up a correct clinical protocol for a particular cell approach, as stem cell therapy is, the development of a complete preclinical study is essential; in fact, only if the criteria chosen for the animal model tests are evaluated in order to be correlated with the future human application, the results obtained can be prognostic.

For this reason, different aspects must be considered to select the most suitable animal model, as these aspects play an important role: (1) the genetic basis of the disease should be the same both for humans and animal models (especially it is an inherited disorder); (2) hallmarks specific of human pathology should be reiterated in animals, so that the study can evaluate eventual improvements through symptoms relief; (3) characteristics of animal itself should be correlated with the aim of the study (e.g. for a genetic screening an animal model with a large number of progeny could be handy, while for a cell therapy preclinical protocol a dog could be better compared with patients' organisms); (4) disease progression should be well characterized, so that

These authors contributed equally to this paper

C. Sitzia • A. Farini
Stem Cell Laboratory, Department of Pathophysiology and Transplantation, Università degli Studi di Milano, Fondazione IRCCS Ca' Granda Ospedale Maggiore Policlinico, Centro Dino Ferrari, via F. Sforza 35, Milan 20122, Italy

S. Erratico
Ystem SRL, via Podgora 7, Milan 20122, Italy

Y. Torrente (✉) • M. Meregalli (✉)
Stem Cell Laboratory, Department of Pathophysiology and Transplantation, Università degli Studi di Milano, Fondazione IRCCS Ca' Granda Ospedale Maggiore Policlinico, Centro Dino Ferrari, via F. Sforza 35, Milan 20122, Italy

Ystem SRL, via Podgora 7, Milan 20122, Italy
e-mail: yvan.torrente@unimi.it; mirella.meregalli@unimi.it

T.A.L. Brevini (ed.), *Stem Cells in Animal Species: From Pre-clinic to Biodiversity*, Stem Cell Biology and Regenerative Medicine, DOI 10.1007/978-3-319-03572-7_1,

reference values are available in scientific literature, in order to allow precise outcome analysis; (5) animal phenotype should be reproducible over generations, so that standard readout parameters can be used and outcomes compared (Willmann et al. 2009). The applications of animal models to clinical research are too spread; in particular, for cell therapy it has a pivotal role, especially in case of pathologies not treatable with pharmacological or conventional approaches, like neurodegenerative diseases.

Muscular dystrophies (MDs) are a heterogeneous group of disorders characterized by the degeneration/regeneration of muscle fibres, associated with progressive muscle weakness due to the development of muscle atrophy or necrosis. Genetically, these pathologies are identified by mutations in different genes of dystrophin-glycoprotein complex (DGC) that act as a bridge across the sarcolemma membrane and guarantee the connection between the extracellular matrix (ECM) and the cytoskeleton actin protein (Ibraghimov-Beskrovnaya et al. 1992).

Depending on the gene where the mutation is localized, muscular dystrophies are classified into Duchenne and Becker muscular dystrophies (DMD, BMD, dystrophin) (Hoffman et al. 1987; Koenig et al. 1987), Emery–Dreifuss (ememin, lamin A/C) (Brown et al. 2001), facioscapulohumeral muscular dystrophy (FSHD, double-homeobox protein 4 (Dux4)); oculo-pharyngeal muscular dystrophy (OPMD, poly(A)-binding protein 2 (Shanmugam et al. 2000)), Miyoshi myopathy (MM, dysferlin) (Liu et al. 1998) and limb-girdle muscular dystrophies (LGMD-2B, dysferlin (Liu et al. 1998); LGMD-2A, calpain III (Richard et al. 1995); LGMD-1C, caveolin (Minetti et al. 1998); LGMD-2C, -2D, -2E, -2F, sarcoglycan complex (Bonnemann et al. 1996)); so far, more than 30 genetically distinct types of MDs have been identified (Kaplan 2011). From a histological point of view, muscle tissue in MDs is characterized by the presence of abundant small-diameters centre-nucleated fibres, indicating regeneration, infiltration of mononucleated cells, accumulated fibrosis and fat infiltration and variation in the size of myofiber cross-sectional areas compared with healthy tissue. All these aspects are reflected in the clinical evolution of the diseases, leading to progressive muscle degeneration. Besides pharmacological approaches, focused on symptoms treatment, different therapeutic perspectives have been explored during the last decade, such as gene therapy or stem cell transplantation; although impressive improvements, nowadays no curative treatment for any MD is available. Nevertheless, to develop new therapeutic strategies, such as to confirm the existing ones or better understand the mechanisms underlining these pathologies, the use of animal models is essential; analysing a whole organism is in fact necessary to identify the effects of a preclinical strategy in a group of systemic disorders, such as MDs.

However, starting from this evidence, it is important to select the most fitting animal model depending on the aims and the characteristics of the research protocol or preclinical study. In fact, different animal models have been developed for MDs in the last years, each one with specific PROs and CONs that must be evaluated, especially in a clinical prospective.

While simple animal models, such as *Drosophila* or zebrafish, are usually used in experiments involving the genetic origin of MDs, mammalian animal models are the most employed in preclinical studies for cell and gene therapies. In particular, compared to vertebrate models, *Drosophila* represents the most rapid and powerful model for genetic screening, thanks to its 10 days long life cycle, small size and large number of progeny; for these reasons, numerous genetic studies on DGC have been performed on *Drosophila*, especially to characterize the neuromuscular junctions (van der Plas et al. 2006; Bogdanik et al. 2008). In a similar way, also zebrafish is commonly used as a model for genetic studies, because of its easily manipulatable translucent embryos and the close resemblance of animal phenotypes to human condition (Lieschke and Currie 2007). An example is represented by the numerous models of DMD, whose application is useful both for the study of physiopathology (Berger et al. 2011; Bassett and Currie 2003; Guyon et al. 2009) and screening of antisense oligonucleotides and other small molecules (Muntoni and Wood 2011; Kawahara et al. 2011).

On the other hand, the mammalian, and in particular rodent, are the most used animal models in the preclinical protocols; even if the origin of the pathological strain is spontaneous, the possibility to genetically engineer the model in order to obtain useful characteristics is an important advantage for an experimental therapy application. In the last years a huge number of murine models for DMD have been developed, starting from the first mdx mutation identified in 1989 as a single base substitution in exon 23 (Sicinski et al. 1989); several allelic variants of the mdx–dystrophin-deficient mouse have been discovered or produced in laboratory (Araki et al. 1997), combining also other approaches to obtain more severe phenotypes, such as MyoD or utrophin lacking mdx mice (Megeney et al. 1999; Deconinck et al. 1997a; Grady et al. 1997; Durbeej and Campbell 2002). Although the lifespan of these animals is shortened, their phenotype is much more benign compared with the one shown by human patients; this aspect must be taken in consideration, especially if mice are used to evaluate the treatment efficacy. Murine models are commonly used also for the dyspherlinopathies: SJL and AJ mice are characterized by mutations in dysferlin gene associated with phenotypical features of progressive muscular dystrophy (Bittner et al. 1999a; Ho et al. 2004). As regards sarcoglycanopathies, BIO 14.6 hamster is the spontaneous animal model (Straub et al. 1998), while several mouse models have been developed with null mutation in each one of the sarcoglycan genes, thanks to homologous recombination technique in embryonic stem cells (Durbeej and Campbell 2002; Duclos et al. 1998; Hack et al. 1998; Araishi et al. 1999).

Although rodents represent important models for studying molecular pathogenesis and the effects of gene manipulation and cell transplantation, other mammalian models, such as cats and dogs, may provide more clinically relevant models of different MDs that will be important for testing new therapies and evaluating benefits in terms of function. As for feline models, the reported cases are few and limited to dystrophinopathies (Carpenter et al. 2003; Gaschen et al. 1992), where a clinical presentation is reported, dogs are well known and relatively common animal model for preclinical studies of different MDs. In fact, for DMD at least three different dog breeds have been developed: the most famous and used, Golden Retriever muscular dystrophy (GRMD) (Sharp et al. 1992; Valentine et al. 1988); the beagle Canine X-linked muscular dystrophy (CXMD1), characterized by a smaller size and a consequent easier maintenance (Shimatsu et al. 2003, 2005) and the Cavalier King Charles Spaniels with Muscular Dystrophy (CKCS-MD), recently obtained by a deletion in exon 50 (Walmsley et al. 2010). Also for sarcoglycanopathies, different canine models have been identified: a young Boston Terrier, Cocker Spaniel and Chihuahua (Walmsley et al. 2010).

In any case, before selecting a good animal model for a preclinical study, the specific PROs and CONs must be taken in consideration, as long as the consciousness that animal models are essential for the first step of in vivo trials; moreover, the results obtained from these models represent a starting point to hypothesize a correlated response in humans, but not always a precise reproduction of the evidence obtained.

In this chapter we will report the animal models for MDs nowadays available for preclinical studies, underlining the specific features of each one; furthermore, we will focus our attention on stem cell preclinical studies performed so far, showing the possible limitations of these approaches in the transition from a preclinical protocol to a clinical application.

2 Stem Cells for Therapy in Animal Models of MDs

Stem cells can be considered the most promising therapeutic approach for the treatment of MDs because of their potential to differentiate directly in the host tissue and give rise to new functional fibres. Despite gene therapy alone, whose limits are represented by the efficiency in delivering the correct gene without being recognized by immune system, the combination of this technique with an allogenic cell transplantation could represent a successful strategy; in fact, the delivery of patient's corrected cells bypasses the problem of an immune response and a rejection in a tissue already characterized by an inflammatory state. To reach this goal, different studies were performed in order to identify the stem cell type that guarantees the best performance in vivo in terms of engrafting and display of functional fibres in the host tissue. Stem cells tested in MDs animal models were derived from different tissues and can be classified into: bone marrow-derived stem cells (BMSC), adipose-derived stem cells (ASC), mesoangioblasts (MABs), satellite cells, blood- and muscle-derived CD133+ cells, side population (SP) cells, umbilical cord stem cells (UCS), Embryonic (ES) and induced pluripotent stem cells (Schuler et al. 1986).

There are advantages and disadvantages in the use of each of these cell types; some references to the specific in vivo studies are reported in Table 1.1.

BMSC, for example, are easily accessible, well characterized and systemically deliverable; on the other hand their participation in muscle regeneration is limited and below therapeutic levels. Experimental protocols were performed in different animal models of DMD (mdx mice and CXMD/GRMD dogs) and LGMD-2B (SJL mice). Also adipose tissue represents a painless and abundant source of mesenchymal stem cells; the multipotency of ASC was demonstrated by Zuk et al. (2001), proposing them as an alternative resource in cell therapy. In addition to the myogenic differentiation potential, the ASC display an immune-modulating behaviour

Table 1.1 List of specific cell-based studies performed in MD animal models

Cell type	Human pathology	Animal model	References
Bone marrow-derived stem cells (BMSC)	DMD	mdx mice	Ferrari et al. (1998), Bittner et al. (1999b), Ferrari and Mavilio (2002), Walsh et al. (2011), Leng et al. (2012)
		CXMD/GRMD dogs	Nitahara-Kasahara et al. (2012)
	LGMD-2B	SJL mice	Vieira et al. (2010)
Adipose-derived stem cells (ASC)	DMD	mdx mice	Rodriguez et al. (2005), Pinheiro et al. (2011)
		GRMD dogs	Vieira et al. (2012)
	LGMD-2B	SJL mice	Vieira et al. (2008)
Umbilical cord stem cells (UCS)	DMD	GRMD dogs	Vieira et al. (2010), Zucconi et al. (2011)
	LGMD-2B	SJL mice	Zucconi et al. (2011), Kong et al. (2004)
Mesoangioblasts	DMD	mdx mice	Berry et al. (2007)
		GRMD dogs	Sampaolesi et al. (2006)
	LGMD-2B	SJL/BIAJ mice	Diaz-Manera et al. (2010)
	LGMD-2D	α-SG null mice	Sampaolesi et al. (2003), Guttinger et al. (2006), Galvez et al. (2006)
Satellite cells	DMD	Nude mdx mice	Montarras et al. (2005) Collins et al. (2005), Boldrin et al. (2009)
		NSG-mdx^{4Cv}	Arpke et al. (2013)
Side population (SP)	DMD	NOD/SCID mdx mice	Gussoni et al. (1999), Motohashi et al. (2008)
	δ-sarcoglycanopathy	Sgcd$^{-/-}$ mice	Lapidos et al. (2004)
Blood and muscle-derived CD133+ cells	DMD	SCID mdx mice	Torrente et al. (2004), Benchaouir et al. (2007), Gavina et al. (2006)
	Dysferlinopathy	SCID/AJ mice	Under submission
Embryonic stem cells (ES) induced pluripotent stem cells (iPS)	DMD	mdx mice	Bhagavati and Xu (2005), Mizuno et al. (2010), Darabi et al. (2012)
	FSHD	FRG-1 mice	Darabi et al. (2009)

that could represent a further advantage in their application to cell transplantation. In this direction several studies were performed in order to evaluate the potential of ASC in treating MDs, in animal models such as mdx mice and GRMD dogs for DMD, and SJL mice for LGMD-2B.

As a third source for mesenchymal stem cells, several groups have addressed umbilical cord vein, meant as blood- or tissue-derived cells. Although in vitro potential of these cells to differentiate into myogenic lineage was demonstrated, in vivo experiments showed contradictory results. As UCS could reach host muscle both in mice (SJL) and dog (GRMD) models, they were unable to completely differentiate into muscle cells and express the lacking protein. Only in a study it was reported that after injection of UCS in SJL mice an apparent therapeutic benefit was shown, but probably due to their immunomodulatory effect.

A particular type of stem cells is represented by blood vessel-associated progenitors called mesoangioblasts (Dellavalle et al. 2007), characterized by the expression of pericyte markers when isolated from postnatal tissues; these cells are able to cross the vessel wall and for this reason they have been used in preclinical models of systemic cell therapy for MDs. Successful results were obtained in different mice models of MDs, such as mdx/utr$^{-/-}$ for DMD, α-SG null mice for LGMD-2D and SJL/BIAJ for dysferlinopathy, further than GRMD dogs. Moreover a Phase I/II clinical trial has been approved and is in progress.

As regards the cell type considered the "real" muscle stem cell, e.g. satellite cell, it was used for several in vivo experiments, demonstrating its ability to contribute to muscle regeneration in young and mature adult muscles. These cells, characterized by the expression of Pax7, once transplanted into an injured muscle, not only generate new muscle fibres, but also contribute to the satellite cells pool; moreover, single cell transplantation gives rise to an extensive proliferation, fibre generation and self-renewal in vivo. On the other hand, only a small number of cells can be derived from muscle and their regenerating efficiency is high only when transplanted fresh-isolated. Nevertheless, numerous studies were performed to evaluate the tissue regeneration and functional recovery in vivo after satellite cells injection, especially in DMD animal models, such as NSG-mdx^{4Cv} and nude mdx mice.

A particular case is represented by side population (SP) cells, a stem cell variant isolated by their ability to rapidly efflux the DNA-binding dye Hoechst 33342; SP cells have been so long isolated from different tissues, such as bone marrow, skin, mammary epithelium, muscle; these stem cells showed to harbour regenerative potential using varying modes of transplantation, including systemic delivery and direct organ injection (Montanaro et al. 2003; Welm et al. 2002; Hierlihy et al. 2002). Although each SP shares the ability exclude Hoechst 33342 and shows overlap of cell surface antigens such as Sca-1, it is unclear whether all tissue-derived SP cells share a common progenitor cell. The most used SP cells in animal models of MDs are the once derived from bone marrow: in fact, positive results were obtained in mdx and Sgcd$^{-/-}$ mice, while a particular subpopulation of SP cells from muscle was transplanted in NOD/SCID mdx mice, all models for DMD.

The last type of somatic stem cells used in animal models of MDs is represented by the muscle and blood-derived CD133+ cells, a particular cell population characterized by the expression of surface marker CD133 and able to differentiate into muscle, hematopoietic and endothelial cell types when exposed to specific cytokines (Torrente et al. 2004). In several articles it was demonstrated that blood-derived CD133+ cells are able to participate to muscle regeneration, generating dystrophin-positive fibres, once injected both intra-arterial and intra-muscle in scid/mdx mice, with best results in the last case (Benchaouir et al. 2007).

A different case is constituted by embryonic stem (ES) cells and induced pluripotent stem cells (Schuler et al. 1986); these cell types, in fact, possess different characteristics if compared with somatic stem cells: their differentiative potential is higher and wider, as their proliferation activity, as long as their teratoma possible induction once transplanted in vivo. This aspect represents the most important limit in the use of ES and iPS as a potential cell source in preclinical studies. Despite that, several studies were performed in order to evaluate the effects of iPS and ES in animal models of MDs: in mdx mice, models of DMD, both the cell populations displayed encouraging results, while in the FRG-1 mice, model of FSHD, the ES-based study demonstrated for the first time how the transplanted cells are capable of generating de novo myofibres with very little or no host contribution in such a dominant type of muscular dystrophy.

3 Animal Models in MDs

Unfortunately, despite many efforts, muscular dystrophies remain fatal. As an example, the identification and characterization of dystrophin gene led to the development of potential treatments for DMD (Bertoni 2008) but currently only the treatment with the steroid drug prednisone/prednisolone was proven to be effective on DMD patient (Backman and Henriksson 1995; Dubowitz et al. 2002). This way, myoblasts therapy obtained promising results in animal models but failed in clinical trials (Miller et al. 1997). To discover the genetic basis of unknown MDs and to develop possible treatments for such diseases, several animal models were established. In this chapter, we would like to give an overview of the applicability in preclinical studies of stem cell therapy in MDs using different animal models.

3.1 Zebrafish

In the last years, the zebrafish (*Danio rerio*) received tremendous attention owing to its advantages as a model system. Since the work of Lieschke and Currie (2007), zebrafish was used as model animal for several human diseases with the great advantage that it closely resembled the human condition. Comparative genomic studies highlighted sequence conservation of dystrophin and members of DGC in vertebrate and invertebrate species (Neuman et al. 2001; Roberts and Bobrow 1998; Gieseler et al. 2000). Moreover, Steffen and colleagues identified the orthologues of the majority of human MD disease in its genome (Steffen et al. 2007) while Catchen et al. showed the conservation among the species of the transcriptional network that activates the myogenesis (Catchen et al. 2011). In 2003, Bassett and collaborators identified a dystrophic mutant of zebrafish carrying a dystrophin null allele (Bassett and Currie 2003) while others were described later by Guyon et al. (2009) and Berger et al. (2011). It was demonstrated that as in human, zebrafish's dystrophin was firstly identified at the peripheral ends of the myofibres and after at the non-junctional sites (Chambers et al. 2001; Clerk et al. 1992). Surprisingly, this model was very similar to DMD patients as, at the onset of the pathology, mutations in dystrophin gene caused the formation of necrotic myofibres, leading to infiltration of mononucleated cells, fibrosis and inflammation (Berger and Currie 2012). Further studies with zebrafish models demonstrated a possible role of sodium channels in preventing the development of fibre detachment. Interestingly, other works showed how dystrophin-deficient zebrafish was a suitable tool to study potential therapies for DMD. Berger et al. assessed that 20–30 % of normal dystrophin transcript levels were needed to recover a severe dystrophic pathology (Berger et al. 2011) while Kawahara and Welch isolated different compounds that ameliorated pathological phenotype of zebrafish (Kawahara et al. 2011; Welch et al. 2007).

α- (and β-) dystroglycan are components of the DGC; when glycosylated, they bind to laminin-2 and anchor the subsarcolemmal cytoskeleton to the ECM. Godfrey et al. described a mutation in the dystroglycan (DAG-1) gene, that caused an out-of-control glycosylation of α-dystroglycan, leading to a reduced binding of α-dystroglycan to laminin-2 (Godfrey et al. 2011). Recently, Hara and colleagues identified a missense mutation in DAG1 associated with limb-girdle muscular dystrophy (Hara et al. 2011). For this disease, commonly known as dystroglycanopathy, different animal models exist. Unfortunately, Dag1 null mice were reported to be embryonic lethal due to the disruption of an extra-embryonic basement membrane (Williamson et al. 1997). However, zebrafish was able to develop normally so that this model was chosen to study the function of the orthologous *dag1* gene in muscle plasticity. Different works described the fundamental role of dystroglycan to maintain muscle integrity (Parsons et al. 2002) and isolated different mutations in dag1 gene, associated with various muscular defects (Gupta et al. 2011; Lin et al. 2011). Importantly, it was demonstrated that in dystroglycan mutants the dystrophin was absent confirming that these proteins were necessary for the assembly and stability of the DGC (Cote et al. 1999; Fritz et al. 1995). Furthermore, in the last years, zebrafish allowed to study several genes directly involved in the glycosylation of α-dystroglycan (Avsar-Ban et al. 2010; Thornhill et al. 2008) and one of those, the ISPD gene, was found to be strictly involved in the development of the congenital MD, the Walker–Warburg syndrome (Roscioli et al. 2012).

The actin cytoskeleton of the myofibres is connected to the ECM by means of integrin, dystroglicans and laminins. Different muscle forms of laminin exist, but the laminin-2 is the most important; mutations in this isoform cause severe congenital MD and merosin deficiency (Helbling-Leclerc et al. 1995). Interesting aspects of pathology were described by using zebrafish mutants: Hall and colleagues demonstrated that myofiber detachment was not due to defects in cell signalling but due to mechanical load of the fibres, that caused apoptosis of muscle fibres, and not necrosis as in DMD patients (Hall et al. 2007). Moreover, Maselli and Jacoby suggested that missense mutation in the lamb2 gene, newly identified in the zebrafish mutant *softy* (*sof*), could be a candidate gene for genetically uncharacterized human MDs (Jacoby et al. 2009; Maselli et al. 2009). Other types of limb-girdle MDs were studied in zebrafish mutants. Roostalu and colleagues knocked down the Dysf gene—that causes dysferlinopathy—and evidenced new steps of the complex sarcolemmal repair of muscle cells mediated by dysferlin, annexin6 and vesicles (Roostalu and Strahle 2012) while Telfer et al. studied the onset of Ullrich congenital MD (UCMD), that is caused by mutations in the collagen VI (Telfer et al. 2010).

3.2 Caenorabditis elegans

Since 1960s, the nematode *C. elegans* was utilized as a precious experimental tool to study development and genetic characteristics of several human diseases. It was largely demonstrated that the large degree of sequence similarity found in genes between *C. elegans* and human genomes resulted in a significant functional similarity of the encoded proteins (Chamberlain and Benian 2000). In particular, as muscle structure and muscle proteins are highly conserved among species, worm was particularly attractive to study muscle. This way, Bessou showed that the nematode was an excellent model to study DMD (Bessou et al. 1998). They identified in the genome of the nematode a gene related to the mammalian dystrophin and utrophin proteins (*dys-1*) whose mutants did not suffer for muscular abnormalities but only hyperactivity. Interestingly, *C. elegans* dystrophin-like transcribed by dys-1 gene shared the same motifs as the human protein. Moreover, it was found that *dys-1* mutants had increased cholinergic activity, opening the intriguing possibility that dystrophin had a role (direct or indirect) in cholinergic signalling (Bessou et al. 1998). This way, Giugia et al. found a significant reduced level of acetylcholine activity in dys-1 mutant (Giugia et al. 1999).

As in mice model of DMD the combined absence of dystrophin and MyoD caused severe myopathy compared with mdx mice (Megeney et al. 1996), Gieseler and collaborators generated a double mutant for dys-1 and hlh-1 that is the *C. elegans* homologue of MyoD. As expected, these mutants suffered for impaired locomotion and uncoordinated movements and their body wall muscles were severely disorganized (Gieseler et al. 2000). They also identified a gene called dyc-1 (dystrophin-like phenotype and Capon related) that reduced the phenotype of double mutants, ameliorating their locomotion. Finally, it was demonstrated that dyc-1 gene was quite similar to human nNOS-binding protein CAPON leading to the possibility that this kind of protein could be a possible therapeutic target for DMD treatment (Gieseler et al. 2000).

As described before for *Drosophila*, also *C. elegans* was successfully used as model organism to study other forms of muscular disease rather than DMD. The nuclear lamina is a protein structure providing structure and stability to the nucleus: it's formed by lamin and its associated proteins (LAPs) and plays a fundamental role in every nuclear function and in controlling transcription factors and histone deacetylases (Broers et al. 2006). Mutations in genes that constitute the nuclear lamina can cause several diseases, such as limb-girdle muscular dystrophy, *LMNA-related* congenital muscular dystrophy (L-CMD) and Emery–Dreifuss muscular dystrophy (EDMD). By means of studies with knockout and mutant strains of *C. elegans*, it was addressed that these pathologies were caused by defects in muscle itself or in the attachment of the muscle to hypodermis and by mis-regulation of neuromuscular junction (Bank et al. 2012).

3.3 *Murine Models*

3.3.1 DMD Animal Models

DMD is the most common form of muscular dystrophy: the discovery of a naturally occurring model, the **mdx mouse** had over 30 years. Mdx mice carried a point mutation in exon 23 of dystrophin, which caused the formation of a premature stop codon and the absence of the protein at the sarcolemma (Sicinski et al. 1989). They showed elevated plasmatic creatine kinase (Cpk) levels, fibres necrosis and calcium overload as observed in DMD patients. Muscle pathology could be evidenced from 2 weeks of age, with a peak in cyclic degeneration and regeneration of myofibres at 4 weeks (Partridge et al. 1989). Mdx myofibres were more prone to contraction-induced damage than wild type mice and showed reduced force per CSA but the observed muscle impairment was milder than in DMD patients. In fact mdx mice conserved quite a normal lifespan. Recently Farini et al. bred the *mdx* mouse with the *scid* immunosuppressed mouse, trying to elucidate the role of immune system in promoting muscle pathology (Farini et al. 2007). Mdx mice, similarly to DMD patients, are characterized by an important inflammatory infiltrates around myofibres. Chronic inflammation is sustained by T lymphocytes and macrophages which promotes collagen deposition and fibrosis. **Scid/mdx mice** lack B and T lymphocyte, so that fibrosis and TGF-β1 expression were dramatically reduced. Moreover scid/mdx, as it is an immuno-permissive system, allowed heterologous cell transplantation, representing a dystrophic model for preclinical studies. The huge problem associated to cell transplantation regards host immune system activation. Although scid/mdx and nude/mdx mice are available, they do not completely overcome this limitation. In fact both models maintain natural killer (NK) cells so that innate immune function is conserved. It was demonstrated that engraftment of human hematopoietic stem cells in scid mice could be improved by the additional deletion of the IL2 receptor common gamma chain (IL2Rg) (Traggiai et al. 2004). This founding was ascribed to lack of host NK cells, especially if donor cells expressed low level of self-MHC class I (Liao et al. 1991; Rideout et al. 2002). As a consequence, Arpke et al. generated the **NSG-mdx^{4Cv} mice** which combined (NOD/SCID; gamma-c) mutation with dystrophin deficiency. Resultant mice were almost identical to mdx mice concerning muscle histology and force assessment. Furthermore they demonstrated the feasibility of xenograft and allograft transplant (Arpke et al. 2013).

The applicability of an animal model correlates with its affinity to the human disease: for this reason, several mutant mice were developed on mdx mice background. Among them double-mutant **mdx/utrophin** mice were generated (Deconinck et al. 1997b). Utrophin, in wild type muscles, is found at the sarcolemma during development, while in mature muscles resides at neuromuscular junctions. In dystrophin-negative myofibres abnormal expression of utrophin was observed at sarcolemma. It has been suggested that mdx mice partially compensate the absence of dystrophin by up-regulating utrophin (a dystrophin homologue)

(Deconinck et al. 1997a). These mice exhibited a severe muscular dystrophy which better recapitulated human disease. As many DMD patients, they developed an important cardiomiopathy and premature death around 3 months of age. Importantly they also developed bone abnormality, scoliosis and kyphosis leading to breathing difficulty (Grady et al. 1997; Isaac et al. 2013).

Similarly, double-mutant **mdx/mTR mice** were generated, which lack both dystrophin and telomerase activity. In fact one hypothesis regarding DMD, postulates that the continuous cycle of fibres degeneration/regeneration leads to exhaustion of stem cell reserve, causing the replacement of lost tissue with fibrosis and adipose tissue. Telomeres are DNA repeats that protect chromosome ends during replication, shortening of telomeres leads to genome instability and cell senescence (Palm and de Lange 2008; Rodier et al. 2005; Sherr and DePinho 2000). It was reported that mdx mouse maintain greater murine satellite cells (MuSC) reserve than DMD patients, thanks to longer telomeres, while a 14-fold greater shortening of telomeres in DMD patients relative to healthy individuals was noticed (Decary et al. 2000). Mdx/mTR mice showed severe dystrophic phenotype which could be ameliorated by transplantation with murine wild type satellite cells (Sacco et al. 2010).

3.3.2 LMGD2B and Miyoshi Myopathy Animal Models

LMGD2B and Miyoshi myopathy arose from a different pathological mechanism than DMD or BMD: in fact the sarcolemma and the DSG complex are preserved. The specific deficit involves dysferlin, a protein fundamental in membrane repair process and vescicular trafficking. Four mice models of dysferlinopathies were described: two are naturally occurring models: SJL/J and AJ, and 2 Dysf–/– models were developed by Brown and Campbell, respectively (Ho et al. 2004; Han and Campbell 2007).

SJ/L mice developed a progressive muscular dystrophy caused by a splice site mutation in the C2 domains of dysferlin gene (Bittner et al. 1999a; Vafiadaki et al. 2001). These mice showed an active myopathy but also an important inflammatory myopathy and they were markedly susceptible to autoimmune diseases and viral infections, a condition that was not confirmed in human patients. These characteristics together with the presence of low level of dysferlin make the SJ/L mice an inappropriate model for human dysferlinopathies (Bittner et al. 1999a).

A progressive muscular dystrophy was observed in **A/J mice** during routinely screening in Jackson Laboratory. A genome-wide analysis revealed a retrotransposon (ETn) insertion within intron 4 of dysferlin gene, causing aberrant splicing and protein deficiency (Ho et al. 2004). Both AJ and SJL mice showed foci of degenerating and regenerating fibres with centrally located nuclei, variation in fibre size and fibre necrosis, together with fibrosis deposition. In A/J mice the onset of muscle pathology appeared at 5 months of age later than in SJL and in dysferlin-deficient Dysf–/– mice and the progression is less severe. Proximal muscle were more severely affected than the distal muscles, especially abdominal muscle were highly compromised, while, differently from mdx mice, the diaphragm were only mild affected.

A new animal model was generated by Bansal et al., by deleting the last three exons of dysferlin: the **Dysf–/– mice** (Bansal et al. 2003). These mice showed the same characteristic of other models of dysferlinopathies. Bansal developed a method to study in vivo dysferlin function. This membrane repair assay allowed in vivo testing of new drugs or cell transplantation efficacy, by dissection of single fibres from treated Dysf–/–mice, thanks to a laser-produced damage and analysis of membrane permeability with a fluorescent dye. In another dysferlin-deficient **Dysf–/– mouse**, the dysferlin gene was inactivated by replacing part of the highly conserved C2E domain with a neomycin gene cassette (Han and Campbell 2007). Recently the importance of immune system in muscular diseases has been deeply investigated. Dysferlinopathic patients were characterized by an important inflammatory infiltrates composed of especially monocytes/macrophages but also of T lymphocytes. Inflammatory cells were both attracted by chronic muscular damage and by specific signalling alteration related to dysferlin deficiency. In fact up-regulation of inflammosome and MHC class I cells and reduced phagocitic activity were reported. Bushby demonstrated that muscular regeneration after notexin injection is attenuated in a mouse model of dysferlinopathy caused by a reduced recruitment of neutrophils in the first phase of regeneration. He suggested that this reduction was consequent to a reduced release of chemotactic agents, secondary to the altered vescicular trafficking in dysferlin-null mice (Chiu et al. 2009).

Moreover Han and Campbell reported over-expression of complement factors in Dysf–/– muscle. To further determine the role of complement activation, they generated ***Dysferlin/C3* double-mutant mice** which lack dysferlin and the central component (C3). They observed an amelioration of muscle pathology in *dysferlin*-deficient mutant mice, but no similar effect in *mdx* mice, proving that complement activation was due primary to dysferlin deficiency, rather than to chronic inflammatory stimuli (Han and Campbell 2007). At the same time they developed ***Dysferlin/C5* double-mutant ko mice** but not beneficial effects were obtained, they suggested that complement damage was not only related to the formation of MAC (membrane attack complex) by C5a-9 but also by the inflammatory environment it provoked. Both double knockout mice did not show significant difference in force generation versus Dysf–/– mice.

Recently Farini and Sitzia showed that immuno-modulation was effective in ameliorating dystrophic feature in **Scid/AJ mice** (Farini et al. 2012). In these mice the absence of B and T lymphocytes reduced inflammatory infiltrates and promoted a switch M1 to M2 in macrophage phenotypes (anti-inflammatory). Consequently a reduction in complement activation and in IL-6 pro-inflammatory cytokine levels was observed. Furthermore scid/AJ mice showed lower levels of plasmatic CpK and increased tetanic force than A/J mice. ScidA/J mice as scid/mdx were also used in preclinical studies involving cell transplantation.

3.3.3 Facioscapulohumeral Muscular Dystrophy Models

The third most common form of MD is FSHD, which is caused by chromatin relaxation of the polymorphic D4Z4 macrosatellite repeat array on chromosome 4 in a repeat array contraction-dependent (FSHD1) or contraction-independent (FSHD2) fashion (van der Maarel et al. 2011). Differently from DMD, FSHD has benign course with heterogeneous phenotypes between patients, both regarding severity and age of onset. Typical affected muscles are facial, shoulder and upper arm muscles, often in asymmetric manner. Recently Gabellini et all generated **FRG-1 transgenic mouse** by over-expressing FRG-1, as this gene is located upstream D4Z4 macrosatellite repeat, which deletion involves a transcriptional silencer of gene located upstream (Gabellini et al. 2002). A role of FRG-1 in pre-mRNA processing was suggested after its identification as a component of purified spliceosomes (Rappsilber et al. 2002). These mice showed a progressive muscular dystrophy comparable to the human disease, which was not imputable to plasma membrane defects as demonstrated by the absence of Evan Blue uptake and normal creatine kinase serum levels and integrity of the DCG complex (Gabellini et al. 2006). In particular the severity of muscle impairment is proportional to the level of FRG-1 over-expression. FRG-1 transgenic mouse showed loss of specific force (Po/CSA) of single muscle fibres and atrophy, which involved particularly fast fibres (D'Antona et al. 2007).

The dystrophin-associated proteins include the dystroglycans, sarcoglycans (Fassati et al. 1997) and syntrophin subcomplexes and with dystrophin and laminins represent the most important membrane cytoskeleton system. Among them, sarcoglycans are glycosylated transmembrane proteins, α-, β-, γ- and δ-, that function synergistically to form a tetrameric complex that is the core dystrophin-associated proteins. Unfortunately, defects in genes encoding for these proteins cause a severe form of muscular dystrophy, known as limb-girdle muscular dystrophy (LGMD-2) D, E, F and C, according to mutations in the α-, β-, γ- and δ-SG, respectively. Affected patients suffer for progressive muscle weakness that leads to death in the second or third decade of life (Angelini et al. 1999). The first animal model for LGMD-2, the so-called inbred Bio 14.6 hamster, was identified in 1962, showing cardiac and muscular problems, such as necrosis and variable size of muscle fibres (Straub et al. 1998; Holt et al. 1998; Li et al. 1999). Only recently it was demonstrated that its dystrophic phenotype was due to a loss-of-function mutation of the δ-SG gene (Sakamoto et al. 1997). Different mouse strains carrying mutations in the SGs were successfully used as model organisms to better elucidate the role of each component in the pathogenesis of muscular dystrophy (Hack et al. 2000). This way, it was described that SGs not only regulated structural integrity of the myofiber plasma membrane but also exerted an important role in vascular smooth muscle function, in inducing ischemic injury in skeletal muscle and in transmitting force in the transverse direction to the long axis of muscle (Watchko et al. 2002). One of the most utilized models is the α-SG-deficient null mutant transgenic mouse; Duclos and colleagues demonstrated that the absence of α-SG protein caused muscle hypertrophy and muscle contractile defects that lead to a progressive muscular dystrophy (Duclos et al. 1998).

3.4 Canine Models

Spontaneous mutations in the dystrophin gene causing X-linked muscular dystrophy similar to DMD were identified the Golden Retriever (Cooper et al. 1988), the Rottweiler (Winand et al. 1994) and the German short-haired pointer (Schatzberg et al. 1999). More recently, Shimatsu and colleagues artificially inseminated a female beagle with spermatozoa from dystrophic Golden Retriever, obtaining the CXMD (Shimatsu et al. 2003). Unfortunately, dogs are not easy to maintain and cannot be genetically manipulable; however, adult dogs have a body mass that is similar to DMD patients and more importantly their muscular pathogenesis resemble human condition (Cooper et al. 1988; Howell et al. 1997). As in humans, Valentine et al. found that young **GRMD** had high concentrations of serum creatine kinase and died for respiratory problem or heart failure, especially in the first 2 weeks after birth (Valentine et al. 1988). In particular, they described various lesions in muscles of GRMD consisting in hyperacidophilic cytoplasm of muscular fibres, enormous accumulation of calcium, signs of muscular regeneration and necrosis of clusters of fibres. As suggested by Hoffman and Gorospe, these kinds of lesions were known as Phase I and were caused by the absence of the dystrophin in every animal species suffering for mutations in this protein. Interestingly, it was noted that muscles of newborn GRMD were affected by maximum two of Phase I lesions while in 2-month-old affected dogs the pathological phenotype was dramatically worsened, with the presence of widespread endomysial fibrosis and myofiber atrophy (Valentine et al. 1988). These lesions, known as Phase II, were only seen in DMD muscles rather than the diaphragm of mdx mice (Gorospe and Hoffman 1992). This way, GRMD dogs were largely used in the last years for preclinical and clinical studies.

As published by Shimatsu, the **CXMDJ** suffered for elevated serum CK level that caused the death of 1–3 of littermates. The first muscle that showed dystrophic phenotype was the diaphragm and only after 3 months of age atrophy and weakness appeared evident in limb muscles. Since 4 months, the clinical course of these dogs rapidly worsen due to macroglossia and joint contracture; however, after 10 months of age, the progression of the disease was retarded (Shimatsu et al. 2005). Similarly to GRMD, these dogs showed fibrosis of the left ventricular wall that was accompanied by degeneration of Purkinje fibres (Valentine et al. 1988; Perloff et al. 1967).

4 Preclinical Studies

Different subpopulations of human stem cells were proven for their efficacy in muscular differentiation: they were injected into animal models for MDs and it was observed whether they ameliorated their dystrophic phenotype. The aim of these studies was to select myogenic stem cells and translate them into clinical trials on human subjects.

4.1 Preclinical Studies for DMD

Since 1989, the group of Partridge showed that myoblast injection into mdx mice ameliorated dystrophin expression (Partridge et al. 1989). According to these exciting results, several clinical trials on human subjects arose but unfortunately failed, as purified myoblasts were not able to proliferate and fuse in host muscular environment (Skuk et al. 2000, 2002). Similarly, Montarras et al. obtained promising results by transplanting into dystrophic mice a pure population of murine-derived satellite cells: they demonstrated that these cells not only formed a pool of satellite cells expressing Pax7 and Pax3 but also possessed an incredible ability of muscular differentiation (Montarras et al. 2005). However, as it was demonstrated for myoblasts, the growth of freshly isolated satellite cells in vitro significantly reduced their in vivo myogenic potential. More recently, Cerletti and colleagues described a newly isolated subpopulation of muscle satellite cells that transplanted into mdx mice not only allowed the re-expression of dystrophin but also regenerated mature muscle fibres and ameliorated host dystrophic phenotype, reducing muscle inflammation and fibrosis (Cerletti et al. 2008). Unfortunately, in vitro expansion of these cells reduced drastically their myogenic potential so that they were not considered for further clinical utilization. Interesting works from several groups identified a population of multipotent muscle-derived stem cell (MDSC), residing in the skeletal muscle (Sarig et al. 2006; Tamaki et al. 2007). It was demonstrated that these cells were different from satellite cells and contained different sub-populations able to efficiently differentiate into skeletal muscle. This way, our group demonstrated that muscle-derived Sca-1+ CD34+ stem cells interacted to endothelium in mdx muscles and participated into muscle regeneration (Torrente et al. 2001) while Qu-Petersen showed that another subpopulation, expressing also c-kit and CD45 could be efficiently transplanted into murine dystrophic muscle (Qu-Petersen et al. 2002). Starting from their demonstration that, under specific stimuli, adult human synovial membrane-derived mesenchymal stem cells (hSM-MSCs) differentiated into muscle (De Bari et al. 2001), de Bari and colleagues assessed that the injection of these cells restored the expression of dystrophin into mdx mice (De Bari et al. 2003). Following the characterization of the immunodeficient dystrophic scid/mdx (Farini et al. 2007), this model was largely used in preclinical studies for muscular diseases. As published by Benchaouir et al. in 2007, two dystrophic CD133+ cell populations, one from blood and one from skeletal muscle, were transduced with a lentivirus carrying a construct designed to skip exon 51 (Benchaouir et al. 2007). As they expressed a short but still functional isoform of dystrophin, engineered CD133+ stem cells were injected into scid/mdx mice. They differentiated into endothelial and muscular cells, took part into muscular regeneration and ameliorated the muscular force in treated animals (Benchaouir et al. 2007).

Following the pioneering work of Partridge regarding the fusion of donor myoblasts in the mdx mouse (Partridge et al. 1989), Prattis e Nakagaki isolated and fully characterized the myoblasts of DMD canine model but never obtained appreciable results (Nakagaki et al. 1994; Prattis et al. 1993). However, Kornegay and colleagues demonstrated that following autologous transplantation myoblasts differentiated in muscle (Kornegay et al. 2012). More recently, Pichavant et al. transfected dystrophic dog muscle precursor cells with a lentivirus carrying canine microdystrophin cDNA and transplanted them into normal dogs (Pichavant et al. 2010). Interestingly Parker and co-workers showed that myoblast transplantation efficiency was higher if the dogs were previously injected with hematopoietic cell, to induce tolerance (Parker et al. 2008). As Dell'Agnola showed that bone marrow stem cell transplantation did not allow dystrophin re-expression in GRMD dogs (Dell'Agnola et al. 2004), Cossu and colleagues transplanted some dystrophic dogs using MABs transduced with a lentiviral vector expressing human micro-dystrophin (Sampaolesi et al. 2006). Treated dogs expressed dystrophin and ameliorated their dystrophic phenotype, improving muscle function and mobility; furthermore, the large majority of muscle fibres showed normal force contraction and absence of defective mobility (Sampaolesi et al. 2006). Recently, Rouger et al. described how the intra-arterial injection of allogenic muscle stem cells into GRMDs contributed to myofiber regeneration, satellite cell replenishment and dystrophin expression. Interestingly, following cellular transplantation, dogs showed long-term dystrophin expression and amelioration of muscle damage course, due to increased regeneration activity (Rouger et al. 2011). The group of Mayana Zatz transplanted GRMDs with different types of stem cells. After injection of human immature dental pulp stem cells, they appreciated low expression of dystrophin (Kerkis et al. 2008) while dystrophin was not identified following intra-arterial injection of umbilical cord mesenchymal stromal cells (Zucconi et al. 2011). However, in this study, it was demonstrated that systemic injections of umbilical cord MSCs was a safe technique; moreover, even if they did not participate directly in formation of muscle, these cells allowed the expression of several molecules that activated the proliferation of stem cells resident in host muscles. As the MABs are similar to human ADSCs regarding size, proliferation and differentiation capacity, they decided to implant these cells in non-immunosuppressed GRMDs. They found that ADSCs reached several muscles and allowed the expression of dystrophin; moreover, they suggested that, as a therapeutically point of view, at least two systemic injections were required for the maintenance of injected cells in the host muscles (Vieira et al. 2012). Nitahara-Kasahara isolated CD271+ MSCs and transduced them with MyoD to induce myogenic differentiation and to allow the formation of myotubes. Combining these findings with the anti-inflammatory role exerted by MSCs, they injected these cells into CXMD and described the formation of clusters of muscular fibres and re-expression of dystrophin (Nitahara-Kasahara et al. 2012). Bhagavati and co-workers demonstrated that ES could have important effects in favouring muscular development as after injection into mdx mice allowed the formation of vascularized skeletal muscle (Bhagavati and Xu 2005). Mizuno et al. demonstrated that iPS cells were quite similar to skeletal muscle stem cells so that after transplantation in murine models of DMD they differentiated into myofibres repairing damaged muscle tissues and replenishing the host satellite cells compartment (Mizuno et al. 2010). More recently, Darabi et al. showed that Pax7-expressing EC/iPS cells regenerated muscle in vivo, improved muscular force of transplanted mdx mice and formed functional satellite cells (Darabi et al. 2012).

4.2 Preclinical Studies for Dysferlinopathies

Thanks to lessons learned from cell transplantation in mdx mice, Guerin et al. injected normal myoblast in SJL mice and demonstrated in vivo rescue of dysferlin expression (Leriche-Guerin et al. 2002). However the number of dysferlin-positive myofibres was lower than the number expected following normal myoblast transplant in mdx mice (Partridge et al. 1989; Kinoshita et al. 1994; Morgan et al. 1990; Vilquin et al. 1995). This result was ascribed to SJL inflammatory background rather than to immunosuppressive treatment, but the need of alternative strategies or different animal models became evident. To overcome immunosuppressive limitation, Kong et al. injected human umbilical cord blood (HUCB) cells intravenous in SJL mice, as these cells were known to be poor immunoreactive (Gluckman et al. 1997; Rocha et al. 2000). They demonstrated the presence of human nuclei within host fibres, but they observed very low frequency of myogenic differentiation (Kong et al. 2004). A step forward was moved in 2008 when the group of Zatz demonstrated the feasibility of human adipose-derived stromal cells transplant in SJL mice, without need of immunosuppression (Vieira et al. 2008). Furthermore they obtained a significant functional improvement in the injected animals although the body-wide distribution of hASCs was widespread.

Promising results obtained using MABs in the field of DMD raised hope for the treatment of dysferlinopathies too. In fact Díaz-Manera demonstrated that adult murine MABs isolated from normal mice, both after intramuscular or intra-arterial administration to SCID/BlAJ mice could reach and survive in the host muscle. They demonstrated that partially restored dysferlin expression allowed normal repairing ability of the membrane after laser-induced lesions (Diaz-Manera et al. 2010). Some tentative were made with mini-gene replacement in animal model of dysferlinopathies as dysferlin messenger RNA is far above the natural packaging size of rAAV: they demonstrated that mini-dysferlin could maintain its functionality (Krahn et al. 2010). Alternative strategy combined the injection of two independent AAV vectors which together contained the full length sequence of dysferlin cDNA. After intramuscular injection of both AAV vectors into muscle of A/J mice, they obtained the expression of full-length dysferlin, the restoration of the membrane repair capacity of myofibres and increased locomotor activity of the mice (Lostal et al. 2010). However recently, controversy arose about the predictive value of membrane repair assay in myofibres isolated from treated dysferlin-deficient mice. In particular Lostal et al. demonstrated that different approaches that rescue membrane repair after laser damage assay (as described by Bansal) failed to alleviate the dysferlin-deficient pathology in mice (Lostal et al. 2010).

4.3 Preclinical Studies for Sarcoglicanopathies

Differently from DMD no experiment involving myoblast transplantation was conducted in sarcoglycanopathies. However sarcoglycan genes are smaller than dystrophin, so that their delivery with recombinant adenovirus or AAV vector is possible. This approach was firstly tempted in BIO14.6 cardiomyopathic hamster, by injection of AAV vector, which carried δ-SG (Holt et al. 1998; Li et al. 1999; Greelish et al. 1999). Allamand et al. demonstrated the rescue of the expression of a-SG at the sarcolemma and the long-term reconstitution of DSG complex after α-sg AAV vector intramuscular injection in *Sgca*-null mice (Allamand et al. 2000). Although these promising results, the therapeutic effect was obtained only when the vector injection is performed before the onset of the dystrophic pathology as inflammatory environment and fibrosis disturbed AAV vectors transduction of myofibres (Allamand et al. 2000; Fougerousse et al. 2007). Different experiments with cell therapy were attempted, using either bone marrow, vessel associated, or UCSs giving disappointing results (Lapidos et al. 2004; Salah-Mohellibi et al. 2006). These results were probably due to inefficient delivery of donor cells to dystrophic muscles. These difficulties were overcome by Sampaolesi et al. thanks to the intra-arterial injection of MABs from dystrophic mice and tranduced with lentiviral vector expressing α-SG in α-Sg null mice. They observed α-sarcoglycan expression, restoration of DGC complex and increased myofibres strength (Sampaolesi et al. 2003). A new intriguing method to treat sarcoglycanopathies arose from experiments of Bartoli's group. They demonstrated that a particular mutation (R77C in a-sarcoglycan gene) which causes protein retention could be bypassed by blocking the protein quality control system. They demonstrated that in vivo delivery of kifunensine (Mannosidase I inhibitor) in living mice could restore protein localization at the sarcolemma without evidence of toxicity (Bartoli et al. 2008).

4.4 Preclinical Studies for FSHD

Unfortunately, pathological mechanisms, underling FSHD development, were only recently elucidated. In particular Gabellini et al. was able to reproduce FSHD in mice model only 10 years ago (Gabellini et al. 2002). For this reason, first studies with cell transplantation in FRG-1 mice started, when limits of myoblasts and other cell type transplantation were well known. Darabi et al. demonstrated that Pax3-induced ES-derived myogenic progenitors were able to engraft into dystrophic FRG-1 mice and produced functional improvement (Darabi et al. 2012). Thanks to these mice the role of over-expression of FRG-1, and DUX4 genes appeared (Wallace et al. 2012). Recently strategies for *FRG-1* knockdown have been developed. In particular through RNA interference (Ferrari et al. 1998) expression cassette Bortolanza et al. showed restoration of muscle function in FRG-1 mice. They demonstrated that intravenous delivery of AAV6-mediated *FRG-1* RNAi was safe and permit a long-term knockdown of FRG-1 mRNA, leading to significant amelioration of muscle function (Bortolanza et al. 2011).

5 Conclusions

Many pathological mechanism theories are developed, thanks to animal models suitable for experimental or genetic manipulation. Several aspects have to be evaluated if researchers want to use an animal model for preclinical studies (Willmann et al. 2009). The following criteria are fundamental: (1) genetic basis of disease must be the same both for humans and animal models; (2) the animal models should have the same phenotypic characteristics of humans; (3) disease progression must be well characterized in order to obtain precise, comparable and reproducible outcome analysis. In case of DMD, suitable animal models have to present muscle weakness, respiratory insufficiency, cardiomyopathy, centralized nuclei myofibres and fibre size variations.

Murine models provide important findings for the basic study of pathogenesis and development of therapies, but generally the clinical phenotype is not completely comparable to that human patient. Mdx animal models have many advantages, firstly the large number of papers published on these mice guaranteed a good understanding of muscle, cardiac and respiratory pathology (Partridge et al. 1989); later the reproducibility of the phenotype and the commercial availability are important advantages. Finally, this animal model has been used to demonstrate efficacy of several potential treatment strategies including small molecule pharmaceuticals as well as gene correction (Goyenvalle et al. 2004). Also scid/mdx animal model, as it is an immuno-permissive system, allowed heterologous cell transplantation, representing a dystrophic model for preclinical studies (Torrente et al. 2004; Benchaouir et al. 2007). GRMD dog displays an inherited degenerative disorder genetically homologous to DMD patients but it shares with DMD patient progressive clinical signs and severe myopathy with contractures and premature death (Thibaud et al. 2007). In human and canine disorders, skeletal muscle shows early fibre necrosis and regeneration together with endomysial and perimysial connective tissue proliferation and severe cardiac defects. The body size and genetic background of canines are more similar to human beings than the murine models. Furthermore, histopathological changes in skeletal muscle appear at preclinical stages in both DMD patient and GRMD dog. GRMD dog represents, thus, the most relevant animal model for DMD, particularly in regard to potential therapeutic approaches and so investigations of stem cells potential in this model are absolutely necessary. One important problem in using animal models is the variability. Moreover, special care must be taken in choosing the control animals and control experiments must be selected with special attention.

Actually, none of the animal models much used totally mimics all the characteristics of the human diseased; the larger animal models, such as dogs, show high variability in disease severity between breeds and among littermates although they all carry the same genetic mutation or non-human primates do not exist for preclinical studies of muscular dystrophy.

In general, methods finalized to increase dystrophin, utrophin, dysferlin and other proteins involved into the onset of MDs are showing great promise in animal models and are beginning to be tested in clinic studies (Goyenvalle et al. 2011). Nevertheless, the optimization of all the possible approaches to treat MDs will be needed before a declaration of success and not before a delicate balancing act between preclinical animal models and human clinical trials.

Up against these considerations, it is very important to increase the number of Phase I clinical trials in human patients when basic research shows the safety of promising therapeutic approaches in animal models. We hope that the achievement of positive results in cycles of animal/human studies will permit a gradual implementation of new therapies that will ameliorate lifespan and life quality of patients affected from different forms of muscular dystrophies.

Acknowledgments This work was supported by Associazione La Nostra Famiglia Fondo DMD Gli Amici di Emanuele, Associazione Amici del Centro Dino Ferrari, EU's 7th Framework programme Optistem 223098 and Ystem s.r.l.

References

Allamand V, Donahue KM, Straub V, Davisson RL, Davidson BL, Campbell KP (2000) Early adenovirus-mediated gene transfer effectively prevents muscular dystrophy in alpha-sarcoglycan-deficient mice. Gene Ther 7:1385–1391

Angelini C, Fanin M, Freda MP, Duggan DJ, Siciliano G, Hoffman EP (1999) The clinical spectrum of sarcoglycanopathies. Neurology 52:176–179

Araishi K et al (1999) Loss of the sarcoglycan complex and sarcospan leads to muscular dystrophy in beta-sarcoglycan-deficient mice. Hum Mol Genet 8:1589–1598

Araki E et al (1997) Targeted disruption of exon 52 in the mouse dystrophin gene induced muscle degeneration similar to that observed in Duchenne muscular dystrophy. Biochem Biophys Res Commun 238:492–497

Arpke RW et al (2013) A new immuno-dystrophin-deficient model, the NSG-Mdx mouse, provides evidence for functional improvement following allogeneic satellite cell transplantation. Stem Cells 31:1611–1620

Avsar-Ban E et al (2010) Protein O-mannosylation is necessary for normal embryonic development in zebrafish. Glycobiology 20:1089–1102

Backman E, Henriksson KG (1995) Low-dose prednisolone treatment in Duchenne and Becker muscular dystrophy. Neuromuscul Disord 5:233–241

Bank EM, Ben-Harush K, Feinstein N, Medalia O, Gruenbaum Y (2012) Structural and physiological phenotypes of disease-linked lamin mutations in C. elegans. J Struct Biol 177:106–112

Bansal D et al (2003) Defective membrane repair in dysferlin-deficient muscular dystrophy. Nature 423:168–172

Bartoli M et al (2008) Mannosidase I inhibition rescues the human alpha-sarcoglycan R77C recurrent mutation. Hum Mol Genet 17:1214–1221

Bassett DI, Currie PD (2003) The zebrafish as a model for muscular dystrophy and congenital myopathy. Hum Mol Genet 12(Spec No 2):R265–R270

Benchaouir R et al (2007) Restoration of human dystrophin following transplantation of exon-skipping-engineered DMD patient stem cells into dystrophic mice. Cell Stem Cell 1:646–657

Berger J, Currie PD (2012) Zebrafish models flex their muscles to shed light on muscular dystrophies. Dis Model Mech 5:726–732

Berger J, Berger S, Hall TE, Lieschke GJ, Currie PD (2011) Dystrophin-deficient zebrafish feature aspects of the Duchenne muscular dystrophy pathology. Neuromuscul Disord 20:826–832

Berry SE, Liu J, Chaney EJ, Kaufman SJ (2007) Multipotential mesoangioblast stem cell therapy in the mdx/utrn–/– mouse model for Duchenne muscular dystrophy. Regen Med 2:275–288
Bertoni C (2008) Clinical approaches in the treatment of Duchenne muscular dystrophy (DMD) using oligonucleotides. Front Biosci 13:517–527
Bessou C, Giugia JB, Franks CJ, Holden-Dye L, Segalat L (1998) Mutations in the Caenorhabditis elegans dystrophin-like gene dys-1 lead to hyperactivity and suggest a link with cholinergic transmission. Neurogenetics 2:61–72
Bhagavati S, Xu W (2005) Generation of skeletal muscle from transplanted embryonic stem cells in dystrophic mice. Biochem Biophys Res Commun 333:644–649
Bittner RE et al (1999a) Dysferlin deletion in SJL mice (SJL-Dysf) defines a natural model for limb girdle muscular dystrophy 2B. Nat Genet 23:141–142
Bittner RE et al (1999b) Recruitment of bone-marrow-derived cells by skeletal and cardiac muscle in adult dystrophic mdx mice. Anat Embryol (Berl) 199:391–396
Bogdanik L et al (2008) Muscle dystroglycan organizes the postsynapse and regulates presynaptic neurotransmitter release at the Drosophila neuromuscular junction. PLoS One 3:e2084
Boldrin L, Zammit PS, Muntoni F, Morgan JE (2009) Mature adult dystrophic mouse muscle environment does not impede efficient engrafted satellite cell regeneration and self-renewal. Stem Cells 27:2478–2487
Bonnemann CG et al (1996) Genomic screening for beta-sarcoglycan gene mutations: missense mutations may cause severe limb-girdle muscular dystrophy type 2E (LGMD 2E). Hum Mol Genet 5:1953–1961
Bortolanza S et al (2011) AAV6-mediated systemic shRNA delivery reverses disease in a mouse model of facioscapulohumeral muscular dystrophy. Mol Ther 19:2055–2064
Broers JL, Ramaekers FC, Bonne G, Yaou RB, Hutchison CJ (2006) Nuclear lamins: laminopathies and their role in premature ageing. Physiol Rev 86:967–1008
Brown SC, Muntoni F, Sewry CA (2001) Non-sarcolemmal muscular dystrophies. Brain Pathol 11:193–205
Carpenter MK, Rosler E, Rao MS (2003) Characterization and differentiation of human embryonic stem cells. Cloning Stem Cells 5:79–88
Catchen JM, Braasch I, Postlethwait JH (2011) Conserved synteny and the zebrafish genome. Methods Cell Biol 104:259–285
Cerletti M et al (2008) Highly efficient, functional engraftment of skeletal muscle stem cells in dystrophic muscles. Cell 134:37–47
Chamberlain JS, Benian GM (2000) Muscular dystrophy: the worm turns to genetic disease. Curr Biol 10:R795–R797
Chambers SP et al (2001) Dystrophin in adult zebrafish muscle. Biochem Biophys Res Commun 286:478–483
Chiu YH et al (2009) Attenuated muscle regeneration is a key factor in dysferlin-deficient muscular dystrophy. Hum Mol Genet 18:1976–1989
Clerk A, Strong PN, Sewry CA (1992) Characterisation of dystrophin during development of human skeletal muscle. Development 114:395–402
Collins CA et al (2005) Stem cell function, self-renewal, and behavioral heterogeneity of cells from the adult muscle satellite cell niche. Cell 122:289–301
Cooper BJ et al (1988) The homologue of the Duchenne locus is defective in X-linked muscular dystrophy of dogs. Nature 334:154–156
Cote PD, Moukhles H, Lindenbaum M, Carbonetto S (1999) Chimaeric mice deficient in dystroglycans develop muscular dystrophy and have disrupted myoneural synapses. Nat Genet 23:338–342
D'Antona G, Brocca L, Pansarasa O, Rinaldi C, Tupler R, Bottinelli R (2007) Structural and functional alterations of muscle fibres in the novel mouse model of facioscapulohumeral muscular dystrophy. J Physiol 584:997–1009
Darabi R, Baik J, Clee M, Kyba M, Tupler R, Perlingeiro RC (2009) Engraftment of embryonic stem cell-derived myogenic progenitors in a dominant model of muscular dystrophy. Exp Neurol 220:212–216

Darabi R et al (2012) Human ES- and iPS-derived myogenic progenitors restore DYSTROPHIN and improve contractility upon transplantation in dystrophic mice. Cell Stem Cell 10:610–619
De Bari C, Dell'Accio F, Tylzanowski P, Luyten FP (2001) Multipotent mesenchymal stem cells from adult human synovial membrane. Arthritis Rheum 44:1928–1942
De Bari C, Dell'Accio F, Vandenabeele F, Vermeesch JR, Raymackers JM, Luyten FP (2003) Skeletal muscle repair by adult human mesenchymal stem cells from synovial membrane. J Cell Biol 160:909–918
Decary S, Hamida CB, Mouly V, Barbet JP, Hentati F, Butler-Browne GS (2000) Shorter telomeres in dystrophic muscle consistent with extensive regeneration in young children. Neuromuscul Disord 10:113–120
Deconinck AE et al (1997a) Utrophin-dystrophin-deficient mice as a model for Duchenne muscular dystrophy. Cell 90:717–727
Deconinck N et al (1997b) Expression of truncated utrophin leads to major functional improvements in dystrophin-deficient muscles of mice. Nat Med 3:1216–1221
Dell'Agnola C et al (2004) Hematopoietic stem cell transplantation does not restore dystrophin expression in Duchenne muscular dystrophy dogs. Blood 104:4311–4318
Dellavalle A et al (2007) Pericytes of human skeletal muscle are myogenic precursors distinct from satellite cells. Nat Cell Biol 9:255–267
Diaz-Manera J et al (2010) Partial dysferlin reconstitution by adult murine mesoangioblasts is sufficient for full functional recovery in a murine model of dysferlinopathy. Cell Death Dis 1:e61
Dubowitz V, Kinali M, Main M, Mercuri E, Muntoni F (2002) Remission of clinical signs in early Duchenne muscular dystrophy on intermittent low-dosage prednisolone therapy. Eur J Paediatr Neurol 6:153–159
Duclos F et al (1998) Progressive muscular dystrophy in alpha-sarcoglycan-deficient mice. J Cell Biol 142:1461–1471
Durbeej M, Campbell KP (2002) Muscular dystrophies involving the dystrophin-glycoprotein complex: an overview of current mouse models. Curr Opin Genet Dev 12:349–361
Farini A et al (2007) T and B lymphocyte depletion has a marked effect on the fibrosis of dystrophic skeletal muscles in the scid/mdx mouse. J Pathol 213:229–238
Farini A et al (2012) Absence of T and B lymphocytes modulates dystrophic features in dysferlin deficient animal model. Exp Cell Res 318:1160–1174
Fassati A et al (1997) Genetic correction of dystrophin deficiency and skeletal muscle remodeling in adult MDX mouse via transplantation of retroviral producer cells. J Clin Invest 100:620–628
Ferrari G, Mavilio F (2002) Myogenic stem cells from the bone marrow: a therapeutic alternative for muscular dystrophy? Neuromuscul Disord 12(suppl 1):S7–S10
Ferrari G et al (1998) Muscle regeneration by bone marrow-derived myogenic progenitors. Science 279:1528–1530
Fougerousse F et al (2007) Phenotypic correction of alpha-sarcoglycan deficiency by intra-arterial injection of a muscle-specific serotype 1 rAAV vector. Mol Ther 15:53–61
Fritz JD, Danko I, Roberds SL, Campbell KP, Latendresse JS, Wolff JA (1995) Expression of deletion-containing dystrophins in mdx muscle: implications for gene therapy and dystrophin function. Pediatr Res 37:693–700
Gabellini D, Green MR, Tupler R (2002) Inappropriate gene activation in FSHD: a repressor complex binds a chromosomal repeat deleted in dystrophic muscle. Cell 110:339–348
Gabellini D et al (2006) Facioscapulohumeral muscular dystrophy in mice overexpressing FRG1. Nature 439:973–977
Galvez BG et al (2006) Complete repair of dystrophic skeletal muscle by mesoangioblasts with enhanced migration ability. J Cell Biol 174:231–243
Gaschen FP et al (1992) Dystrophin deficiency causes lethal muscle hypertrophy in cats. J Neurol Sci 110:149–159
Gavina M, Belicchi M, Camirand G (2006) VCAM-1 expression on dystrophic muscle vessels has a critical role in the recruitment of human blood-derived CD133+ stem cells after intra-arterial transplantation. Blood 108:2857–2866

Gieseler K, Grisoni K, Segalat L (2000) Genetic suppression of phenotypes arising from mutations in dystrophin-related genes in Caenorhabditis elegans. Curr Biol 10:1092–1097

Giugia J, Gieseler K, Arpagaus M, Segalat L (1999) Mutations in the dystrophin-like dys-1 gene of Caenorhabditis elegans result in reduced acetylcholinesterase activity. FEBS Lett 463:270–272

Gluckman E et al (1997) Outcome of cord-blood transplantation from related and unrelated donors. Eurocord Transplant Group and the European Blood and Marrow Transplantation Group. N Engl J Med 337:373–381

Godfrey C, Foley AR, Clement E, Muntoni F (2011) Dystroglycanopathies: coming into focus. Curr Opin Genet Dev 21:278–285

Gorospe JR, Hoffman EP (1992) Duchenne muscular dystrophy. Curr Opin Rheumatol 4:794–800

Goyenvalle A et al (2004) Rescue of dystrophic muscle through U7 snRNA-mediated exon skipping. Science 306:1796–1799

Goyenvalle A, Seto JT, Davies KE, Chamberlain J (2011) Therapeutic approaches to muscular dystrophy. Hum Mol Genet 20:R69–R78

Grady RM, Teng H, Nichol MC, Cunningham JC, Wilkinson RS, Sanes JR (1997) Skeletal and cardiac myopathies in mice lacking utrophin and dystrophin: a model for Duchenne muscular dystrophy. Cell 90:729–738

Greelish JP et al (1999) Stable restoration of the sarcoglycan complex in dystrophic muscle perfused with histamine and a recombinant adeno-associated viral vector. Nat Med 5:439–443

Gupta V et al (2011) The zebrafish dag1 mutant: a novel genetic model for dystroglycanopathies. Hum Mol Genet 20:1712–1725

Gussoni E et al (1999) Dystrophin expression in the mdx mouse restored by stem cell transplantation. Nature 401:390–394

Guttinger M, Tafi E, Battaglia M, Coletta M, Cossu G (2006) Allogeneic mesoangioblasts give rise to alpha-sarcoglycan expressing fibres when transplanted into dystrophic mice. Exp Cell Res 312:3872–3879

Guyon JR et al (2009) Genetic isolation and characterization of a splicing mutant of zebrafish dystrophin. Hum Mol Genet 18:202–211

Hack AA et al (1998) Gamma-sarcoglycan deficiency leads to muscle membrane defects and apoptosis independent of dystrophin. J Cell Biol 142:1279–1287

Hack AA, Groh ME, McNally EM (2000) Sarcoglycans in muscular dystrophy. Microsc Res Tech 48:167–180

Hall TE et al (2007) The zebrafish candyfloss mutant implicates extracellular matrix adhesion failure in laminin alpha2-deficient congenital muscular dystrophy. Proc Natl Acad Sci U S A 104:7092–7097

Han R, Campbell KP (2007) Dysferlin and muscle membrane repair. Curr Opin Cell Biol 19:409–416

Hara Y et al (2011) A dystroglycan mutation associated with limb-girdle muscular dystrophy. N Engl J Med 364:939–946

Helbling-Leclerc A et al (1995) Mutations in the laminin alpha 2-chain gene (LAMA2) cause merosin-deficient congenital muscular dystrophy. Nat Genet 11:216–218

Hierlihy AM, Seale P, Lobe CG, Rudnicki MA, Megeney LA (2002) The post-natal heart contains a myocardial stem cell population. FEBS Lett 530:239–243

Ho M et al (2004) Disruption of muscle membrane and phenotype divergence in two novel mouse models of dysferlin deficiency. Hum Mol Genet 13:1999–2010

Hoffman EP, Brown RH Jr, Kunkel LM (1987) Dystrophin: the protein product of the Duchenne muscular dystrophy locus. Cell 51:919–928

Holt KH et al (1998) Functional rescue of the sarcoglycan complex in the BIO 14.6 hamster using delta-sarcoglycan gene transfer. Mol Cell 1:841–848

Howell JM, Fletcher S, Kakulas BA, O'Hara M, Lochmuller H, Karpati G (1997) Use of the dog model for Duchenne muscular dystrophy in gene therapy trials. Neuromuscul Disord 7:325–328

Ibraghimov-Beskrovnaya O, Ervasti JM, Leveille CJ, Slaughter CA, Sernett SW, Campbell KP (1992) Primary structure of dystrophin-associated glycoproteins linking dystrophin to the extracellular matrix. Nature 355:696–702
Isaac C et al (2013) Dystrophin and utrophin "double knockout" dystrophic mice exhibit a spectrum of degenerative musculoskeletal abnormalities. J Orthop Res 31:343–349
Jacoby AS et al (2009) The zebrafish dystrophic mutant softy maintains muscle fibre viability despite basement membrane rupture and muscle detachment. Development 136:3367–3376
Kaplan JC (2011) The 2011 version of the gene table of neuromuscular disorders. Neuromuscul Disord 20:852–873
Kawahara G, Karpf JA, Myers JA, Alexander MS, Guyon JR, Kunkel LM (2011) Drug screening in a zebrafish model of Duchenne muscular dystrophy. Proc Natl Acad Sci U S A 108:5331–5336
Kerkis I et al (2008) Early transplantation of human immature dental pulp stem cells from baby teeth to golden retriever muscular dystrophy (GRMD) dogs: local or systemic? J Transl Med 6:35
Kinoshita I, Vilquin JT, Guerette B, Asselin I, Roy R, Tremblay JP (1994) Very efficient myoblast allotransplantation in mice under FK506 immunosuppression. Muscle Nerve 17:1407–1415
Koenig M, Hoffman EP, Bertelson CJ, Monaco AP, Feener C, Kunkel LM (1987) Complete cloning of the Duchenne muscular dystrophy (DMD) cDNA and preliminary genomic organization of the DMD gene in normal and affected individuals. Cell 50:509–517
Kong KY, Ren J, Kraus M, Finklestein SP, Brown RH Jr (2004) Human umbilical cord blood cells differentiate into muscle in sjl muscular dystrophy mice. Stem Cells 22:981–993
Kornegay JN et al (2012) Canine models of Duchenne muscular dystrophy and their use in therapeutic strategies. Mamm Genome 23:85–108
Krahn M et al (2010) A naturally occurring human minidysferlin protein repairs sarcolemmal lesions in a mouse model of dysferlinopathy. Sci Transl Med 2:50ra69
Lapidos KA et al (2004) Transplanted hematopoietic stem cells demonstrate impaired sarcoglycan expression after engraftment into cardiac and skeletal muscle. J Clin Invest 114:1577–1585
Leng Y, Zheng Z, Zhou C, Zhang C, Shi X, Zhang W (2012) A comparative study of bone marrow mesenchymal stem cell functionality in C57BL and mdx mice. Neurosci Lett 523:139–144
Leriche-Guerin K, Anderson LV, Wrogemann K, Roy B, Goulet M, Tremblay JP (2002) Dysferlin expression after normal myoblast transplantation in SCID and in SJL mice. Neuromuscul Disord 12:167–173
Li J, Dressman D, Tsao YP, Sakamoto A, Hoffman EP, Xiao X (1999) rAAV vector-mediated sarcogylcan gene transfer in a hamster model for limb girdle muscular dystrophy. Gene Ther 6:74–82
Liao NS, Bix M, Zijlstra M, Jaenisch R, Raulet D (1991) MHC class I deficiency: susceptibility to natural killer (NK) cells and impaired NK activity. Science 253:199–202
Lieschke GJ, Currie PD (2007) Animal models of human disease: zebrafish swim into view. Nat Rev Genet 8:353–367
Lin YY, White RJ, Torelli S, Cirak S, Muntoni F, Stemple DL (2011) Zebrafish Fukutin family proteins link the unfolded protein response with dystroglycanopathies. Hum Mol Genet 20:1763–1775
Liu J et al (1998) Dysferlin, a novel skeletal muscle gene, is mutated in Miyoshi myopathy and limb girdle muscular dystrophy. Nat Genet 20:31–36
Lostal W et al (2010) Efficient recovery of dysferlin deficiency by dual adeno-associated vector-mediated gene transfer. Hum Mol Genet 19:1897–1907
Maselli RA et al (2009) Mutations in LAMB2 causing a severe form of synaptic congenital myasthenic syndrome. J Med Genet 46:203–208
Megeney LA, Kablar B, Garrett K, Anderson JE, Rudnicki MA (1996) MyoD is required for myogenic stem cell function in adult skeletal muscle. Genes Dev 10:1173–1183
Megeney LA, Kablar B, Perry RL, Ying C, May L, Rudnicki MA (1999) Severe cardiomyopathy in mice lacking dystrophin and MyoD. Proc Natl Acad Sci U S A 96:220–225
Miller RG et al (1997) Myoblast implantation in Duchenne muscular dystrophy: the San Francisco study. Muscle Nerve 20:469–478

Minetti C et al (1998) Mutations in the caveolin-3 gene cause autosomal dominant limb-girdle muscular dystrophy. Nat Genet 18:365–368
Mizuno Y et al (2010) Generation of skeletal muscle stem/progenitor cells from murine induced pluripotent stem cells. FASEB J 24:2245–2253
Montanaro F, Liadaki K, Volinski J, Flint A, Kunkel LM (2003) Skeletal muscle engraftment potential of adult mouse skin side population cells. Proc Natl Acad Sci U S A 100:9336–9341
Montarras D et al (2005) Direct isolation of satellite cells for skeletal muscle regeneration. Science 309:2064–2067
Morgan JE, Hoffman EP, Partridge TA (1990) Normal myogenic cells from newborn mice restore normal histology to degenerating muscles of the mdx mouse. J Cell Biol 111:2437–2449
Motohashi N et al (2008) Muscle CD31(–) CD45(–) side population cells promote muscle regeneration by stimulating proliferation and migration of myoblasts. Am J Pathol 173:781–791
Muntoni F, Wood MJ (2011) Targeting RNA to treat neuromuscular disease. Nat Rev Drug Discov 10:621–637
Nakagaki K, Ozaki J, Tomita Y, Tamai S (1994) Alterations in the supraspinatus muscle belly with rotator cuff tearing: evaluation with magnetic resonance imaging. J Shoulder Elbow Surg 3:88–93
Neuman S, Kaban A, Volk T, Yaffe D, Nudel U (2001) The dystrophin/utrophin homologues in Drosophila and in sea urchin. Gene 263:17–29
Nitahara-Kasahara Y et al (2012) Long-term engraftment of multipotent mesenchymal stromal cells that differentiate to form myogenic cells in dogs with Duchenne muscular dystrophy. Mol Ther 20:168–177
Palm W, de Lange T (2008) How shelterin protects mammalian telomeres. Annu Rev Genet 42:301–334
Parker MH, Kuhr C, Tapscott SJ, Storb R (2008) Hematopoietic cell transplantation provides an immune-tolerant platform for myoblast transplantation in dystrophic dogs. Mol Ther 16:1340–1346
Parsons MJ, Campos I, Hirst EM, Stemple DL (2002) Removal of dystroglycan causes severe muscular dystrophy in zebrafish embryos. Development 129:3505–3512
Partridge TA, Morgan JE, Coulton GR, Hoffman EP, Kunkel LM (1989) Conversion of mdx myofibres from dystrophin-negative to -positive by injection of normal myoblasts. Nature 337:176–179
Perloff JK, Roberts WC, De Leon AC Jr, O'Doherty D (1967) The distinctive electrocardiogram of Duchenne's progressive muscular dystrophy. An electrocardiographic-pathologic correlative study. Am J Med 42:179–188
Pichavant C et al (2010) Expression of dog microdystrophin in mouse and dog muscles by gene therapy. Mol Ther 18:1002–1009
Pinheiro CH et al (2011) Local injections of adipose-derived mesenchymal stem cells modulate inflammation and increase angiogenesis ameliorating the dystrophic phenotype in dystrophin-deficient skeletal muscle. Stem Cell Rev 8:363–374
Prattis SM, Gebhart DH, Dickson G, Watt DJ, Kornegay JN (1993) Magnetic affinity cell sorting (MACS) separation and flow cytometric characterization of neural cell adhesion molecule-positive, cultured myogenic cells from normal and dystrophic dogs. Exp Cell Res 208:453–464
Qu-Petersen Z et al (2002) Identification of a novel population of muscle stem cells in mice: potential for muscle regeneration. J Cell Biol 157:851–864
Rappsilber J, Ryder U, Lamond AI, Mann M (2002) Large-scale proteomic analysis of the human spliceosome. Genome Res 12:1231–1245
Richard I et al (1995) Mutations in the proteolytic enzyme calpain 3 cause limb-girdle muscular dystrophy type 2A. Cell 81:27–40
Rideout WM III, Hochedlinger K, Kyba M, Daley GQ, Jaenisch R (2002) Correction of a genetic defect by nuclear transplantation and combined cell and gene therapy. Cell 109:17–27
Roberts RG, Bobrow M (1998) Dystrophins in vertebrates and invertebrates. Hum Mol Genet 7:589–595

Rocha V et al (2000) Graft-versus-host disease in children who have received a cord-blood or bone marrow transplant from an HLA-identical sibling. Eurocord and International Bone Marrow Transplant Registry Working Committee on Alternative Donor and Stem Cell Sources. N Engl J Med 342:1846–1854
Rodier F, Kim SH, Nijjar T, Yaswen P, Campisi J (2005) Cancer and aging: the importance of telomeres in genome maintenance. Int J Biochem Cell Biol 37:977–990
Rodriguez AM et al (2005) Transplantation of a multipotent cell population from human adipose tissue induces dystrophin expression in the immunocompetent mdx mouse. J Exp Med 201:1397–1405
Roostalu U, Strahle U (2012) In vivo imaging of molecular interactions at damaged sarcolemma. Dev Cell 22:515–529
Roscioli T et al (2012) Mutations in ISPD cause Walker-Warburg syndrome and defective glycosylation of alpha-dystroglycan. Nat Genet 44:581–585
Rouger K et al (2011) Systemic delivery of allogenic muscle stem cells induces long-term muscle repair and clinical efficacy in Duchenne muscular dystrophy dogs. Am J Pathol 179:2501–2518
Sacco A et al (2010) Short telomeres and stem cell exhaustion model Duchenne muscular dystrophy in mdx/mTR mice. Cell 143:1059–1071
Sakamoto A et al (1997) Both hypertrophic and dilated cardiomyopathies are caused by mutation of the same gene, delta-sarcoglycan, in hamster: an animal model of disrupted dystrophin-associated glycoprotein complex. Proc Natl Acad Sci U S A 94:13873–13878
Salah-Mohellibi N et al (2006) Bone marrow transplantation attenuates the myopathic phenotype of a muscular mouse model of spinal muscular atrophy. Stem Cells 24:2723–2732
Sampaolesi M et al (2003) Cell therapy of alpha-sarcoglycan null dystrophic mice through intra-arterial delivery of mesoangioblasts. Science 301:487–492
Sampaolesi M et al (2006) Mesoangioblast stem cells ameliorate muscle function in dystrophic dogs. Nature 444:574–579
Sarig R, Baruchi Z, Fuchs O, Nudel U, Yaffe D (2006) Regeneration and transdifferentiation potential of muscle-derived stem cells propagated as myospheres. Stem Cells 24:1769–1778
Schatzberg SJ et al (1999) Molecular analysis of a spontaneous dystrophin 'knockout' dog. Neuromuscul Disord 9:289–295
Schuler W et al (1986) Rearrangement of antigen receptor genes is defective in mice with severe combined immune deficiency. Cell 46:963–972
Shanmugam V, Dion P, Rochefort D, Laganiere J, Brais B, Rouleau GA (2000) PABP2 polyalanine tract expansion causes intranuclear inclusions in oculopharyngeal muscular dystrophy. Ann Neurol 48:798–802
Sharp NJ et al (1992) An error in dystrophin mRNA processing in golden retriever muscular dystrophy, an animal homologue of Duchenne muscular dystrophy. Genomics 13:115–121
Sherr CJ, DePinho RA (2000) Cellular senescence: mitotic clock or culture shock? Cell 102:407–410
Shimatsu Y et al (2003) Canine X-linked muscular dystrophy in Japan (CXMDJ). Exp Anim 52:93–97
Shimatsu Y et al (2005) Major clinical and histopathological characteristics of canine X-linked muscular dystrophy in Japan, CXMDJ. Acta Myol 24:145–154
Sicinski P, Geng Y, Ryder-Cook AS, Barnard EA, Darlison MG, Barnard PJ (1989) The molecular basis of muscular dystrophy in the mdx mouse: a point mutation. Science 244:1578–1580
Skuk D, Goulet M, Roy B, Tremblay JP (2000) Myoblast transplantation in whole muscle of non-human primates. J Neuropathol Exp Neurol 59:197–206
Skuk D, Goulet M, Roy B, Tremblay JP (2002) Efficacy of myoblast transplantation in nonhuman primates following simple intramuscular cell injections: toward defining strategies applicable to humans. Exp Neurol 175:112–126
Steffen LS et al (2007) Zebrafish orthologs of human muscular dystrophy genes. BMC Genomics 8:79

Straub V et al (1998) Molecular pathogenesis of muscle degeneration in the delta-sarcoglycan-deficient hamster. Am J Pathol 153:1623–1630
Tamaki T et al (2007) Clonal multipotency of skeletal muscle-derived stem cells between mesodermal and ectodermal lineage. Stem Cells 25:2283–2290
Telfer WR, Busta AS, Bonnemann CG, Feldman EL, Dowling JJ (2010) Zebrafish models of collagen VI-related myopathies. Hum Mol Genet 19:2433–2444
Thibaud JL, Monnet A, Bertoldi D, Barthelemy I, Blot S, Carlier PG (2007) Characterization of dystrophic muscle in golden retriever muscular dystrophy dogs by nuclear magnetic resonance imaging. Neuromuscul Disord 17:575–584
Thornhill P, Bassett D, Lochmuller H, Bushby K, Straub V (2008) Developmental defects in a zebrafish model for muscular dystrophies associated with the loss of fukutin-related protein (FKRP). Brain 131:1551–1561
Torrente Y et al (2001) Intraarterial injection of muscle-derived CD34(+)Sca-1(+) stem cells restores dystrophin in mdx mice. J Cell Biol 152:335–348
Torrente Y et al (2004) Human circulating AC133(+) stem cells restore dystrophin expression and ameliorate function in dystrophic skeletal muscle. J Clin Invest 114:182–195
Traggiai E et al (2004) Development of a human adaptive immune system in cord blood cell-transplanted mice. Science 304:104–107
Vafiadaki E et al (2001) Cloning of the mouse dysferlin gene and genomic characterization of the SJL-Dysf mutation. Neuroreport 12:625–629
Valentine BA, Cooper BJ, de Lahunta A, O'Quinn R, Blue JT (1988) Canine X-linked muscular dystrophy. An animal model of Duchenne muscular dystrophy: clinical studies. J Neurol Sci 88:69–81
van der Maarel SM, Tawil R, Tapscott SJ (2011) Facioscapulohumeral muscular dystrophy and DUX4: breaking the silence. Trends Mol Med 17:252–258
van der Plas MC, Pilgram GS, Plomp JJ, de Jong A, Fradkin LG, Noordermeer JN (2006) Dystrophin is required for appropriate retrograde control of neurotransmitter release at the Drosophila neuromuscular junction. J Neurosci 26:333–344
Vieira NM et al (2008) SJL dystrophic mice express a significant amount of human muscle proteins following systemic delivery of human adipose-derived stromal cells without immunosuppression. Stem Cells 26:2391–2398
Vieira NM et al (2010) Human multipotent mesenchymal stromal cells from distinct sources show different in vivo potential to differentiate into muscle cells when injected in dystrophic mice. Stem Cell Rev 6:560–566
Vieira NM et al (2012) Human adipose-derived mesenchymal stromal cells injected systemically into GRMD dogs without immunosuppression are able to reach the host muscle and express human dystrophin. Cell Transplant 21:1407–1417
Vilquin JT, Wagner E, Kinoshita I, Roy R, Tremblay JP (1995) Successful histocompatible myoblast transplantation in dystrophin-deficient mdx mouse despite the production of antibodies against dystrophin. J Cell Biol 131:975–988
Wallace LM et al (2012) RNA interference inhibits DUX4-induced muscle toxicity in vivo: implications for a targeted FSHD therapy. Mol Ther 20:1417–1423
Walmsley GL et al (2010) A Duchenne muscular dystrophy gene hot spot mutation in dystrophin-deficient Cavalier King Charles Spaniels is amenable to exon 51 skipping. PLoS One 5:e8647
Walsh S, Nygren J, Ponten A, Jovinge S (2011) Myogenic reprogramming of bone marrow derived cells in a W(4)(1)Dmd(mdx) deficient mouse model. PLoS One 6:e27500
Watchko JF, O'Day TL, Hoffman EP (2002) Functional characteristics of dystrophic skeletal muscle: insights from animal models. J Appl Physiol 93:407–417
Welch EM et al (2007) PTC124 targets genetic disorders caused by nonsense mutations. Nature 447:87–91
Welm BE, Tepera SB, Venezia T, Graubert TA, Rosen JM, Goodell MA (2002) Sca-1(pos) cells in the mouse mammary gland represent an enriched progenitor cell population. Dev Biol 245:42–56

Williamson RA et al (1997) Dystroglycan is essential for early embryonic development: disruption of Reichert's membrane in Dag1-null mice. Hum Mol Genet 6:831–841
Willmann R, Possekel S, Dubach-Powell J, Meier T, Ruegg MA (2009) Mammalian animal models for Duchenne muscular dystrophy. Neuromuscul Disord 19:241–249
Winand NJ, Edwards M, Pradhan D, Berian CA, Cooper BJ (1994) Deletion of the dystrophin muscle promoter in feline muscular dystrophy. Neuromuscul Disord 4:433–445
Zucconi E et al (2011) Preclinical studies with umbilical cord mesenchymal stromal cells in different animal models for muscular dystrophy. J Biomed Biotechnol 2011:715251
Zuk PA et al (2001) Multilineage cells from human adipose tissue: implications for cell-based therapies. Tissue Eng 7:211–228

Chapter 2
Understanding Tissue Repair Through the Activation of Endogenous Resident Stem Cells

Iolanda Aquila, Carla Vicinanza, Mariangela Scalise, Fabiola Marino, Christelle Correale, Michele Torella, Gianantonio Nappi, Ciro Indolfi, and Daniele Torella

1 The Clinical Need of Stem Cells for Cardiovascular Diseases

Heart failure (HF) has been singled out as an epidemic and is a staggering clinical and public health problem, associated with significant mortality, morbidity, and healthcare expenditures, particularly among those aged ≥65 years (Roger 2013). In particular, HF has a prevalence of roughly six million in the United States (a similar number is reached in Europe) and more than 23 million worldwide. After the diagnosis of HF, survival remains quite poor with estimates of 50 % and 10 % at 5 and 10 years, respectively. Despite modest progress in reducing HF-related mortality, hospitalizations for HF remain frequent and rates of readmissions continue to rise (Roger 2013). It is predicted that the constant progress in the primary prevention of

I. Aquila • M. Scalise • F. Marino • C. Correale • C. Indolfi
Molecular and Cellular Cardiology, Department of Medical and Surgical Sciences, Magna Graecia University, Catanzaro, Italy

C. Vicinanza
Molecular and Cellular Cardiology, Department of Medical and Surgical Sciences, Magna Graecia University, Catanzaro, Italy

Stem Cell and Regenerative Biology Unit (BioStem), Research Institute for Sport & Exercise Sciences (RISES), Liverpool John Moores University, Liverpool, UK

M. Torella • G. Nappi
Department of Cardio-Thoracic and Respiratory Sciences, Second University of Naples, Naples, Italy

D. Torella (✉)
Molecular and Cellular Cardiology, Department of Medical and Surgical Sciences, Magna Graecia University, Catanzaro, Italy

Department of Cardio-Thoracic and Respiratory Sciences, Second University of Naples, Naples, Italy
e-mail: dtorella@unicz.it

T.A.L. Brevini (ed.), *Stem Cells in Animal Species: From Pre-clinic to Biodiversity*, Stem Cell Biology and Regenerative Medicine, DOI 10.1007/978-3-319-03572-7_2,

HF will eventually lead to decreasing incidence of the disease. Accordingly, the constant improvement in modern medical care of the acute cardiac diseases will result in improved survival. However, the latter in turn will anyway increase the prevalence of HF. This is so because modern management for HF is mainly a symptomatic treatment that does not fight against its cause, leave organ transplantation as the only alternative available to restore function, with all the logistic, economic and biological limitations associated with this intervention (Kahan 2011). Indeed, the root problem responsible for the poor outcome of the CHF is a deficit of functional myocardial contractile cells (cardiomyocytes) and adequate coronary circulation to nurture them resulting in pathological cardiac remodelling, which, in turn, triggers the late development of cardiac failure in these patients (Jessup and Brozena 2003). For this reason, it has been a goal of cardiovascular research for the past decade to find methods to replace the cardiomyocytes lost as a consequence of the MI in order to prevent or reverse the pathological cardiac remodelling. Overall, the need to identify new therapies has become a key research area in regenerative cardiovascular medicine and stem cell-based therapies are fast becoming an attractive and highly promising experimental treatment for heart disease and failure (Terzic and Nelson 2010).

2 Adult Heart Self-Renewal and Tissue-Specific Endogenous Cardiac Stem/Progenitor Cells

Out of the limelight and apart from the cultural and philosophical wars, over the past 15 years there has been a slow but steady re-evaluation of the prevalent paradigm about adult mammalian—including human—tissue cellular homeostasis. It has been slowly appreciated that the parenchymal cell population of most, if not all, adult tissues is in a continuous process of self-renewal with cells continuously dying and new ones being born. Once cell turnover was accepted as a widespread phenomenon in the adult organs, it was rapidly surmised that in order to preserve tissue mass, each organ constituted mainly of terminally differentiated cells needed to have a population of tissue-specific regenerating cells. Not surprisingly, this realization was rapidly followed by the progressive identification of stem cells in each of the adult body tissues (Yamanaka 2007; Robinton and Daley 2012; Rountree et al. 2012; Reule and Gupta 2011; Kopp et al. 2011; Kotton 2012; Buckingham and Montarras 2008; Suh et al. 2009).

For a long time, the cardiovascular research community has treated the adult mammalian heart as a post-mitotic organ without intrinsic regenerative capacity. The prevalent notion was that the >20-fold increase in cardiac mass from birth to adulthood and in response to different stimuli in the adult heart, results exclusively from the enlargement of pre-existing myocytes (Hunter and Chien 1999; Soonpaa and Field 1998; Laflamme and Murry 2011). It was accepted that this myocyte hypertrophy, in turn, was uniquely responsible for the initial physiological adaptation and subsequent deterioration of the overloaded heart. This belief was based on two generally accepted notions: (a) all myocytes in the adult heart were formed during fetal life or shortly thereafter, were terminally differentiated and could not be recalled into the cell cycle (Nadal-Ginard 1978; Chien and Olson 2002); therefore, all cardiac myocytes have to be of the same chronological age as the individual (Oh et al. 2001); (b) the heart has no intrinsic parenchymal regenerative capacity because it lacks a stem/progenitor cell population able to generate new myocytes. Despite published evidence that this prevalent view was incorrect (Beltrami et al. 2001; Quaini et al. 2002; Urbanek et al. 2003, 2005; Anversa and Nadal-Ginard 2002; Nadal-Ginard et al. 2003), it took the publication of Bergmann et al. in 2009, based on ^{14}C dating in human hearts showing that during a lifetime the human heart renews ~50 % of its myocytes (Bergmann et al. 2009), to produce a significant switch in the prevalent opinion. After this, intensive research and still much controversy on the adult mammalian heart's capacity for self-renewal has finally brought a consensus that new cardiomyocytes are indeed formed throughout adult mammalian life (Bergmann et al. 2009; Hsieh et al. 2007). However, the physiological significance of this myocyte renewal, the origin of the new myocytes as well as the rate of adult myocyte turnover are still highly debated. Indeed, while Bergmann et al. (2009) have calculated a yearly cardiomyocyte turnover of about 1 % (Bergmann et al. 2009) others calculated 4–10 % (Senyo et al. 2013) and some as high as 40%/year (Kajstura et al. 2012). This very high spread on the "measured" values of such an important phenomenon raises questions about the conceptual and methodological approaches used in these studies. If Bergmann's calculations were the one more close to the reality, because this "measured" self-renewal is far from being very robust, its physiological significance would be highly doubtful. However, the conclusion of Bergmann et al. (2009) on the rate of turnover depends on the validity of a complex mathematical formula, whose impact on the results dwarfs that of the measured data. Their calculations identify the highest turnover rate during youth and early adulthood followed by a steady decrease with age. The latter conclusion, which is contrary to most or all the turnover values measured for all other human tissues, including the heart (Nadal-Ginard et al. 2003), has passed without a ripple. Similar discrepancy exists with respect to the origin of the new myocytes and their physiological significance. Three main sources of origin of the new myocytes have been claimed: (a) circulating progenitors, which through the bloodstream home to the myocardium and differentiate into myocytes (Quaini et al. 2002); (b) mitotic division of the pre-existing myocytes (Boström et al. 2010; Bersell et al. 2009; Kühn et al. 2007) and (c) a small population of resident myocardial and/or epicardial multipotent stem cells able to differentiate into the main cell types of the heart: myocytes, smooth and endothelial vascular and connective tissue cells (Torella et al. 2007; Rasmussen et al. 2011). It is clear now that the blood borne precursors, although well documented as a biological phenomenon (Eisenberg et al. 2006), might be limited to very special situation (Orlic et al. 2001) and their direct regenerative import is very limited, if any (Loffredo et al. 2011). Myocyte replacement, particularly after injury, was originally attributed to differentiation of a stem-progenitor cell compartment (Beltrami et al. 2003) a source confirmed by genetic cell fate mapping (Hsieh et al. 2007). However, more recently this last investigator group, using the same genetic tools, claims that myocytes in the border zone of an infarct are actually replaced by the division of pre-existing post-mitotic myocytes (Senyo et al. 2013). Pre-existing cardiomyocyte division has not been convincingly documented and/or remains to be confirmed by different authors. This result, in addition to being contrary to most known biology of terminally differentiated cells, would shift the target of regenerative therapy towards boosting mature cardiomyocyte cell cycle re-entry. However, Senyo et al. (2013) only document that a very small fraction of cardiomyocyte DNA replication occurs in cells that have already activated the αMHC gene, a well-documented part of myocyte development, which falls short of documenting mature myocyte re-entry into the cell cycle. This evidence is indeed equally compatible with new myocyte formation from the pool of multipotent cardiac progenitor cells because it is a well-documented fact that newly born myocytes are not yet terminally differentiated and are capable of a few rounds of mitosis before irreversibly withdrawing from the cell cycle (Nadal-Ginard et al. 2003). Undoubtedly, the best documented source of the regenerating myocardial cells in the adult mammalian heart, including the human, is a small population of cells distributed throughout the atria and ventricles of the young, adult and senescent mammalian myocardium, that have the phenotype, behaviour and regenerative potential of bona fide cardiac stem cells (eCSCs) (Torella et al. 2007; Rasmussen et al. 2011; Srivastava and Ivey 2006). In 2003, we identified the first population of eCSCs in the adult mammalian rat heart (Beltrami et al. 2003). These cells express the stem cell marker c-kit (c-kit^{pos}), are positive for Sca-1 and MDR-1 (ABCG2), yet are negative for markers of the blood cell lineage, CD31, CD34 and CD45 (described as Lin^{neg}). They are self-renewing, clonogenic and multipotent and exhibit significant regenerative potential when injected into the adult rat heart following a myocardial infarction (MI), forming new myocytes and vasculature and restoring cardiac function (Beltrami et al. 2003). c-kit^{pos} eCSCs with similar properties to those originally identified in the rat have been identified and characterized in the mouse (Messina et al. 2004; Fransioli et al. 2008), dog (Linke et al. 2005), pig (Ellison et al. 2011) and human (Messina et al. 2004; Torella et al. 2006a, b; Bearzi et al. 2007; Arsalan et al. 2012). These cells are present at a similar density in all species (~1 eCSC per 1,000 cardiomyocytes or 45,000 human eCSCs per gram of tissue) (Torella et al. 2007). Similar to the rodent heart, the distribution of c-kit^{pos} eCSCs in the pig and human heart varies with cardiac chamber. Although a variety of markers have been proposed to identify eCSCs in different species and throughout development (Messina et al. 2004; Oh et al. 2003; Matsuura et al. 2004; Martin et al. 2004; Laugwitz et al. 2005; Moretti et al. 2006; Kattman et al. 2006; Wu et al. 2006; Smart et al. 2011), it still remains to be determined whether these markers identify different populations of eCSCs or, more likely, different developmental and/or physiological stages of the same cell type (Ellison et al. 2010). Recently, another multipotent cell type, present in the epicardium and derived from the pro-epicardial organ has been described (Ellison et al. 2007a). The role of these cells in normal or pathological myocyte turnover remains to be elucidated.

The progeny of a single eCSC is able to differentiate into cardiac myocytes, smooth muscle and endothelial vascular cells and when transplanted into the border zone of an infarct regenerates functional contractile muscle and the microvasculature of the tissue (Beltrami et al. 2003). In a normal adult myocardium, at any given time, most of the eCSCs are quiescent and only a small fraction is active to replace the myocytes and vascular cells lost by wear and tear. In response to stress (hypoxia, exercise, work overload, or other damage), however, a proportion of the resident eCSCs are rapidly activated, they multiply and generate new muscle and vascular cells (Ellison et al. 2007a, 2012a), contributing to cardiac remodelling. Recently, two studies have questioned the in situ myogenic potential of c-kit$^{(pos)}$ cardiac cells in adult life as being significantly reduced compared to their neonatal counterparts (Zaruba et al. 2010), and in a model of myocardial cryo-injury (Jesty et al. 2012). Thus, whether the c-kit^{pos} eCSCs are necessary and/or sufficient for the adult cardiac regenerative response to damage/injury remained unproven. However, recently, using mouse and rat experimental protocols of severe diffuse myocardial damage which unlike an experimental infarct spares the eCSCs (Ellison et al. 2007b) combined with several genetic murine models and cell transplantation approaches, we have shown that the eCSCs, in the presence of a patent coronary circulation, fulfil the criteria as the cell type necessary and sufficient for myocyte regeneration, leading to complete cellular, anatomical and functional myocardial recovery (Fig. 2.1) (Ellison et al. 2013). To follow c-kit^{pos}eCSC physiological response to cardiac injury, we induced severe diffuse myocardial damage in adult rats with a single high dose of isoproterenol (ISO) (Ellison et al. 2007b). This treatment—in the presence of a patent coronary circulation—produces a Takotsubo-like cardiomyopathy (Akashi et al. 2008) (a clinical syndrome affecting up to 2 % of patients with symptoms and signs of acute myocardial infarction (AMI), characterized by catecholamine overdrive and reversible cardiomyopathy) killing-10 % of the LV myocytes and resulting in overt acute heart failure (Ellison et al. 2007b). Interestingly, the myocardial damage and heart failure spontaneously reverses anatomically and functionally by 28 days. The anatomical and functional recovery is met through a robust c-kit$^{(pos)}$ eCSC activation and ensuing new myocyte formation, the latter completely balancing the myocyte loss by ISO injury. Using inducible double-transgenic reporter mice to track the fate of adult cardiomyocytes in a "pulse-chase" fashion we compellingly show that new myocytes after diffuse myocardial injury are not generated through the division of pre-existing terminally differentiated myocytes but rather from non-myocyte cells, with the characteristics of a stem-progenitor compartment. Then, to directly identify whether c-kit^{pos} eCSCs replenish cardiomyocytes lost by myocardial damage, we genetically tagged in situ the resident c-kit^{pos}eCSCs and their committed progeny. Through these in vivo genetic cell-fate mapping experiments, we have eventually proved that new myocytes after myocardial injury in the adult mammalian heart originate from resident c-kit^{pos}eCSCs. Furthermore, we have shown in a rat model of severe cardiomyopathy induced by ISO injury and 5-Fluoro Uracil that the ablation of the eCSCs abolishes regeneration and functional recovery. The regenerative process however is completely restored by replacing the ablated eCSCs with the tagged progeny of one eCSC. These eCSCs recovered from the primary host, and re-cloned, retain

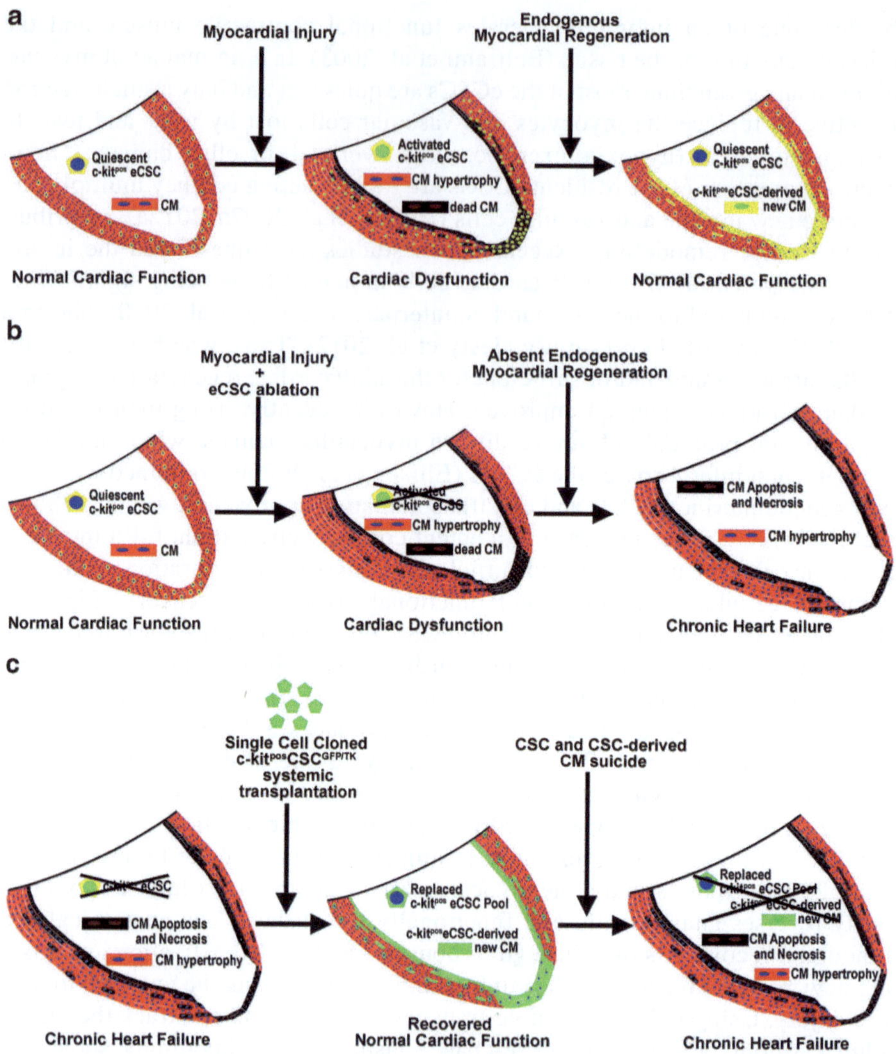

Fig. 2.1 eCSCs are necessary and sufficient for myocardial repair and regeneration. (**a**) Schematic of myocyte damage and cardiac recovery through eCSC activation and new myocyte formation. (**b**) When eCSC are ablated, cardiac regeneration is absent with the development of a severe cardiomyopathy. (**c**) If CSCs are exogenously transplanted, cardiac anatomy and function is restored. However, the selective suicide of the transplanted CSCs and their progeny sets back the animal in heart failure

their regenerative potential in vivo and in vitro. Finally, after regeneration, selective suicide of these exogenous eCSCs and their progeny abolishes the regeneration, severely impairing ventricular performance. Thus, overall these data have provided the ultimate and solid documentation that the resident tissue-specific eCSCs are necessary and sufficient for the regeneration of the adult myocardium and establish these cells as true cardiac regenerative agents.

3 Autologous Cardiac Stem Cell Therapy for Heart Failure

Many questions about eCSC basic biology still remain unanswered, particularly their long-term effectiveness and regenerative potential as well as their origin. It is imperative that such issues be addressed quickly if the full potential of these cells is to be realized, manipulated and applied clinically. In particular, it is imperative to document whether the teratogenic and neoplastic potential of the in vitro expanded eCSCs is low enough to make their use in humans safe. However, clinical trials using autologous cardiac stem/progenitor cells are already underway (Bolli et al. 2011; Makkar et al. 2012). In the SCIPIO (Stem Cell Infusion in Patients with Ischemic Cardiomyopathy; NCT00474461) trial, 16 patients with ischaemic cardiomyopathy with post-infarction LV dysfunction (ejection fraction ≤40 %) who had undergone coronary artery bypass grafting, had 500,000–1 million of autologous c-kit positive, lineage negative, cardiac progenitor cells infused intracoronary, ~4 months after surgery (Bolli et al. 2011). The control group was not given any treatment. LVEF increased by 8 EF points at 4 months after infusion, whereas the LVEF did not change in the control patients, during the corresponding time interval. Moreover, LVEF increased by 12 EF points in eight of the treated patients at 1-year follow-up. cMRI of seven of the treated patients showed that infarct size decreased at 4 and 12 months (Bolli et al. 2011). Furthermore, the interim analysis of myocardial function and viability by magnetic resonance in SCIPIO on a total of 33 patients (20 CSC-treated and 13 control subjects) confirmed the improvement in both global and regional LV function, and a reduction in infarct size at 1 year (Chugh et al. 2012). In the prospective, randomized cardiosphere-derived autologous stem cells to reverse ventricular dysfunction (CADUCEUS) trial, 17 patients (with left ventricular ejection fraction of 25–45 %) were infused into the infarct-related artery with up to 25 million, CD105-positive, autologous cardiosphere-derived cells (CDCs), 1.5–3 months after myocardial infarction (Makkar et al. 2012). Eight patients received standard care and acted as the control group. Compared with controls at 6 months, MRI analysis of patients treated with CDCs showed significant reductions in scar size and mass, increased viable heart mass, regional contractility and systolic wall thickening. However, changes in end-diastolic volume, end-systolic volume, and LVEF did not differ between groups at 6 months (Makkar et al. 2012). Recently, the ALCADIA clinical trial was initiated, which will focus on a hybrid biotherapy approach for treating chronic ischemic cardiomyopathy. This translational study is focusing on the safety and efficacy of autologous clonally amplified CSCs, which have shown to be enriched for embryonic stem cell markers and have mesenchymal cell characteristics (Matsubara and Kyoto Prefectural University School of Medicine 2012). This trial is also investigating cell therapy with the controlled release of basic fibroblast growth factor (bFGF) from a gelatin hydrogel sheet. While this innovative work has proved promising in the respect that transplantation of autologous CSCs has not resulted in any adverse health effects, we now await further studies, which focus on the efficacy of eCSC-based therapies and compare these to results obtained with BMDCs. Indeed, because of the high cost and the long wait for the availability of the cells for autologous cardiac stem/

precursor cell therapy, it will become imperative to compare the beneficial effects of this approach to that obtained with BMDCs because of their easier availability, accessibility and lower cost of the procedure. Furthermore, the widespread use and applicability of autologous cardiac stem cell therapy is highly debatable. Firstly, the procedure for cell acquisition, scale-up and transplantation is complex, time consuming and very expensive. The isolation and expansion of eCSCs to the number needed from catheter and surgical biopsies takes 1–3 months. Therefore, the cells are not available to be administered when they would be most effective, that is when a patient with an AMI in progress arrives at the hospital. Furthermore, the cost of the procedure in human and material resources would make it unavailable to patients beyond those few required to establish proof-of-concept for the therapy and to a small group of individuals with abundant economical resources. Finally, eCSCs undergo senescence with severe pathological consequences (Torella et al. 2004, 2006a; Matsubara and Kyoto Prefectural University School of Medicine 2012; Chimenti et al. 2003). Accordingly, for the cohort of patients (the aged population) most likely candidates for the regenerative therapy, >50 % of their eCSCs can be senescent and unable to participate in the regenerative process (Torella et al. 2004, 2006a; Matsubara and Kyoto Prefectural University School of Medicine 2012; Chimenti et al. 2003). Thus, if eCSC "aging" is an age or cell cycle dependent process, which affects all or most of the eCSC population, most or all regenerative therapies based on eCSC isolation and expansion will likely result in further exhaustion of the self-renewal capability of these cells with an accelerated loss of their regenerative capacity.

4 Stimulation of the Myocardial Endogenous Capacity for Repair and Regeneration

As above mentioned, while autologous eCSCs undoubtedly hold great promise for cardiac repair, their isolation and expansion prior to cell transplantation can be complex, time consuming and costly. This has raised the question of whether it may be advantageous to target the activation and regenerative capacity of the resident eCSCs to reconstitute damaged myocardium in the absence of cell therapy. One potential therapeutic mechanism of action by which the different forms of transplantation cell therapy are now receiving a lot of attention, is the activation, through a paracrine mechanism, of survival pathways in the cohort of cells at risk together with the endogenous regeneration compartment, represented by the eCSCs. A corollary of this hypothesis is that the identification of the molecules secreted by the transplanted cells should make possible the design of therapies, which eliminate the use of the cells and concentrate on the administration of the principal effector molecules these cells had identified. Myocardial regenerative cell-free therapies effective on the in situ activation, multiplication and differentiation of the resident eCSCs should have many advantages over those based on cell transplantation. First, therapeutic components should be available as "off-the-shelf" and ready to use at all times without the lag time required for the cell therapy approaches; second, they should be affordable, in terms of the production costs of the medicinal product; third, such a therapy should be easy to apply and compatible with current clinical standard of care for AMI, including the widespread use of percutaneous coronary interventions (PCI); and fourth, because of the robustness of the regenerative response produced it should be able to produce and/or recover ~50–60 g of functional myocardial tissue, which is the minimum needed to change the course of the disease in a seriously ill patient. In an attempt to move towards cell-free, protein-based therapies, various growth factors and cytokines have been identified as potential candidates for therapeutic cardiac regeneration and as this list expands so too does our awareness of growth factor-mediated regenerative potential. Vascular endothelial factor (VEGF) is one such factor, which has been identified as central in promoting neo-vascularization post-MI (Crottogini et al. 2003). Initially phase II clinical trials suggested that limited functional benefits were observed upon direct administration of VEGF post-MI (Henry et al. 2003). However, this is now known to be due to the short half-life of VEGF and goes some way to demonstrate how important initial experimental studies are when designing clinical trials. Recent studies have focused on delivering VEGF in combination with various scaffolds and have achieved much greater success in stimulating angiogenesis and restoring cardiac function (Wu et al. 2011; Formiga et al. 2010). Neuregulin 1 (NRG-1) is another key factor implicated in stimulating cardiac repair and regeneration (Wadugu and Kühn 2012; Waring et al. 2012). An Ig-domain containing form of NRG-1β, also known as glial growth factor 2 (GG2) has been shown to improve LVEF and remodelling in pigs post-MI, compared to controls (Kasasbeh et al. 2011). It is thought that NRG-1 imparts functional benefits by activating and increasing c-kit^{pos} eCSC proliferation (Waring et al. 2012), inducing cardiomyocyte replacement (Bersell et al. 2009), protecting cardiomyocytes from apoptosis and improving mitochondrial function (Kasasbeh et al. 2011). Testing regenerative therapies in mouse models of human diseases, although a necessary step in pre-clinical assays, is not an accurate predictor of their human effectiveness. This is so not only because of the potential biological differences between the two species but because of the three order of magnitude difference in mass between the two organisms, which make the challenges not only quantitatively but qualitatively different. Therefore, it is necessary that pre-clinical testing of therapies be carried out in a model, which is more similar in tissue biology, size and physiology to the human than the rodent models commonly used. The pig, because of its size, rapid growth rate, well-known physiology and availability, has proven a very useful and frequently used pre-clinical large animal model for many pathologies, particularly those involving tissue regeneration. Thus, we have recently tested the regenerative effects of intracoronary administration of two growth factors known to be involved in the paracrine effect of the transplanted cells (Ellison et al. 2011). Insulin-like growth factor I (IGF-1) and hepatocyte growth factor (HGF), in doses ranging from 0.5 to 2 μg HGF and 2 to 8 μg IGF-1, were intracoronary administered, just below the site of left anterior descendent occlusion, 30 min after AMI during coronary reperfusion in the pig. This growth factor cocktail triggers a regenerative response from the c-kit^{pos} eCSCs, which is potent and able to produce anatomically, histologically and physiologically significant regeneration of the damaged myocardium without the need for cell transplantation (Ellison et al. 2011). IGF-1 and HGF induced eCSC migration, proliferation and functional cardiomyogenic and microvasculature differentiation. Furthermore, IGF-1/HGF, in a dose-dependent manner, improved cardiomyocyte survival, and reduced fibrosis and cardiomyocyte reactive hypertrophy. Interestingly, the effects of a single administration of IGF-1/HGF are still measurable 2 months after its application, suggesting the existence of a feedback loop triggered by the external stimuli that activates the production of growth and survival factors by the targeted cells, which explains the persistence and long duration of the regenerative myocardial response. These histological changes were correlated with a reduced infarct size and an improved ventricular segmental contractility and ejection fraction at the end of the follow-up assessed by cMRI (Ellison et al. 2011). Despite their effectiveness, the administration of IGF-1 and HGF has a significant drawback. Although it is very effective in the regeneration of the myocytes and micro-vessels lost, the rate of maturation of the newly formed myocytes is heterogeneous and quite slow. While the newly formed myocytes which are in contact with spared ones mature rapidly and can reach a diameter close to a normal pig cardiomyocyte, there is an inverse correlation between new myocyte size and their distance from the small islands of spared myocardium scattered within the ischemic zone (Ellison et al. 2011). With the exception of those new myocytes in close proximity to spared micro-islands of surviving pre-existing myocytes within the ischemic tissue or those in the border region, at 3 weeks after treatment the length and diameter of the remaining new myocytes (~85 % of those regenerated) is between 1/2 and 1/5, respectively, of an adult myocyte, which means that their volume is significantly less than 1/10th of their mature counterparts (Ellison et al. 2011). Because of this slow maturation process, although the therapy is very effective in restoring the number of myocytes lost by the AMI this is not the case as to the regeneration of the lost ventricular mass which lags behind very significantly. In consequence, the myocardial generation of force capacity, that is the meaningful functional recovery, also lags significantly behind the regeneration of the cell numbers to the pre-AMI state. Despite the beneficial effect of the therapy in reducing the scar area, pathological remodelling and partial recovery of ventricular function, there is little doubt that it would be desirable to obtain a more rapid recovery of the ventricular mass and the capacity to generate force. All the currently proposed autologous cell approaches are very attractive from the theoretical and biological standpoint. For those rare diseases with chronic and long-term evolution affecting hundreds or even thousands of potential patients to be treated, these personalized therapies, despite their high cost in medical and material resources, might even make sense from an economic standpoint. Unfortunately, this is not the case for diseases of high prevalence, such as the consequences of ischaemic heart disease, with millions of patients/candidates for regenerative therapy. Not even the developed world has the resources needed to start a program of personalized regenerative medicine for the patients already in CHF who presently are left with heart transplantation as the only realistic option for recovery. Therefore, although the cell transplantation approaches outlined are very valuable as proof-of-concept and as research tools with the possibility of greatly improving a narrow subset of patients in need of therapy, we believe that all of the autologous cell strategies taken together, now and in the foreseeable future, are and will continue to be ineffective to favourably impact the societal healthcare problem posed by the consequences of CHF post-AMI (Ellison et al. 2012b). Moreover, as outlined above, a consensus is gaining ground that most of the favourable effects of cell transplantation protocols used until now exert their beneficial effect by a paracrine mechanism of the transplanted cells over the surviving myocardial cells at risk and/or through the activation of the endogenous myocardial regenerative capacity represented by the eCSCs. If this is correct, then there seems to be little advantage in the use of autologous cells because a similar, and perhaps enhanced, effect can be obtained by the administration of the proper cell type isolated from allogeneic sources. These can be produced in large amounts beforehand, kept stored frozen before their use, and remain available at all times, which would allow their use not only for the treatment of the pathological remodelling once it has developed but soon after the acute insult in order to induce early regeneration of the cells lost in order to prevent or diminish the pathological remodelling. Mesenchymal stem cells (MSC) have a broad repertoire of secreted trophic and immunodulatory cytokines, however they also secrete factors that negatively modulate cardiomyocyte apoptosis, inflammation, scar formation and pathological remodelling (Ranganath et al. 2012). Medicetty and colleagues (2012) used a porcine model of AMI and delivered 20–200 million allogeneic, multipotent, adult BMDCs (MultiStem; that are nonimmunogeneic and can suppress activated T-cell proliferation and have anti-inflammatory and angiogenic properties as well), directly to the myocardium via the infarct related vessel using a transarterial microsyringe catheter-based delivery system, 2 days after AMI. Echocardiography showed significant improvements in regional and global LV function and remodelling at 30 and 90 days after myocardial injury (Medicetty et al. 2012). Rapidly following on from this pre-clinical study, Penn et al. (2012) conducted a multi-centre phase I trial of the effects of adventitial delivery of MultiStem in patients 2–5 days after primary PCI. In patients with EF determined to be <45 % before the MultiStem injection, at 4 months after AMI, a 1, 4, 14, and 11 % absolute increase in EF was observed following injection of 20, 50, and 100 million cells, respectively (Penn et al. 2012). Recently, Marban and colleagues (2012) have tested the safety and efficacy of using allogeneic, mismatched cardiosphere-derived Cells (CDCs) in infarcted rats. Rats underwent permanent ligation of the LAD coronary artery and two million CDCs or vehicle were intramyocardially injected at four sites in the peri-infarct zone. Three weeks post-MI, animals that received allogeneic CDCs exhibited smaller scar size, increased infarcted wall thickness and attenuation of LV remodelling. Allogeneic CDC transplantation resulted in a robust improvement of fractional area change (~12 %), ejection fraction (~20 %), and fractional shortening (~10 %), and this was sustained for at least 6 months. Furthermore, allogeneic CDCs stimulated endogenous regenerative mechanisms (cardiomyocyte cycling, recruitment of c-kit^{pos} eCSCs, angiogenesis) and increased myocardial VEGF, IGF-1 and HGF (Malliaras et al. 2012). Unlike other cell types (Janssens 2010;

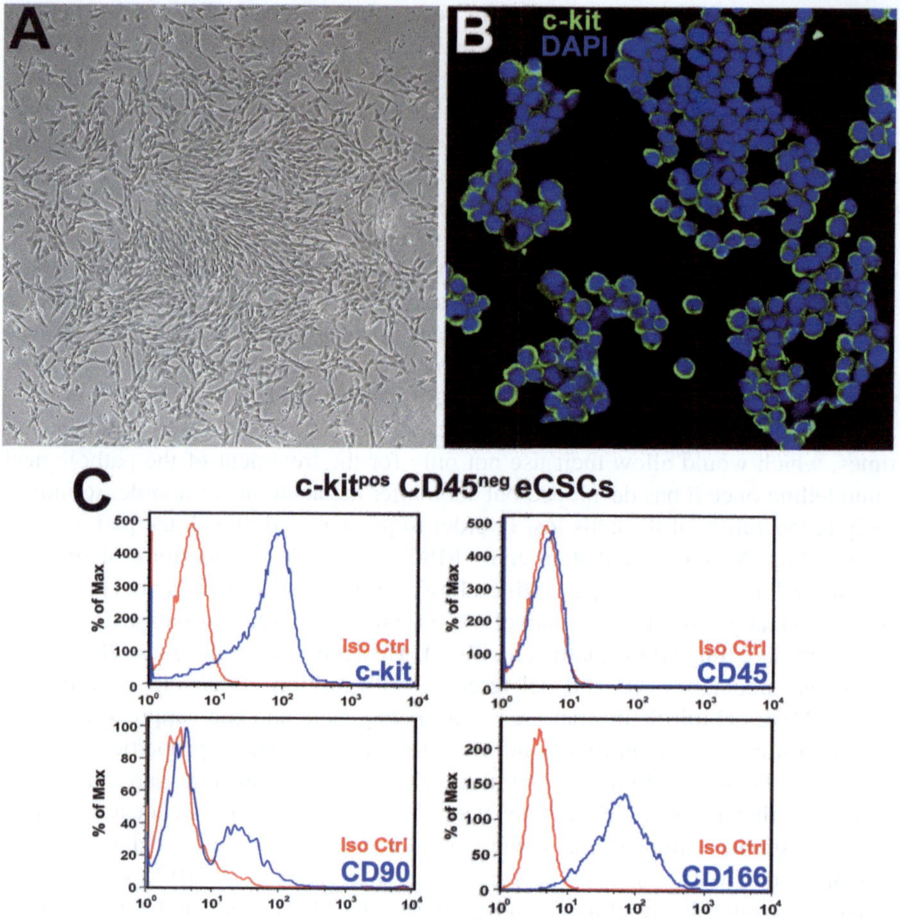

Fig. 2.2 Endogenous cardiac stem cell essential phenotype. (**a**) Light microscopy representative image of long-term cultured pig eCSCs. (**b**) Cytospin preparation and c-kit immunofluorescence of cloned eCSCs. (**c**) Essential CD phenotype of a typical CSC preparation (**b**, **c**) are adapted from Ellison et al. (45)

Hofmann et al. 2005), eCSCs have a very high tropism for the myocardium (Ellison et al. 2013). Under proper culture conditions it is possible to clone and expand a single rodent, porcine or human eCSC to up to 1×10^{10} cells without detectable alteration of karyotype, loss of differentiating properties or the phenotype of the differentiated progeny (Fig. 2.2) (Ellison et al. 2011). These cloned cells produce a repertoire of pro-survival and cardiovascular regenerative growth factors [Our unpublished data]. For this reason, we decided to test whether these in vitro expanded cells, when administered into allogeneic animals, would be the source of a more complex and physiologic mixture of growth and differentiating factors which, through a paracrine effect would produce a robust activation of the eCSCs with more rapid maturation of their progeny. It was expected that once their short-term effect had been produced and the auto/paracrine feedback loop of growth factor production has been activated in the eCSCs, the allogeneic cells would be eliminated (presumably by apoptosis) and that the regeneration triggered by activated eCSCs would be completely autologous. c-kit^{pos} eCSCs do not express either MHC-I locus or co-activator molecules and have strong immunomodulatory properties in vitro when tested in the mixed lymphocyte reaction [Our unpublished Data]. We therefore expected the expanded cells to survive long enough in the allogeneic host to produce their paracrine effect before being eliminated by the host immune system. Allogeneic, non-matched, cloned male EGFP-transduced porcine eCSCs, were administered intracoronary in white Yorkshire female pigs, 30 min after MI and coronary reperfusion (Ellison et al. 2009). Pig serum was injected to control pigs after MI (CTRL). The cells or sera were injected through a percutaneous catheter into the anterior descending coronary artery just below the site of balloon occlusion used to produce the AMI. We found a high degree of EGFP^{pos}/c-kit^{pos} heterologous HLA non-matched allogeneic porcine CSCs nesting in the damaged pig myocardium at 30 min through to 1 day after MI. At 3 weeks post-AMI, all the injected allogeneic cells had disappeared from the myocardium and peripheral tissues (i.e. spleen). There was significant activation of the endogenous GFP^{neg} c-kit^{pos} CSCs (eCSCs) following allogeneic CSC treatment (Fig. 2.3), so that by 3 weeks after MI, there was increased new cardiomyocyte and capillary formation, which was not evident in the control hearts (Fig. 2.3). Moreover, through paracrine mechanisms, c-kit^{pos} heterologous HLA non-matched allogeneic CSC treatment preserved myocardial wall structure and attenuated remodelling by reducing myocyte hypertrophy, apoptosis and scar formation (fibrosis) (Ellison et al. 2009). In summary, intracoronary injection of allogeneic CSCs after MI in pigs, which is a clinically relevant MI model, activates the eCSCs through a paracrine mechanism resulting in improved myocardial cell survival, function, remodelling and regeneration. A possible risk of using large numbers of in vitro expanded CSCs is the appearance of transformed cells with the potential to form abnormal growths. This risk is completely eliminated by the use of allogeneic cells, with a different HLA allele from the recipient, because they all get eliminated by the immune system without immunosuppression. Claims that some of the transplanted allogeneic cells have a long-term survival in the host, have not been reproduced or thoroughly documented (Malliaras et al. 2012; Quevedo et al. 2009; Huang et al. 2010). If their survival proves to be correct, many of the immunology concepts, which have ruled transplant biology until now, will need to be revised. Furthermore, despite thorough pathological examination and contrary to many iPS- and ECS-derived cell lines, the adult tissue-specific eCSCs have a very low or non-existent capacity to form tumours and/or teratomas in syngeneic or immunodeficient animals [our unpublished data and (Chong et al. 2011)]. Allogeneic CSC therapy is conceptually and practically different from any presently in clinical use. The proposed cell therapy is only a different form of growth factor therapy able to deliver a more complex mixture of growth factors than our present knowledge permits us to prepare. The factors

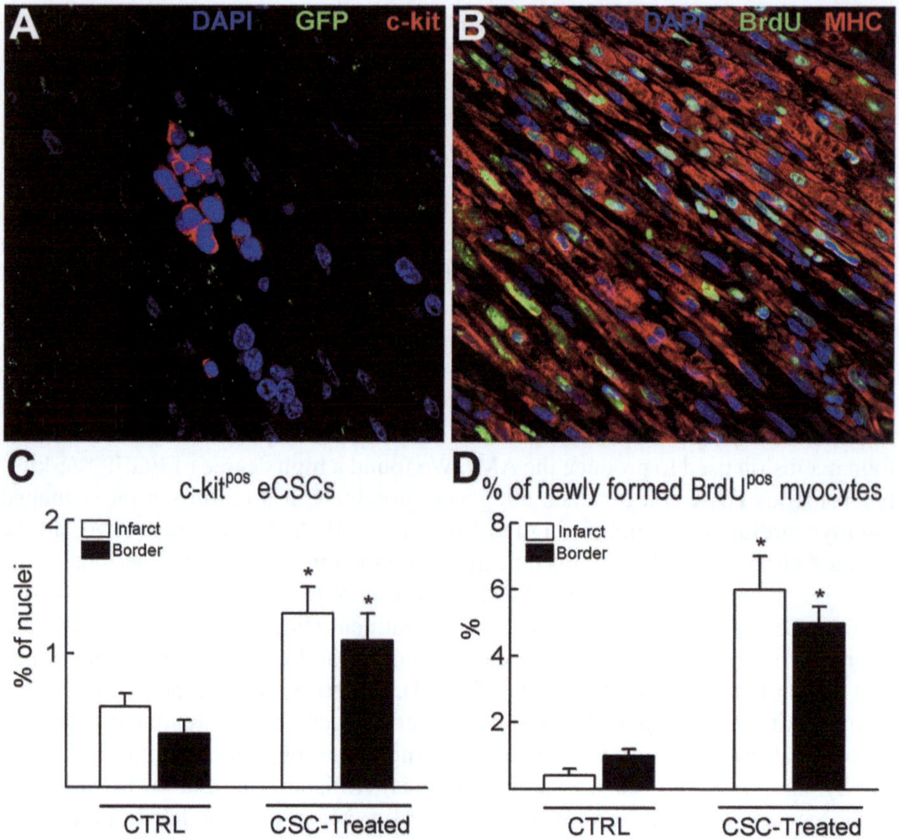

Fig. 2.3 Activation of tissue-specific endogenous resident c-kit^{pos} CSCs following intracoronary injection of c-kit^{pos} allogeneic porcine CSCs after acute myocardial infarction in pigs. (**a**) A cluster of activated GFP^{neg}, c-kit^{pos} (*red*) endogenous CSCs in the 3-week-old infarcted region of the allogeneic CSC-treated porcine myocardium. Nuclei are stained by DAPI in *blue*. (**b**) Regenerating band of newly formed BrdU^{pos} (*green*) cardiomyocytes (*red*, MHC) in the infarct region, 3 weeks following allogeneic CSC treatment. Nuclei are stained by DAPI in *blue*. (**c**) The number of c-kit^{pos} endogenous CSCs significantly increased following intracoronary allogeneic CSC treatment. $*P<0.05$ vs. CTRL. (**d**) New BrdU^{pos} myocyte formation significantly increased following allogeneic CSC therapy. $*P<0.05$ vs. CTRL. Adapted from Ellison et al. J Cardiovasc Transl Res. 2012;5:667–77

produced by the allogeneic cells are designed to stimulate the endogenous stem cells of the target tissue but the transplanted cells themselves survive only transiently and do not directly participate in the production of progeny that contributes to the regenerated tissue. Once more information is available, the allogeneic cells could be used either alone or in combination with the available factor therapy to improve the activation of the eCSCs and the maturation of their progeny.

5 Summary and Conclusions

The findings that the adult heart harbours a regenerative multipotent cell population composed by eCSCs and that mammalian, including human, cardiomyocytes are replaced throughout adulthood represents a paradigm shift in cardiovascular biology. The presence of this regenerative agent within the adult heart supports the view that the heart has the potential to repair itself if the eCSCs can be properly stimulated. Indeed, it is predicted that in the near future it should be possible to replace cell transplantation-based myocardial regeneration protocols with an "off-the-shelf", readily available, unlimited and effective regenerative/reparative therapy based on specific growth factor administration or on the paracrine secretion by allogeneic CSC transplantation able to produce the activation in situ of the resident eCSCs. However, before reaching this optimistic clinical scenario, it is mandatory to obtain a better understanding of eCSC biology in order to fully exploit their regeneration potential. The latter will ultimately lead to developing realistic and clinically applicable myocardial regeneration strategies. Cardiac regenerative medicine is set to revolutionize the treatment of cardiac diseases and such research will have significant and long-term impact on socio-economics and patient well-being. Indeed, therapies which are based on findings from high quality research will undoubtedly cut deaths from cardiovascular disease, reduce recovery times, increase life expectancy and quality of care and save money.

References

Akashi YJ, Goldstein DS et al (2008) Takotsubo cardiomyopathy: a new form of acute, reversible heart failure. Circulation 118:2754–2762

Anversa P, Nadal-Ginard B (2002) Myocyte renewal and ventricular remodelling. Nature 415: 240–243

Arsalan M, Woitek F et al (2012) Distribution of cardiac stem cells in the human heart. ISRN Cardiol 2012:483407

Bearzi C, Rota M et al (2007) Human cardiac stem cells. Proc Natl Acad Sci U S A 104(35): 14068–14073

Beltrami AP, Urbanek K et al (2001) Evidence that human cardiac myocytes divide after myocardial infarction. N Engl J Med 344:1750–1757

Beltrami AP, Barlucchi L et al (2003) Adult cardiac stem cells are multipotent and support myocardial regeneration. Cell 114:763–776

Bergmann O, Bhardwaj RD et al (2009) Evidence for cardiomyocyte renewal in humans. Science 324:98–102

Bersell K, Arab S et al (2009) Neuregulin1/ErbB4 signaling induces cardiomyocyte proliferation and repair of heart injury. Cell 138:257–270

Bolli R, Chugh AR et al (2011) Cardiac stem cells in patients with ischaemic cardiomyopathy (SCIPIO): initial results of a randomised phase 1 trial. Lancet 378:1847–1857

Boström P, Mann N et al (2010) C/EBPβ controls exercise-induced cardiac growth and protects against pathological cardiac remodeling. Cell 143:1072–1083

Buckingham M, Montarras D (2008) Skeletal muscle stem cells. Curr Opin Genet Dev 18: 330–336

Chien KR, Olson EN (2002) Converging pathways and principles in heart development and disease: CV@CSH. Cell 110:153–162

Chimenti C, Kajstura J et al (2003) Senescence and death of primitive cells and myocytes lead to premature cardiac aging and heart failure. Circ Res 93:604–613

Chong JJ, Chandrakanthan V et al (2011) Adult cardiac-resident MSC-like stem cells with a proepicardial origin. Cell Stem Cell 9:527–540

Chugh AR, Beache GM et al (2012) Administration of cardiac stem cells in patients with ischemic cardiomyopathy: the SCIPIO trial: surgical aspects and interim analysis of myocardial function and viability by magnetic resonance. Circulation 126:S54–S64

Crottogini A, Meckert PC et al (2003) Arteriogenesis induced by intramyocardial vascular endothelial growth factor 165 gene transfer in chronically ischemic pigs. Hum Gene Ther 14(14): 1307–1318

Eisenberg CA, Burch JB et al (2006) Bone marrow cells transdifferentiate to cardiomyocytes when introduced into the embryonic heart. Stem Cells 24:1236–1245

Ellison GM, Torella D et al (2007a) Myocyte death and renewal: modern concepts of cardiac cellular homeostasis. Nat Clin Pract Cardiovasc Med 4(suppl 1):S52–S59

Ellison GM, Torella D et al (2007b) Acute beta-adrenergic overload produces myocyte damage through calcium leakage from the ryanodine receptor 2 but spares cardiac stem cells. J Biol Chem 282:11397–11409

Ellison GM, Torella D et al (2009) Use of heterologous non-matched cardiac stem cells (CSCs) without immunosuppression as an effective regenerating agent in a porcine model of acute myocardial infarction. Eur Heart J 30(Abstract Supplement):495

Ellison GM, Galuppo V et al (2010) Cardiac stem and progenitor cell identification: different markers for the same cell? Front Biosci 2:641–652

Ellison GM, Torella D et al (2011) Endogenous cardiac stem cell activation by insulin-like growth factor-1/hepatocyte growth factor intracoronary injection fosters survival and regeneration of the infarcted pig heart. J Am Coll Cardiol 58(9):977–986

Ellison GM, Waring CD et al (2012a) Physiological cardiac remodelling in response to endurance exercise training: cellular and molecular mechanisms. Heart 98:5–10

Ellison GM, Nadal-Ginard B et al (2012b) Optimizing cardiac repair and regeneration through activation of the endogenous cardiac stem cell compartment. J Cardiovasc Transl Res 5(5): 667–677

Ellison GM, Vicinanza C et al (2013) Adult c-kit(pos) cardiac stem cells are necessary and sufficient for functional cardiac regeneration and repair. Cell 154:827–842

Formiga FR, Pelacho B et al (2010) Sustained release of VEGF through PLGA microparticles improves vasculogenesis and tissue remodeling in an acute myocardial ischemia-reperfusion model. J Control Release 147(1):30–37

Fransioli J, Bailey B et al (2008) Evolution of the c-kit-positive cell response to pathological challenge in the myocardium. Stem Cells 26(5):1315–1324

Henry TD, Annex BH et al (2003) The VIVA trial: vascular endothelial growth factor in ischemia for vascular angiogenesis. Circulation 107(10):1359–1365

Hofmann M, Wollert KC et al (2005) Monitoring of bone marrow cell homing into the infarcted human myocardium. Circulation 111:2198–2202

Hsieh PC, Segers VF et al (2007) Evidence from a genetic fate-mapping study that stem cells refresh adult mammalian cardiomyocytes after injury. Nat Med 13(8):970–974

Huang XP, Sun Z et al (2010) Differentiation of allogeneic mesenchymal stem cells induces immunogenicity and limits their long-term benefits for myocardial repair. Circulation 122: 2419–2429

Hunter JJ, Chien KR (1999) Signaling pathways for cardiac hypertrophy and failure. N Engl J Med 341:1276–1283

Janssens S (2010) Stem cells in the treatment of heart disease. Annu Rev Med 61:287–300

Jessup M, Brozena S (2003) Heart failure. N Engl J Med 348:2007–2018

Jesty SA, Steffey MA et al (2012) c-kit+ precursors support postinfarction myogenesis in the neonatal, but not adult, heart. Proc Natl Acad Sci U S A 109(33):13380–13385

Kahan BD (2011) Fifty years in the vineyard of transplantation: looking back. Transplant Proc 43:2853–2859

Kajstura J, Rota M et al (2012) Cardiomyogenesis in the aging and failing human heart. Circulation 126:1869–1881

Kasasbeh E, Murphy A et al (2011) Neuregulin-1β improves cardiac remodeling after myocardial infarction in swine. Circulation 124:Abstract 15531

Kattman SJ, Huber TL et al (2006) Multipotent flk-1+ cardiovascular progenitor cells give rise to the cardiomyocyte, endothelial, and vascular smooth muscle lineages. Dev Cell 11:723–732

Kopp JL, Dubois CL et al (2011) Progenitor cell domains in the developing and adult pancreas. Cell Cycle 10:1921–1927

Kotton DN (2012) Next generation regeneration: the hope and hype of lung stem cell research. Am J Respir Crit Care Med 185(12):1255–1260

Kühn B, Del Monte F et al (2007) Periostin induces proliferation of differentiated cardiomyocytes and promotes cardiac repair. Nat Med 13:962–969

Laflamme MA, Murry CE (2011) Heart regeneration. Nature 473:326–335

Laugwitz KL, Moretti A et al (2005) Postnatal isl1+ cardioblasts enter fully differentiated cardiomyocyte lineages. Nature 433:647–653

Linke A, Müller P et al (2005) Stem cells in the dog heart are self-renewing, clonogenic, and multipotent and regenerate infarcted myocardium, improving cardiac function. Proc Natl Acad Sci U S A 102(25):8966–8971

Loffredo FS, Steinhauser ML et al (2011) Bone marrow-derived cell therapy stimulates endogenous cardiomyocyte progenitors and promotes cardiac repair. Cell Stem Cell 8:389–398

Makkar RR, Smith RR et al (2012) Intracoronary cardiosphere-derived cells for heart regeneration after myocardial infarction (CADUCEUS): a prospective, randomised phase 1 trial. Lancet 379:895–904

Malliaras K, Li TS et al (2012) Safety and efficacy of allogeneic cell therapy in infarcted rats transplanted with mismatched cardiosphere-derived cells. Circulation 125:100–112

Martin CM, Meeson AP et al (2004) Persistent expression of the ATP-binding cassette transporter, Abcg2, identifies cardiac SP cells in the developing and adult heart. Dev Biol 265:262–275

Matsubara H, Kyoto Prefectural University School of Medicine (2012) AutoLogous human cardiac-derived stem cell to treat ischemic cardiomyopathy (ALCADIA). ClinicalTrials.gov. Available from: http://clinicaltrials.gov/ct2/show/NCT 00981006

Matsuura K, Nagai T et al (2004) Adult cardiac Sca-1-positive cells differentiate into beating cardiomyocytes. J Biol Chem 279:11384–11391

Medicetty S, Wiktor D et al (2012) Percutaneous adventitial delivery of allogeneic bone marrow derived stem cells via infarct related artery improves long-term ventricular function in acute myocardial infarction. Cell Transplant 21(6):1109–1120

Messina E, De Angelis L et al (2004) Isolation and expansion of adult cardiac stem cells from human and murine heart. Circ Res 95(9):911–921

Moretti A, Caron L et al (2006) Multipotent embryonic isl1+ progenitor cells lead to cardiac, smooth muscle, and endothelial cell diversification. Cell 127:1151–1165

Nadal-Ginard B (1978) Commitment, fusion and biochemical differentiation of a myogenic cell line in the absence of DNA synthesis. Cell 15:855–864

Nadal-Ginard B, Kajstura J et al (2003) Myocyte death, growth, and regeneration in cardiac hypertrophy and failure. Circ Res 92:139–150

Oh H, Taffet GE et al (2001) Telomerase reverse transcriptase promotes cardiac muscle cell proliferation, hypertrophy, and survival. Proc Natl Acad Sci U S A 98:10308–10313

Oh H, Bradfute SB et al (2003) Cardiac progenitor cells from adult myocardium: homing, differentiation, and fusion after infarction. Proc Natl Acad Sci U S A 100:12313–12318

Orlic D, Kajstura J et al (2001) Bone marrow cells regenerate infarcted myocardium. Nature 410:701–705

Penn MS, Ellis S et al (2012) Adventitial delivery of an allogeneic bone marrow-derived adherent stem cell in acute myocardial infarction: phase I clinical study. Circ Res 110:304–311

Quaini F, Urbanek K et al (2002) Chimerism of the transplanted heart. N Engl J Med 346:5–15
Quevedo HC, Hatzistergos KE et al (2009) Allogeneic mesenchymal stem cells restore cardiac function in chronic ischemic cardiomyopathy via trilineage differentiating capacity. Proc Natl Acad Sci U S A 106:14022–14027
Ranganath SH, Levy O et al (2012) Harnessing the mesenchymal stem cell secretome for the treatment of cardiovascular disease. Cell Stem Cell 10:244–258
Rasmussen TL, Raveendran G et al (2011) Getting to the heart of myocardial stem cells and cell therapy. Circulation 123:1771–1779
Reule S, Gupta S (2011) Kidney regeneration and resident stem cells. Organogenesis 7:135–139
Robinton DA, Daley GQ (2012) The promise of induced pluripotent stem cells in research and therapy. Nature 481:295–305
Roger VL (2013) Epidemiology of heart failure. Circ Res 113:646–659
Rountree CB, Mishra L et al (2012) Stem cells in liver diseases and cancer: recent advances on the path to new therapies. Hepatology 55:298–306
Senyo SE, Steinhauser ML et al (2013) Mammalian heart renewal by pre-existing cardiomyocytes. Nature 493:433–436
Smart N, Bollini S et al (2011) De novo cardiomyocytes from within the activated adult heart after injury. Nature 474:640–644
Soonpaa MH, Field LJ (1998) Survey of studies examining mammalian cardiomyocyte DNA synthesis. Circ Res 83:15–26
Srivastava D, Ivey KN (2006) Potential of stem-cell-based therapies for heart disease. Nature 441:1097–1099
Suh H, Deng W et al (2009) Signaling in adult neurogenesis. Annu Rev Cell Dev Biol 25:253–275
Terzic A, Nelson TJ (2010) Regenerative medicine advancing health care 2020. J Am Coll Cardiol 55:2254–2257
Torella D, Rota M et al (2004) Cardiac stem cell and myocyte aging, heart failure, and insulin-like growth factor-1 overexpression. Circ Res 94:514–524
Torella D, Ellison GM et al (2006a) Resident human cardiac stem cells: role in cardiac cellular homeostasis and potential for myocardial regeneration. Nat Clin Pract Cardiovasc Med 3(suppl 1): S8–S13
Torella D, Ellison GM et al (2006b) Biological properties and regenerative potential, in vitro and in vivo, of human cardiac stem cells isolated from each of the four chambers of the adult human heart. Circulation 114:87
Torella D, Ellison GM et al (2007) Resident cardiac stem cells. Cell Mol Life Sci 64:661–673
Urbanek K, Quaini F et al (2003) Intense myocyte formation from cardiac stem cells in human cardiac hypertrophy. Proc Natl Acad Sci U S A 100:10440–10445
Urbanek K, Torella D et al (2005) Myocardial regeneration by activation of multipotent cardiac stem cells in ischemic heart failure. Proc Natl Acad Sci U S A 102:8692–8697
Wadugu B, Kühn B (2012) The role of neuregulin/ErbB2/ErbB4 signaling in the heart with special focus on effects on cardiomyocyte proliferation. Am J Physiol Heart Circ Physiol 302(11): H2139–H2147
Waring CD, Vicinanza C et al (2012) The adult heart responds to increased workload with physiologic hypertrophy, cardiac stem cell activation, and new myocyte formation. Eur Heart J. Oct 25. [Epub ahead of print] PMID: 23100284
Wu SM, Fujiwara Y et al (2006) Developmental origin of a bipotential myocardial and smooth muscle cell precursor in the mammalian heart. Cell 127:1137–1150
Wu J, Zeng F et al (2011) Infarct stabilization and cardiac repair with a VEGF-conjugated, injectable hydrogel. Biomaterials 32(2):579–586
Yamanaka S (2007) Strategies and new developments in the generation of patient-specific pluripotent stem cells. Cell Stem Cell 1:39–49
Zaruba MM, Soonpaa M et al (2010) Cardiomyogenic potential of C-kit(+)-expressing cells derived from neonatal and adult mouse hearts. Circulation 11:121(18)

Chapter 3
Large Animal Induced Pluripotent Stem Cells as Models of Human Diseases

Anjali Nandal and Bhanu Prakash V.L. Telugu

Abbreviations

EpiSC	Epiblast stem cells
ESC	Embryonic stem cells
FGF	Fibroblast growth factor
iPSC	Induced pluripotent stem cells
LIF	Leukemia inhibiting factor

1 Pluripotent Stem Cells

Pluripotency is defined by the ability of cells to self renew indefinitely, and differentiate into cells representing all the three germ layers: ectoderm, mesoderm, and endoderm. In mouse, pluripotent stem cells have been established either from the inner cell mass (ICM) of early blastocyst-the embryonic stem cells (ESC); a more mature epiblast-the epiblast stem cells (EpiSC); germline tumors or teratocarcinomas-the embryonic carcinoma cells (ECC); or the primordial germ cells (PGC) of the fetus-the embryonic germ cells (EGC). Among the different pluripotent cell types, ESC are of particular interest because they are relatively more stable in culture and can be modified with desired mutations, and transferred to the next generation. To be considered as a genuine ESC, the stem cells have to pass standard pluripotency tests and be capable of: (1) forming embryoid bodies in vitro; (2) giving

A. Nandal • Bhanu Prakash V.L.Telugu (✉)
Department of Animal and Avian Sciences, University of Maryland,
2121 ANSC Building, College Park, MD 20742, USA

Animal Bioscience and Biotechnology Laboratory, USDA Agricultural Research Service (USDA-ARS), Beltsville, MD 20705, USA
e-mail: btelugu@umd.edu

T.A.L. Brevini (ed.), *Stem Cells in Animal Species: From Pre-clinic to Biodiversity*,
Stem Cell Biology and Regenerative Medicine, DOI 10.1007/978-3-319-03572-7_3,

rise to teratomas upon transplantation into an immunocompromised animal; (3) derivation of chimeric offspring in vivo with a demonstration that the ESC contributed to the tissues derived from the endoderm, mesoderm, and ectoderm, in addition to the germline; and (4) succeeding in the most stringent of the pluripotency tests, which is tetraploid complementation, where the ESC alone gives rise to the embryo proper (Nagy et al. 1993).

Among the ESC established from the preimplantation embryos so far, the ESC from mouse (m) is the only cell type that has met all the stringent measures of pluripotency outlined above, indicating that all pluripotent stem cells from the early embryos are not identical. This is evident from the human and primate ESC that differ in morphology, growth factor requirements, and other aspects of their phenotype from mESC. The existence of at least two, if not more distinct pluripotent stem cells was further corroborated by the establishment of a different kind of ESC, also from the mouse, the EpiSC (Brons et al. 2007; Tesar et al. 2007). The mEpiSC are similar to the primate ESC in colony morphology and require activin A and FGF2 for pluripotency, which is in stark contrast to mESC that are dependent on LIF. Because the EpiSC are established from an advanced differentiated epiblast—stage of mouse conceptuses in comparison to the "naïve" ESC from the ICM, they are classified as "primed" ESC. Besides the obvious differences in phenotype and growth factor requirements, the mEpiSC also differ from mESC in their inability to tolerate dispersion into single cells, responsiveness to BMP4, and relatively poor contribution to chimerism, confirming their distinctiveness from the naïve ESC. It can therefore be hypothesized that primate ESC that share properties of EpiSC, presumably originated from a naïve epiblast, but spontaneously progressed to the primed epiblast stage in vitro before giving rise to a stable cell line. Arguably, the primate ESC are actually EpiSC though they have been classified under ESC. Regardless, at least in mouse, the primed cells can be converted into naïve cells by incorporating small molecule compounds activating WNT signaling pathway, and by using kinase inhibitors inhibiting the MEK signaling pathway (Rodriguez et al. 2012). This interchangeability is of particular interest to stem cell biologists because, the naïve stemness, a hallmark of mESC is coveted for their ease of passage, cryopreservation, rapid proliferation, stability in culture, and high competence for producing germline chimeras (Brons et al. 2007; Tesar et al. 2007).

A little less than a decade after the first publication of mESC, reports of ESC/ESC-like lines from in vivo- or in vitro-derived embryos were reported from pigs (Notarianni et al. 1990; Piedrahita et al. 1990) and sheep (Notarianni et al. 1991). These ESC-like cells bore characteristics of primed ESC in their morphology but failed to maintain self-renewal, or remain in an undifferentiated state, or differentiate efficiently into various cell types of the three germ layers in vitro or in vivo. Although chimera formation was reported, tissue analysis was minimal or ESC-derived tissue types were limited. Similarly, the origin of the differentiated teratoma tissue was assumed to be ESC-derived, but it was not proven. Despite two decades of additional effort, proven ESC lines from any domestic large animal species has yet to be reported. This failure could largely be attributable to differences in the developmental biology of ungulate (u) species. In human and mouse embryos, morula stage is generally similar to ungulates but one significant difference is that human and mouse embryos reach blastocyst stage 24–48 h earlier than cattle and swine. In human and mouse, the epiblast develops in vivo following implantation but ungulates have an extended peri-implantation period following hatching during which, epiblast is formed (Vejlsted et al. 2006), extraembryonic membranes expand (Geisert et al. 1982); gastrulation begins and Rauber's Layer is shed, exposing underlying epiblast to the uterine luminal fluid (Flechon et al. 2004). These distinct phenomena suggest that the ungulate embryos are possibly exposed to unique developmental cues and hence, the optimal time to establish uESC lines could be different from human and mouse. Analysis of recently established putative bovine (Lim et al. 2011) and porcine ESC (Alberio et al. 2010) showed that although they expressed OCT4, SOX2, and NANOG; only the bovine ESC were AP and REX1 positive. It is noteworthy to point out that both AP and REX1 expressions were observed in the initial culture of the porcine epiblasts but were lost during cell line establishment (Alberio et al. 2010). This indicates that the expression of some other crucial factors may have been compromised with in vitro culture as the three core pluripotency factors were present in both bovine and porcine ESC. Other reasons could be the non-availability of factors that define pluripotency in uESC, and the lack of appropriate in vitro culture system(s) to maintain pluripotent and self-renewing uESC. Since robust uESC are not available, one possibility of bypassing the need of uESC is the derivation of induced pluripotent stem cells (iPSC) as described below.

2 Induced Pluripotent Cells

Six years ago, in the initial landmark publication, Takahashi and Yamanaka (2006) reported reprogramming of somatic cells to iPSC by the ectopic expression of four transcription factors Oct4, Sox2, Klf4, and c-Myc. The mouse iPSC that was reported in this and subsequent studies had properties common with ESC, including their ability to form teratomas, generate germline chimeras and live offspring (Kang et al. 2009; Zhao et al. 2009). The technology has since revolutionized the field and has been adopted to successfully establish iPSC from humans (Takahashi et al. 2007) and several other primate and non-primate species. Genome-integrating lentiviral (Winkler et al. 2010) and retroviral vectors (Okita et al. 2007) were widely used initially for the generation of iPSC. However, a spate of recent publications were aimed at creating iPSC lines by either modifying the vectors, including adenoviruses (Stadtfeld et al. 2008), plasmid vectors (Okita et al. 2010), episomal vectors (Yu et al. 2009), minicircle DNA (Jia et al. 2010); or modes of delivery, such as those required for delivery of small molecules/chemicals (Desponts and Ding 2010; Hou et al. 2013), mRNA (Tavernier et al. 2012) and recombinant proteins (Zhou et al. 2009). Various mechanisms to generate iPSC are briefly discussed in Fig. 3.2. Nevertheless, the aim is to improve reprogramming efficiency of the somatic cells by using minimal or no reprogramming factors as chemical reprogramming could efficiently generate cells with the least genomic variations. The realization that

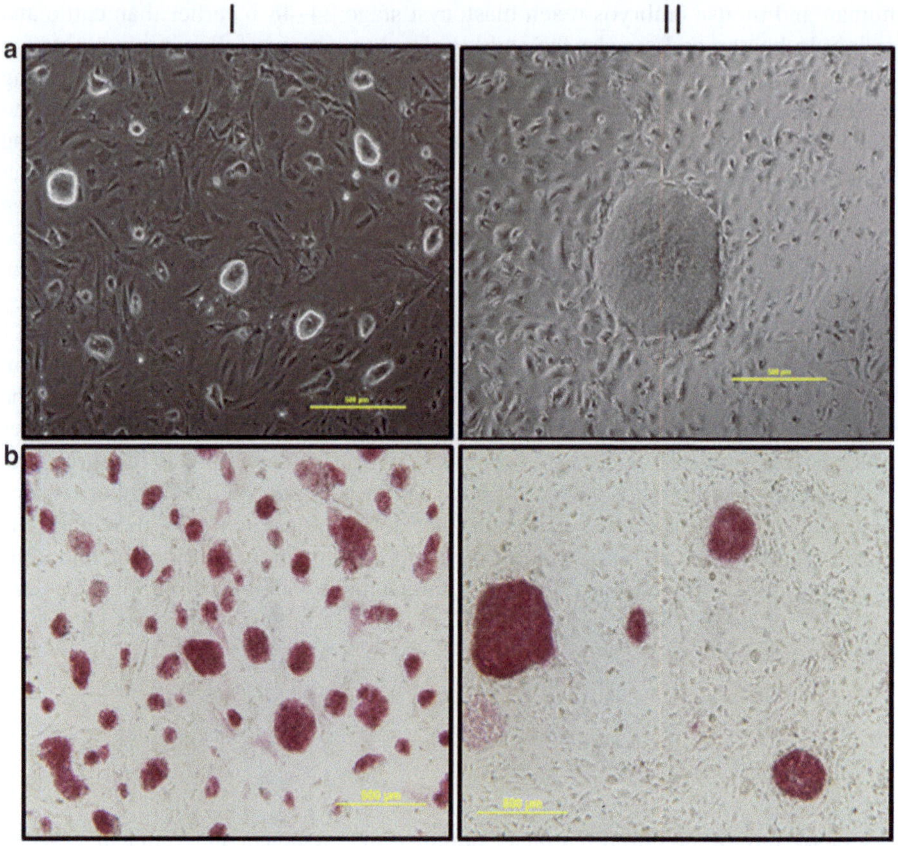

Fig. 3.1 Typical morphological features of 'naïve' and 'primed' induced pluripotent cells (**a**). (**I**) The colonies of naïve-type caprine iPSC reprogrammed from fetal fibroblasts with six factors (OSKMLN) through lentivial transduction and 2i/LIF (Leukemia Inhibitory Factor) supplemented medium on a feeder layer of mouse embryonic fibroblasts (MEF) at day 3. (**II**) Colonies of primed-type goat iPSC cells derived from fibroblasts by the delivery of four factors and addition of bFGF to the medium on a feeder layer of MEF at 3–4 days. (**b**) Alkaline phosphatase (AP) stained naïve and primed caprine colonies to further emphasize the size and morphology difference between the colonies (unpublished data). The brightfield and AP stained images were magnified using 4X and all the iPSC were maintained on feeders

iPSC could be readily created from these animal models prompted investigation into the establishment of similar lines from other important animal models such as the dog and pig, and from endangered species and disappearing breeds (Roberts et al. 2009; Telugu et al. 2010b). It is believed that somatic cells transformed into iPSC maintain the culture requirements typical of the species of origin with mouse cells giving rise to the naïve type of stem cells, and human cells generating cell lines with the characteristics of the primed state (Telugu et al. 2010a). This indicates that once somatic cells are reprogrammed to pluripotency, it follows the

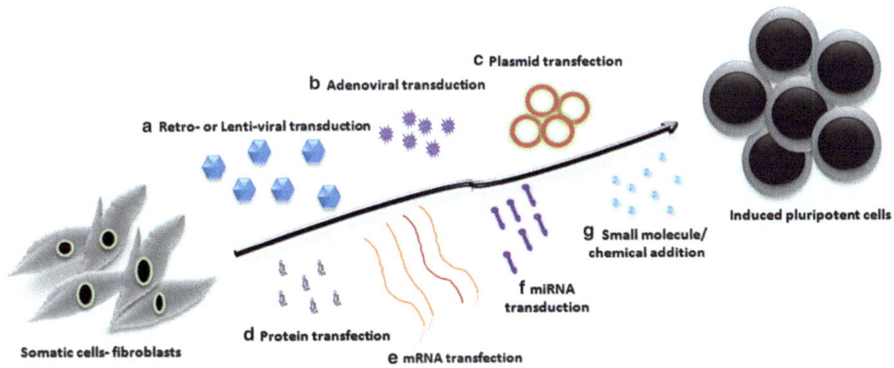

Fig. 3.2 Various reprogramming strategies to induce pluripotent stem cells from adult somatic cells. Earlier methods generally involved the use of efficient viral delivery system by (A) retroviruses and lentiviruses, which integrate into the genome at high frequency or (B) adenoviral transduction, where virus does not integrate. (C) Transfection of comparatively less efficient but non-viral and non-integrating DNA-based methods like plasmids, minicircles, and transposons have been successfully utilized, albeit with less frequency. Similarly, non-integrating (D) reprogramming factor proteins and (E) in vitro transcribed RNA (also capped and ply A-tailed) can be transfected in the cells for efficient reprogramming. Additionally, (F) miRNA-mediated reprogramming and (G) efficient induction of pluripotency by small molecules/chemicals has also been recently reported

default behavior that is typical of the epiblast of its species. It would be therefore very useful to dissect the properties of ungulate iPSC and, as a consequence, their typical default stemness state. Since ESC and iPSC are very similar, iPSC also either have a naïve or primed stemness state. The differences between naïve and primed iPSC are illustrated by brightfield and Alkaline Phosphatase (AP) stained images for goats (unpublished data) in Fig. 3.1.

In many studies with pig, sheep, and bovine iPSC, it is difficult to classify the cells either as naïve or primed as they were propagated in media supplemented with both LIF and FGF2 (Montserrat et al. 2012; Sumer et al. 2011; Liu et al. 2012). However, in two of these studies, morphology was described as similar to that of the human ESC which suggests that they could be primed cells. In some other studies, the iPSC generated were possibly wrongly classified as naïve as the morphological characteristics described were similar to the human primed cells (Han et al. 2011). In another study, true pluripotency was observed and chimera formation at least at the blastocyst stage, was reported for sheep iPSC (Montserrat et al. 2011). Therefore, the whole picture is unclear but suggests that currently in most cases, ungulate iPSC belong mainly to the primed category. It would be a significant study in the future to develop novel methods to reprogram ungulate primed iPSC into naïve iPSC. The establishment of naïve iPSC by Telugu et al. (2010a, b) is certainly a good beginning. In spite of various advances in generating iPSC from ungulates and other large animals, there are still many issues that need to be resolved before we can efficiently generate robust iPSC from these animals.

3 Difficulties in Making Completely Reprogrammed Large Animal iPSC

In large animals, iPSC have been from the somatic cells of nonhuman primates (Liu et al. 2008), dogs (Shimada et al. 2010), pigs (Esteban et al. 2009, 2010; Ezashi et al. 2009; Wu et al. 2009; Telugu et al. 2010a, b; West et al. 2010; Montserrat et al. 2012; Hall et al. 2012), cattle (Huang et al. 2011), buffalo (Deng et al. 2012), sheep (Liu et al. 2012), goat (Ren et al. 2011) and horse (Nagy et al. 2011). A summary of the different studies that reported the establishment of iPSC from various farm animals is outlined in Table 3.1. In majority of these studies, however, the established iPSCs were plagued by deficiencies such as the failure to turn-off expression of exogenous genes or factors responsible for reprogramming, failure to self-sustain for a longer period of time, or more importantly meet all the criteria for pluripotency, specifically the in vivo tests.

One major challenge in establishing robust farm animal iPSC from the outset has been the difficulty in identifying the right set of conditions and factors required for each large animal species to propagate and sustain iPSC in culture. This has been a rather arduous task, since some experiments to replicate the culture conditions that are successful with the mouse or human have failed to yield self-sustaining large animal iPSC lines. The search has been made complex by the species-specific requirements of ungulates, which led many investigators down an unsuccessful path of trial and error to identify the "missing" factors. Different combinations of genes or factors have been introduced in somatic cells with varying efficiencies, although OCT4 is usually included (Montserrat et al. 2011) as a reprogramming factor in most animal species, and NANOG is necessary for complete reprogramming in bovines (Malaver-Ortega et al. 2012). In some species like bovine, it is imperative that the reprogramming factors should be from the same species as the somatic cells but in others like ovine, the origin of the reprogramming genes does not seem critical. It could be due to conservation of reprogramming factors in various animal species, although a comparison of efficiencies of these factors from different species for reprogramming a particular somatic cell would be beneficial. It is also very important, according to some studies, to use plasmid-based vectors rather than ones based on a virus (Huang et al. 2011) as the use of integrating retroviral vectors leads to the continued expression of the transgenes in the reprogrammed cells. In addition to the risk of mutagenesis due to the random integration of transgenes, continued expression of pluripotent genes could also limit the differentiation potential of iPSC. For iPSC to have a better utility, the inserted transgenes should either be deleted or effectively silenced after the cells have been reprogrammed. Alternatively, reprogramming must be achieved through the use of non-integrating vectors or plasmids (Okita et al. 2008; Yu et al. 2009), introduction of "stemness" proteins rather than genes (Zhou et al. 2009), or pharmaceutically with a suitable combination of small molecules or chemicals. The disadvantage of using non-integration methods of iPS cell generation is that their induction efficiencies are quite low and as a result, they inefficiently generate completely reprogrammed iPSC.

Table 3.1 Summary of large animal iPSC reports

Animal species	Year/author	Parental cells	Culture conditions	Tests	Markers	Factors	iPSC type
Porcine	Fan et al. (2013)	Ear fibroblasts, Bone marrow cells, fetal fibroblasts	DMEM with LIF, bFGF	Embryoid bodies, Teratomas	Oct4, Nanog, Sox2, c-Myc, Klf4	Human OSKM	Naive
–	Fujishiro et al. (2013)	Embryo fibroblasts	bFGF, LIF, pLIF, FK	Embryoid bodies, chimera	Oct4, Nanog, ERas, Sox2, Lin28, SSEA 1, SSEA-3, SSEA-4	Human OSKM	Naive
–	Park et al. (2013)	Embryo fibroblasts	DMEM, FBS	Embryoid bodies, in vitro differentiation	Oct4, Sox2, Nanog, TDGF1, REX1, bFGF, FGFR1, FGFR2, Nodal	Mouse OSKM	Primed
–	Cheng et al. (2012)	Embryo fibroblasts	KSR, KO-DMEM, LIF, bFGF	Embryoid bodies, Teratomas	Oct4, Sox2, Nanog, Klf4, SSEA-1 and SSEA-4, Tra-1-60, Tra-1-81, TERT	Mouse OSKM	Naive
–	Hall et al. (2012)	Fetal fibroblasts	KO-DMEM, KSR, bFGF	Embryoid bodies	OSKM, Nanog, SSEA-1, SSEA-4	Human OSKM, Nanog	Primed
–	Montserrat et al. (2012)	Ear fibroblasts	hbFGF, LIF	Embryoid bodies, Teratomas	Nanog, SSEA-4, TRA-1-60	Human SKM	Primed
–	Telugu et al. (2010a)	Fetal fibroblasts	KSR, hLIF, 2 inhibitors PD03259 CHIR99021	Teratomas	Oct4, Sox2, Nanog, Cdh1, Lin28, Gcnf, Gnl3, Zfp42, Utf1, Tdgf1, Acvr2b,	Human OSKM Nanog, Lin28	Naive

(continued)

Table 3.1 (continued)

Animal species	Year/author	Parental cells	Culture conditions	Tests	Markers	Factors	iPSC type
–	West et al. (2011)	Mesenchymal stem cells	KSR, hbFGF, mTeSR1 medium	Embryoid bodies, Chimera	AP, Oct4, Sox2	Human OSKM Nanog, Lin28	Primed
–	Ezashi et al. (2009)	Porcine fetal fibroblasts	FGF2	Embryoid bodies, Teratomas	AP, SSEA-1, Oct4, Nanog, Sox2,	Human OSKM	Primed
–	Esteban et al. (2009)	Miniature pig fibroblast	FGF2, LIF	Teratomas	AP, Nanog, Sox2, Klf4, Oct4, Rex1, Lin28, SSEA4	Mouse and Human OKSM	Primed
–	Wu et al. (2009)	Ear fibroblast and bone marrow cells	hESC medium with doxycycline	Embryoid bodies, Teratomas	AP, SSEA3, SSEA4, Tra-1-60, Tra-1-81, Oct4, Nanog, Sox2, Rex1 and Cdh1.	Human OSKM Nanog, Lin28	Primed
Cattle	Cao et al. (2012)	Fetal fibroblasts	LIF, bFGF	Embryoid bodies, Teratomas	Oct4, Nanog, AP, SSEA-1	Human Oct4, Porcine SKM	Primed
–	Huang et al. (2011)	Fetal fibroblasts	2i/LIF medium, PD0325901, CHIR99021	Embryoid bodies, Teratomas	Oct4, Sox2, Cdh1, DPPA-3, Nanog, ZFP42, Tra-1-60, Tra-1-81, SSEA3 and SSEA4	Bovine OSKM	Not clear
–	Han et al. (2011)	Fetal fibroblasts	KSR, bFGF	Embryoid bodies, Teratomas	DPPA-3, DPPA-4, Esrrb, TERT, E-Cadherin	Bovine OSKM, Lin 28, Nanog Human OSKM	Naive?

–	Sumer et al. (2011)	Adult fibroblasts	bFGF, LIF	Embryoid bodies, Teratomas	OSKM,, Nanog, Rex1, ALP1, AP, SSEA1, SSEA4	Human OSKM, Nanog	Not clear
Buffalo	Deng et al. (2012)	Embryonic fibroblasts	Pifithrin	Embryoid bodies, Teratomas	AP, Oct4, SOX2, Nanog, SSEA1, SSEA4, Tra-1-81, E-Cadherin	Buffalo OSKM	Primed
Sheep	Liu et al. (2012)	Fetal fibroblasts	bFGF, LIF	Embryoid bodies, Teratomas, tetraploid embryo	AP, Oct4, Nanog	Human OSKM	Naive
–	Sartori et al. (2012)	Fetal fibroblasts	Human and mouse ESC media	Embryoid bodies, Chimera, Teratomas	Nanog, SSEA1, SSEA4	Mouse OSKM	Not clear
–	Bao et al. (2011)	Primary ear fibroblasts	KSR	Embryoid bodies, Teratomas	AP, Oct4, Sox2, Nanog, SSEA1, DPPA4, Tra-1-81, Rex1, E-Cadherin	Human OSKM, Nanog, Lin28, SV40 Large T, hTERT	Naive
–	Li et al. (2011)	Fetal fibroblasts	hFGF2, LIF, KO-DMEM	Embryoid bodies, Teratomas	AP, Oct4, Sox2, Nanog, SSEA4	Mouse OSKM	Primed
Goat	Ren et al. (2011)	Primary ear fibroblasts	ES media	Embryoid bodies, Teratomas	AP, Oct4, Sox2, Nanog, SSEA1, DPPA4, Tra-1-60, TRA-1-81, Rex1, SALLA, E-Cadherin	Mouse OSKM, SV40 Large T, hTERT	Naive
Horse	Breton et al. (2013)	Fetal and adult fibroblasts	KSR, Knockout DMEM, bFGF, LIF	Embryoid bodies, Teratomas	Oct4, Sox2, Nanog, Rex1, Lin28, SSEA1, SSEA4, Tra1-60	Mouse OSKM	Primed

(continued)

Table 3.1 (continued)

Animal species	Year/author	Parental cells	Culture conditions	Tests	Markers	Factors	iPSC type
–	Khodadadi et al. (2012)	Adult fibroblasts	LIF, bFGF, EGF	Embryoid bodies, Teratomas	AP, Oct4, Sox2, Nanog, SSEA1, SSEA 4, STAT3	Human OSK	Primed
–	Nagy et al. (2011)	Fetal fibroblasts	bFGF, LIF, GSK; MEK; TGF; ALK receptor inhibitor, Thiazovivin	Embryoid bodies, Teratomas	AP, Oct4, Klf4, Nanog, SSEA1, SSEA4, DPPA4, Tra-1-60, Tra-1-81	Mouse OSKM	Primed
Dog	Whitworth et al. (2012)	Dermal fibroblasts	Konckout DMEM, KSR, LIF, bFGF, GSK; MEK and TGF-β antagonist	Embryoid bodies	AP, Nanog, Oct4, Telomerase, SSEA1, SSEA4, TRA1-60, TRA1-81, Rex1.	Human OSKM, Lin28, Nanog	Naive
–	Lee et al. (2011)	Fibroblasts, adipose stromal cells	hbFGF, hLIF	Embryoid bodies, Teratomas	AP, Oct4, Sox2, Nanog, SSEA4, Tra-1-60	Human OSKM	Primed
–	Shimada et al. (2010)	Embryonic fibroblasts	hbFGF, hLIF, VPA, PD0325901, CHIR99021, A83-01	Embryoid bodies, Teratomas	AP, Oct4	Canine OSKM	Primed
Nonhuman primates	Emborg et al. (2013)	Skin fibroblasts from female macaca	Primate ES cell medium, bFGF	Teratoma	SOX2, Nanog	Human OSKM	Primed
–	Deleidi et al. (2011)	Macaque skin fibroblasts	Human stem cell media with KSR	Neurons, Teratomas	AP, OSKM Nanog, SSEA 4, Tra-1-60, Tra-1-81	Human OSKM	Primed

–	Zhong et al. (2011)	Macaca oral fibroblasts	hbFGF, PD0325901, VPA, Y-27632	Embryoid bodies, neural precursors, cardiomyocytes, hepatocytes, Teratomas	Oct4, Nanog, SSEA3 and 4, Tra-1-60, Tra-1-81	Human OSKM	Primed
–	Liu et al. (2008)	Rhesus macaque ear fibroblast	KSR	Embryoid bodies, Teratomas	AP, Oct4, Sox2, Nanog, SSEA4, Tra-1-60, Tra-1-81	Monkey OSKM	Primed

The culture media used also has a significant effect on the quality of iPSC generated and should be standardized for each animal species. The presence of FBS in the culture medium used in the reprogramming stage was implicated in generating pre-iPSC in mice (Chen et al. 2012), because it contains growth factors of the BMP family that cause alterations of H3K9 methyltransferase and demethylase activities. The alternative use of Dulbecco's Modified Eagle's Medium (DMEM) supplemented with knockout serum replacement (KSR) medium (Chen et al. 2010) improved the efficiency of "true" iPSC generation (Chen et al. 2012). These findings from mouse cells may not apply to improving reprogramming condition for porcine iPSC, but it seems possible that the differences of DNA methylation status of the endogenous *OCT4* promoter observed between porcine iPSC reprogrammed in KSR (Wu et al. 2009) and FBS-supplemented media is due to FBS interference with reprogramming of the latter (Fujishiro et al. 2013). Secondary to the choices of the growth factor(s) and culture media, another technical consideration is the concentration of the supplement as the commercially available mouse or human ligands do not necessarily carry similar potency in farm animals. Another unresolved issue which could affect the quality of iPSC generated is the choice of cell lines for use as feeders, as well as their density in culture. Yet another cause could be incomplete resetting of epigenetic profile. Variations in fidelity of epigenetic reprogramming, which can be different in different cells, can be quite significant even in iPSC that otherwise meet established pluripotency criteria. The donor cell-type-specific epigenetic patterns ("somatic memory" of the cells) and other unknown factors can affect complete reprogramming into, and later differentiation of iPSCs.

Although human and mouse iPSC are well defined, most other animal models also suffer from a general lack of reliable iPSC markers. For instance, reports of canine pluripotent cell derivation have shown that in different studies the surface marker expression (such as SSEA-1, SSEA-3, SSEA-4) varies, despite the common expression of OCT4, SOX2, and NANOG (Shimada et al. 2010; Whitworth et al. 2012). Ungulates, such as pig, sheep, and cow, also show inconsistencies in surface marker expression (Huang et al. 2011; Li et al. 2011), and there is a further complication in that porcine, and bovine blastocysts express primary pluripotency genes and surface markers, such as OCT4 and SSEA-4, in both the ICM and the trophectoderm. Furthermore, Ezashi et al. (2009) reported that the process that results in pig iPSC also produces the by-product of trophectoderm-like cells. Like iPSC, these cells can grow in iPSC culture conditions, have high expression of telomerase, and a subset of pluripotency genes, making it difficult to distinguish them from iPSC following reprogramming. The lack of appropriate markers for different cells in different animals makes it impossible to distinguish animal-specific iPSC from other cells and judge the extent of their reprogramming.

In addition to problems with characterization, another consideration is the accumulation of chromosomal abnormalities following routine passages of human and mouse iPSC within a few generations (Ben-David et al. 2010; Mayshar et al. 2010; Ronen and Benvenisty 2012). This finding suggests that the long-term culture of large animal iPSCs may result in similar abnormalities, and therefore should be monitored carefully for culture-induced genetic changes. Another limitation has been the limited number of studies demonstrating that the uiPSC can be directed in vitro to transform into more specialized sub-lineages that might be tested for their ability to form functional grafts in large animals like pigs. In one example, primed iPSC derived from pig fetal fibroblasts (Ezashi et al. 2009) were directed along the ectoderm lineage to form a mixture of cells that included rod photoreceptor lineage cells (Zhou et al. 2011). These cells when injected into the eye were able to integrate into the retina, differentiate into photoreceptors, and generate outer segment-like projections (Zhou et al. 2011). In other studies, due to lack of further differentiation, the state of reprogramming could never be tested experimentally. A limited number of other papers have demonstrated analogous directed differentiation in vitro from pig iPSC (Bui et al. 2012; Yang et al. 2013) but the work carried out in other species has so far been very limited in scope.

Bonafide iPSC from ungulates are not here yet, but we are closer than ever before in making this a reality. Above all, a defined conceptual framework is beginning to emerge due to the recognition of the difference between naïve and primed epiblast iPSC and signaling networks involved in maintaining them. More in vivo studies on the factors, signaling pathways, and their interactions for the generation of iPSC would make the generation of fully reprogrammed cells easy and possible. The best strategy for future iPSCs derivation would be to use chemicals or small molecules for reprogramming, reduce the number of transcription factors used, and use non-genome-integrating methods of DNA or protein delivery. Since integration-free methods are very inefficient, they can be improved by better combination of factors, selection of an easily reprogrammable parental cell type, and by determining optimal culture conditions. After production of robust iPSC from large animal species, an important step for the translation of basic research into clinical applications is the testing of these cells using large animal models that are intermediate between laboratory rodents and humans.

4 Potential Utility of Large Animal iPSC to Human Disease Therapy

The ability to propagate in an undifferentiated state and differentiate into target cell types under defined conditions make the ESC and iPSC valuable tools for investigating early embryonic developmental events and regenerative medicine applications. For example, miPSC lines have been generated from animals with a specific genetic disease, the genetic defect was corrected in vitro, and the cells were transplanted back into a preclinical mouse model (Hanna et al. 2007; Wu et al. 2011). Likewise, hematopoietic progenitors derived from genetically corrected iPSC were transplanted into the mouse model and this led to the improved production of human β-globin (Wang et al. 2012). Another recent and more significant observation was the establishment of functional liver buds from iPSC-derived human hepatic progenitors. The liver buds had confirmed functionality as determined by detoxification of drugs, and secretion of human albumin (Takebe et al. 2013). These significant albeit proof-of-principle studies in mice need to be thoroughly evaluated for their safety and efficacy on a larger recipient host prior to clinical application in humans. Although, nonhuman primates are the best candidates for these studies, they are also incredibly expensive to maintain and perform research on due to various ethical concerns. This precludes their use for promising "upcoming" therapies. Conceivably, in this regard, large animal species specifically ungulates are better suited to bridge the gap. Although relatively more expensive to maintain than the mouse, ungulates enable long-term evaluation of the safety and efficacy of potential stem cell therapies (Dall et al. 2002; Vodicka et al. 2005). Specifically, pig has several potential advantages over mouse for predicting whether or not stem-cell-based therapy is likely to be safe when considering outcomes, such as toxicity, immune responses, and tumorigenicity. Their large size facilitates the expansion of implanted human or animal cells to numbers sufficient for biochemical analysis, standardization of dosage, the decision on the type of surgical equipment required for humans, and also allows multiple samplings over time. Additional key characteristics such as the similarity in physiology, anatomy, and pathology to humans as compared to mice, and above all similarity in their perceived primed stemness to humans make them attractive model systems (Ourednik et al. 2001; Brevini et al. 2008).

Pigs and other ungulates are genetically well standardized, and their genomes are sequenced, making transgenic approaches feasible (Campbell et al. 1996; Wilmut et al. 1997). At the cellular level, many of the transcriptional regulators, genes and proteins share a high degree of similarity to humans and can be used interchangeably (Zanjani et al. 1994a, b; Verfaillie 2000) in various species. For these reasons, many of the domestic animals have already made inroads into transplantation and preclinical research. Sheep and goats have been used extensively in preclinical studies of bone marrow transplantation, and for the xenotransplantation of human stem cells into sheep and goat fetuses for more than 50 years (Zanjani et al. 1991a, b; Zeng et al. 2005; Narayan et al. 2006; Almeida-Porada et al. 2007; Michelini et al. 2008). Similar to sheep and goats, pigs have also been widely recognized for their potential as donors for xenotransplantation research (Klymiuk et al. 2010). They are also used as models for studying the pathophysiology of human diseases, such as cystic fibrosis (Rogers et al. 2008), where mice fail to develop the relevant symptoms encountered in human patients. Pigs also have an important role in studying cardiovascular disease, lipoprotein metabolism, diabetes, wound, burn repair, intestine and immune system development (Swindle 2007). They already play a significant role in developing surgical techniques and cell delivery procedures, and for optimizing the number and type of cells to be used for a particular type of graft. For instance, pigs have already been treated with a variety of "adult" stem cells to determine whether cardiac function can be improved after induced ischemia (Amado et al. 2006; Gandolfi et al. 2011; Mazhari and Hare 2012). Pig iPSC, after conversion to endothelial cell precursors, have been successfully transplanted into mice with myocardial infarctions that appear to promote neovascularization in the ischemic regions (Gu et al. 2012). Presumably, the next step will be to use iPSC to test disease therapies and mechanisms involved in different disease conditions to confirm the regenerative potential of the iPSC approach. Porcine iPSC, for example, could be matched to specific pigs, making them essentially "patient-specific," and then could be differentiated into specific lineages. These qualities make it possible to use pigs to test transplantation therapies with iPSC for safety and efficacy before applying the procedures to human patients. Another ungulate, equine, can serve as an excellent model for a range of human degenerative diseases, especially those involving joints, bones, tendons and ligaments, such as arthritis. Another large animal, dog, can serve as a clinically relevant model of human hereditary disease as compared to rodents. Canine iPSC have also been used to test the relevance of human heart therapies as dog's heart is more similar to humans as compared to the mouse (Lee et al. 2011).

Currently, many groups are using large animal iPSC for transplantation in the more traditional small animal mouse model. Zhu et al. (2011) reported the generation of insulin-producing pancreatic cells from rhesus monkey iPSC but tested their efficacy in a diabetic mouse model. Similarly, Zhong et al. (2011) used genetically modified nonhuman primate iPSC in a mouse model to show that suicide genes can be conditionally activated to eliminate pluripotent cells in vivo. Although such studies provide clues regarding how iPSC behave following transplantation in humans, transplantation of autologous iPSC or their derivatives using large animal models would be more insightful by providing better preclinical safety data needed to progress towards human clinical trials. Since no single animal model is ideal, a rational approach to their use for chimeric studies of human stem cells would arguably involve related studies in more than one model. Initial experiments in small, relatively inexpensive and versatile animal models provide a good platform for moving to larger animal models with greater relevance for preclinical studies. Clinical researchers generally gravitate toward the use of large animal models due to their similarity to humans, whereas basic researchers lean towards the small diverse animal models as they are comparatively inexpensive and easy to maintain. Regenerative medicine can benefit enormously from joining of these two schools of thought and in designing studies that would use many different models to accelerate the translation of animal research to human applications. As the use and efficiency of iPSC generation increases, so would the range of applications for these diverse animal models. It is likely that the increasingly sophisticated applications involving human/animal chimeric models will emerge through advances in genetic manipulation, naïve iPSC generation and differentiation, efficient cell delivery, and in vivo imaging, all immensely significant developments in their own right. In the context of human/animal chimeric studies, they represent a means to a greater end: the eventual use of human-induced pluripotent cells to treat human diseases.

Besides the promise of stem cell therapies, large animal iPSC also offer greater potential for use in basic research, biotechnology, agriculture, pharmacy, and medicine. In ungulates where the ESC are hard to isolate and maintain, iPSC have a much greater role to play in genetic engineering applications as well, such as the generation of transgenic animals with special production characteristics or resistance against diseases, biopharming human recombinant proteins (Nowak-Imialek et al. 2011), and reviving endangered or extinct breeds and species (Ben-Nun et al. 2011). As the field of large animal iPSC advances, it is anticipated that the findings will enable new stem-cell based regenerative therapies in human and veterinary medicine, and aid in the development of preclinical models leading to human applications.

5 Conclusions

Numerous studies have revealed that iPSC can be established using a variety of methods, cell sources, induction methods, reprogramming factors, chemicals, and culture conditions. The iPSC are quite similar to ESC in morphology, proliferative capacity, surface antigens, overall gene expression, epigenetic status of pluripotent cell-specific genes and specific gene loci. Therefore, the recent establishment of iPSC from medically relevant large animals such as pigs, cattle and sheep, for which authentic ESC are not currently available was a valuable step forward for testing the potential utility of stem cells in human medicine and agriculture. In addition to the advantages associated with ESC, iPSC will also allow the establishment of "personalized" stem cells for transplantation. If iPSC can be created from the desired animal and successfully directed along particular pathways of differentiation, it will be possible to transplant the cells into the same animal from which the founder cells originated. This would protect against the immune-rejection and promote better survival of the graft. The ability of these transplanted cells to be incorporated into damaged organs or tissues can be easily measured and the functionality and stability of these grafts can be assessed over a longer life span of large animals. Of equal importance, a study on the large animals will provide confidence that stem cell transplantation can be performed safely without the risk of cancer occurring over a period of years from adolescence to more mature age.

Despite the rapid progress of the field, several key concerns remain. Completely reprogrammed iPSCs from most large animals or humans are difficult to derive on a routine basis, and there is a general lack of effective and reproducible reprogramming protocols. Therefore, the need of the hour is to develop reliable differentiation protocols and reagents capable of driving differentiation into targeted lineages such as neuronal, cardiac, endothelial and hepatic cells. Although no animal study can truly compare to a human study, every effort should be made to ensure that the model system is as close to the human system as possible, particularly when translational medical research is the goal. With this in mind, large animal models should and can play a more significant role in translational research, but they are often overlooked due to perceived difficulties and/or costs.

References

Alberio R, Croxall N et al (2010) Pig epiblast stem cells depend on activin/nodal signaling for pluripotency and self-renewal. Stem Cells Dev 19(10):1627–1636

Almeida-Porada G, Porada C et al (2007) The human-sheep chimeras as a model for human stem cell mobilization and evaluation of hematopoietic grafts' potential. Exp Hematol 35(10): 1594–1600

Amado LC, Schuleri KH et al (2006) Multimodality noninvasive imaging demonstrates in vivo cardiac regeneration after mesenchymal stem cell therapy. J Am Coll Cardiol 48(10): 2116–2124

Bao L et al (2011) Reprogramming of ovine adult fibroblasts to pluripotency via drug-inducible expression of defined factors. Cell Res 21:600–608
Ben-David U, Benvenisty N et al (2010) Genetic instability in human induced pluripotent stem cells: classification of causes and possible safeguards. Cell Cycle 9(23):4603–4604
Ben-Nun IF, Montague SC et al (2011) Induced pluripotent stem cells from highly endangered species. Nat Methods 8(10):829–831
Breton A et al (2013) Derivation and characterization of induced pluripotent stem cells from equine fibroblasts. Stem Cells Dev 22:611–621
Brevini TA, Antonini S et al (2008) Recent progress in embryonic stem cell research and its application in domestic species. Reprod Domest Anim 43(suppl 2):193–199
Brons IG, Smithers LE et al (2007) Derivation of pluripotent epiblast stem cells from mammalian embryos. Nature 448(7150):191–195
Bui HT, Kwon DN et al (2012) Epigenetic reprogramming in somatic cells induced by extract from germinal vesicle stage pig oocytes. Development 139(23):4330–4340
Campbell KH, McWhir J et al (1996) Sheep cloned by nuclear transfer from a cultured cell line. Nature 380(6569):64–66
Cao H, Yang P, Pu Y, Sun X, Yin H, Zhang Y, Zhang Y, Li Y, Liu Y, Fang F, Zhang Z, Tao Y, Zhang X (2012) Characterization of bovine induced pluripotent stem cells by lentiviral transduction of reprogramming factor fusion proteins. Int J Biol Sci 8(4):498–511
Chen J, Liu J et al (2010) Towards an optimized culture medium for the generation of mouse induced pluripotent stem cells. J Biol Chem 285(40):31066–31072
Chen J, Liu H et al (2012) H3K9 methylation is a barrier during somatic cell reprogramming into iPSCs. Nat Genet. In press
Cheng D et al (2012) Porcine induced pluripotent stem cells require LIF and maintain their developmental potential in early stage of embryos. PLoS One 7:e51778
Dall AM, Danielsen EH et al (2002) Quantitative [18F]fluorodopa/PET and histology of fetal mesencephalic dopaminergic grafts to the striatum of MPTP-poisoned minipigs. Cell Transplant 11(8):733–746
Deleidi M, Cooper O, Hargus G, Levy A, Isacson O (2011) Oct4-induced reprogramming is required for adult brain neural stem cell differentiation into midbrain dopaminergic neurons. PLoS One 6:e19926
Deng Y, Liu Q et al (2012) Generation of induced pluripotent stem cells from buffalo (Bubalus bubalis) fetal fibroblasts with buffalo defined factors. Stem Cells Dev 21(13):2485–2494
Desponts C, Ding S (2010) Using small molecules to improve generation of induced pluripotent stem cells from somatic cells. Methods Mol Biol 636:207–218
Emborg ME et al (2013) Induced pluripotent stem cell-derived neural cells survive and mature in the nonhuman primate brain. Cell Rep 3:646–650
Esteban MA, Xu J et al (2009) Generation of induced pluripotent stem cell lines from tibetan miniature pig. J Biol Chem 284(26):17634–17640
Esteban MA, Peng M et al (2010) Porcine induced pluripotent stem cells may bridge the gap between mouse and human iPS. IUBMB Life 62(4):277–282
Ezashi T, Telugu BP et al (2009) Derivation of induced pluripotent stem cells from pig somatic cells. Proc Natl Acad Sci U S A 106(27):10993–10998
Fan N, Lai L (2013) Genetically modified pig models for human diseases. J Genet Genomics 40:67–73
Flechon JE, Degrouard J et al (2004) Gastrulation events in the prestreak pig embryo: ultrastructure and cell markers. Genesis 38(1):13–25
Fujishiro SH et al (2013) Generation of naive-like porcine-induced pluripotent stem cells capable of contributing to embryonic and fetal development. Stem Cells Dev 22:473–482
Gandolfi F, Vanelli A et al (2011) Large animal models for cardiac stem cell therapies. Theriogenology 75(8):1416–1425
Geisert RD, Brookbank JW et al (1982) Establishment of pregnancy in the pig: II. Cellular remodeling of the porcine blastocyst during elongation on day 12 of pregnancy. Biol Reprod 27(4):941–955

Gu M, Nguyen PK et al (2012) Microfluidic single-cell analysis shows that porcine induced pluripotent stem cell-derived endothelial cells improve myocardial function by paracrine activation. Circ Res 111(7):882–893
Hall VJ, Kristensen M et al (2012) Temporal repression of endogenous pluripotency genes during reprogramming of porcine induced pluripotent stem cells. Cell Reprogram 14(3):204–216
Han X et al (2011) Generation of induced pluripotent stem cells from bovine embryonic fibroblast cells. Cell Res 21:1509–1512
Hanna J, Wernig M et al (2007) Treatment of sickle cell anemia mouse model with iPS cells generated from autologous skin. Science 318(5858):1920–1923
Hou P et al (2013) Pluripotent stem cells induced from mouse somatic cells by small-molecule compounds. Science 341:651–654
Huang B, Li T et al (2011) A virus-free poly-promoter vector induces pluripotency in quiescent bovine cells under chemically defined conditions of dual kinase inhibition. PLoS One 6(9): e24501
Jia F, Wilson KD et al (2010) A nonviral minicircle vector for deriving human iPS cells. Nat Methods 7(3):197–199
Kang L, Wang J et al (2009) iPS cells can support full-term development of tetraploid blastocyst-complemented embryos. Cell Stem Cell 5(2):135–138
Khodadadi K et al (2012) Induction of pluripotency in adult equine fibroblasts without c-MYC. Stem Cells Int 2012:429160
Klymiuk N, Aigner B et al (2010) Genetic modification of pigs as organ donors for xenotransplantation. Mol Reprod Dev 77(3):209–221
Lee AS, Xu D et al (2011) Preclinical derivation and imaging of autologously transplanted canine induced pluripotent stem cells. J Biol Chem 286(37):32697–32704
Li Y, Cang M et al (2011) Reprogramming of sheep fibroblasts into pluripotency under a drug-inducible expression of mouse-derived defined factors. PLoS One 6(1):e15947
Lim ML, Vassiliev I et al (2011) A novel, efficient method to derive bovine and mouse embryonic stem cells with in vivo differentiation potential by treatment with 5-azacytidine. Theriogenology 76(1):133–142
Liu H, Zhu F et al (2008) Generation of induced pluripotent stem cells from adult rhesus monkey fibroblasts. Cell Stem Cell 3(6):587–590
Liu J, Balehosur D et al (2012) Generation and characterization of reprogrammed sheep induced pluripotent stem cells. Theriogenology 77(2):338–346.e331
Malaver-Ortega LF, Sumer H et al (2012) The state of the art for pluripotent stem cells derivation in domestic ungulates. Theriogenology 78(8):1749–1762
Mayshar Y, Ben-David U et al (2010) Identification and classification of chromosomal aberrations in human induced pluripotent stem cells. Cell Stem Cell 7(4):521–531
Mazhari R, Hare JM (2012) Translational findings from cardiovascular stem cell research. Trends Cardiovasc Med 22(1):1–6
Michelini M, Papini S et al (2008) Prolonged human/sheep cellular chimerism following transplantation of human hemopoietic stem cells into the ewe celomic cavity. Int J Dev Biol 52(4):365–370
Montserrat N, Garreta E et al (2011) Simple generation of human induced pluripotent stem cells using poly-beta-amino esters as the non-viral gene delivery system. J Biol Chem 286(14): 12417–12428
Montserrat N, de Onate L et al (2012) Generation of feeder-free pig induced pluripotent stem cells without Pou5f1. Cell Transplant 21(5):815–825
Nagy A, Rossant J et al (1993) Derivation of completely cell culture-derived mice from early-passage embryonic stem cells. Proc Natl Acad Sci U S A 90(18):8424–8428
Nagy K, Sung HK et al (2011) Induced pluripotent stem cell lines derived from equine fibroblasts. Stem Cell Rev 7(3):693–702
Narayan AD, Chase JL et al (2006) Human embryonic stem cell-derived hematopoietic cells are capable of engrafting primary as well as secondary fetal sheep recipients. Blood 107(5): 2180–2183

Notarianni E, Laurie S et al (1990) Maintenance and differentiation in culture of pluripotential embryonic cell lines from pig blastocysts. J Reprod Fertil Suppl 41:51–56
Notarianni E, Galli C et al (1991) Derivation of pluripotent, embryonic cell lines from the pig and sheep. J Reprod Fertil Suppl 43:255–260
Nowak-Imialek M, Kues W et al (2011) Pluripotent stem cells and reprogrammed cells in farm animals. Microsc Microanal 17(4):474–497
Okita K, Ichisaka T et al (2007) Generation of germline-competent induced pluripotent stem cells. Nature 448(7151):313–317
Okita K, Nakagawa M et al (2008) Generation of mouse induced pluripotent stem cells without viral vectors. Science 322(5903):949–953
Okita K, Hong H et al (2010) Generation of mouse-induced pluripotent stem cells with plasmid vectors. Nat Protoc 5(3):418–428
Ourednik V, Ourednik J et al (2001) Segregation of human neural stem cells in the developing primate forebrain. Science 293(5536):1820–1824
Park KM, Cha SH, Ahn C, Woo HM (2013) Generation of porcine induced pluripotent stem cells and evaluation of their major histocompatibility complex protein expression in vitro. Vet Res Commun 37:293–301
Piedrahita JA, Anderson GB et al (1990) Influence of feeder layer type on the efficiency of isolation of porcine embryo-derived cell lines. Theriogenology 34(5):865–877
Ren J, Pak Y et al (2011) Generation of hircine-induced pluripotent stem cells by somatic cell reprogramming. Cell Res 21(5):849–853
Roberts RM, Telugu BP et al (2009) Induced pluripotent stem cells from swine (Sus scrofa): why they may prove to be important. Cell Cycle 8(19):3078–3081
Rodriguez A, Allegrucci C et al (2012) Modulation of pluripotency in the porcine embryo and iPS cells. PLoS One 7(11):e49079
Rogers CS, Stoltz DA et al (2008) Disruption of the CFTR gene produces a model of cystic fibrosis in newborn pigs. Science 321(5897):1837–1841
Ronen D, Benvenisty N (2012) Genomic stability in reprogramming. Curr Opin Genet Dev 22(5):444–449
Sartori C et al (2012) Ovine-induced pluripotent stem cells can contribute to chimeric lambs. Cell Reprogram 14:8–19
Shimada H, Nakada A et al (2010) Generation of canine induced pluripotent stem cells by retroviral transduction and chemical inhibitors. Mol Reprod Dev 77(1):2
Stadtfeld M, Nagaya M et al (2008) Induced pluripotent stem cells generated without viral integration. Science 322(5903):945–949
Sumer H, Liu J et al (2011) NANOG is a key factor for induction of pluripotency in bovine adult fibroblasts. J Anim Sci 89(9):2708–2716
Swindle MM (2007) Swine in the laboratory: surgery, anesthesia, imaging, and experimental techniques. CRC Press, Boca Raton, FL
Takahashi K, Tanabe K et al (2007) Induction of pluripotent stem cells from adult human fibroblasts by defined factors. Cell 131(5):861–872
Takahashi K, Yamanaka S (2006) Induction of pluripotent stem cells from mouse embryonic and adult fibroblast cultures by defined factors. Cell 126(4):663–676
Takebe T et al (2013) Vascularized and functional human liver from an iPSC-derived organ bud transplant. Nature 499:481–484
Tavernier G, Wolfrum K et al (2012) Activation of pluripotency-associated genes in mouse embryonic fibroblasts by non-viral transfection with in vitro-derived mRNAs encoding Oct4, Sox2, Klf4 and cMyc. Biomaterials 33(2):412–417
Telugu BP, Ezashi T et al (2010a) Porcine induced pluripotent stem cells analogous to naive and primed embryonic stem cells of the mouse. Int J Dev Biol 54(11–12):1703–1711
Telugu BP, Ezashi T et al (2010b) The promise of stem cell research in pigs and other ungulate species. Stem Cell Rev 6(1):31–41
Tesar PJ, Chenoweth JG et al (2007) New cell lines from mouse epiblast share defining features with human embryonic stem cells. Nature 448(7150):196–199

Vejlsted M, Du Y et al (2006) Post-hatching development of the porcine and bovine embryo–defining criteria for expected development in vivo and in vitro. Theriogenology 65(1):153–165

Verfaillie CM (2000) Meeting report on an NHLBI workshop on ex vivo expansion of stem cells, July 29, 1999, Washington, D.C. National Heart Lung and Blood Institute. Exp Hematol 28(4):361–364

Vodicka P, Smetana K Jr et al (2005) The miniature pig as an animal model in biomedical research. Ann N Y Acad Sci 1049:161–171

Wang Y, Zheng CG et al (2012) Genetic correction of beta-thalassemia patient-specific iPS cells and its use in improving hemoglobin production in irradiated SCID mice. Cell Res 22(4): 637–648

West FD, Terlouw SL et al (2010) Porcine induced pluripotent stem cells produce chimeric offspring. Stem Cells Dev 19(8):1211–1220

West FD et al (2011) Brief report: chimeric pigs produced from induced pluripotent stem cells demonstrate germline transmission and no evidence of tumor formation in young pigs. Stem Cells 29:1640–1643

Whitworth DJ, Ovchinnikov DA et al (2012) Generation and characterization of LIF-dependent canine induced pluripotent stem cells from adult dermal fibroblasts. Stem Cells Dev 21(12): 2288–2297

Wilmut I, Schnieke AE et al (1997) Viable offspring derived from fetal and adult mammalian cells. Nature 385(6619):810–813

Winkler T, Cantilena A et al (2010) No evidence for clonal selection due to lentiviral integration sites in human induced pluripotent stem cells. Stem Cells 28(4):687–694

Wu Z, Chen J et al (2009) Generation of pig induced pluripotent stem cells with a drug-inducible system. J Mol Cell Biol 1(1):46–54

Wu G, Gentile L et al (2011) Efficient derivation of pluripotent stem cells from siRNA-mediated Cdx2-deficient mouse embryos. Stem Cells Dev 20(3):485–493

Yang JY, Mumaw JL, Liu Y, Stice SL, West FD (2013) SSEA4-positive pig induced pluripotent stem cells are primed for differentiation into neural cells. Cell Transplant 22:945–959

Yu J, Hu K et al (2009) Human induced pluripotent stem cells free of vector and transgene sequences. Science 324(5928):797–801

Zanjani ED, Mackintosh FR et al (1991a) Hematopoietic chimerism in sheep and nonhuman primates by in utero transplantation of fetal hematopoietic stem cells. Blood Cells 17(2):349–363, discussion 364–366

Zanjani ED, Pallavicini MG et al (1991b) Successful stable xenograft of human fetal hemopoietic cells in preimmune fetal sheep. Trans Assoc Am Physicians 104:181–186

Zanjani ED, Flake AW et al (1994a) Long-term repopulating ability of xenogeneic transplanted human fetal liver hematopoietic stem cells in sheep. J Clin Invest 93(3):1051–1055

Zanjani ED, Silva MR et al (1994b) Retention and multilineage expression of human hematopoietic stem cells in human-sheep chimeras. Blood Cells 20(2–3):331–338, discussion 338–340

Zeng F, Chen M et al (2005) Identification and characterization of engrafted human cells in human/goat xenogeneic transplantation chimerism. DNA Cell Biol 24(7):403–409

Zhao XY, Li W et al (2009) iPS cells produce viable mice through tetraploid complementation. Nature 461(7260):86–90

Zhong B, Trobridge GD et al (2011) Efficient generation of nonhuman primate induced pluripotent stem cells. Stem Cells Dev 20(5):795–807

Zhou H, Wu S et al (2009) Generation of induced pluripotent stem cells using recombinant proteins. Cell Stem Cell 4(5):381–384

Zhou L, Wang W et al (2011) Differentiation of induced pluripotent stem cells of swine into rod photoreceptors and their integration into the retina. Stem Cells 29(6):972–980

Zhu FF, Zhang PB et al (2011) Generation of pancreatic insulin-producing cells from rhesus monkey induced pluripotent stem cells. Diabetologia 54(9):2325–2336

Chapter 4
Fetal Adnexa-Derived Stem Cells Application in Horse Model of Tendon Disease

Anna Lange-Consiglio and Fausto Cremonesi

1 Structure of Tendon

Healthy tendon is a highly specialized tissue designed to resist enormous unidirectional forces. Tendons are responsible not only for transmitting mechanical forces from skeletal muscle to bone but also as joint stabilizers and as "shock absorbers" to limit muscle damage. The functional properties of the tissue are derived from its structure and its components.

Tendon is composed predominantly of water (approximately 70 %). Of the remaining 30 % dry matter, the major constituents are collagen and a noncollagenous matrix representing the extracellular matrix (ECM) that is synthesized by a few residing tendon fibroblasts. The main component of the ECM is *type I* collagen (Williams et al. 1980). Other collagens (for example: collagen *-II*, *-III*, *-IV*, and *-V*) are present although in smaller quantities and in specific locations. *Type II* collagen is found within enthesious insertions and regions where the tendon changes direction around a bony prominence, reflecting the fibrocartilage-like nature of the matrix in this region designed to withstand compressional as well as tensional forces. *Types III*, *IV*, and *V* are confined to basement membranes and endotendon (Goodship et al. 1994; Perez-Castro and Vogel 1999).

Collagen molecules are arranged hierarchically into fibrils, fibers, and fascicles, according to the direction of the force application (Fig. 4.1). Endotendinous connective tissue (endotenon) surrounds the fiber bundles and contains blood vessels, lymphatics, and nerves. The whole tendon, composed of multiple bundles and endotenon, is surrounded by the epitenon, a thin layer of connective tissue that is contiguous with the endotenon. Intermolecular cross-links between the *type I* collagen molecules reinforce the tensile strength of the tendon matrix (Thorpe et al. 2010).

A. Lange-Consiglio (✉) • F. Cremonesi
Reproduction Unit, Large Animal Hospital, Università degli Studi di Milano,
Polo Veterinario di Lodi, via dell'Università, 6, 26900 Lodi, Italy
e-mail: anna.langeconsiglio@unimi.it

T.A.L. Brevini (ed.), *Stem Cells in Animal Species: From Pre-clinic to Biodiversity*,
Stem Cell Biology and Regenerative Medicine, DOI 10.1007/978-3-319-03572-7_4,

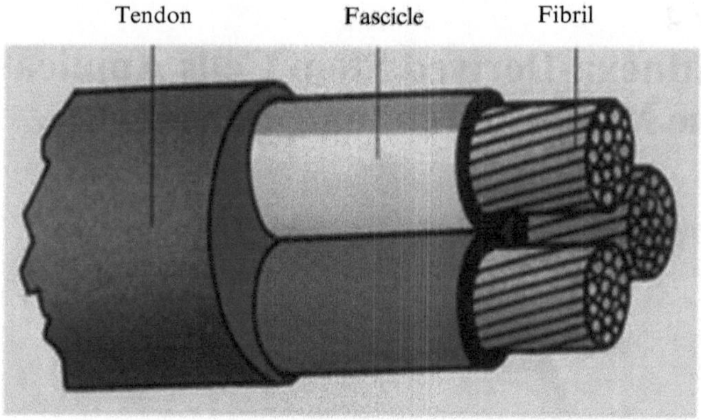

Fig. 4.1 Schematic diagram representing the architectural hierarchy of the tendon

The noncollagenous substance is composed of tenocytes and glycoproteins. Three distinct cell types (*I*, *II*, and *III*), have been identified on light microscopy within the fascicles of normal equine tendon, based on the morphology of their nuclei. The distribution of cell types varies with age, between tendons and within tendons (Goodship et al. 1994; Smith and Webbon 1996). The functions of these different cell types are unclear but it is postulated that *type II* and *III* cells have higher metabolic activity because their nuclei are larger and contain nucleoli. These cells may be concerned primarily with ECM synthesis (Smith and Webbon 1996) but considerably more information is needed on the nature of these different cell types and their function. There is a further population of cells resident within the endotenon septa of the tendon whose functions are largely unknown.

Of the glycoproteins, cartilage oligomeric matrix protein (COMP) is one of the most abundant (Smith et al. 1997). Proteoglycans consist of glycosaminoglycan side-chain(s) attached to a protein core. Various glycosaminoglycans have been demonstrated within the equine superficial digital flexor tendon (SDFT) including chondroitin sulfate, dermatan sulfate, keratan sulfate, heparin, heparin sulfate, and hyaluronic acid (Smith and Webbon 1996), but this knowledge is only of limited benefit as this does not indicate the nature of the proteins to which they are attached. The metacarpophalangeal regions of normal SDFT contain more of the larger, cartilage-like proteoglycans, and experience higher rates of proteoglycan synthesis compared to the midmetacarpal region. These differences probably reflect functional and metabolic variance that exists between regions of tension and compression (Smith and Webbon 1996; Perez-Castro and Vogel 1999). The small proteoglycans decorin, fibromodulin, and biglycan have been thought to influence tenocyte functions, collagen fibrillogenesis, and the spatial organization of fibers, thereby influencing tendon strength (Gu and Wada 1996).

Tendons have high mechanical strength and elasticity necessary to perform their function. Tendons also exhibit viscoelasticity as they display properties of stress relaxation and creep (Sharma and Maffulli 2005a).

2 Mechanical Properties of Tendon

Biomechanical properties of tendons during their repair and regeneration have been studied extensively and their properties compared with normal tendon. Mechanical testing involves separate clamps to grip the muscle–tendon complex at one end and the bone at the other end ensuring that the ends are held firmly without slippage, the tendon is loaded along its longitudinal axis, and the force and displacement are recorded until the tissue fails (Riemersa and Schamhardt 1982) The in vitro mechanical properties of normal tendon have been characterized and shown to approximate a sigmoidal curve when force is plotted against elongation (Goodship et al. 1994) (Fig. 4.2).

There is an initial lax phase, termed the "toe" region, where the crimp is eliminated followed by a linear phase due to progressive stretching of the straightened collagen fibers, until tendon failure occurs at extreme forces. The methodology of in vitro biomechanical testing has been shown to be repeatable, and, to provide objective data on the biomechanical properties of tendons in response to various treatments. The mechanical properties of tendon may be correlated at the molecular level with the mechanical characteristics of the collagen fibers (Cohen et al. 1974). Tendon viscoelastic behavior (Kastelic and Baer 1980; Hooley et al. 1980), which is the result of complex interactions between various components, is dependent on age and activity. Tendons do not usually fail or rupture under normal conditions; therefore, it is more appropriate to quantify the physical properties within the linear region, including stress relaxation, creep, hysteresis loop, and viscoelasticity.

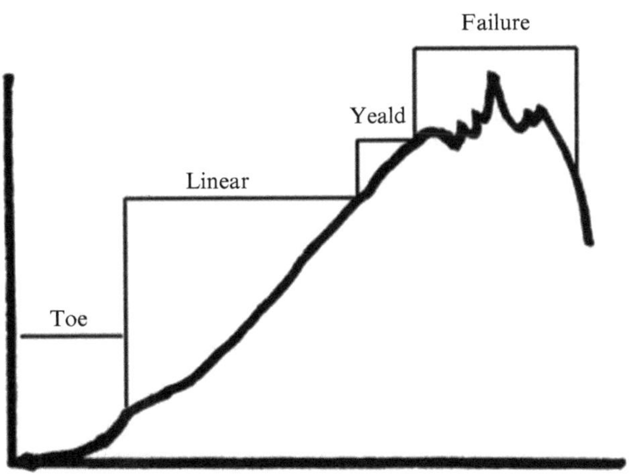

Fig. 4.2 Simplified stress–strain curve for the superficial digital flexor tendon. (1) "Toe" region; (2) linear deformation; (3) yield; (4) rupture

3 Tendons and Their Diseases in Man

A variety of tendons in man may be affected by overstrain injury, including elements of the rotator cuff and supraspinatus tendon in the shoulder, the forearm extensors, biceps brachii, the Achilles tendon in the lower limb, the tibialis posterior tendon in the foot, and the patella tendon in the knee (Kader et al. 2002; Rees et al. 2006). Tendon injuries, clinically appearing either as acute or chronic disease, can be caused by both extrinsic (trauma) and intrinsic (overstrain) factors (Rees et al. 2006). Whilst spontaneous acute injuries do occur in tendon diseases, there is evidence that such injuries are often preceded by degeneration of the tendon matrix in the period prior to onset of clinical signs (Kannus and Jozsa 1991; Astrom and Rausing 1995). There is a growing body of evidence that proteolytic enzymes, such as matrix metalloproteinases, in part mediate such degeneration (Jones et al. 2006). In recent decades, the incidence of tendon injury has risen due to both an increase in the elderly population, and a rise in participation to recreational and competitive sporting activities (Kader et al. 2002).

In man, loading of a tendon up to 2 % strain leads to flattening of the crimp pattern in tendon. Between 2 and 4 % strain, tendons deform in a linear fashion as a result of intramolecular sliding of collagen triple helices and fibers become more parallel. At strains less than 4 % tendons behave in an elastic fashion and return to their original length when unloaded. Microscopic failure occurs when the strain exceeds 4 %, and between 8 and 10 % strain, macroscopic failure occurs (Sharma and Maffulli 2005b).

4 The Importance of Injuries to Energy-Storing Tendons in Large Athletic Animals

It is widely accepted that there is no model in which all potential therapies may be tested appropriately; however, it has become essential to have an animal model matched to the clinical pathology prior to initiating a clinical trial. When selecting an appropriate animal model, the salient criteria to consider are a closely matched anatomic structure and the ease with which the treatment can be translated to a clinical setting (Smith et al. 2005). A study about adhesions formed after injury to the flexor tendon in the primate was found to have the closest resemblance to humans in both surgical technique and physical attributes of the matching tissues and organs (Pruzansky 1987). Use of dogs and chickens has been ruled out primarily because their physical and anatomic features do not match the human anatomy. This is essential for research translation from an animal model to the human situation. In short, testing materials, biological factors, and surgical techniques in an animal model that mimics the pathogenic condition in humans is ideal because it will facilitate research translation to humans. Murine models that simulate various disease conditions, such as genetic disorders, growth factor deficiencies, cancers, and immunodeficiency are available to many investigators. Animal models have given a major boost to the design of research tools and are available for the development and validation of novel therapies prior to human trials. These models are used to test procedures quantitatively and to determine the ease and effectiveness of surgical procedures, and they allow for statistically significant studies in mice, rats, and rabbits. In addition, larger animals, such as goats, sheep, as well as porcine and equine models, have been utilized in appropriate circumstances. The choice of the animal model is also determined by the outcomes sought by the investigators. Most animal models are appropriate for the determination of histology, cell proliferation, and biochemical analysis in order to determine the effectiveness of surgical or therapeutic treatment procedures. The need to have a technically reproducible and mechanically stable tendon injury model is dominant to the ultimate utility of the animal model. In addition, the availability of techniques that will measure the strength of the regenerate tissue is necessary within most musculoskeletal components. These considerations set the lower limit to the anatomical size and morphological features within the experimental model. In this context a large animal as the equine is a good model for the study of human tendon disease (Roshan et al. 2008).

As in human, among equine there are athletes, too. Tendon injuries associated with athletic activity and aging are most frequent and important both in people and horses. Due to its clinical importance and to the easier attainability of tissue specimens not achievable in the medical field, the most completely investigated pathologies of tendons are those of the distal limb of the horse, in particular the injury to SDFT. The SDFT in the horse is a weight bearing tendon and has many similarities to the human Achilles tendon in both its structure and matrix composition. This tendon acts as a spring, absorbing and releasing elastic energy during different phases of the stride, contributing to both the high efficiency of locomotion and, along with its associated muscle, also acting as a shock absorber for the limb (Wilson et al. 2001). The horse has maximized its ability to store energy for efficient locomotion by having maximal strains in the SDFT far higher than in humans, recorded at 16 % at the gallop in horses (Minetti et al. 1999). This value is in close agreement with measured in vitro strains at rupture of 12–21 %. These findings suggest that at maximal exercise the SDFT operates close to its physiological limits with a relatively narrow safety margin. Consequently, minor disruption of the tendon matrix composition and arrangement dramatically increase the incidence of tendonitis. Conversely, improving the "quality," or strength, of the tendon matrix reduces incidence.

The specialized nature of equine digital tendons and certain human tendons essential for the efficiency of high-speed locomotion, and the large size of both species, bring the clinical relevance of laboratory rodent and rabbit models of exercise-induced tendon damage into question.

The risk of SDFT injury is one of the most frequent causes of lameness of thoroughbred horses internationally (Kasashima et al. 2004; Ely et al. 2009) and it has been shown to increase with horse age in a number of surveys (Kasashima et al. 2004; Perkins et al. 2005). Re-injury rates of 23–67 % have been reported in various equine sports disciplines; 19–70 % are ultimately retired due to the original or the subsequent re-injury (Dyson 2004; Perkins et al. 2005; O'Meara et al. 2010). All of these factors cause significant economic loss and animal welfare concerns.

The risk of Achilles tendon (AT) injury in people, is associated with elite athletic activity and with age, as for SDFT injury in horses; incidences of up to 29 %, 18 %, and 52 % have been reported in elite male runners, elite gymnasts, and running athletes, respectively, with the latter figure correlated with hours spent training (Knobloch et al. 2008; Emerson et al. 2010). Up to 29 % of patients with AT injuries may require surgery (Zafar et al. 2009); time for a return to activity varies from 6 weeks to almost 10 months depending on the complexity and chronicity of the injury, and re-rupture rates of 2–12 % have been reported (Saxena et al. 2011).

The functional equivalence of the SDFT and AT is largely due to their pivotal role in saving energy during high-speed locomotion, by reducing muscular work (Malvankar and Khan 2011). Most tendons ("positional" tendons) simply connect skeletal muscle and bone, stabilizing and moving joints. In the horse, recovery of mechanical work during tendon elastic recoil (i.e., with the SDFT acting as a "biological spring") has been estimated at 36 % at gallop; in the human AT, energy returns contribute to over 50 % of positive work done at the ankle (Farris et al. 2011). Energy-storing tendons must undergo high strains (i.e., changes in tendon length as a percentage of initial length) to store sufficient amounts of energy.

5 Evidence of Microdamage and Aetiological Factors in Tendons Diseases

The majority of clinical tendonitis lesions have, as a precursor, some undetectable subclinical change that precedes clinical tendonitis and this change is age-related and accelerated by exercise. Degenerative changes described in human tendons are similar (but not identical) to those found in equine exercise study and include evidence of new collagen synthesis with increased *type III* collagen and reduced *type I* collagen, increases in levels of fibronectin, tenascin C, GAGs, and proteoglycans and histological observations of apparently irreversible matrix change including irregular arrangement, altered crimping, disruption, and reductions in density of collagen (Riley et al. 1994a, b; Jarvinen et al. 1997; Cook et al. 2004). Fibronectin and tenascin C are glycoproteins that are thought to modulate cellular activity and migration (Riley 2005). Increased amounts of *type III* collagen at AT rupture sites in one study were not associated with evidence of recent synthesis (cleaved propeptides), supporting the theory that this accumulates prior to clinical injury (Eriksen et al. 2002). Histological degenerative lesions seen in human AT, but not noted in horses, include mucoid degeneration, fatty degeneration (accumulation of adipocytes), calcification, and neovascularization; this may to some extent relate to the fact that they are not identical anatomical structures. In both asymptomatic and painful degenerate human tendons, various histological changes have been noted in tenocyte populations including increased or decreased cellularity, plump ovoid or rounded nuclei with ultrastructural evidence of increased production of collagen and proteoglycans, and chondroid change in more extreme cases; tendon cells cultured from lesions have maintained a higher proliferation rate and produced greater quantities of *type III* collagen (Rolf et al. 2001; Cook et al. 2004). Ultrastructural changes in the tenocytes have been interpreted as indicative of hypoxic degeneration, including alterations in sizes and shapes of mitochondria and nuclei, increased numbers and sizes of lysosomal vacuoles, "hypoxic" and lipid vacuolation and intracytoplasmic or mitochondrial calcification (Kannus and Jozsa 1991; Leadbetter 1992).

In equine, when the surface of a tendon is viewed at an angle, the presence of a wave form or crimp within the fascicles can be detected (Goodship and Birch 1996). This crimp pattern plays a role in imparting elasticity to the tendon during the early stages of loading (up to 2–3 % of strain). Crimp angles and lengths were found to be 19–20° and 17–19 μm respectively in the midmetacarpal region of the young SDFT (Wilmink et al. 1992) in equine species, reducing to 12–17° and 11–15 μm with aging. The elimination of the crimp with age is thought to contribute partly to the increased stiffness of tendon with age.

Most tendons in mature horses exhibit patchy acellularity, although there is no overall decrease in cell number with age in the equine SDFT (Birch et al. 1999). Further degenerative signs become more evident in all horses age >3 years (Pool 1996) and these include matrix fibrillation, chondroid metaplasia, chondrone formation, neovascularization, and fibroplasia (Pool 1996). All these changes have the effect of inducing relative tendon weakness and, although an association between elastic modulus and age has been demonstrated, the correlation between decreasing mechanical strength and age has not been proven (Gillis et al. 1997).

Recent investigations have indicated that the equine SDFT attains maturity at approximately 2 years of age. Collagen fibril diameter, mature collagen cross-links, and crimp morphology have stabilized by this age (Patterson-Kane et al. 1997a, b). Tendons from horses over 2 years of age have significantly stiffer mechanical properties (Gillis et al. 1995b). This increased stiffness is thought to be associated with the reduction of crimp, the presence of increasing numbers of nonreducible cross-links and decreasing fascicle size in older tendons (Gillis et al. 1997).

Exercise also causes a reduction in the mass average diameter of collagen fibrils within the central region of the SDFT compared to the peripheral region fibrils (Patterson-Kane et al. 1997c). As glycation levels are unaltered, which indicates that the smaller fibrils are not new collagen, these results suggest that microdamage, with splitting of the collagen fibers, is occurring.

In the human being, Dowling et al. (2000) report about investigations of specific pathogenetic mechanisms of tendonitis. These can be classified into two broad categories: (a) physical and (b) vascular.

(a) Physical mechanisms
Fatigue, poor conformation, lack of fitness, incoordinate muscle activity (McIlwraith 1987) in nonathletic people will all act to produce excessive biomechanical forces on the tendon (Silver et al. 1983; Crevier-Denoix and Pourcelot 1997) which may accelerate the degenerative change by physically disrupting the matrix, or are sufficient to induce full clinical tendonitis by exceeding the

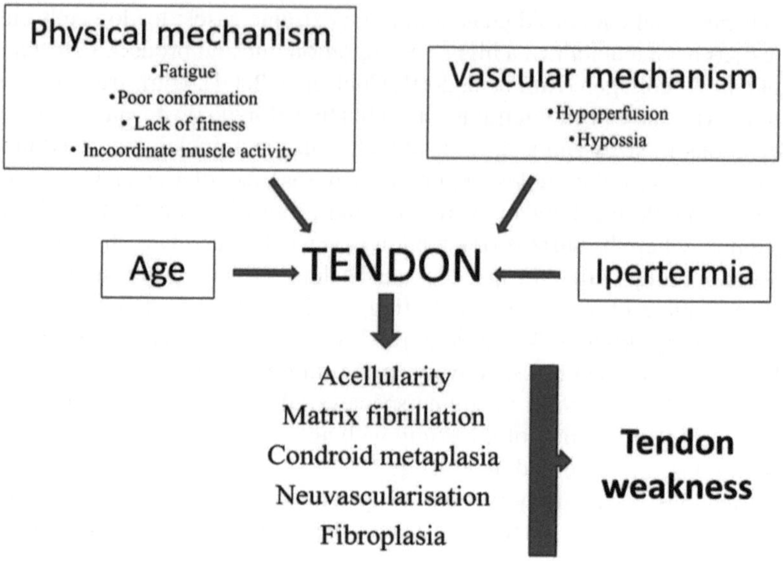

Fig. 4.3 Aetiological factors in tendons diseases

mechanical properties of the tendon. On the other hand, it is generally agreed that the levels and most likely also the patterns, types, and durations of physical stress on tendons are the most important aetiological factors in athletic injuries (Riley 2005). In addition to the mechanical environment, the only other directly measured or calculated stress factor specific to energy-storing tendons is hyperthermia. Due to loss of some strain energy as heat (hysteresis), which is not effectively dispersed by the vascular system, temperatures of at least 43–45 °C have been measured in the SDFT core of the galloping horse (Wilson and Goodship 1994). Conservative estimates of hyperthermia in the human AT, calculated by mathematical modeling, predicted a core temperature of at least 41 °C during 30 min of treadmill running, with the potential to rise to up to 44 °C in some individuals (Farris et al. 2011). Such temperatures would be expected to be lethal or at the very least highly stressful to fibroblastic cells.

(b) Vascular mechanisms

Other damaging conditions that may be experienced during and/or following exercise within the tendon tissue include hypoxia due to tendon hypoperfusion and increased levels of reactive oxygen species (ROS) induced by ischaemia-reperfusion and/or hyperthermia (Millar et al. 2012).

However, the equine SDFT has good vascular network (Kraus-Hansen et al. 1992) and blood flow in this tendon has been recorded at a level similar to that of resting skeletal muscle and is increased by exercise (Jones 1993). At present there is little experimental evidence to support or refute these potential mechanisms, but it seems logical that each may interact or summate to cause tendon injury (Fig. 4.3).

6 Tendon Healing

The restoration of normal tendon function after injury requires reestablishment of tendon fibers and the gliding mechanism between tendon and its surrounding structures (Abrahamsson and Gelberman 1994). The initial stage of repair involves formation of scar tissue that provides continuity at the injury site (Dunphy 1967) however, lack of mechanical stimulus on the tendon will cause proliferation of scar tissue and subsequent adhesions that are undesirable and harmful because they impede normal tendon function, particularly in the hand. Although stability to the injury site is necessary, mobility is critical, and mechanical loading that is associated with motion of the healing tendon decreases the formation of postoperative adhesions and increases the strength. After tendon injury, the body initiates a cascade of distinct events or phases distinguishable by the cellular and biochemical processes that occur. The sequence of repair involves a progression through three stages: tissue inflammation, cell proliferation, and remodeling (Ingraham et al. 2003; Sharma and Maffulli 2005b) (Fig. 4.4).

During inflammatory stage, vasodilators and proinflammatory molecules attract inflammatory cells from the surrounding tissue. Erythrocytes, platelets, neutrophils, monocytes, and macrophages migrate to the wound site where the clot, cellular debris, and foreign body matter are engulfed and resorbed by phagocytosis. Fibroblasts recruited to the site begin to synthesize various components of the ECM (Lindsay and Birch 1964) Angiogenic factors are released during this phase and initiate the formation of a vascular network (Myers and Wolf 1974) These processes include an increase in DNA and in ECM, which establishes continuity and partial stability at the site of injury.

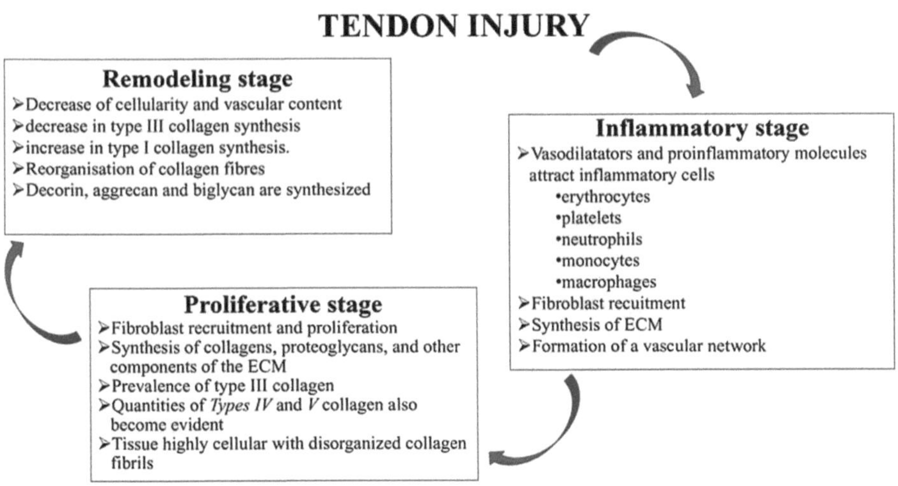

Fig. 4.4 Stages of tendon healing

In the proliferative stage, fibroblasts and their rapid proliferation at the wound site are responsible for the synthesis of collagens, proteoglycans, and other components of the ECM. These components are initially arranged in a random manner within the ECM, which at this point is composed largely of type III collagen (Garner et al. 1989). *Type III* collagen forms interfibrillar cross-links conferring early stability and mechanical strength to the site of injury (Williams et al. 1980; Cheung et al. 1983). Increased quantities of *Types IV* and *V* collagen also become evident (Williams et al. 1980; Silver et al. 1983). At the end of the proliferative stage, the repair tissue is highly cellular and contains relatively large amounts of water and an abundance of ECM components.

The remodeling stage begins 6–8 weeks after injury. This phase is characterized by a decrease in cellularity, reduced matrix synthesis, decrease in *type III* collagen, and an increase in *type I* collagen synthesis (Roshan et al. 2008). Loose fibrils of *Type I* and *III* predominate until 6 months post injury. After this period, linearly arranged *Type I* collagen fibrils prevail, indicating progressive remodeling and normalization of the healing tissue (Watkins et al. 1985). *Type I* collagen fibers are organized longitudinally along the tendon axis and are responsible for the mechanical strength of the regenerate tissue (Liu et al. 1995). During the later phases of remodeling, interactions between the collagen structural units lead to higher tendon stiffness and consequently greater tensile strength; however, the repair tissue never achieves the characteristics of normal tendon. Indeed, abnormal quantities of *Type III* collagen, smaller collagen fibrils, and lack of fiber bundles and linear arrangement may still be detected up to 14 months post injury (Silver et al. 1983) and probably persist even longer. It is the abnormal composition and arrangement of matrix in fibrous scar tissue, which has poor biomechanical properties in comparison to normal tendon, and the slow rate of healing that are believed to be responsible for the high incidence of reinjure (Pool 1996).

7 Diagnosis of Tendonitis

Ultrasonography is presently the most common technique used to diagnose tendonitis and monitor healing. The commonly measured variables include tendon and lesion cross-sectional area (CSA), lesion type and location, and fiber alignment (Smith et al. 1994; Gillis et al. 1995a, b; Genovese et al. 1996). An increase in tendon CSA is reportedly the most sensitive indicator of fiber damage (Genovese et al. 1996) and for detecting re-injury during the convalescent period when the reintroduction of exercise may be excessive. Additional measurement includes changes in echogenicity. Also magnetic resonance could be applied for diagnosis of tendinitis. Kasashima et al. (2002) demonstrated that magnetic resonance imaging is the better imaging modality for the objective detection of chronic scar tissue in live horses when the chronic scar tissue is difficult to discern by ultrasonography.

8 Classical Therapies in Tendon Diseases

Injury initiates several signaling events that recruit fibroblasts and stimulate the local tenocyte population to synthesize collagen and other extracellular components, establishing physical continuity at the site. In most cases of tendon laceration or rupture, surgical intervention is required to direct the natural process of healing, and occasionally the damage exceeds the natural ability to repair even with existing treatment modalities. Tendon repair is a slow process and characterized by a scar with mechanically inferior tissue, at least initially and it is complicated by the need to provide appropriate and timely tension to the repair tissue (Roshan et al. 2008).

In the past, repair of tendon injuries were performed by different methods. In particular, clinical therapy was based on ice application, bandage, box rest, and controlled exercise. An alternative approach consisted on the use of corticosteroid (inflammation reduction) and other drugs (sodium hyaluronate, polysulphated glycosaminoglycans, beta aminoproprionitrile fumarate). Furthermore, surgical treatments like accessory ligament desmotomy, local irritation by line firing or pin firing were commonly used. More recently ultrasound, laser therapy, and electromagnetic field therapy have been considered. Unfortunately, they did not allow complete tissue healing and quite often people or animals did not regain competitiveness.

9 New Treatments in Tendinopathies: Regenerative Medicine

During rehabilitation with controlled exercise, there is an ideal mechanical stimulation allowing the newly created tissue to organize itself in the direction of the force application, hence this approach can be referred to as "in vivo tissue engineering." But a better understanding of the mechanisms that regulate the function and the differentiation of tendon cells is essential to developing new treatments for tendinopathy such as tendon rupture or ectopic ossification resulting from injury caused by overuse or trauma. In recent years, research on tendinopathy therapies has focused on regenerative medicine and tissue engineering approaches, which are considered to be promising tools to prospect normal restore, or close to normal, structure and function to an injured organ. Regenerative medicine requires an exogenous cell source and this approach comes from the knowledge that most tissues have a sub- or side-population of precursor cells (tissue-specific progenitor cells) used to replenish cells due to natural turnover and aid in repair post injury (da Silva Meirelles et al. 2006). Several lines of evidence suggest that multipotential stem cells are present also in tendons and ligaments. First, both human and mouse tendons develop fibrocartilage and ossification in response to injury (Fenwick et al. 2002). Second, tendon-derived immortalized cell lines or human tendon-derived fibroblasts express genes of adipogenic, osteogenic, and chondrogenic differentiation pathways, suggesting that they possess multiple differentiation capacities in vitro (Salingcarnboriboon et al. 2003; de Mos et al. 2007). Finally, postnatal stem cells capable of differentiating into adipocytes and osteoblastic cells have been identified in human periodontal ligaments (Seo et al. 2004) while human and mouse tendons harbor a unique cell population, termed tendon stem/progenitor cells (TSPCs), that has universal stem cell characteristics such as clonogenicity, multipotency, and self-renewal capacity (Bi et al. 2007). Recently, Lovati et al. (2011) identified TSPCs specifically in the horse SDFT with the ability to be highly clonogenic, to grow fast and to differentiate in different induced-cell lineages as well as bone marrow-derived progenitor cells (BM-MSCs). The hypothesis that TSPCs possess a mesenchymal stem cell (MSC) behavior opens a new prospective for tendon regenerative medicine approaches because TSPCs could represent an important tool to study basic tendon biology. The exact site for TPSCs cells within tendon is not known, but they are most likely to reside in the endotenon tissue between the collagen fascicles and adjacent to the vasculature (da Silva Meirelles et al. 2006). Although this might be true in young growing tendon, mature equine tendon, however, does not appear to possess a substantial population of these cells capable of differentiating into multiple cell lines and this may explain why this component of the repair process is limited and hence natural repair is inferior to normal tendon.

During the repair process, there is a large influx of cells into the lesion. Kajikawa et al. (2007) showed that at 24 h after the injury, the wound contained circulation-derived cells but not tendon-derived cells. Tendon-derived cells appeared in the injured area at 3 days after the wound, and significantly increased in number with time and maintained a high level of proliferative activity until 7 days after the injury, whereas the circulation-derived cells decreased in number and are replaced by the tendon-derived cells. These findings suggest that circulation-derived and tendon-derived cells contribute to the healing of tendons in different periods as part of a biphasic process but that the cells mainly involved in the synthesis of new tissue are believed to be tendon-derived cells (Kajikawa et al. 2007). For these reasons some authors hypothesized that the implantation of far greater numbers of progenitor stem cells, than are present normally within tendon tissue, would have the potential of regenerating or improving the repair of the tendon. Fibroblasts derived from tendon or other sources could be used (Kryger et al. 2007), but the removal of sections of tendon to recover cells leads to the formation of a secondary lesion in the horse that is unacceptable. Alternative cell sources under investigation include dermal fibroblasts, which were shown to be capable of functionally bridging a tendon defect and to have similar histological and tensile properties to the tenocyte-seeded scaffold (Liu et al. 2006) although in vitro these cells behave differently from tenocytes (Evans and Trail 2001). By contrast, the implantation of MSCs of different sources in far greater numbers than are present normally within tendon tissue would have the potential for regenerating or repairing tendon. MSCs have been implanted into surgical defects in tendons in multiple in vivo experiments in laboratory animals with mostly positive outcomes. Most of these models used surgically created defects in rabbit or rat tendons and have variously shown some improvement in structure and strength of defects implanted with MSCs in a biodegradable scaffold (collagen gel, Vicryl knitted mesh or fibrin glue) over controls implanted with just the scaffold, as assessed by histology or simple biochemical assays (Awad et al. 2003; Juncosa-Melvin et al. 2007; Butler et al. 2008). In other studies using a rat patellar defect model, MSCs implantation has been associated with both greater ultimate tensile stress and improved quality of reparative tissue determined by an increased collagen *I/III* ratio (Hankemeier et al. 2005, 2007). Thus, MSCs-seeded constructs implanted in vivo have shown the ability to integrate into the tissue and induce the synthesis of tissue-specific ECM. Implantation has been associated with both greater ultimate tensile stress and improved quality of reparative tissue determined by an increased collagen *I/III* ratio (Hankemeier et al. 2005, 2007).

Unfortunately, it is still unclear whether the major contribution of the MSCs to the healing process is to differentiate into tenocytes and thus produce ECM molecules, whether it is rather to supply growth factors and thus stimulate the residing cells within the tendon (Richardson et al. 2007; Chong et al. 2009) or whether a combination of the two mechanisms occurs (Yagi et al. 2010; Fortier and Travis 2011). In addition, recent studies have suggested an anti-inflammatory role of implanted stem cells. Animal model studies have demonstrated that MSCs are hypo-immunogenic and inhibit the activation of T and B lymphocytes and natural killer cells (Ren et al. 2008; Herrero and Pérez-Simón 2010). The precise mechanism of the anti-inflammatory effect of these cells is largely unknown, although a combination of this activity along with an antiapoptotic effect, additional recruitment of local multipotent stem cells, stimulation of vascular ingrowths, and the liberation of growth factors could all contribute to tissue repair (Alves et al. 2011). It is not known which of these actions occur after stem cell implantation, although current opinion favors the paracrine actions as being most important (Murphy et al. 2003).

The clinical nature of the employment of MSCs for horse tendinopathies preclude the routine *post mortem* analyses that can answer to these questions, but some experimental works has been carried out to monitor the fate of injected MSCs in horses and the structural aspect of the healing. Guest et al. (2008) studied the fate of autologous and allogeneic MSCs transfected with green fluorescent protein (GFP) following injection into the SDFT and revealed that GFP labeled cells located mainly within injected lesions, but with a small proportion integrated into healthy tendon. Furthermore, the authors showed that both autologous and allogeneic MSCs may be used without stimulating an undesirable cell-mediated immune response from the host. Other postmortem examinations have shown that MSCs application improved the ECM structure of damaged tendons. In histological sections of MSCs-treated tendon lesions, compared to non-treated tendon lesions, increased tendon fiber densities, increased organization of the collagen fibers, and a reduced vascularity have been found (Smith et al. 2003, 2009; Nixon et al. 2008). However, statistically significant differences have not been detected either in DNA, proteoglycan, or total collagen content, or in the expressions of *type I* collagen, *type III* collagen, or tissue remodeling enzymes such as matrix metalloproteinases (Schnabel et al. 2009; Guest et al. 2008), suggesting that the beneficial effect of MSCs is due to improvement of structural organization rather than of matrix composition. However, it has been shown that MSCs treatment can enhance expression levels of COMP (Guest et al. 2008; Crovace et al. 2010), a glycoprotein that is known to be important for tendon elasticity and stiffness. Ultrasonographic follow-up examinations showed significant improvements in fiber alignment and echogenicity scores at 1, 3, and 6 months after MSCs treatment (Leppänen et al. 2009), supporting the histological findings in the above-mentioned studies.

The optimal in vivo regenerative response due to MSCs could be achieved by the combination of three key factors (Butler et al. 2008):

1. A cell source capable of the formation of an optimal matrix.
2. A scaffold capable of promoting the survival of implanted cells by mechanical protection and/or nutritional support.
3. An anabolic stimulus usually combining, for musculoskeletal tissues, growth factors, and appropriate mechanical load to promote optimal ECM synthesis and organization.

9.1 Cell Source

Today, there are multiple choices for the selection of a cell source for regenerative medicine and at this moment it is not clear which source will prove to be therapeutically optimal (Table 4.1). The cells carrying the greatest hope for effective different cell lines are MSCs that have been isolated from bone marrow and fat but, recently, other sources of MSCs potentially useful for tendon and ligament therapies include extra-fetal stem cells that can be easily recovered at birth and stored for future use (Cremonesi et al. 2011; Lange-Consiglio et al. 2012) and embryonic or fetal embryonic-like stem cells (ESC or fdESC, respectively) (Guest et al. 2010; Watts et al. 2011).

9.1.1 Adult MSCs

The stromal compartment of bone marrow was the first source reported to contain multipotent progenitor cells (Fortier et al. 1998; Pittenger et al. 1999). For this reason, bone marrow is currently the best investigated origin of MSC. Bone marrow is generally collected from the sternum and it is favored for cell-based therapies in equine regenerative medicine probably to the reliable isolation success of bone marrow-derived MSC following an easy preparation procedure and separation of MSC via plastic adherence and cell culture (Vidal et al. 2006) Looking at differentiation capacities, bone marrow-derived MSC can be directed towards the osteogenic, chondrogenic, and adipogenic lineage (Pittenger et al. 1999), with differences in the potency of differentiation compared to MSCs derived from other sources. For equine MSC, Kisiday et al. (2008) reported a higher chondrogenic differentiation potential in response to TGF-β1 compared to MSC derived from adipose tissue. In contrast, compared to umbilical cord blood-derived MSC, chondrogenic differentiation of bone marrow MSC was inferior (Berg et al. 2009). Osteogenic differentiation potential of bone marrow MSC was reported to be highest among

Table 4.1 Sources for cell therapy of tendinopathies

Cell source		Advantages	Disadvantages	References
Embryo	Embryonic stem cells (ESC)	– Pluripotent	– Teratoma formation	Paris and Stout (2010)
Extra-fetal tissues	Amnion-derived cells	– No invasive collection – High plasticity and proliferative capacity – High number of immediately available cells for therapy – Well-tolerated by horses	– Strict surveillance of parturition	Lange-Consiglio et al. (2012, 2013a)
	MSCs from umbilical cord tissue	– No invasive collection – Greater multipotent than BM-MSCs – Possibility to obtain more rapidly proliferating cells by cell sorting – No immune response	– Strict surveillance of parturition	Corradetti et al. (2011) Carrade et al. (2011)
Adult tissues	Concentrated bone marrow aspirate (BMC)	– Minimal manipulation – No cell expansion	– Invasive aspiration procedure with risk of pneumopericardium – No reports on the use of BMC on tendonitis	Fortier et al. (2010)
	Stromal vascular fraction from adipose tissue	– Minimal manipulation – No cell expansion – Well-tolerated by horse	– Invasive collection	Alexander (2012)

(continued)

Table 4.1 (continued)

Cell source		Advantages	Disadvantages	References
Adult stem/progenitor cells	MSCs from bone marrow (BM-MSCs)	– Multipotent	– Invasive aspiration procedure with risk of pneumopericardium	Durando et al. (2006), Vidal et al. (2012)
		– No immune response	– Limited potential than ESC in terms of expansion (delay of 2–4 weeks to obtain a sufficient number of cells to in vivo implant)	Butler et al. (2008)
	MSCs from adipose tissue	– Higher proliferative potential and less senescence of BM-MSCs	– Invasive collection	Vidal et al. (2007)
		– Multipotent		Colleoni et al. (2009)
	Tendon stem/progenitor cells	– Possible activation of this endogenous population	– Invasive collection (removal of sections of tendons leads to the formation of secondary lesion)	Lovati et al. (2011)
		– Multipotent	– Mature equine tendon do not posses a substantial population of these cells	
Adult differentiated cells	Tenocytes	– Appropriate tendon matrix synthesis	– Invasive collection – Age-related reduction in synthesis of matrix ability	Kryger et al. (2007)
	Fibroblasts derived from tendon	– Appropriate tendon matrix synthesis	– Invasive collections	Kryger et al. (2007)
	Dermal fibroblasts	– Easy to recover, with acceptable donor site lesion – Similar histological and tensile properties than tenocyte	– Different protein-matrix synthesis than tenocytes	Liu et al. (2006)

MSC derived from several other sources (Toupadakis et al. 2010) but lowest compared to amniotic-derived cells (Lange-Consiglio et al. 2012, 2013a). Although BM-MSCs represent the most widely investigated cells for application in veterinary regenerative medicine to date, it is important to underline that procurement of BM from horses requires an invasive aspiration procedure, which occasionally has been associated with pneumopericardium (Durando et al 2006). In addition, BM-MSCs represent a very small fraction (0.001–0.01 %) of the total population of mononuclear cells isolated from a Ficoll density gradient of bone marrow aspirate, so, any clinical effect of bone marrow aspirate might be attributed to the numerous bioactive substances in the acellular fraction such as growth factors produced by cells or platelets. The BM-MSCs, moreover, have a more limited potential than embryonic stem cells in terms of in vitro proliferation ability (about 32 days for expansion between isolation and implantation) and this is a true limitation to use autologous BM-MSCs because there is a delay of 2–4 weeks while MSCs are multiplied in vitro before sufficient numbers are available for use (Guest et al. 2010; Lange-Consiglio et al. 2013a). Furthermore, BM-MSCs show reduced plasticity and growth with increasing donor age and in vitro passage number (Vidal et al. 2012) and do not appear to noticeably improve long-term functionality (Paris and Stout 2010).

The concentrated bone marrow aspirate (BMC) can overcame these problems avoiding the lag time from diagnosis to treatment when culture-expanded BM-MSCs are used. In addition to the concentration of stem cells, the concentrations of platelets and therefore anabolic growth factors are increased (Fortier et al. 2010), in addition it allows the concentration of the relatively few stem cells, and the production of a cell pool with minimal cell manipulation and no risk of cell transformation during the growth in vitro. When combined with thrombin, the fibrinogen present in BMC is converted to fibrin and a solid scaffold forms to retain the cells and growth factors in a given location. However, the precursors of bone and fat cells contained into BM may be detrimental to the healing by triggering the formation of dystrophic mineralization or metaplasia at the site of inoculation.

No peer-reviewed preclinical or clinical reports on the use of BMC for tendonitis have been published while an acute cartilage injury (15-mm-diameter lesions) was treated with BMC (Fortier et al. 2010). At 8 months, all macroscopic, histologic, and magnetic resonance imaging data indicated sustained improvement in BMC-grafted repair tissue compared with the control.

Another source of MSC commercially used in humans and animals is adipose tissue (Vidal et al. 2007; Del Bue et al. 2008; Jezierska-Woźniak et al. 2010; Martinello et al. 2011). The quick and successful recovery of adipose-derived MSC via lipectomy (horses) (Vidal et al. 2007) or collection of visceral fat during ovariohysterectomies (dogs) (Martinello et al. 2011), followed by enzymatic digestion, makes adipose tissue a promising MSC source for clinical applications. Adipose-derived MSCs seem to display a higher proliferation potential and less senescence compared to MSCs from other sources (Wagner et al. 2005; Colleoni et al. 2009; Vidal and Lopez 2011). Several studies report a good differentiation potential into mesodermal tissue lineages (Jezierska-Woźniak et al. 2010; Martinello et al. 2011), whereas other groups observed a limited chondrogenic and osteogenic potential in rat or horse adipose-derived MSC (Yoshimura et al. 2007; Vidal et al. 2008). However, the collection of this tissue in the horse remains invasive.

In any case, since MSCs treatment regime was first published by Hertel (2001) there have been several experimental and clinical studies with encouraging results, giving evidence of the benefit and safety of adult MSCs application for tendon regeneration.

In these studies, autologous adult progenitor cells have been used, either expanded bone marrow-derived MSCs (Pacini et al. 2007; Smith 2008, 2009; Burk and Brehm 2011), or adipose-derived MSCs (Del Bue et al. 2008; Schnabel et al. 2009) or adipose-derived nuclear cells (Nixon et al. 2008; Dahlgren 2009; Leppänen et al. 2009). Furthermore, the effects of autologous bone marrow-derived expanded MSCs and bone marrow-derived mononuclear cells on tendon healing have been compared, revealing a similar improvement in both treatment groups compared to the control group, which was demonstrated by significantly improved ultrasonography and histology scores, higher COMP expressions and relatively lower *type III* collagen contents (Crovace et al. 2007, 2010; Lacitignola et al. 2008).

9.1.2 Extra-Fetal MSCs

To overcome the invasive collection of BM and adipose tissue, progenitor cells derived from extra-fetal sources could represent attractive alternative candidates, with the further potential to circumvent many of the limitations of BM-MSCs and to open new perspectives for developmental biology and regenerative medicine (Lange-Consiglio et al. 2012).

Extra-fetal sources of MSCs are the umbilical cord (both, matrix and blood) and amnion.

Use of adnexal tissue has many potential advantages including the noninvasive nature of the isolation procedure and the large tissue mass from which cells can be harvested. Moreover, the fact that the placenta is fundamental for maintaining feto-maternal tolerance during pregnancy suggests that cells present in placental tissue may have immunomodulatory characteristics (Evangelista et al. 2008) that could decrease the risk of the recipient rejecting transplanted stem cells immunologically.

With regard to amnion, this membrane is the innermost layer of the fetal adnexa and consists of a thin epithelial layer, a thick basement membrane, and an avascular stroma. Since the amniotic epithelium develops prior to gastrulation, which is a "tipping point" at which cell fate is specified, it has been suggested that pluripotent stem cells, derived from the contiguity of embryonic epiblast (Fig. 4.5), may be retained in the amnion even at full term pregnancy. The fibroblast layer of the amnion originates from the extraembryonic mesoderm that develops from the hypoblast around 2–3 weeks post-fertilization.

In equine species, and for the first time in veterinary medicine, amnion-derived mesenchymal cells (AMCs) were characterized (Lange-Consiglio et al. 2012) and afterward compared to equine BM-MSCs (Lange-Consiglio et al. 2013a). Results of these studies demonstrated that AMSCs and BM-MSCs both exhibited adult stromal

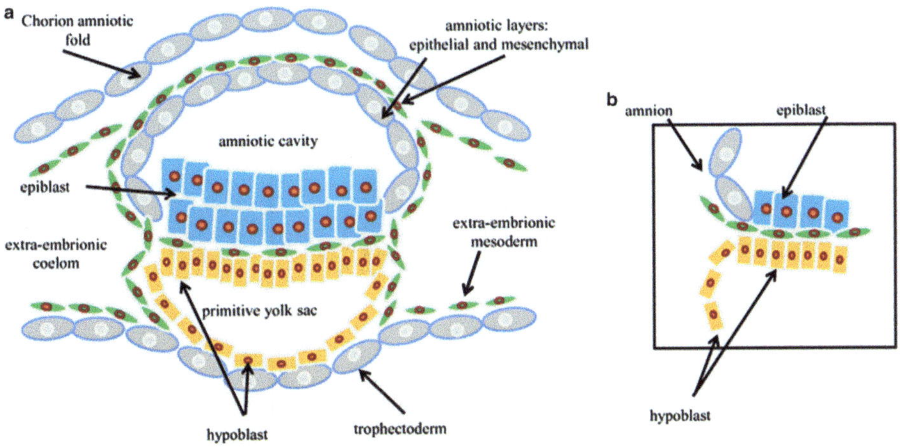

Fig. 4.5 Illustration of embryo at early gastrulation. (**a**) During gastrulation, an early amniotic cavity appears as chorion amniotic folds develop. The epithelial layer of the amnion, at the embryonic disk, is continuous with the epiblast. The mesenchymal layer is composed of extraembryonic mesoderm. (**b**) Detail of the amnion that is continuous with epiblast, outlined in (**a**)

cell-specific gene expression and immunocytochemical properties but showed substantial differences in the density of collection, and in their proliferative and differentiation potential.

BM-MSCs were collected at low density: on average, 8×10^6 MSCs were obtained from 30 mL of sternal aspirated from each analyzed sample, with a maximum of 25×10^6 and a minimum of 5.2×10^6 MSCs, while 100×10^6 stromal cells from about 24 g of tissue were obtained after digestion of amnion tissue.

For AMCs the mean doubling time (DT) was significantly lower ($P<0.05$) than that observed for BM-MSCs (1.17 ± 0.15 vs. 3.27 ± 0.19 days, respectively). The average DT value of AMCs was 2 days shorter than the value observed in BM-MSCs (1.17 vs. 3.27 days). We calculated that when 5×10^5 AMCs are plated, it is possible to obtain an average of 8×10^6 cells in 6 days, while, by plating 5×10^5 of BM-MSCs, it takes 14 days to obtain the same number of cells. Also Fortier and Smith (2008) reported that in the equine species at least 2–3 weeks are necessary to expand cells isolated from BM and to get 10×10^6 cells available for implantation into a tendon injury, even if this therapeutic dose is empirically derived and other doses could potentially be efficacious. Moreover, BM-MSCs, compared to AMCs, also demonstrated a significantly ($P<0.05$) lower clonogenic capability expresses as colony forming unit (CFU), with a mean value of 1 CFU-F each 590.15 cells seeded for BM-MSCs and 1 CFU-F each 242.73 cells seeded for AMSCs.

About differentiation potential, after differentiation in osteogenic, adipogenic, chondrogenic, and neurogenic lines, BM-MSCs did not differentiate into glial cells and the osteogenic differentiation process was longer compared to AMCs (Table 4.2). In addition, Lange-Consiglio et al. (2012) showed that equine amnion-derived cells can also be frozen and recovered without loss of their functional integrity in terms of morphology, presence of specific stemness markers, and differentiation ability.

Table 4.2 Characteristics of the MSCs isolated from bone marrow and amniotic membrane

MSCs-source	Phenotype	Efficiency of isolation	Differentiation potential	Doubling time	Clonogenic capability	In vivo results % re-injuries
Bone marrow	Mesenchymal and hematopoietic markers: CD44+, CD29+, CD105+,CD34–,HLA-ABC+,HLA-DR– Pluripotent markers: Oct-4+, SSEA-4+, TRA-1-60+	From 30 ml about 8×10^6 MSCs	Osteogenic Adipogenic Chondrogenic	3.27 ± 0.19 days	1 CFU-F each 590.15 cells seeded	23.08
Amnion membrane	Mesenchymal and hematopoietic markers: CD44+, CD29+, CD105+,CD34–,HLA-ABC+,HLA-DR– Pluripotent markers: Oct-4+, SSEA-4+, TRA-1-60+	From 24 g about 100×10^6 MSCs	Osteogenic Adipogenic Chondrogenic Neurogenic	1.17 ± 0.15 days	1 CFU-F each 242.73 cells seeded	4

+ and – signs respectively indicate presence or absence of gene expression

The noninvasive nature and low cost of collection, the rapid proliferation along with a greater differentiation potential, and the "off the shelf" preparation potential could make AMCs useful for cell therapy.

Cells obtained from other extra-fetal adnexa in domestic animal share a relatively consistent set of surface markers, which appears to support the hypothesis that they are MSC-like (for review see Cremonesi et al. 2011).

Extra-Fetal MSCs and Clinical Application

In a preclinical study, Carrade et al. (2011) injected autologous and allogeneic equine MSCs derived from either umbilical cord blood or umbilical cord tissue in non-injured horse joints to test the immunological properties of these cells. Their data suggest that there are no differences in the type or degree of inflammation elicited by self, related, or nonself cells. Moreover, the inflammation was clinically mild, not associated with gait abnormalities, self-limiting, allowing to conclude that, with the mild to moderate self-limiting inflammation and the absence of a systemic immune response, placental-derived MSCs do not elicit any hyperacute or acute (MHC-mediated) immune response.

To date, in the horse, the only data about the first clinical approaches for the treatment of horse spontaneous tendon diseases employing equine amniotic cells are from our laboratory (Lange-Consiglio et al. 2012, 2013b). After comparative study employing cryopreserved heterologous AMCs and fresh autologous BM-MSCs in spontaneous tendon diseases in vivo, our results showed that AMCs were well-tolerated by horses, and all of the clinical findings reported (i.e., the quick reduction in gross tendon size, palpation sensitivity, and ultrasonographic CSA measurements), provide compelling evidence to support the exertion of beneficial effects by the injected cells. Moreover, the ultrasonographic evolution showed for tendon and ligament architecture is similar to that reported after injecting autologous BM-MSCs. Comparing this innovative therapy with the conventional treatment with BM-MSCs, we observed that, although the recovery time is very similar (only a minimum period of about 4 months is required in both groups), the rate of re-injury in amniotic-treated horse was lower (4.00 % vs. 23.08 % for amniotic and BM-MSCs, respectively). The good result with AMSCs may be due to the opportunity to inject these cells in real time, before any ultrastructural change occurs within the injured tendon. Indeed, BM-MSCs, used as fresh autologous cells, required prolonged in vitro culture, limiting the time frame for implantation and probably, in this situation, animals were prone to re-injury due to a not optimal regeneration. The regenerated tissue could be less elastic and therefore functionally inferior to normal or native tendon. The AMCs, instead, were used as cryopreserved cells, then immediately available. The timing of the implantation should be carefully considered as well. Using AMSCs it could be hypothesized that the earlier implantation would offer greater benefits to the healing tendon by reducing inflammation, recruiting native stem cells, and promoting production of collagen and other ECM proteins (Alves et al. 2011). The good time of implantation, together with the higher plasticity and proliferative capacity of the amniotic-derived cells compared to BM-MSCs (Lange-Consiglio et al. 2013a), represent the main features of interest for this novel approach for the treatment of equine tendon diseases. In addition, recent studies of in vivo transplantation of ovine amniotic membrane-derived epithelial cells into ovine Achilles tendon (Barboni et al. 2012) or into spontaneous equine tendon injuries (Muttini et al. 2013), demonstrated production of ovine collagene supporting the possibility to apply amniotic epithelial cells for tenocytes regeneration. Moreover, according to previous works (Smith et al. 2003; Dyson 2004) for equine embryonic stem cells and fetal-derived equine embryonic-like stem cells, it is possible to speculate that AMSCs, being most primitive compared to adult cells, exhibit higher level of engraftment with respect to BM-MSCs.

9.1.3 Embryonic or Fetal Embryonic-Like Stem Cells

To overcome the problem related to the use of adult stem cells, in addition to amniotic-derived cells, very recently, the first results concerning the use of ESC or fdESC, respectively for treatment of equine tendinopathy have been published (Guest et al. 2010; Watts et al. 2011). Guest et al. (2010) injected labeled ESC or labeled MSCs into artificially created tendon lesions, and horses were subjected to euthanasia after 20, 30, 60, or 90 days. ECS showed a high survival rate in vivo, which remained constant over 90 days and also migrated to other damaged areas within the tendon, whereas MSCs showed a lower and decreasing survival rate and did not migrate. Furthermore, there was no evidence of any immune response or tumor formation following cell application. Watts et al. (2011) reported that, over a period of 8 weeks after fdESC application, no differences in tendon-specific gene expression, collagen, or DNA content could be detected compared to the controls, while a significant improvement in tissue architecture, tendon size, lesion size, and tendon linear fiber pattern was shown. However, while these results are very encouraging, the long-term success and safety of ESC treatment remain to be proven.

9.2 Scaffold

Apart from blood cells, most, if not all other, normal cells in human and animal tissues are anchorage-dependent residing in a solid matrix called ECM. There are numerous types of ECM in human and animal tissues, which usually have multiple components and tissue-specific composition.

ECM provides structural support and physical environment for cells residing in that tissue to attach, grow, migrate, and respond to signals. Moreover, ECM gives the tissue its structural and therefore mechanical properties, such as rigidity and elasticity that is associated with the tissue functions. For example, well-organized thick bundles of collagen type I in tendon are highly resistant to stretching and are responsible for the high tensile strength of tendons. On the other hand, randomly

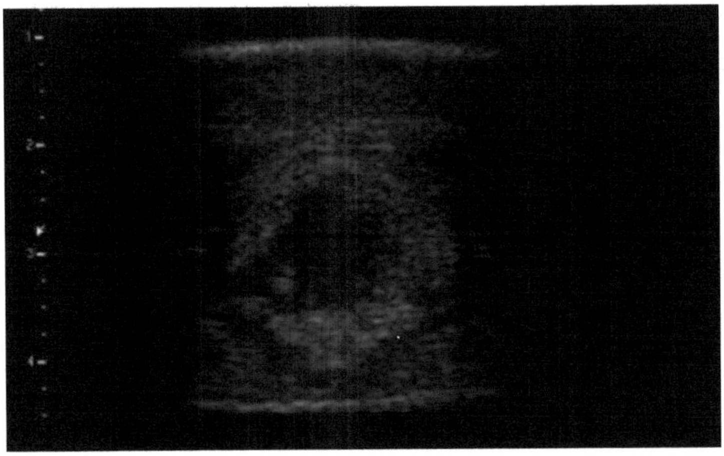

Fig. 4.6 Tendon lesion manifests within the central core of the tissue

distributed collagen fibrils and elastin fibers of skin are responsible for its toughness and elasticity. In addition, ECM may act as reservoir of growth factors and potentiates their bioactivities and provides a degradable physical environment so as to allow neovascularization and remodeling in response to developmental, physiological, and pathological challenges during tissue dynamic processes namely morphogenesis, homeostasis, and wound healing, respectively.

Intuitively, the best scaffold for an engineered tissue should be the ECM of the target tissue in its native state. Nevertheless, the multiple functions, the complex composition and the dynamic nature of ECM in native tissues make it difficult to mimic exactly (Chan and Leong 2008).

Scaffolds, typically made of polymeric biomaterials, provide the structural support for cell attachment and subsequent tissue development. As pointed out previously, in the horse, tendon injuries are mostly located in the SDFT, which represent the strongest tendon in the equine body. The SDFT displays several similarities to the human Achilles tendon concerning anatomy, biomechanics, and pathogenesis of tendinopathy. Yet, in most cases, pathomorphology of tendinopathy differs in lesion size. In the horse, one typical so-called "core lesion" is usually centrally located within the tendon, extended in length and still surrounded by intact tendon tissue. The equine SDFT injury lends itself to cell therapy because it provides many of the additional elements required for tendon tissue engineering. The lesion manifests within the central core of the tissue provides a natural enclosure for implantation that, at the time of stem cell implantation, is filled of granulation tissue which acts as a scaffold (Alves et al. 2011). This enables the application of MSCs without any artificial scaffold material, merely by injecting a cell suspension directly into the lesion; thereby the MSCs are exposed to a natural environment providing collagen fibers and growth factors (Fig. 4.6).

Table 4.3 Some growth factors involved during tendon repair

Growth factor	Activity	Repair phase
PDGF	• Induces expression and attraction of other growth factors (e.g., IGF-1)	Inflammatory
	• Increases cell proliferation	
	• Stimulates the synthesis of collagen and other ECM components	
IGF-1	• Stimulates recruitment of fibroblasts and inflammatory cells to the injury site	Inflammatory
	• Cell proliferation (DNA synthesis)	Proliferative
	• ECM remodeling	Remodeling
TGF-β	• Regulation of cell migration	Inflammatory
	• Cell proliferation (DNA synthesis)	Proliferative
	• Stimulates cell–matrix interactions	
	• Collagen type III synthesis	

9.3 Growth Factors

Growth factors are peptide signaling molecules that regulate many aspects of cellular metabolism including the cell cycle, cell growth, and differentiation, and the production and destruction of ECM products. Their effects are mediated primarily via autocrine and paracrine mechanisms, which provide the rationale for local administration of exogenous growth factors to influence cellular metabolism (Schnabel et al. 2009). Of the growth factors influencing tendon metabolism, platelet-derived growth factor (PDGF), insulin-like growth factor-I (IGF-I), and transforming growth factor β (TGF-β) show the most promise for enhancing tendon healing (Dahlgren et al. 2002). In Table 4.3 are summarized some effects of these factors involved during tendon repair.

PDGF induces the expression of other growth factors such as IGF-1 during the initial repair phase. In addition, the delivery of PDGF to tendon injuries in animal models increases cell proliferation and stimulates the synthesis of collagen and other ECM components in a dose-dependent manner during the remodeling phase (Wang et al. 2004; Thomopoulos et al. 2005). During the initial repair process and the inflammatory phase, upregulation of growth factors and cytokines such as IGF-1 stimulates the migration and proliferation of fibroblasts and inflammatory cells to the wound site (Kurtz et al. 1999; Tsuzaki et al. 2000). IGF-1 may be stored as an inactive precursor protein in normal tendon and, upon injury, enzymes release the growth factor to exert its biological activity. During the later phases such as remodeling, IGF-1 stimulates synthesis of collagen and other ECM components; studies in vitro have shown that the effects of IGF-1 on matrix metabolism are dose dependent (Abrahamsson et al. 1991; Abrahamsson and Lohmander 1996). Investigations with equine flexor tendon injury models have shown that both cell proliferation and collagen content increase on treatment with IGF-1. These changes are accompanied by increased stiffness in the treated tendon (Dahlgren et al. 2002).

Transforming growth factor-β (TGF-β) is a product of most cells that are involved in the healing process; During the initial inflammatory phase after trauma, TGF-β expression is elevated and stimulates cellular migration and proliferation, as well as interactions within the repair zone (Kashiwagi et al. 2004). Synthesis of collagen type I and collagen type III is increased greatly during the later phases. One of the isoforms of the growth factor, TGF-β1, is responsible for the initial scar tissue that forms to establish tissue continuity at the wound site. In the later phases of wound healing, increased expression of TGF-β1 leads to scar proliferation and reduced functionality. Transforming growth factor-β3 acts as a negative regulator of scarring at the wound site (Klein et al. 2002; Campbell et al. 2004). TGF-β1 also serves to regulate the synthesis of collagen in tendon mechanically during physical exercise (Heinemeier et al. 2003).

Although exogenous IGF-I has been shown to stimulate tendon healing in vivo in an equine model (Dahlgren et al. 2002) it has a short half-life, which necessitates repeated dosing, making clinical application challenging and costly. For this reason Schnabel et al. (2009) examined the effects of MSCs, as well as IGF-I gene-enhanced MSCs (AdIGF-MSC) on tendon healing in vivo finding that both MSC and AdIGF-MSC injection resulted in significant histological tendon healing and a trend towards improved biomechanical characteristics of healing tendon in the early period after injury, with minimal added value of IGF-I gene-enhanced MSC implantation compared to native MSCs injection. This minimal added value underlines the possibility that the MSCs can act mainly to supply immunomodulatory and trophic factors.

Current knowledge of the biological characteristics of MSCs also focus much attention on the role of paracrine factors released from these cells, on the mechanisms of their regenerative/reparative potential and on their ability to control the inflammatory process.

To test the hypothesis that the beneficial effects observed with transplantation of AMCs to promote horse tendon repair (Lange-Consiglio et al. 2012, 2013b) are mediated via AMC-secreted paracrine-acting molecules and by their immunomodulatory properties, we evaluated the immunomodulatory capacity of horse amniotic membrane-derived cells and of the conditioned medium derived from the culture of these cells (AMCs-CM), and we investigated the effects of injecting AMCs-CM in spontaneous horse tendon and ligament injuries in vivo (Lange-Consiglio et al. 2013c Stem Cell Development accepted in press).

To our knowledge for the first time in veterinary medicine, we have demonstrated that horse AMCs are capable of inhibiting peripheral blood mononuclear cells (PBMC) proliferation not only when cultured in cell–cell contact with responder cells, but also when separated from them by a transwell membrane. The inhibitory effects were more pronounced when increasing numbers of AMCs were added to the cultures, suggesting a dose-dependent effect. The finding that inhibition of T-cell proliferation was induced by AMCs in the transwell system suggests that soluble factors are implicated in this phenomenon. This hypothesis was further supported by our findings that PBMC proliferation was inhibited by the conditioned medium derived from AMCs (AMCs-CM). Our results are similar to findings that have been reported using human term placenta, which clearly demonstrate that cells isolated from the mesenchymal portion of human amniotic membrane, as well as CM derived from the culture of these cells, could both inhibit lymphocyte proliferation (Wolbank et al. 2007; Magatti et al. 2008; Rossi et al. 2012).

Even though it is generally accepted that MSC from amniotic membrane or other sources act in a paracrine fashion, mechanisms underlying the immunosuppressive actions of these cells remain to be fully elucidated (for review see: Nauta and Fibbe 2007; Herrero and Pérez-Simón 2010; Engela et al. 2012). The identity of these soluble factors remains unknown, even though TGF-β1, hepatocyte growth factor, prostaglandin E2, interleukin-10, heme oxygenase-1, interleukin-6, and human leukocyte antigen-G5 are all known to be constitutively produced by MSC (Aggarwal and Pittenger 2005; Chabannes et al. 2007; Selmani et al. 2008). Recently Rossi et al. (2012) suggested prostaglandins as one of the key effector molecules of the antiproliferative activity of human amniotic membrane-derived cells.

An important role for soluble factors in stem/progenitor cell-mediated reparative effects has been supported by animal experiments using CM as an effective treatment for different tissue injuries. Indeed, in a porcine model of myocardial ischemia and reperfusion, intravenous and intracoronary injection of CM obtained from human MSC cultures reduced infarct size and improved cardiac performance (Timmers et al. 2007). Furthermore, systemic infusion of human MSC-CM has been shown to reduce apoptosis and stimulate proliferation of hepatocytes in a rat model of acute liver injury (van Poll et al. 2008). In support of a key role for soluble factors in the suppressive action of human amniotic membrane-derived cells (Cargnoni et al. 2009, 2012), used a mouse model of lung fibrosis to show that a reduction in severity and progression of the disease could be obtained both after transplantation of amniotic membrane-derived cells and also after the injection of their conditioned medium, thereby underlining the anti-inflammatory effect of these media.

All of the important results described above were obtained using CM in induced disease, while in our study, for the first time, CM derived from cultured horse amniotic membrane-derived cells was used to treat spontaneous horse tendon and ligament injuries (unpublished data). We treated 13 sports horses with different tendon or ligament injuries that were treated with 2 mL of AMCs-CM ultrasonografically injected in the lesions and followed up for 2 years. Clinical and ultrasonographical evaluation of the involved tendon or ligament did not reveal evidence of inappropriate tissue or tumor formation, independently of the type and severity of diseases. In the absence of any perioperative collateral treatment, it was possible to attest that no local or focal reaction has occurred. These results are suggestive of a correlated anti-inflammatory effect of AMCs-CM. Indeed, besides a demonstrated lack of worsening of the lesion straight after treatment, a marked reduction in swelling and CSA in the very early postoperative period was also noticed. Moreover, interestingly, using power doppler US, neo-vessels were imaged within the treated area as soon as 1 month after AMCs-CM injection. Vessel size and quantity decreased and eventually disappeared by approximately 2 months after treatment, while improvement in echogenicity and fiber architecture were also observed.

This is clearly correlated to the positive tissue healing process, probably induced by angio-acting cytokines, and must be considered as an important timing predictor in the rehabilitation program. Clinical outcomes after AMCs-CM injection can be considered favorable after 2 years of follow-up where the rate of re-injuries was 9 %. Considering that increasing evidence has recently highlighted the ability of MSC isolated from various sources to produce bioactive molecules, which are potentially able to exert several types of paracrine effects (e.g., anti-scarring, anti-inflammatory, antiapoptotic) on target cells (Meirelles et al. 2009), our observations lead us to hypothesize that soluble factors released by AMCs may interact with and stimulate tendon-resident cells to initiate an anti-inflammatory and angiogenic response that leads to a regenerative/reparative response. In particular, factors derived from AMCs-CM may counteract the action of inflammatory cells accumulated at the site of injury, and may also exert trophic effects, as documented by Tang et al. (2004) in ischemic myocardium and in bleomycin-induced lung fibrosis (Cargnoni et al. 2012). These results are consistent and overlapping (in terms of time interval needed to reach complete healing) with the beneficial effects of transplanted horse amniotic membrane-derived cells that we have previously observed in the same animal model (Lange-Consiglio et al. 2012, 2013b), and therefore suggest that cell treatment could be substituted by the use of CM derived from the culture of these cells.

Our results may open the way to consideration of cell-free treatment approaches for tendon and ligament diseases, offering a series of added advantages with respect to other current treatments. In particular, AMCs-CM can be produced easily and in large quantities; it can be stored efficiently because it maintains its efficacy after the lyophilization process; as a cell-free treatment, it can drastically reduce the risk of adverse immunological reactions, infectious risks, and other potential long-term negative effects caused by the presence of exogenous cells; finally, it is also conceivable that AMCs-CM could be administered safely via intravenous injection, avoiding clot formation and lung capillary entrapment (Cargnoni et al. 2009). Even though this study identifies AMCs-CM as a novel therapeutic biological cell-free product in spontaneous tendon and ligament diseases, elucidation of the nature and characteristics of the paracrine soluble molecules underlying the efficacy of AMCs-CM therapy, and an understanding of the mechanisms responsible for the beneficial effects of the paracrine actions performed by stem cells in general, remain ongoing challenges for all researchers in the field.

10 Debate

Although our understanding of tendon biology and the biological processes that regulate repair have progressed tremendously, many challenges need to be addressed to bring about a successful treatment strategy. Recently, amniotic-derived stromal cells have been shown to differentiate into multiple lineages under appropriate chemical cues and to heal tendon diseases. In addition, growth factors derived from these cells (amniotic cells) can improve the healing of tendon injured lowering the recurrence rate. Treatment options involving growth factors, alone or in combination with a stromal population, may form functional repair or regenerate tendon faster than existing treatment modalities.

In any case, irrespective of the type of stem cell transplanted or of injected CM, there was no evidence of any significant adverse effects after MSCs treatment in the horses studied in different trials. Evaluation of tendons clinically, ultrasonographically, scintigraphically, and histologically showed no evidence of inappropriate tissue or tumor formation and clinical outcomes after MSCs treatment were favorable. Whereas high re-injury rates of 23 % up to 80 % following conventional treatment regimes have been reported (Dowling et al. 2000; Dyson 2004), re-injury rates following MSCs treatment are decreased. After a period variable between 48 weeks and 3 years of adult stem cell transplantation, the average of re-injured varied between 13 and 36 % (Smith 2008; Leppänen et al. 2009; Burk and Brehm 2011; Godwin et al. 2012), while with amniotic-derived cells, varied between 0 % at the beginning of the study on three treated cases (Lange-Consiglio et al. 2012) and 4.0 % on 48 cases treated after a follow-up of 2 years (Lange-Consiglio et al. 2013b) and increased to 9 % with conditioned medium, but in this study the clinical record was very limited.

Even in spite of the best results obtained with the AMSCs, it is very difficult to demonstrate improved healing in treated animals and the comparison between different sources of cells because no control animals, generally, are included. Difficulties to determine a control population are correlated to the treatments given in the equine industry. Moreover, the interpretation of the results can be misleaded by the variety of athletic sports in which the horses were involved, the nature of injury, the severity of the initial injury, and the age of the horse. In addition, the statistical analysis is difficult to standardize considering the treatment of different equine tendons that are highly variable. All these factors are influencing the prognosis. The optimal time of treatment and the optimal dose and route of administration of stem cells have not been determined yet for any stem cell type, lesion, or animal species. Many authors agree with Richardson (2007) suggesting that the optimal timing of cell implantation is after the initial inflammatory phase but before fibrous-tissue formation. It has been hypothesized that the presence of mature fibrous tissue within the tendon would make implantation more difficult and reduce the benefits of the stem cell therapy because of its persistence. Other issues are debated and concern the amount of stem cells needed to stimulate regeneration or the period of follow-up to allow a direct comparison. To date, there are no dose response-based studies: in literature the range varies between 1 and 50×10^6 cells depending on the extent of the lesion while the follow-up ranges from 1 month to 3 years.

Further research studies will be required to answer these questions.

It is hoped that experience gained from treating naturally occurring tendon injury in horses will provide sufficient supportive data to encourage the translation of this technology into the human field where large randomized controlled trials will lead to a higher level of clinical evidence (Godwin et al. 2012).

11 Conclusion

Exercise-induced injury to energy-storing tendons of horses and people occurs at a similarly high frequency, with failure of tissue regeneration contributing to high re-injury rates or retirement from athletic activity. There is now increasing research focus on the role of the tenocyte in accumulation of the matrix and cellular micro-damage that predisposes to tendon rupture. Access to equine tissues and the ability to make measurements in vivo have already significantly improved our knowledge of the composition, structure, and mechanical environment of injury-prone and non-injury-prone digital tendons over a wide age range. The imperative now is to develop tractable cell culture models using this species in order to study stress-induced damage and dysfunction; these models will facilitate a more authentic replication of the in vivo microenvironment (Patterson-Kane et al. 2012).

Acknowledgments The Authors thank Dr S. Tassan for providing ultrasound image of tendon core lesion.

References

Abrahamsson SO, Gelberman R (1994) Maintenance of the gliding surface of tendon autografts in dogs. Acta Orthop Scand 65:548–552. doi:10.1016/j.jhsa.2007.09.007

Abrahamsson SO, Lohmander S (1996) Differential effects of insulin-like growth factor-I on matrix and DNA synthesis in various regions and types of rabbit tendons. J Orthop Res 14: 370–376. doi:10.1002/jor.1100140305

Abrahamsson SO, Lundborg G, Lohmander LS (1991) Long-term explant culture of rabbit flexor tendon: effects of recombinant human insulin-like growth factor-I and serum on matrix metabolism. J Orthop Res 9:503–515

Aggarwal S, Pittenger MF (2005) Human mesenchymal stem cells modulate allogeneic immune cell responses. Blood 105:1815–1822. doi:10.1182/blood-2004-04-1559

Alexander RW (2012) Understanding adipose-derived stromal vascular fraction (AD-SVF) cell biology and use on the basis of cellular, chemical, structural and paracrine components: a concise review. J Prolother 4:e855–e869

Alves A, Stewart A, Dudhia J (2011) Cell-based therapies for tendon and ligament incurie. Vet Clin North Am Equine Pract 27:315–333. doi:10.1016/j.cveq.2011.06.001

Astrom M, Rausing A (1995) Chronic Achilles tendinopathy. A survey of surgical and histopathologic findings. Clin Orthop Relat Res 316:151–164

Awad HA, Boivin GP, Dressler MR et al (2003) Repair of patellar tendon injuries using a cell-collagen composite. J Orthop Res 21:420–431

Barboni B, Russo V, Curini V et al (2012) Achilles tendon regeneration can be improved by amniotic epithelial cell allotransplantation. Cell Transplant 21:2377–2395

Berg L, Koch T, Heerkens T et al (2009) Chondrogenic potential of mesenchymal stromal cells derived from equine bone marrow and umbilical cord blood. Vet Comp Orthop Traumatol 22:363–370. doi:10.3415/VCOT-08-10-0107

Bi Y, Ehirchiou D, Kilts TM et al (2007) Identification of tendon stem/progenitor cells and the role of the extracellular matrix in their niche. Nat Med 13:1219–1227

Birch HL, Bailey JVB, Bailey AJ et al (1999) Age-related changes to the molecular and cellular components of equine flexor tendons. Equine Vet J 31:391–396

Burk J, Brehm W (2011) Stammzellentherapie von Sehnenverletzungen—klinische Ergebnisse von 98 Fällen. Pferdeheilkunde 27:153–161

Butler DL, Juncosa-Melvin N, Boivin GP et al (2008) Functional tissue engineering for tendon repair: a multidisciplinary strategy using mesenchymal stem cells, bioscaffolds, and mechanical stimulation. J Orthop Res 26:1–9. doi:10.1002/jor.20456

Campbell BH, Agarwal C, Wang JH (2004) TGF-beta1, TGFbeta3, and PGE(2) regulate contraction of human patellar tendon fibroblasts. Biomech Model Mechanobiol 2:239–245. doi:10.1007/s10237-004-0041-z

Cargnoni A, Gibelli L, Tosini A et al (2009) Transplantation of allogeneic and xenogeneic placenta-derived cells reduces bleomycin-induced lung fibrosis. Cell Transplant 18:405–422. doi:10.3727/096368909788809857

Cargnoni A, Ressel L, Rossi D et al (2012) Conditioned medium from amniotic mesenchymal tissue cells reduces progression of bleomycin-induced lung fibrosis. Cytotherapy 14:153–161. doi:10.3109/14653249.2011.613930

Carrade DD, Owens SD, Galuppo LD et al (2011) Clinicopathologic findings following intra-articular injection of autologous and allogeneic placentally derived equine mesenchymal stem cells in horses. Cytotherapy 13:419–430. doi:10.3109/14653249.2010.536213

Chabannes D, Hill M, Merieau E et al (2007) A role for heme oxygenase-1 in the immunosuppressive effect of adult rat and human mesenchymal stem cells. Blood 110:3691–3694. doi:10.1182/blood-2007-02-075481

Chan BP, Leong K (2008) Scaffolding in tissue engineering: general approaches and tissue-specific considerations. Eur Spine J 17(suppl 4):467–479. doi:10.1007/s00586-008-0745-3

Cheung DT, DiCesare P, Benya PD et al (1983) The presence of inter molecular cross-links in type III collagen. J Biol Chem 12:7774–7778

Chong AK, Chang J, Go JC (2009) Mesenchymal stem cells and tendon healing. Front Biosci 14:4598–4605

Cohen RE, Hooley CJ, McCrum NG (1974) Mechanism of the viscoelastic deformation of collagenous tissue. Nature 247:59–61

Colleoni S, Bottani E, Tessaro I et al (2009) Isolation, growth and differentiation of equine mesenchymal stem cells: effect of donor, source, amount of tissue and supplementation with basic fibroblast growth factor. Vet Res Commun 33:811–821. doi:10.1007/s11259-009-9229-0

Cook JL, Feller JA, Bonar SF et al (2004) Abnormal tenocyte morphology is more prevalent than collagen disruption in asymptomatic athletes' patellar tendons. J Orthop Res 22:334–338

Corradetti B, Lange-Consiglio A, Barucca M et al (2011) Size-sieved subpopulations of mesenchymal stem cells from intervascular and perivascular equine umbilical cord matrix. Cell Prolif 44:330–342. doi:10.1111/j.1365-2184.2011.00759.x

Cremonesi F, Corradetti B, Lange Consiglio A (2011) Fetal adnexa derived stem cells from domestic animal: progress and perspectives. Theriogenology 75:1400–1415. doi:10.1016/j.theriogenology.2010.12.032

Crevier-Denoix N, Pourcelot P (1997) Additional research on tendon strains and stresses. Am J Vet Res 58:569–570

Crovace A, Lacitignola L, De Siena R, Rossi G, and Francioso E. (2007) Cell therapy for tendon repair in horses: an experimental study. Vet Res Commun 31: supplement 1, 281–283

Crovace A, Lacitignola L, Rossi G et al (2010) Histological and immunohistochemical evaluation of autologous cultured bone marrow mesenchymal stem cells and bone marrow mononucleated cells in collagenase-induced tendinitis of equine superficial digital flexor tendon. Vet Med Int 2010:250978. doi:10.4061/2010/250978

da Silva Meirelles L, Chagastelles PC, Nardi NB (2006) Mesenchymal stem cells reside in virtually all post-natal organs and tissues. J Cell Sci 119:2204–2213

Dahlgren LA (2009) Fat-derived mesenchymal stem cells for equine tendon repair. Regen Med 4: Suppl.2, S14

Dahlgren LA, van der Meulen MC, Bertram JE et al (2002) Insulin-like growth factor-I improves cellular and molecular aspects of healing in a collagenase induced model of flexor tendinitis. J Orthop Res 20:910–919 de Mos M, Koevoet WJ, Jahr H et al (2007) Intrinsic differentiation potential of adolescent human tendon tissue: an in-vitro cell differentiation study. BMC Musculoskelet Disord 8:16

Del Bue M, Ricco S, Ramoni R et al (2008) Equine adipose-tissue derived mesenchymal stem cells and platelet concentrates: their association in vitro and in vivo. Vet Res Commun 32(suppl 1): S51–S55. doi:10.1007/s11259-008-9093-3

Dowling BA, Dart AJ, Hodgson DR et al (2000) Superficial digital flexor tendonitis in the horse. Equine Vet J 32:369–378

Dunphy JE (1967) The healing of wounds. Can J Surg 10:281–287

Durando MM, Zarucco L, Schaer TP et al (2006) Pneumopericardium in a horse secondary to sternal bone marrow aspiration. Equine Vet Educ 18:75–79

Dyson SJ (2004) Medical management of superficial digital flexor tendonitis: a comparative study in 219 horses (1992-2000). Equine Vet J 36:415–419

Ely ER, Avella CS, Price JS et al (2009) Descriptive epidemiology of fracture, tendon and suspensory ligament injuries in National Hunt racehorses in training. Equine Vet J 41:372–378

Emerson C, Morrissey D, Perry M et al (2010) Ultrasonographically detected changes in Achilles tendons and self reported symptoms in elite gymnasts compared with controls—an observational study. Man Ther 15:37–42. doi:10.1016/j.math.2009.05.008

Engela AU, Baan CC, Dor FJ et al (2012) On the interactions between mesenchymal stem cells and regulatory T cells for immunomodulation in transplantation. Front Immunol 3:126. doi:10.3389/fimmu.2012.00126

Eriksen HA, Pajala A, Leppilahti J et al (2002) Increased content of type III collagen at the rupture site of human Achilles tendon. J Orthop Res 20:1352–1357

Evangelista M, Soncini M, Parolini O (2008) Placenta-derived stem cells: new hope for cell therapy? Cytotechnology 58:33–42

Evans CE, Trail IA (2001) An in vitro comparison of human flexor and extensor tendon cells. J Hand Surg Br 26:307–313

Farris DJ, Trewartha G, McGuigan MP (2011) Could intra-tendinous hyperthermia during running explain chronic injury of the human Achilles tendon? J Biomech 44:822–826. doi:10.1016/j.jbiomech.2010.12.015

Fenwick S, Harral R, Hacney R et al (2002) Endochondral ossification in Achilles and patella tendinopathy. Rheumatology (Oxford) 41:474–476

Fortier LA, Smith RK (2008) Regenerative medicine for tendinous and ligamentous injuries of sport horses. Vet Clin North Am Equine Pract 24:191–201. doi:10.1016/j.cveq.2007.11.002

Fortier LA, Travis AJ (2011) Stem cells in veterinary medicine. Stem Cell Res Ther 2:9. doi:10.1186/scrt50

Fortier LA, Nixon AJ, Williams J et al (1998) Isolation and chondrocytic differentiation of equine bone marrow-derived mesenchymal stem cells. Am J Vet Res 59:1182–1187

Fortier LA, Potter HG, Rickey EJ et al (2010) Concentrated bone marrow aspirate improves full-thickness cartilage repair compared with microfracture in the equine model. J Bone Joint Surg Am 92:1927–1937. doi:10.2106/JBJS.I.01284

Garner WL, McDonald JA, Koo M (1989) Identification of the collagen-producing cells in healing flexor tendons. Plast Reconstr Surg 83:875–879

Genovese RL, Reef VB, Longo KL et al (1996) Superficial digital flexor tendonitis long term sonographic and clinical study of racehorses. In: Rantanen NW, Hauser ML (eds) Proceedings of the Dubai equine international symposium, Dubai, pp 187–205

Gillis C, Meagher DM, Cloninger A et al (1995a) Ultrasonographic cross-sectional area and mean echogenicity of the superficial and deep digital flexor tendons in 50 trained Thoroughbred racehorses. Am J Vet Res 56:1265–1269

Gillis C, Sharkey N, Stover S et al (1995b) Effect of maturation and aging on material and ultrasonographic properties of equine superficial digital flexor tendon. Am J Vet Res 56:1345–1350

Gillis C, Pool RR, Meagher DM et al (1997) Effect of maturation and aging on the histomorphometric and biochemical characteristics of equine superficial digital flexor tendon. Am J Vet Res 58:425–430

Godwin EE, Young NJ, Dudhia J et al (2012) Implantation of bone marrow-derived mesenchymal stem cells demonstrates improved outcome in horses with overstrain injury of the superficial digital flexor tendon. Equine Vet J 44:25–32. doi:10.1111/j.2042-3306.2011.00363.x

Goodship AE, Birch HL (1996) The pathophysiology of the flexor tendons in the equine athlete. In: Rantanen NW, Hauser ML (eds) Proceedings of the Dubai equine international symposium, Dubai, pp 83–107

Goodship AE, Birch HL, Wilson AM (1994) The pathobiology and repair of tendon and ligament injury. Vet Clin North Am Equine Pract 10:323–348

Gu J, Wada Y (1996) Effect of exogenous Decorin on cell morphology and attachment of Decorin-deficient fibroblasts. J Biochem 119:743–748

Guest DJ, Smith MR, Allen WR (2008) Monitoring the fate of autologous and allogeneic mesenchymal progenitor cells injected into the superficial digital flexor tendon of horses: preliminary study. Equine Vet J 40:178–181. doi:10.2746/042516408X276942

Guest DJ, Smith MR, Allen WR (2010) Equine embryonic stem-like cells and mesenchymal stromal cells have different survival rates and migration patterns following their injection into damaged superficial digital flexor tendon. Equine Vet J 42:636–642. doi:10.1111/j.2042-3306.2010.00112.x

Hankemeier S, Keus M, Zeichen J et al (2005) Modulation of proliferation and differentiation of human bone marrow stromal cells by fibroblast growth factor 2: potential implications for tissue engineering of tendons and ligaments. Tissue Eng 11:41–49

Hankemeier S, van Griensven M, Ezechieli M et al (2007) Tissue engineering of tendons and ligaments by human bone marrow stromal cells in a liquid fibrin matrix in immunodeficient rats: results of a histologic study. Arch Orthop Trauma Surg 127:815–821

Heinemeier K, Langberg H, Olesen JL et al (2003) Role of TGF-beta1 in relation to exercise-induced type I collagen synthesis in human tendinous tissue. J Appl Physiol 95:2390–2397

Herrero C, Pérez-Simón JA (2010) Immunomodulatory effect of mesenchymal stem cells. Braz J Med Biol Res 43:425–430

Hertel D (2001) Enhanced suspensory ligament healing in 100 horses by stem cells and other bone marrow components. Proc Am Assoc Equine Pract 47:319–321

Hooley CJ, McCrum NG, Cohen RE (1980) The viscoelastic deformation of tendon. J Biomech 13:521–528

Ingraham JM, Hauck RM, Ehrlich HP (2003) Is the tendon embryogenesis process resurrected during tendon healing? Plast Reconstr Surg 112:844–854

Jarvinen M, Jozsa L, Kannus P et al (1997) Histopathological findings in chronic tendon disorders. Scand J Med Sci Sports 7:86–95

Jezierska-Woźniak K, Nosarzewska D, Tutas A et al (2010) Use of adipose tissue as a source of mesenchymal stem cells. Postepy Hig Med Dosw (Online) 64:326–332

Jones AJ (1993) Normal and diseased equine digital flexor tendon: blood flow, biochemical and serological studies. PhD Thesis, University of London, London

Jones G, Corps A, Pennington C et al (2006) Expression profiling of metalloproteinases and tissue inhibitors of metalloproteinases in normal and degenerate human Achilles tendon. Arthritis Rheum 54:832–842. doi:10.1002/art.21672

Juncosa-Melvin N, Matlin KS, Holdcraft RW et al (2007) Mechanical stimulation increases collagen type I and collagen type III gene expression of stem cell-collagen sponge constructs for patellar tendon repair. Tissue Eng 13:1219–1226

Kader D, Saxena A, Movin T et al (2002) Achilles tendinopathy: some aspects of basic science and clinical management. Br J Sports Med 36:239–249

Kajikawa Y, Morihara T, Watanabe N et al (2007) GFP chimeric models exhibited a biphasic pattern of mesenchymal cell invasion in tendon healing. J Cell Physiol 210:684–691

Kannus P, Jozsa L (1991) Histopathological changes preceding spontaneous rupture of a tendon. A controlled study of 891 patients. J Bone Joint Surg Am 73:1507–1525

Kasashima Y, Kuwano A, Katayama Y et al (2002) Magnetic resonance imaging application to live horse for diagnosis of tendinitis. J Vet Med Sci 64:577–582

Kasashima Y, Takahashi T, Smith RK et al (2004) Prevalence of superficial digital flexor tendonitis and suspensory desmitis in Japanese Thoroughbred flat racehorses in 1999. Equine Vet J 36: 346–350

Kashiwagi K, Mochizuki Y, Yasunaga Y et al (2004) Effects of transforming growth factor-beta 1 on the early stages of healing of the Achilles tendon in a rat model. Scand J Plast Reconstr Surg Hand Surg 38:193–197

Kastelic J, Baer E (1980) Deformation in tendon collagen. Symp Soc Exp Biol 34:397–435

Kisiday JD, Kopesky PW, Evans CH et al (2008) Evaluation of adult equine bone marrow- and adipose-derived progenitor cell chondrogenesis in hydrogel cultures. J Orthop Res 26:322–331

Klein MB, Yalamanchi N, Pham H et al (2002) Flexor tendon healing in vitro: effects of TGF-beta on tendon cell collagen production. J Hand Surg Am 27A:615–620

Knobloch K, Yoon U, Vogt PM (2008) Acute and overuse injuries correlated to hours of training in master running athletes. Foot Ankle Int 29:671–676

Kraus-Hansen AE, Fackelman GE, Becker C et al (1992) Preliminary studies on the vascular anatomy of the equine superficial digital flexor tendon. Equine Vet J 24:46–51

Kryger GS, Chong AK, Costa M et al (2007) A comparison of tenocytes and mesenchymal stem cells for use in flexor tendon tissue engineering. J Hand Surg Am 32:597–605

Kurtz CA, Loebig TG, Anderson DD et al (1999) Insulin-like growth factor I accelerates functional recovery from Achilles tendon injury in a rat model. Am J Sports Med 27:363–369

Lacitignola L, Crovace A, Rossi G, Francioso E (2008) Cell therapy for tendinitis, experimental and clinical report.Vet Res Commun:S33–S38.

Lange-Consiglio A, Corradetti B, Bizzaro D et al (2012) Characterization and potential applications of progenitor-like cells isolated from horse amniotic membrane. J Tissue Eng Regen Med 6:622–635. doi:10.1002/term.465

Lange-Consiglio A, Corradetti B, Meucci A et al (2013a) Characteristics of equine mesenchymal stem cells derived from amnion and bone marrow: in vitro proliferative and multilineage potential assessment. Equine Vet J. 45:737–744. doi:10.1111/evj.12052

Lange-Consiglio A, Tassan S, Corradetti B et al (2013b) Investigating the potential of equine mesenchymal stem cells derived from amnion and bone marrow in equine tendon diseases treatment in vivo. Cytotherapy 15:1011–1020. doi: 10.1016/J.JCYT.2013.03.002

Lange-Consiglio A, Rossi D, Tassan S et al (2013c) Conditioned medium from horse amniotic membrane-derived multipotent progenitor cells: immunomodulatory activity *in vitro* and first clinical application in tendon and ligament injuries *in vivo*. Stem Cell Dev 22:3015–3024. doi:10.1089/scd.2013.0214

Leadbetter WB (1992) Cell-matrix response in tendon injury. Clin Sports Med 11:533–578

Leppänen M, Miettinen S, Mäkinen S et al (2009) Management of equine tendon & ligament injuries with expanded autologous adipose-derived mesenchymal stem cells: a clinical study. Regen Med 4(suppl 2):21

Lindsay WK, Birch JR (1964) The fibroblast in flexor tendon healing. Plast Reconstr Surg 34: 223–232

Liu SH, Yang RS, al-Shaikh R, Lane JM (1995) Collagen in tendon, ligament, and bone healing. A current review. Clin Orthop Relat Res 318:265–278

Liu W, Chen B, Deng D et al (2006) Repair of tendon defect with dermal fibroblast engineered tendon in a porcine model. Tissue Eng 12:775–788

Lovati AB, Corradetti B, Lange Consiglio A et al (2011) Characterization and differentiation of equine tendon-derived progenitor cells. J Biol Regul Homeost Agents 25(2 suppl):S75–S84

Magatti M, De Munari S, Vertua E et al (2008) Human amnion mesenchyme harbors cells with allogeneic T-cell suppression and stimulation capabilities. Stem Cells 26:182–192

Malvankar S, Khan WS (2011) Evolution of the Achilles tendon: the athlete's Achilles heel? Foot (Edinburgh) 21:193–197

Martinello T, Bronzini I, Maccatrozzo L et al (2011) Canine adipose-derived-mesenchymal stem cells do not lose stem features after a long-term cryopreservation. Res Vet Sci 91:18–24. doi:10.1016/j.foot.2011.08.004

McIlwraith CW (1987) Diseases of joints, tendons, ligaments, and related structures. In: Stashak TS (ed) Adams' lameness in horses. Lea & Febiger, Philadelphia, pp 339–485

Meirelles LS, Fontes AM, Covas DT et al (2009) Mechanisms involved in the therapeutic properties of mesenchymal stem cells. Cytokine Growth Factor Rev 20:419–427
Millar NL, Reilly JH, Kerr SC et al (2012) Hypoxia: a critical regulator of early human tendinopathy. Ann Rheum Dis 71:302–310. doi:10.1136/ard.2011.154229
Minetti AE, ArdigO LP, Reinach E et al (1999) The relationship between mechanical work and energy expenditure of locomotion in horses. J Exp Biol 202:2329–2338
Murphy JM, Fink DJ, Hunziker EB et al (2003) Stem cell therapy in a caprine model of osteoarthritis. Arthritis Rheum 48:3464–3474
Muttini A, Valbonetti L, Abate M et al (2013) Ovine amniotic epithelial cells: in vitro characterization and transplantation into equine superficial digital flexor tendon spontaneous defects. Res Vet Sci 94:158–169. doi:10.1016/j.rvsc.2012.07.028
Myers B, Wolf M (1974) Vascularization of the healing wound. Am Surg 40:716–722
Nauta AJ, Fibbe WE (2007) Immunomodulatory properties of mesenchymal stromal cells. Blood 110:3499–3506
Nixon AJ, Dahlgren LA, Haupt JL et al (2008) Effect of adipose-derived nucleated cell fractions on tendon repair in horses with collagenase-induced tendinitis. Am J Vet Res 69:928–937. doi:10.2460/ajvr.69.7.928
O'Meara BO, Bladon B, Parkin TDH et al (2010) An investigation of the relationship between race performance and superficial digital flexor tendonitis in the thoroughbred racehorse. Equine Vet J 42:322–326. doi:10.1111/j.2042-3306.2009.00021.x
Pacini S, Spinabella S, Trombi L et al (2007). Suspension of bone marrow-derived undifferentiated mesenchymal stromal cells for repair of superficial digital flexor tendon in race horses. Tissue Eng 13:2949–2955
Paris DB, Stout TA (2010) Equine embryos and embryonic stem cells: defining reliable markers of pluripotency. Theriogenology 74:516–524. doi:10.1016/j.theriogenology.2009.11.020
Patterson-Kane JC, Firth EC, Goodship AE et al (1997a) Age related differences in collagen crimp patterns in the superficial digital flexor tendon core region of untrained horses. Aust Vet J 75:39–44
Patterson-Kane JC, Parry DA, Birch HL et al (1997b) An age-related study of morphology and cross-link composition of collagen fibrils in the digital flexor tendons of young Thoroughbred horses. Connect Tissue Res 36:253–260
Patterson-Kane JC, Wilson AM, Firth EC et al (1997c) Comparison of collagen fibril populations in the superficial flexor tendon of exercised and nonexercised Thoroughbreds. Equine Vet J 29:121–125
Patterson-Kane JC, Becker DL, Rich T (2012) Spontaneously arising disease: review article. The pathogenesis of tendon microdamage in athletes: the horse as a natural model for basic cellular research. J Comp Pathol 147:227–247. doi:10.1016/j.jcpa.2012.05.010
Perez-Castro AV, Vogel KG (1999) In situ expression of collagen and proteoglycan genes during development of fibrocartilage in bovine deep flexor tendon. J Orthop Res 17:139–148
Perkins NR, Reid SW, Morris RS (2005) Risk factors for injury to the superficial digital flexor tendon and suspensory apparatus in Thoroughbred racehorses in New Zealand. N Z Vet J 53:184–192
Pittenger MF, Mackay AM, Beck SC et al (1999) Multilineage potential of adult human mesenchymal stem cells. Science 284:143–147
Pool RR (1996) Pathologic changes in tendonitis of athletic horses. In: Rantanen NW, Hauser ML (eds) Proceedings of the Dubai equine international symposium, Dubai, pp 109–117
Pruzansky ME (1987) A primate model for the evaluation of tendon adhesions. J Surg Res 42:273–276
Rees JD, Wilson AM, Wolman RL (2006) Current concepts in the management of tendon disorders. Rheumatology 45:508–521
Ren G, Zhang L, Zhao X et al (2008) Mesenchymal stem cell-mediated immunosuppression occurs via concerted action of chemokines and nitric oxide. Cell Stem Cell 2:141–150. doi:10.1016/j.stem.2007.11.014

Richardson LE, Dudhia J, Clegg PD et al (2007) Stem cells in veterinary medicine-attempts at regenerating equine tendon after injury. Trends Biotechnol 25:409–416, PubMed PMID: 17692415.eng

Riemersa DJ, Schamhardt HC (1982) The cryo-jaw, a clamp designed for in vitro rheology studies of horse digital flexor tendons. J Biomech 15:619–620

Riley GP (2005) Gene expression and matrix turnover in overused and damaged tendons. Scand J Med Sci Sports 15:241–251

Riley GP, Harrall RL, Constant CR et al (1994a) Tendon degeneration and chronic shoulder pain: changes in the collagen composition of the human rotator cuff tendons in rotator cuff tendinitis. Ann Rheum Dis 53:359–366

Riley GP, Harrall RL, Constant CR et al (1994b) Glycosaminoglycans of human rotator cuff tendons: changes with age and in chronic rotator cuff tendinitis. Ann Rheum Dis 53:367–376

Rolf CG, Fu BSC, Pau A et al (2001) Increased cell proliferation and associated expression of PDGFRb causing hypercellularity in patellar tendinosis. Rheumatology 40:256–261

Roshan J, Kesturu G, Balian G et al (2008) Tendon: biology, biomechanics, repair, growth factors, and evolving treatment options. Hand Surg 33A:102–112

Rossi D, Pianta S, Magatti M et al (2012) Characterization of the conditioned medium from amniotic membrane cells: prostaglandins as key effectors of its immunomodulatory activity. PLoS One 7:e46956. doi:10.1371/journal.pone.0046956

Salingcarnboriboon R, Yoshitake H, Tsuji K et al (2003) Establishment of tendon-derived cell lines exhibiting pluripotent mesenchymal stem cell-like property. Exp Cell Res 287:289–300

Saxena A, Ewen B, Maffulli N (2011) Rehabilitation of the operated Achilles tendon: parameters for predicting a return to activity. J Foot Ankle Surg 50:37–40. doi:10.1053/j.jfas.2010.10.008

Schnabel LV, Lynch ME, van der Meulen MC et al (2009) Mesenchymal stem cells and insulin-like growth factor-I gene-enhanced mesenchymal stem cells improve structural aspects of healing in equine flexor digitorum superficialis tendons. J Orthop Res 27:1392–1398. doi:10.1002/jor.20887

Selmani Z, Naji A, Zidi I et al (2008) Human leukocyte antigen-G5 secretion by human mesenchymal stem cells is required to suppress T lymphocyte and natural killer function and to induce CD4+CD25highFOXP3+ regulatory T cells. Stem Cells 26:212–222

Seo BM, Miura M, Gronthos S et al (2004) Investigation of multipotent postnatal stem cells from human periodontal ligament. Lancet 364:149–155

Sharma P, Maffulli N (2005a) Basic biology of tendon injury and healing. Surgeon 3:309–316

Sharma P, Maffulli N (2005b) Tendon injury and tendinopathy: healing and repair. J Bone Joint Surg 87:187–202

Silver IA, Brown PM, Goodship AE (1983) A clinical and experimental study of tendon injury, healing and treatment in the horse. Equine vet J Suppl 1:1–43

Smith RK (2008) Mesenchymal stem cell therapy for equine tendinopathy. Disabil Rehabil 30:1752–1758. doi:10.1080/09638280701788241

Smith RKW, Webbon PM (1996) The physiology of normal tendon and ligament. In: Rantanen NW, Hauser ML (eds) Proceedings of the Dubai equine international symposium, Dubai, pp 55–81

Smith RKW, Jones R, Webbon PM (1994) The cross-sectional areas of normal equine digital flexor tendons determined ultrasonographically. Equine Vet J 26:460–465

Smith RK, Zunino L, Webbon PM et al (1997) The distribution of cartilage oligomeric matrix protein (COMP) in tendon and its variation with tendon site, age and load. Matrix Biol 16:255–271

Smith RK, Korda M, Blunn GW et al (2003) Isolation and implantation of autologous equine mesenchymal stem cells from bone marrow into the superficial digital flexor tendon as a potential novel treatment. Equine Vet J 35:99–102

Smith AM, Forder JA, Annapureddy SR et al (2005) The porcine forelimb as a model for human flexor tendon surgery. J Hand Surg Br 30B:307–309

Smith R, Young N, Dudhia J et al (2009) Effectiveness of bone-marrow-derived mesenchymal progenitor cells for naturally occurring tendinopathy in the horse. Regen Med 4(suppl 2): 25–26

Tang YL, Zhao Q, Zhang YC et al (2004) Autologous mesenchymal stem cell transplantation induce VEGF and neovascularization in ischemic myocardium. Regul Pept 117:3–10

Thomopoulos S, Harwood FL, Silva MJ et al (2005) Effect of several growth factors on canine flexor tendon fibroblast proliferation and collagen synthesis in vitro. J Hand Surg Am 30A:441–447

Thorpe CT, Clegg PD, Birch HL (2010) A review of tendon injury: why is the equine superficial digital flexor tendon most at risk? Equine Vet J 42:174–180. doi:10.2746/042516409X480395

Timmers L, Lim SK, Arslan F et al (2007) Reduction of myocardial infarct size by human mesenchymal stem cell conditioned medium. Stem Cell Res 1:129–137. doi:10.1016/j.scr.2008.02.002

Toupadakis CA, Wong A, Genetos DC et al (2010) Comparison of the osteogenic potential of equine mesenchymal stem cells from bone marrow, adipose tissue, umbilical cord blood, and umbilical cord tissue. Am J Vet Res 71:1237–1245

Tsuzaki M, Brigman BE, Yamamoto J et al (2000) Insulin-like growth factor-I is expressed by avian flexor tendon cells. J Orthop Res 18:546–556

van Poll D, Parekkadan B, Cho CH et al (2008) Mesenchymal stem cell-derived molecules directly modulate hepatocellular death and regeneration in vitro and in vivo. Hepatology 47:1634–1643. doi:10.1002/hep.22236

Vidal MA, Lopez MJ (2011) Adipogenic differentiation of adult equine mesenchymal stromal cells. Methods Mol Biol 702:61–75. doi:10.1007/978-1-61737-960-4_6

Vidal MA, Kilroy GE, Johnson JR et al (2006) Cell growth characteristics and differentiation frequency of adherent equine bone marrow-derived mesenchymal stromal cells: adipogenic and osteogenic capacity. Vet Surg 35:601–610

Vidal MA, Kilroy GE, Lopez MJ et al (2007) Characterization of equine adipose tissue-derived stromal cells: adipogenic and osteogenic capacity and comparison with bone marrow-derived mesenchymal stromal cells. Vet Surg 36:613–622

Vidal MA, Robinson SO, Lopez MJ et al (2008) Comparison of chondrogenic potential in equine mesenchymal stromal cells derived from adipose tissue and bone marrow. Vet Surg 37: 713–724. doi:10.1111/j.1532-950X.2008.00462.x

Vidal MA, Walker NJ, Napoli E et al (2012) Evaluation of senescence in mesenchymal stem cells isolated from equine bone marrow, adipose tissue, and umbilical cord tissue. Stem Cells Dev 21:273–283. doi:10.2460/ajvr.73.9.1435

Wagner W, Wein F, Seckinger A et al (2005) Comparative characteristics of mesenchymal stem cells from human bone marrow, adipose tissue, and umbilical cord blood. Exp Hematol 33: 1402–1416

Wang XT, Liu PY, Tang JB (2004) Tendon healing in vitro: genetic modification of tenocytes with exogenous PDGF gene and promotion of collagen gene expression. J Hand Surg Am 29A: 884–890

Watkins JP, Auer JA, Gay S, Morgan SJ (1985) Healing of surgically created defects in the equine superficial digital flexor tendon: Collagen-type transformation and tissue morphologic reorganisation. Am J Vet Res 46:2091–2096

Watts AE, Yeager AE, Kopyov OV et al (2011) Fetal derived embryonic-like stem cells improve healing in a large animal flexor tendonitis model. Stem Cell Res Ther 2:4. doi:10.1186/scrt45

Williams IF, Heaton A, McCullagh KG (1980) Cell morphology and collagen types in equine tendon scar. Res Vet Sci 28:302–310

Wilmink J, Wilson AM, Goodship AE (1992) Functional significance of morphology and micromechanics of collagen fibres in relation to partial rupture of the superficial digital flexor tendon in racehorses. Res Vet Sci 53:354–359

Wilson AM, Goodship AE (1994) Exercise-induced hyperthermia as a possible mechanism for tendon degeneration. J Biomech 27:899–905

Wilson AM, McGuigan MP, Su A et al (2001) Horses damp the spring in their step. Nature 414: 895–899

Wolbank S, Peterbauer A, Fahrner M et al (2007) Dose-dependent immunomodulatory effect of human stem cells from amniotic membrane: a comparison with human mesenchymal stem cells from adipose tissue. Tissue Eng 13:1173–1183

Yagi H, Soto-Gutierrez A, Parekkadan B et al (2010) Mesenchymal stem cells: mechanisms of immunomodulation and homing. Cell Transplant 19:667–679. doi:10.3727/096368910X508753
Yoshimura H, Muneta T, Nimura A et al (2007) Comparison of rat mesenchymal stem cells derived from bone marrow, synovium, periosteum, adipose tissue, and muscle. Cell Tissue Res 327: 449–462
Zafar MS, Mahmood A, Maffulli N (2009) Basic science and clinical aspects of Achilles tendinopathy. Sports Med Arthrosc Rev 17:190–197. doi:10.1097/JSA.0b013e3181b37eb7

Part II
Stem Cells in “Non-conventional” Species

Chapter 5
Using Stem Cells to Study and Preserve Biodiversity in Endangered Big Cats

Rajneesh Verma and Paul John Verma

1 Introduction

One of every four animal species on the planet is threatened by extinction. Historically, strategies for preserving biodiversity have focused on saving habitat and, by default, species in these native environments (in situ) (Ben-Nun et al. 2011).

One support strategy is managing 'assurance' populations ex situ for hundreds of mammals, birds, reptiles, amphibians and fish. This preserves genetic integrity, allows basic research to be conducted and is a source of animals for reintroduction programmes. Ex situ species propagation approaches are complex and demand expertise and resources, including specialized animal space and facilities. Even then, achieving reproduction is challenging. Animals can and often do have preferences for sexual partners, and it is expensive and complicated to transfer wild animals between locations to make ideal genetic matches to retain maximal heterozygosity. As a result, assisted reproductive technologies (ART) have been explored for helping manage ex situ wildlife populations.

Much progress has been made during the last decade in the development of assisted reproductive techniques, not only in the economically driven fields of human infertility and domestic animal breeding, but also in the area of animal conservation. Endangered felids are often difficult to breed either in captivity or even

R. Verma
Monash Institute of Medical Research, Monash University, Melbourne, VIC, Australia

Institute of Molecular Bioscience, Mahidol University, Bangkok, Thailand

P.J. Verma (✉)
Stem Cells and Reprogramming Group, Biological Engineering Laboratories, Faculty of Engineering, Monash University, Melbourne, VIC 3800, Australia

South Australian Research and Development Institute (SARDI), Adelaide, Australia

Turretfield Research Centre, Holland Road, Rosedale, Australia
e-mail: Paul.Verma@monash.edu

T.A.L. Brevini (ed.), *Stem Cells in Animal Species: From Pre-clinic to Biodiversity*, Stem Cell Biology and Regenerative Medicine, DOI 10.1007/978-3-319-03572-7_5,

under natural conditions. One of the most important reasons for infertility or sub-fertility in this group is decreased genetic diversity caused by inbreeding, due to genetic bottlenecks as a consequence of geographical isolation and population contraction. Because of this, there has been increasing interest in maintaining genetic diversity for the conservation of wild felids and preservation of valuable cat breeds. ART comprising of techniques, such as artificial insemination (AI), intracytoplasmic sperm injection (ICSI), in vitro fertilization (IVF) and somatic cell nuclear transfer (SCNT), has been promoted over the past two decades as a potential means to conserve and manage threatened wildlife populations. But one of the major problems with the implementation of in situ and ex situ conservation programmes is the lack of availability of suitable biological material, which is required for a better understanding of reproductive patterns as well to maximize reproductive efficiency. This constraint arises from the strict procedures adopted for restraining or anaesthetizing free-living animals for collection of biological/reproductive samples due to inherent risks (Verma et al. 2012).

2 SCNT, an Alternative to ART

SCNT also called cloning represents an alternative for the production of animals genetically (chromosomally) identical to the donor somatic cell, albeit with mitochondria being contributed by the recipient oocyte, and offers the possibility of preventing the extinction of wild species. However, owing to the limited availability of oocytes from wild animals, the cloning of endangered species requires the use of donor oocytes from a related domestic species. Interspecies SCNT consists of the reconstruction of a cloned embryo by use of a donor somatic cell and a recipient oocyte from a different species, but from the same genus; whereas inter-generic SCNT consists of the reconstruction of a cloned embryo in which the donor nucleus and recipient cytoplast differ both in species and in genus, with oocyte-specific mitochondria in both cases. Several studies have demonstrated that it is possible to produce embryos from endangered species by interspecies or inter-generic SCNT (Gomez et al. 2008) (Fig. 5.1); however; few live cloned wild mammals have resulted and these animals were derived from embryos reconstructed with donor oocytes of the same genus. Viable offspring from inter-generic SCNT have not been produced in any mammalian species, although pregnancies have been established with inter-generic cloned embryos after transfer into sheep or domestic cat recipient oocytes (Gomez et al. 2000). The successful development of interspecies and inter-generic cloned embryos is dependent on a variety of factors, which are similar to those reported for intra-species SCNT including source of oocyte cytoplasts, cell cycle synchronization, mitochondrial compatibility and genotype of the donor cells. Moreover, increasing evidence shows that aberrant epigenetic alterations that arise during SCNT may be associated with perinatal and neonatal losses and the production of abnormal offspring (Wu et al. 2010).

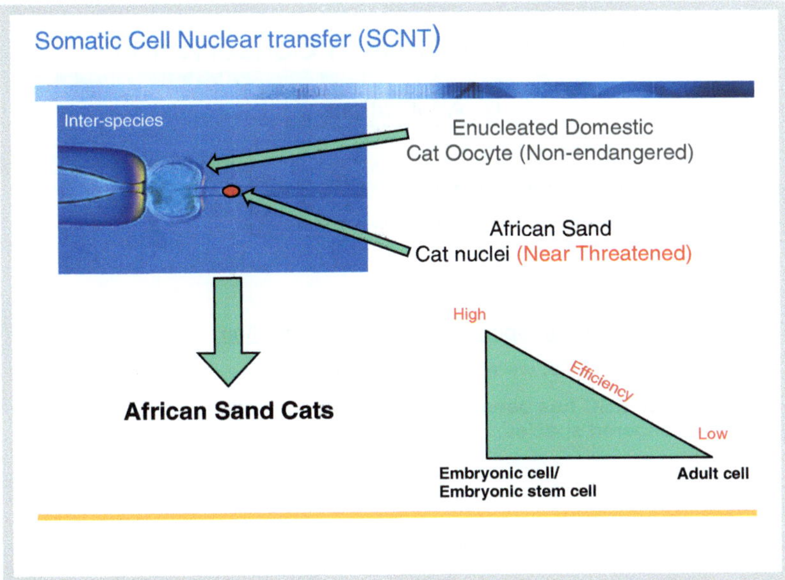

Fig. 5.1 Inter-species cloning and efficiency parameter

3 Pluripotent Stem Cells

Pluripotent stem cells differentiate into all the cell types in the body including gametes in vivo, while retaining the capacity for indefinite self-renewal in vitro. The best known example of pluripotent cells are embryonic stem cells (ESC) and the potential application of these cells are that they (1) can contribute to embryos for, e.g. chimeras, (2) can be differentiated to gametes in vitro and (3) can be used as a donor cell for nuclear transfer (Fig. 5.2). ESC are typically derived from inner cell mass (ICM) of blastocysts, which are destroyed in the process, raising ethical and logistical concerns for the derivation of stem cell lines from endangered species. Therefore for species in which embryos are particularly difficult to obtain or those, which are endangered, this approach was not particularly useful and feasible. In addition, to date no true ESC have been reported for any species for other than rodents (Malaver-Ortega et al. 2012; Verma et al. 2012).

4 Induced Pluripotent Stem Cells

Despite this reality check on what is possible today, it is essential to consider what might be promising for future intensive management of endangered species, especially as embryo and molecular technologies advance. So it is imperative to

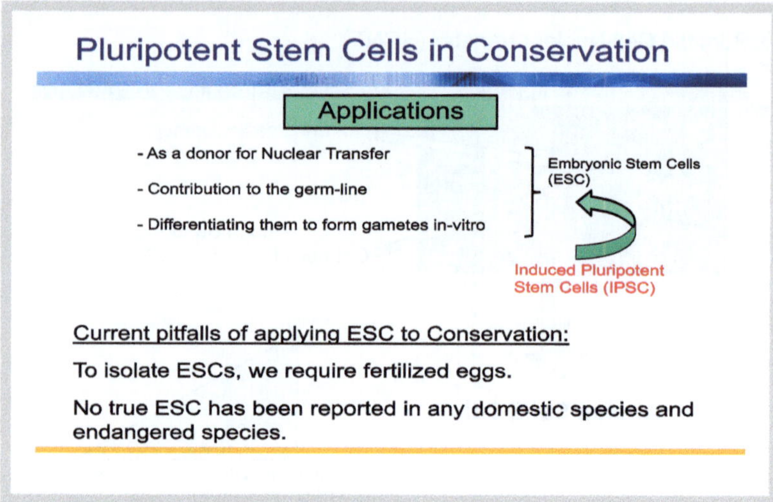

Fig. 5.2 Pluripotent stem cells and its applications in conservation

consider the potential benefits offered by induced pluripotent stem cells (iPSC) generated by reprogramming adult or differentiated somatic cells to a pluripotent stem cell-like state using defined transcription factors.

In a seminal study, Japanese scientists (Takahashi and Yamanaka 2006) used viral transduction of mouse fibroblasts to screen a combination of 24 candidate genes with putative roles in pluripotency and remarkably found that four previously known transcription factors (Oct3/4, Sox2, Klf4 and cMyc) could reprogram mouse embryonic fibroblasts (MEFs) and tail tip fibroblasts into ES-like cells, which were almost indistinguishable from mouse ESC in terms of pluripotency and coined the term iPSC.

iPSC have now been isolated from rodents (mouse and rats) (Takahashi et al. 2007; Honda et al. 2010), primates (human and monkeys) (Liu et al. 2008; Park et al. 2008), livestock (pigs, sheep, horse and cattle) (Ezashi et al. 2009; Nagy et al. 2011; Sumer et al. 2011; Khodadadi et al. 2012; Liu et al. 2012) and endangered species (Ben-Nun et al. 2011; Verma et al. 2012). The use of iPSC technology to provide a source of pluripotent cells for use in felid conservation was considered to be likely a more successful approach than isolating ESC from endangered felids embryos because it is a non-invasive technique, which only requires somatic cells. In mice, iPSC are similar to ESC and can form chimeric embryos when injected into blastocysts. This opened a new chapter, offering the possibility to convert cells from skin to ESC regardless of age and gender of donor and to use them for ART in various forms.

5 Snow Leopard

The snow leopard (*Uncia uncia*) is a moderately large cat native to the mountain ranges of Central Asia. They live between 3,000 and 5,500 m (9,800 and 18,000 ft) above sea level in the rocky mountain ranges of Central Asia. However, their secretive nature means that their exact numbers are unknown, although it has been estimated that between 3,500 and 7,000 snow leopards exist in the wild and between 600 and 700 in zoos worldwide (Kleihman and Garman 1978).

6 First Snow Leopard iPSC

Before attempting to produce iPSC from snow leopard, we came across a number of unknowns as listed below (Fig. 5.3).

Later, we reported the successful derivation of iPSC from ear fibroblasts of the snow leopard transfected with five human factors *OCT4*, *SOX2*, *KLF4*, *cMYC* and *NANOG*. We believe the high transduction efficiency (96 %) for this study, measured using a pMx-GFP construct, was important in achieving a successful outcome (Fig. 5.4).

Of particular importance was the observation that the three key exogenous pluripotency transgenes (OCT4, SOX2, NANOG) were silenced at later passages. Interestingly, we observed a similar requirement of transcription factors for induction of pluripotency and an identical pattern of subsequent silencing of transgenes for three globally diverse endangered felids in a subsequent study (Verma et al. 2013). During the course of our study, we observed that snow leopard iPSC at P14, when tagged with mCherry reporter and injected into the peri-vitelline space of mouse morulae, did not compromise development of the mouse embryos. We next examined whether these iPSC were able to contribute to developing mouse embryos by incorporation into the ICM (Fig. 5.5). We propose this as a novel in vivo assay to assess the embryo contribution ability of iPSC and pluripotent cells from exotic

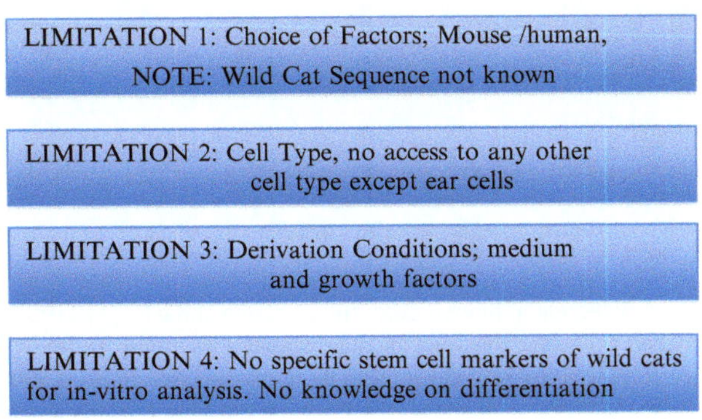

Fig. 5.3 Limitations for producing snow leopard iPSC

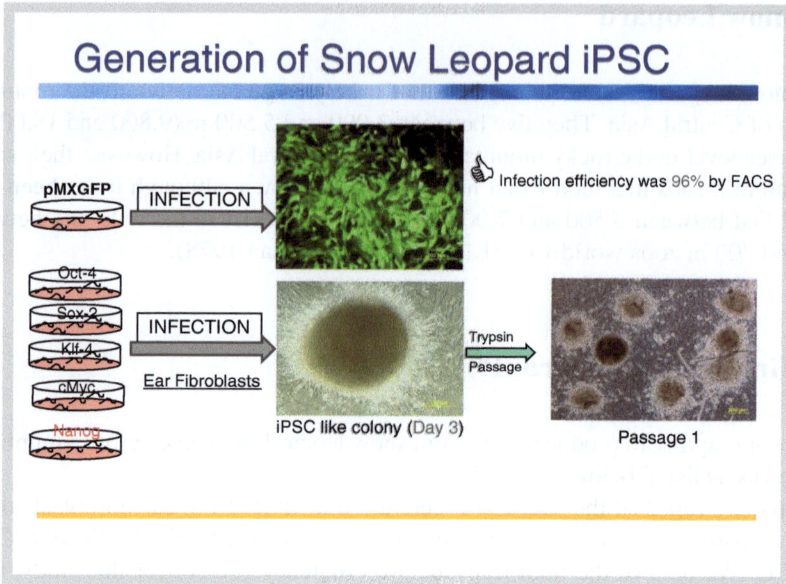

Fig. 5.4 Experimental diagram to show the procedure for producing snow leopard iPSC

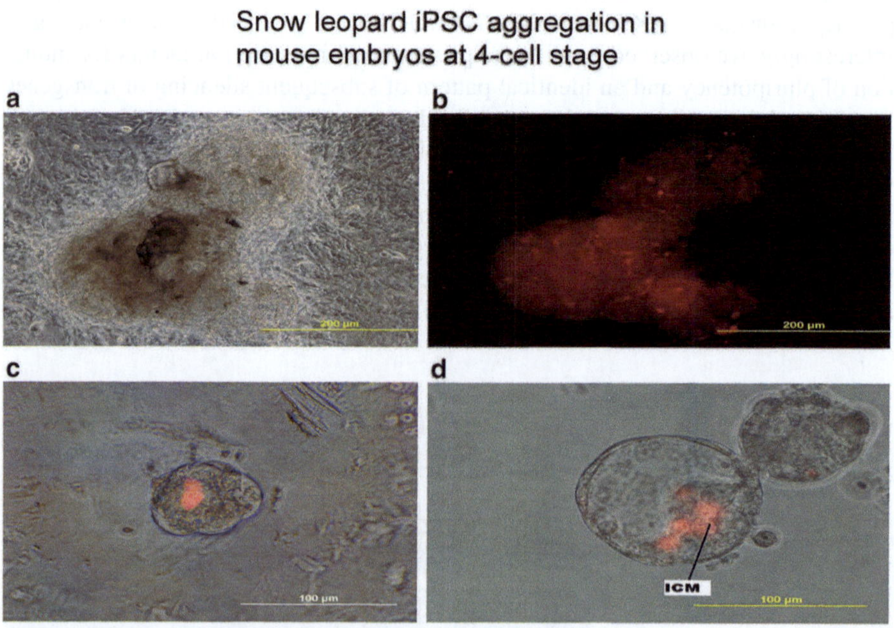

Fig. 5.5 Potential in vivo assay for assessing snow leopard iPSC. Snow leopard iPSC tagged with mCherry reporter construct and aggregated with mouse embryos at the 4-cell stage (**a**) Snow leopard iPSC colony (Bright field), (**b**) Snow leopard iPSC colony expressing mCherry reporter, (**c**) Snow leopard iPSC localized into a morula and (**d**) Snow leopard iPSC seen to appear or localize in the ICM of hatching mouse blastocyst

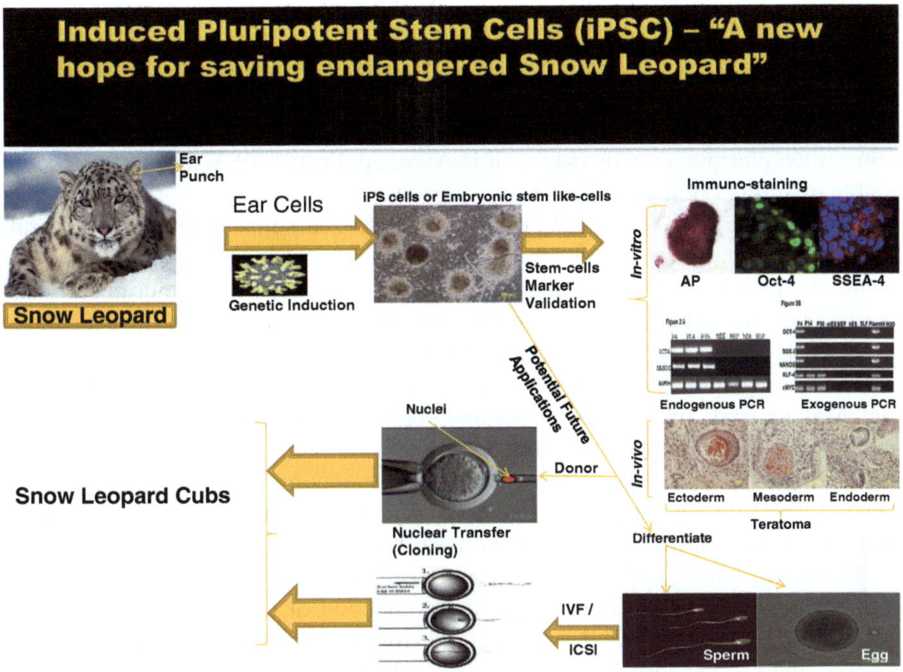

Fig. 5.6 Experimental layout of generating snow leopard iPSC and future applications

species where species-specific testing is impossible because of the often extremely limited access to gametes, especially oocytes and embryos.

In conclusion, we believe this is the first report on the derivation of iPSC from both a felid and an endangered species. This is also the first report on the induction of pluripotency in a large animal with concomitant silencing of the pluripotency associated transgenes. The iPSC technology has the potential to impact on conservation of endangered species at a number of levels. It can provide insights into pluripotency and development in species where embryos are difficult to access. Furthermore, iPSC generated from the endangered species can be easily expanded for banking of genetic material or used as a reprogrammed donor cell to improve NT outcomes. They may also create opportunities to prevent extinction in a wide range of threatened animals in the future. For example, it may eventually be possible to differentiate cell lines with proven pluripotency in vitro to produce gametes or use these cell lines in vivo in conjunction with tetraploid complementation to produce whole animals (Fig. 5.6). This report has relevance to understanding pluripotency in big cats and also has application in domestic cats, which are companion animals and are unique biomedical models to study genetic diseases (e.g. HIV, arthritis and diabetes; (Verma et al. 2013)).

7 Future of iPSC in Endangered Species

If the many challenges associated with advanced embryo culture, finding the appropriate surrogate mother and synchronizing the embryo to the uterus can be overcome, then pluripotent cells may well improve application of SCNT for producing viable offspring (Holt et al. 2004).

Generating sperm from iPSC derived from frozen somatic cell samples from long-dead animals would provide a way to infuse much needed genetic diversity using already proven AI methods. An analogous approach using pluripotent cell-derived oocytes could provide an endless resource for fundamental investigations into IVF, ICSI and SCNT (Holt et al. 2004).

There is also potential for these cells as a scalable resource of rare genetic material. Making iPSC for diverse species and populations available to every interested institution would accelerate research progress on analysing phylo-geographic structure, delineating subspecies, tracing paternities, evaluating gene flow and assessing genetic variation—information critical for decision-making in managing both ex situ and in situ wildlife populations (Pope 2000).

Therefore, wildlife-related studies will need to focus on fidelity measures for the reprogramming process to ensure the production of legitimate and 'fit' iPSC. As these are rare species and some methods for generating the cells reply on viral vectors, the process also has to be proven indisputably safe for offspring (Verma et al. 2013).

References

Ben-Nun IF, Montague SC, Houck ML, Tran HT, Garitaonandia I, Leonardo TR, Wang YC, Charter SJ, Laurent LC, Ryder OA, Loring JF (2011) Induced pluripotent stem cells from highly endangered species. Nat Methods 8:829–831

Ezashi T, Telugu BPVL, Alexenko AP, Sachdev S, Sinha S, Roberts RM (2009) Derivation of induced pluripotent stem cells from pig somatic cells. Proc Natl Acad Sci U S A 106: 10993–10998

Gomez MC, Pope CE, Harris R, Davis A, Mikota S, Dresser BL (2000) Births of kittens produced by intracytoplasmic sperm injection of domestic cat oocytes matured in vitro. Reprod Fertil Dev 12:423–433

Gomez MC, Pope CE, Kutner RH, Ricks DM, Lyons LA, Ruhe M, Dumas C, Lyons J, Lopez M, Dresser BL, Reiser J (2008) Nuclear transfer of sand cat cells into enucleated domestic cat oocytes is affected by cryopreservation of donor cells. Cloning Stem Cells 10:469–483

Holt WV, Pickard AR, Prather RS (2004) Wildlife conservation and reproductive cloning. Reproduction 127:317–324

Honda A, Hirose M, Hatori M, Matoba S, Miyoshi H, Inoue K, Ogura A (2010) Generation of induced pluripotent stem cells in rabbits: potential experimental models for human regenerative medicine. J Biol Chem 285:31362–31369

Khodadadi K, Sumer H, Pashaiasl M, Lim S, Williamson M, Verma PJ (2012) Induction of pluripotency in adult equine fibroblasts without c-MYC. Stem Cells International: 429160. doi: 10.1155/2012/429160

Kleihman MS, Garman RH (1978) An endoscopic approach to a snow leopard. Gastroenterology 74:1348

Liu H, Zhu F, Yong J, Zhang P, Hou P, Li H, Jiang W, Cai J, Liu M, Cui K, Qu X, Xiang T, Lu D, Chi X, Gao G, Ji W, Ding M, Deng H (2008) Generation of induced pluripotent stem cells from adult rhesus monkey fibroblasts. Cell Stem Cell 3:587–590

Liu J, Balehosur D, Murray B, Kelly JM, Sumer H, Verma PJ (2012) Generation and characterization of reprogrammed sheep induced pluripotent stem cells. Theriogenology 77:338–346

Malaver-Ortega LF, Sumer H, Liu J, Verma PJ (2012) The state of the art for pluripotent stem cells derivation in domestic ungulates. Theriogenology 78:1749–1762

Nagy K, Sung H-K, Zhang P, Laflamme S, Vincent P, Agha-Mohammadi S, Woltjen K, Monetti C, Michael IP, Smith LC, Nagy A (2011) Induced pluripotent stem cell lines derived from equine fibroblasts. Stem Cell Rev 7:693–702

Park IH, Lerou PH, Zhao R, Huo H, Daley GQ (2008) Generation of human-induced pluripotent stem cells. Nat Protoc 3:1180–1186

Pope CE (2000) Embryo technology in conservation efforts for endangered felids. Theriogenology 53:163–174

Sumer H, Liu J, Malaver-Ortega LF, Lim ML, Khodadadi K, Verma PJ (2011) NANOG is a key factor for induction of pluripotency in bovine adult fibroblasts. J Anim Sci 89:2708–2716

Takahashi K, Okita K, Nakagawa M, Yamanaka S (2007) Induction of pluripotent stem cells from fibroblast cultures. Nat Protoc 2:3081–3089

Takahashi K, Yamanaka S (2006) Induction of pluripotent stem cells from mouse embryonic and adult fibroblast cultures by defined factors. Cell 126:663–676

Verma R, Holland MK, Temple-Smith P, Verma PJ (2012) Inducing pluripotency in somatic cells from the snow leopard (Panthera uncia), an endangered felid. Theriogenology 77:220–8, 228.e1-2

Verma R, Liu J, Holland MK, Temple-Smith P, Williamson M, Verma PJ (2013) Nanog is an essential factor for induction of pluripotency in somatic cells from endangered felids. Biores Open Access 2:72–76

Wu Z, Chen J, Ren J, Bao L, Liao J, Cui C, Rao L, Li H, Gu Y, Dai H, Zhu H, Teng X, Cheng L, Xiao L (2010) Generation of pig-induced pluripotent stem cells with a drug-inducible system (vol 1, pg 46, 2009). J Mol Cell Biol 2:104

Chapter 6
Pluripotent and Multipotent Domestic Cat Stem Cells: Current Knowledge and Future Prospects

Martha C. Gómez and C. Earle Pope

1 Introduction

The domestic cat (*Felis catus*) is a mammalian species of particular importance due to their evolutionary history, role as companion animals and as a research model. According to The Humane Society, in the United States 86.4 million cats are kept as companion animals. Veterinary care can be a major expense for cat owners and development of new health-related technologies has become a priority for veterinarians. Recently, stem-cell-based therapies have been developed to improve health and well-being in domestic animals (Ribitsch et al. 2010; Fortier and Travis 2011). Hundreds of cats, including a Bengal tiger and a Florida panther, have received "adult stem cells" for tissue replacement therapy. Although most of the stem-cell-based therapies have been applied by veterinarians without the support of research studies demonstrating their efficacy or safety (Cyranoski 2013), a recent pilot study in which adult stem cells were injected into the kidney, as a treatment for chronic kidney disease (CKD), induced a mild improvement in renal function of domestic cats (Quimby et al. 2011). Thus, transplantation of adult stem cells is a promising tool for cell replacement therapy and regenerative medicine.

Also, stem-cell-based technologies have been proposed as an alternative approach for the conservation of genetically valuable animals and preservation of endangered animals (Travis et al. 2009). The domestic cat has been demonstrated to be a valuable model for developing assisted reproductive technologies that can be applied to the conservation of endangered felids (Pope et al. 1993, 2006a, b, 2012). Although the feasibility of producing domestic cats and endangered felids by in vitro fertilization (Pope et al. 1993, 2006a, b, 2012) and somatic cell nuclear transfer (SCNT) (Gómez et al. 2004, 2008) has been demonstrated, practical applications depend on

M.C. Gómez (✉) • C.E. Pope
Audubon Nature Institute, Center for Research of Endangered Species,
14001 River Road, New Orleans, LA 70131, USA
e-mail: info@mcgomez.com

T.A.L. Brevini (ed.), *Stem Cells in Animal Species: From Pre-clinic to Biodiversity*,
Stem Cell Biology and Regenerative Medicine, DOI 10.1007/978-3-319-03572-7_6,

improving efficiency, particularly that of interspecies SCNT. Potential applications of using stem cells for enhancing reproduction in endangered felids range from the generation of stable characterized stem cells for use as donor nuclei to improve efficiency of interspecies SCNT (Jaenish et al. 2002), in vitro differentiation into germ cells (oocytes and spermatozoa; Nayernia et al. 2006; Kee et al. 2009; Yu et al. 2009b), and restoration of spermatogenesis with the haplotype of the endangered felid donor by spermatogonial stem cell transplantation (Silva et al. 2012).

Sequencing of the domestic cat genome revealed that 90 % of the 20,285 putative genes are homologous to human genes (Pontius et al. 2007). The cat provides the only naturally occurring model for human AIDS pathogenesis in its endemic fatal transmissible feline immunodeficiency virus (Siebelink et al. 1990; Willet et al. 1997). Moreover, feline Alzheimer's disease in aged animals was recognized as a feline disorder with pronounced amyloid and phosphorylated tau deposition (Gunn-Moore et al. 2006, 2007), indicative of an identical disease mechanism as seen in Alzheimer's disease in humans. Thus, the domestic cat as a model for certain human diseases may more closely resemble the human conditions than does the current mouse model for that disease. Interest in the generation of genetically modified domestic cat stem cells to produce genetically identical cats that carry genes for specific human disorders has gradually increased after recent reports on production of transgenic cats (Yin et al. 2008; Gómez et al. 2009; Wongsrikeao et al. 2011). Cat somatic cells exhibit limited proliferation, become aneuploid during in vitro culture (Gómez et al. 2006), and have an extremely low frequency of homologous recombination (Gómez et al. unpublished data). In contrast, pluripotent stem cells have shown the capability of long-term proliferation, which allows longer in vitro genetic manipulations and possibly increases the homologous recombination efficiency. The latter could represent an efficient method to produce genetically engineered cats as a model for human diseases.

Successful application of stem-cell-based therapies in cats is dependent on development of robust methods to efficiently isolate fully pluripotent stem cells and a better understanding of the mechanisms that control stem cell fate to differentiate these cells into specific phenotypes. In this chapter, we will describe recent progress on the isolation, culture, and characterization of pluripotent and multipotent stem cells and preliminary trials in regenerative medicine in domestic cats and prospects for applications to conservation and veterinary care of endangered felids.

2 Stem Cell Definition

Stem cells are characterized by their functional capacity to: (1) self-renew, either by symmetric cell division generating two stem cells, or by asymmetric cell division resulting in one stem cell and one other cell with restricted differentiation capacity, and (2) differentiate into other cell types (Weissman et al. 2001). The latter ability is known as totipotency; and can be further classified according to differentiation pattern. Stem cells are classified as either pluripotent or multipotent.

Pluripotent stem cells have the ability to proliferate extensively in vitro in a morphologically undifferentiated state while retaining a normal diploid karyotype during extended culture. Also, the cells are able to differentiate into all derivates of the three embryonic layers (Murry and Keller 2008). In contrast, multipotent stem cells, isolated from tissues or organs (adult stem cells), are characterized by having limited proliferation and can only differentiate into specific cell types within the particular lineage from which they were originated.

Both types of stem cells have different potential applications for stem-cell-based therapies. Although pluripotent stem cells can differentiate into all cell types, direct injection into a host results in formation of tumors, limiting clinical use in their undifferentiated state. Thus, cell differentiation must be induced before transplantation. In contrast, multipotent stem cells are not tumorigenic after transplantation into the recipient; however, their limited ability to undergo differentiation reduces their potential applications. Regardless of the stem cell potency, both types of cells offer several advantages for their use in regenerative medicine in domestic cats and endangered felids.

3 Domestic Cat Pluripotent Embryonic Stem Cells

In the domestic cat, putative embryonic stem cells (cESC) isolated from the ICM of in vivo-derived and in vitro-produced blastocysts are the only pluripotent cells that have been generated (Yu et al. 2008; Gómez et al. 2010). In these studies, 70 % of plated embryos formed outgrowth colonies, while 50 % of the outgrowths developed into primary colonies, depending on the culture conditions (Table 6.1). Cat ESC colonies proliferated for several passages (6–12), but progressively lost their self-renewal capacity and spontaneously differentiated during in vitro culture. Continuous proliferation of ESC requires optimal in vitro culture conditions to maintain self-renewal and pluripotent characteristics. In this section, we consider the culture conditions that significantly influenced the derivation of cat ESC, the methods used for stem cell characterization and present data, acquired in the author's laboratory, which will help to reveal the intracellular signaling pathways that contribute to self-renewal in cat ESC.

3.1 Culture Conditions for Cat ESC

3.1.1 Feeder Cell Layers

Previous reports on culturing mouse and human ESC have shown that both the derivation and maintenance of ESC require the support of a feeder cell layer (Evans and Kaufman 1981; Thomson et al. 1995, 1998). Mouse feeder cell layers inhibit differentiation of human ESC by supplying growth factors and extracellular matrix

(Yu et al. 2008). Similarly, we have observed that cell growth rate is significantly reduced when cat fibroblast cells are cultured in a defined commercial medium containing human albumin, instead of FBS (Gómez et al. unpublished data). Although it is not possible to make a direct comparison between fibroblasts and ESC, due to differences in cell type and cell cycle pathway, these results corroborate that factors in FBS positively influence cell growth in cat cells. It is not clear why supplementing the culture medium of cat ESC with KOSR did not maintain prolonged self-renewal in an undifferentiated state as did FBS, but it appears that differences in lipid content between the two supplements may be an important influence. In one study with human ESC, it was found that lipid-rich bovine serum albumin was the active ingredient in KOSR and that lipids, not albumin apoprotein, were responsible for stimulating the self-renewal effect (Garcia-Gonzalo and Izpisua 2008). In a later study, it was found that retinol, conjugated linoleic acid, and eicosapentaenoic acid were some of the lipids present in lipid-rich bovine serum albumin that were actively involved in self-renewal of human ESC; however, the authors clearly specified that the optimal response of cells to lipid-rich serum albumin may be due to synergism between the different lipids (Rajala et al. 2011). In contrast, relatively little is known about the complete lipid content of FBS and about the impact of lipids on cell growth, but it is generally accepted that the composition of FBS may directly influence the outcome of cell culture. FBS supported spontaneous differentiation of human ESC into beating cardiomyocytes during in vitro culture (Bettiol et al. 2007). Similar effects of prolonged supplementation of human ESC with human serum have been reported. In fact, differentiation rates of human ESC were significantly increased when cells were cultured with human serum for more than ten passages, possibly due to the presence of molecules that may accelerate cell differentiation (Richards et al. 2003). Similarly, we observed that cat ESC cultured in medium supplemented with 15 % FBS and 5 % KOSR can spontaneously differentiate into beating cardiomyocytes and differentiation is significantly increased after passage six (Gómez et al. 2010). Additional studies are required to determine which active lipids and additional factors have a positive effect on self-renewal of cat ESC to develop a supplement that could be used as a replacement for FBS.

3.1.3 Cell Dissociation

The method of dissociating cat ESC colonies for further passages is another factor that affects their stability. Enzymes such as trypsin are used to dissociate mouse ESC into single-cell suspension during routine passage without evident negative effects on cell survival and their pluripotency. In contrast, monkey, pig, and human ESC died after enzymatic single-cell dissociation (Thomson and Marshall 1998; Brevini et al. 2007; Xu et al. 2010). Cat ESC differentiate within one to two passages after dissociation with trypsin-EDTA (Yu et al. 2008). In a recent human ESC study, it was demonstrated that cell death and low plating efficiency after enzymatic dissociation was caused by downregulation of the E-cadherin signaling pathway, as a consequence of the disruption of cell–cell interaction (Xu et al. 2010). E-cadherin is a molecule, highly expressed in human ESCs, that is involved in maintaining cell–cell adhesion and cortical actin cytoskeletal arrangement (Eastham et al. 2007); thus, loss of its activity is associated with cell death. Xu et al. (2010) also demonstrated, that mouse ESC are more tolerant to single-cell dissociation due to their capacity to synthesize new E-cadherin after enzymatic dissociation. The authors suggest that distinct mechanisms are used by cells of each species to stabilize E-cadherin and this may be influenced by their specific culture conditions. Although little is known about the mechanism of self-renewal in cat ESC, it is possible that they are using a similar E-cadherin signaling pathway as human ESC; thus, susceptibility to enzymatic dissociation may be due to the loss of cell–cell contact.

Rho-associated coiled-coil forming protein, serine/threonine kinase (ROCK), is a downstream effector of the Rho signaling involved in cytoskeleton remodeling of actin stress fibers and focal adhesion (Leung et al. 1996) and Rho-ROCK signaling pathway is stabilized by E-cadherin (Xu et al. 2010). When human ESCs were disassociated with trypsin, the activity of Rho-ROCK signaling was up-regulated, and this up-regulation induced cell hypercontraction and consequently, lower plating efficiency. When ESC were treated with a ROCK inhibitor, the stress on the cell fibers disappeared and focal adhesion was observed with an increase in cell plating, and the levels of E-cadherin were significantly increased. These results not only demonstrated that Rho-ROCK signaling is involved in stabilization of E-cadherin, but also confirmed that inhibition of the signaling pathway, when levels of E-cadherin are low, is beneficial to increase cell–cell adhesion and survival rates. ROCK inhibition also attenuates induced apoptosis in human ESC and significantly reduces cell death after single-cell dissociation without affecting their pluripotent ability (Watanabe et al. 2007). Accordingly, based on these results, and the fact that cat ESC must be passaged as aggregates due to their sensitivity to clonal expansion, for further passages the author's currently pretreat cat ESC with ROCK inhibitor and mechanically disaggregate colonies by cutting them with an ultra-sharp splitting blade.

3.2 *Characterization of Cat ESC*

Several methods have been used to characterize the "stemness" and potency of pluripotent stem cells. Traditionally, the pluripotent characteristic of a stem cell is measured by their cell morphology, expression of pluripotent markers, self-renewal capacity, and in vitro differentiation potential into several cell types. However, pluripotent cells must meet additional criteria, with more rigorous assays, to establish true pluripotency. These additional criteria include the generation of teratomas that express cells from the three germinal cell layers, production of chimeras by injection of cells into a blastocyst and integration of the cells into the host embryo, and germline transmission. The majority of domestic mammalian species ESC are defined as putative or-like ESC because they do not fulfill all criteria to be defined as true pluripotent cells. Similarly, true pluripotency in cat ESC has not been accomplished to date.

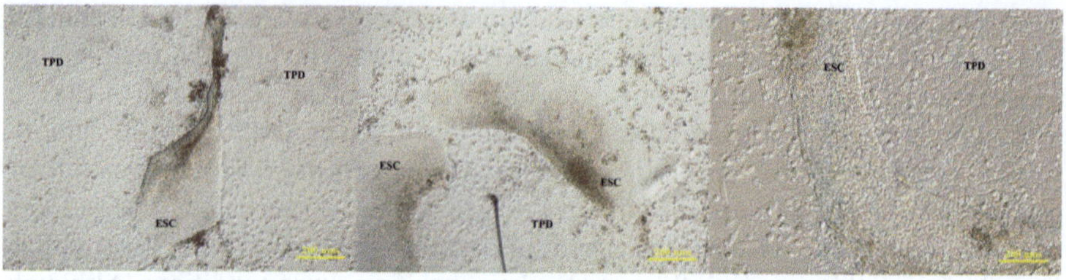

Fig. 6.1 Morphology of cat ESC at 6 days after plating. Primary colonies can be clearly identified by marked borders, as uniformly dense cluster of cells that have a large nucleus: cytoplasm ratio, prominent nucleoli, and a distinct dome shape. A flat monolayer of TPD cells can be also visualized around the ESC colony. The TPD cells contain lipid droplets, and grow usually as a flat monolayer

3.2.1 Morphological Characteristics of Cat ESC

The cell morphology of cat ESC can be described as a uniformly dense cluster of cells that have a large nucleus:cytoplasm ratio, prominent nucleoli, and a distinct dome shape (Fig. 6.1). When a whole embryo is plated, a flat monolayer of trophectoderm (TPD) cells can be also visualized around the ESC colony. The TPD cells contain lipid droplets, and grow usually as a flat monolayer (Fig. 6.1). Morphology of cat ESC colonies can be heterogeneous. Most colonies, whether cultured with leukemia inhibitory factor (LIF), or a combination of LIF and basic fibroblast growth factor (bFGF), displayed a compact dome shape similar to that of mouse ESC (Yin et al. 2008; Gómez et al. 2010). However, large flattened colonies similar to those of human ESC were also seen in cat ESC colonies cultured in stem cell medium supplemented with 20 % FBS (Yu et al. 2008).

3.2.2 Expression of Pluripotent Markers

The pluripotent state of ESC is mainly regulated by a "core" of transcription factors, octamer-binding transcription factor Oct-4 (*OCT-4*), SRY-box 2 (*SOX-2*), and *NANOG*, which act in conjunction with other transcription factors to establish and maintain a robust pluripotent state (see review; Young 2011). Cat ESC have been further characterized by the expression of the core *OCT-4*, *SOX-2* and *NANOG* transcription factors and other markers involved in pluripotency. Alkaline phosphatase activity was observed with different degrees of intensity, depending on the undifferentiating state of the colony. Expression of cell-surface stage-specific embryonic antigens *SSEA-1*, *SSEA-3*, and *SSEA-4*, as well as the expression of *OCT-4* and *NANOG*, was also confirmed by immunocytochemical staining (Yu et al. 2008; Gómez et al. 2010). On the other hand, the expression of pluripotent associated markers *OCT-4, NANOG* (Yu et al. 2009a; Gómez et al. 2010), and *SOX-2* and the proto-oncogene *C-MYC* (Gómez et al. 2010) were detected at the mRNA level, suggesting that cat ESC colonies express the "core" of transcription factors reported in other mammalian ESC. Interestingly, we found that up-regulation of *NANOG* without the expression of *OCT-4* is capable of maintaining cat ESC colonies in an undifferentiated state. Instead, at the onset of differentiation, *NANOG* is downregulated (Gómez et al. 2010). These results clearly demonstrated the significance of *NANOG* for maintaining pluripotency in cat ESC. Our observation was recently confirmed when *NANOG* was shown to be an essential factor for reprogramming and maintaining prolonged self-renewal of iPSC from endangered felids (Verma et al. 2013).

3.2.3 In Vitro Differentiation of Cat ESC

The ability of cat ESC to differentiate into derivates of the three embryonic layers has been tested with the embryoid body (EB) formation assay. In vitro culture of EBs showed that cat ESC were able to differentiate, when cultured without LIF and without a feeder cell layer, into cells morphologically similar to neurons and epithelial cells (Yu et al. 2008). However, proper characterization of the three germ cell layers by expression of specific protein markers for each cell type was not reported. Also, cat ESC exhibited spontaneous differentiation into well-developed contractile functional cardiomyocytes when grown in a mass and in the presence of a feeder cell layer and growth factors which expressed alpha actinin (Yu et al. 2008) and desmin (Gómez et al, unpublished data) confirming mesoderm differentiation.

3.3 *Signaling for Self-Renewal in Cat ESC*

Developmental signaling pathways control the self-renewal and differentiation of stem cells. As mentioned in the previous section, feeder cell layers play an important role in derivation and proliferation of cat ESC. However, the undefined nature of feeder cell layers posed a significant challenge in determining the signaling pathways involved in self-renewal in cat ESC. Nonetheless, there are additional factors other than feeder cell layers that selectively activate self-renewal signaling pathways. Several studies have clearly suggested that stem cells from different species use diverse self-renewal mechanisms. Mouse ESC self-renewal is regulated by the gp130/JAK/STAT pathway as a result of LIF-dependent activation (Niwa et al. 1998), as well as Bone morphogenic protein 4 signaling pathway (BMP4; Ying et al. 2003). In contrast, human ESC self-renewal is regulated through fibroblast growth factor (FGF), and transforming growth factor-β (TGF-β)/Activin/nodal pathways (Thomson et al. 1998; Laping et al. 2002). Cat ESC have been cultured on MEF feeder cell layers with cytokines and growth factors that were shown to be beneficial for mouse and human ESC. However, none of the factors were able to promote prolonged self-renewal (Yu et al. 2008; Gómez et al. 2010). The signaling pathway(s) for self-renewal of cat ESC have not been elucidated.

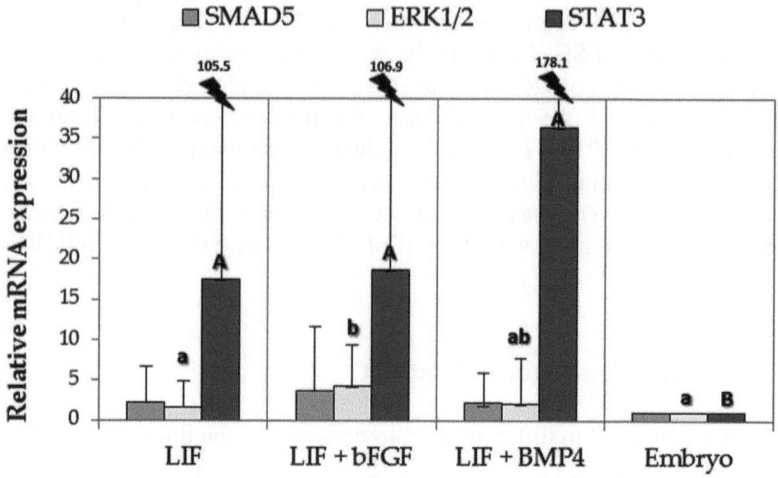

Fig. 6.2 Relative transcript abundance of SMAD5, ERK1/2, and STAT3 genes in in vitro produced cat blastocysts at day 8 and cat ESC derived on cat feeder cell layers and LIF, and cultured in: (1) LIF only, (2) a combination of LIF and bFGF, and (3) a combination of LIF and BMP4. mRNA transcript levels for each gene were determined by qRT-PCR and signals were normalized with their corresponding 18SrRNA signal. Relative mRNA expression was expressed as n-fold differences relative to embryos which were set at 1. Different lower and capital letters denoted samples that differ significantly in the relative expression of ERK 1/2 and STAT3 within treatments and embryos ($P<0.05$). *Jagged arrow* values are greater than the ones shown in the graph

In the author's laboratory, preliminary experiments were performed to identify if downstream transcription factors of the intracellular signaling pathway that are involved in self-renewal of pluripotent mouse and human ESC are also involved in self-renewal of cat ESC. We first designed primers for transcription detection of signal transducer and activator of transcription 3 (STAT3), the extracellular regulated kinase 1/2 (ERK1/2), and the receptor-regulated SMAD 5 (SMAD5), molecules which are phosphorylated and activated by different signaling pathways. Primer sets were tested for efficiency and amplification by standard curve analysis and sequencing of PCR products in in vitro-produced cat blastocysts. qRT-PCR analysis showed the presence of the three transcripts (Fig. 6.2), suggesting that downstream transcription factors involved in self-renewal of mouse and human ESC are expressed in cat embryos.

Mouse ESC require the presence of LIF to maintain the cells in an undifferentiated state (Niwa et al. 1998), while human ESC do not (Thomson et al. 1998). In the mouse, the mechanism of self-renewal is dependent on LIF activation of STAT3 through the heterodimerization of gp130 and the LIF receptor (LIFR), and activation of the Janus-associated tyrosine kinase (JAK) pathway (Niwa et al. 1998). However, LIF also up-regulates ERK1/2 activity, which leads to the loss of pluripotency and the onset of differentiation (Burdon et al. 1999). Similarly, our study showed that LIF signaling resulted in significant up-regulation of STAT3 transcripts in cat ESC cultured in LIF or in a combination of LIF and bFGF (Fig. 6.2).

However, the transcript levels of ERK1/2 were not increased in cat ESC cultured with LIF only, but expression levels of ERK1/2 were greater when cat ESC were cultured in the combination of LIF and bFGF (Fig. 6.2). It is not clear why combined treatment of LIF and bFGF induced higher levels of ERK1/2 transcripts, but since both factors induced ERK1/2 activation through different pathways, it is possible that accumulation of ERK1/2 occurs and consequently higher transcript levels were present, as recently reported in rabbit ESC (Hsieh et al. 2011).

FGF signaling activates the mitogen-activated protein kinase (MEK)/ ERK1/2 and the PI3L/AKT pathways. Activation of these pathways plays a dissimilar role in ESC self-renewal of different mammalian species. For example, inhibition of the PI3L/AKT pathway in mouse ESC cultured in LIF only results in an increased expression of ERK1/2 and cell differentiation. However, inhibition of both the PI3L/AKT and MEK/ERK1/2 pathways results in self-renewal (Paling et al. 2004). In contrast, activation of both PI3L/AKT and MEK/ERK1/2 pathways in rabbit ESC cultured with a combination of LIF and bFGF results in robust self-renewal (Hsieh et al. 2011). In our study, although we did not measure the expression of AKT, we similarly observed that activation of ERK1/2 with a combination of LIF and bFGF results in prolonged self-renewal and enhanced cell proliferation (Gómez et al. 2010). The biological significance of this effect in rabbit ESC was reported to be improved colony morphology and stronger expression of some pluripotent markers. Nonetheless, similar effects in colony morphology were not observed in cat ESC and expression of pluripotent markers was not tested, although supplementation with higher doses of bFGF (20 ng/mL) promoted cell proliferation and prolonged self-renewal (Gómez et al. 2010). These combined results suggest that the JAK/STAT signaling pathway is activated by LIF, and that cat ESC use this pathway for self-renewal. The beneficial self-renewal effect observed when bFGF was added to the culture medium suggests that the FGF signaling pathway is also involved in self-renewal of cat ESC. However, the two pathways were not sufficient to maintain cat ESC in their pluripotent state.

Inhibition of the MEK/ERK pathway with BMP4 enhanced self-renewal of mouse ESC (Ying et al. 2003; Qi et al. 2004). Thus, we did an experiment to see if culturing cat ESC with a combination of BMP4 and LIF enhanced self-renewal. Our results showed that BMP4 did not have an inhibitory effect on ERK1/2. In fact, transcript levels of ERK1/2 in cat ESC cultured with both BMP4 and LIF were not downregulated, and the levels of expression were similar to that of cat ESC cultured in LIF only or a combination of LIF and bFGF (Fig. 6.2). Moreover, BMP4 induced cell differentiation of cat ESC, as previously reported in humans (Schuldiner et al. 2000). In human ESC, the BMP-signaling pathway activates SMAD1/5/8, which promotes trophectodermal differentiation (Lyssiotis et al. 2011). To determine if BMP4 induced cell differentiation in cat ESC through the BMP-signaling pathway, we measured expression levels of SMAD 5 in cat ESC cultured in both BMP4 and LIF. Interestingly, an increase in SMAD5 was not observed upon stimulation of the BMP-signaling pathway (Fig. 6.2). It is possible that paracrine molecules secreted by the CEF feeder cells were interacting indirectly with the ESC and reducing expression levels of SMAD5, but not enough to compensate for the differentiation

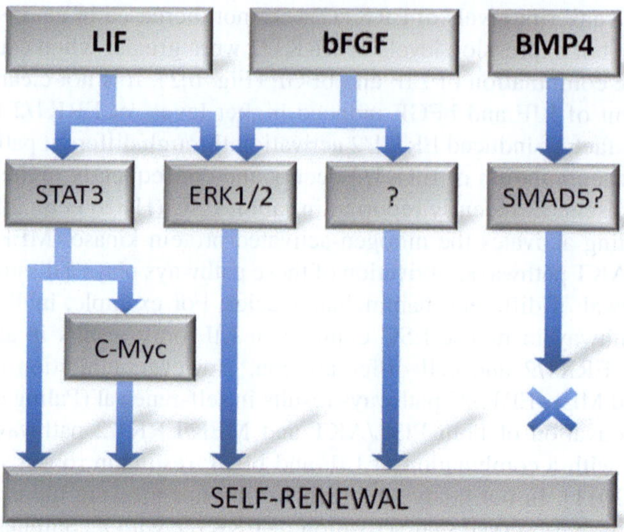

Fig. 6.3 An interpretation of the effect of LIF, bFGF, and BMP4 on activation of downstream transcription factors associated with self-renewal signaling pathways in cat ESC. LIF signaling activates STAT3 pathway and STAT-3 induces c-MYC expression, as detected by RT-PCR in cat ESC colonies. The STAT3 pathway acts as one of the self-renewal pathways in cat ESC. bFGF signaling in combination with LIF activates the downstream protein ERK 1/2 which is involved in self-renewal in cat ESC. bFGF may also activate other signaling pathways that are involved in self-renewal and maintaining cat ESC in an undifferentiated state. BMP4-signaling was responsible for inducing cell differentiation. BMP4 did not have an inhibitory effect on ERK 1/2 and did not increase SMAD 5. It is possible that paracrine molecules secreted by CEF feeder cells were interacting indirectly with ESC and reducing expression levels of SMAD5, but not enough to compensate for the differentiation effect induced by BMP4. Cat ESC do not use the BMP-signaling pathway for inhibiting ERK 1/2 activity and self-renewal

effect induced by BMP4. Thus, our results indicate that cat ESC do not use the BMP-signaling pathway for inhibiting ERK1/2 activity and self-renewal.

An interpretation of the effect of LIF, bFGF, and BMP4 on activation of downstream transcription factors associated with self-renewal signaling pathways in cat ESC is given in Fig. 6.3. In summary, the JAK/STAT pathway is activated by LIF signaling and possibly acts as one of the self-renewal pathways in cat ESC. In addition, the MERK/ERK pathway activated by bFGF signaling may be involved in self-renewal through activation of the downstream protein ERK and other signaling pathways. The BMP-signaling was responsible for inducing cell differentiation, while it is not clear if ERK activation by a combination of LIF and bFGF is also involved in cell differentiation. Additional signaling pathways are involved in self-renewal and maintaining ESC in an undifferentiated state. Ongoing studies detecting downstream transcription factors that are influenced by specific regulatory pathways will provide a means to identify modulators of ESC fate. Ensuring proper culture conditions is required for the maintenance of robust and stable cat ESC lines.

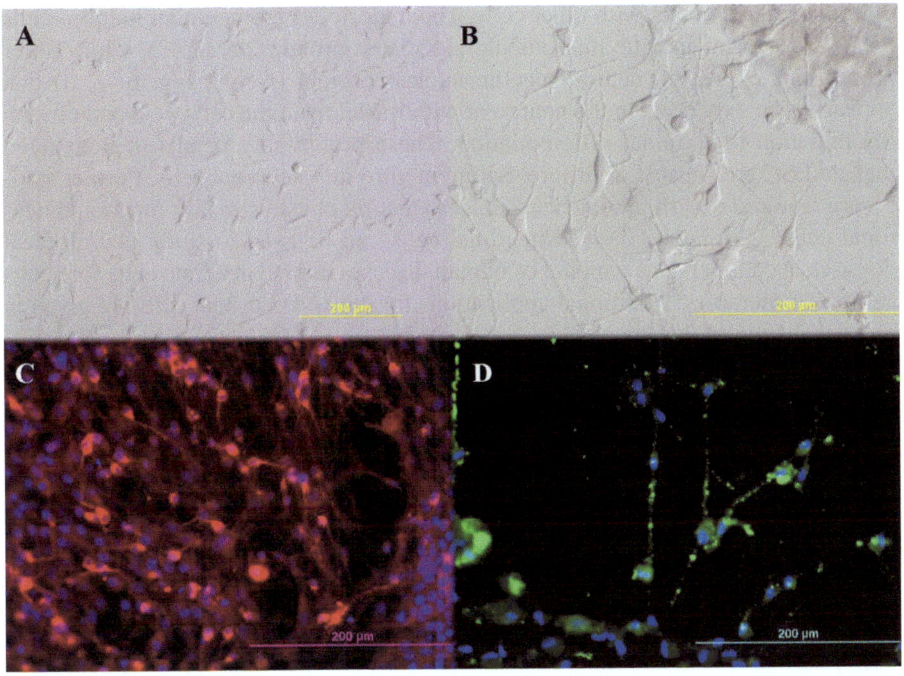

Fig. 6.4 Morphological characterization and neuronal differentiation of cat ESC. (**a**) Pre-induction: Single cat ESC in growth medium 48 h after colony dissociation. Single cat ESC appear with a fibroblastic morphology. (**b**) After pre-induction treatment with bFGF followed by addition of neuronal induction medium. Cat ESC change to a neuronal morphology with long, thin processes and round cell bodies. Neuronal phenotype cells showing positive expression for neuronal specific markers (**c**) MAP2 (red), and (**d**) NeuN (green). Nuclei are stained with DAPI (blue)

3.4 Induced Differentiation of Cat ESC

Clearly, spontaneous in vitro differentiation of cat ESC is not a viable method to obtain a pure culture of a desired cell type; even less so when true pluripotency has not been confirmed. Thus, efficient methods that direct the differentiation of cat ESC and produce a homogeneous population of a specific cell type, before cells spontaneously differentiate, would benefit the production of cat cells for future stem-cell-based therapies.

3.4.1 Neuronal Differentiation

In the author's laboratory, a preliminary experiment was done to determine the ability of cat ESC to differentiate into neuronal cells. Cat ESC colonies were induced with neuronal induction medium, using the same protocol effectively tested for the induction of feline bone marrow into neuron-like cells (Martin et al. 2002). We found that high numbers of cat ESC rapidly differentiated into neuronal phenotype cells that possessed a round cell body with long and thin processes (Fig. 6.4b), some of which made contact with other cells. The majority of the neuronal induced cells expressed neuron-specific microtubule associated protein 2 (Map-2), while lower proportions expressed neuron-specific nuclear protein (NeuN; Fig. 6.4c, d) that appears to be expressed in the neuron at withdrawal from the cell cycle and/or with the initiation of terminal differentiation. These preliminary results demonstrated that cat ESC are capable of differentiating in vitro into neuronal cells. Further studies are required to differentiate cat ESC into mature physiologically functional neuronal cells. Ultimately, derived neuronal cells can be used to study neurological diseases in cats, serve as a model of human diseases, and to generate cells for stem-cell-based therapies. Additional applications for derived neuronal cells are given in another section of this chapter.

3.5 *Current and Future Applications of Stem-Cell-Based Therapies Using Cat ESC*

There are no reports indicating that cat ESC are being used currently for stem-cell-based therapies for regenerative medicine, possibly due to the limited stem cell proliferation of the cells in vitro. However, our recent results confirming that cat ESC can be induced to differentiate in vitro into neuronal cells opens the door for future studies focusing on neuronal diseases. One possible application of derived neuronal stem cells could be for the study and treatment of Alzheimer's disease (AD). Histopathological features of AD in humans consist of extracellular deposition of amyloid-β peptide, an enzymatic product of amyloid precursor protein (App), and intracellular neurofibrillary tangles (NFT). Aged cats have a similar AD that is recognized as a feline disorder with pronounced amyloid and phosphorylated tau deposition (Gunn-Moore et al. 2006, 2007) that is indicative of an identical disease mechanism to that of AD in humans. Moreover, the human brain and the cat brain both have cerebral cortex with similar lobes, sharing some neurological aptitudes (Mann 1979). These properties and conditions are indicative that cats could be a potential model for human AD.

4 Domestic Cat Multipotent Stem Cells

In the domestic cat, two types of multipotent stem cells, mesenchymal and spermatogonial, have been isolated, partially characterized, and initial transplantation trials have been done. One drawback to developing stem-cell-based technologies is that most isolated multipotent adult stem cells have more limited life-spans in culture than do ESC. Understanding self-renewal and how these cells commit to differentiation has progressed slowly, in part due to the absence of molecular markers that reliably define the target cell and the lack of optimal in vitro culture conditions. In this section, we will give an overview of current knowledge for the molecular characterization of mesenchymal and spermatogonial stem cells and describe some preliminary transplantation trials performed in the domestic cat using these adult stem cells.

4.1 Cat Mesenchymal Stem Cells

Mesenchymal stem cells (MSC) are a heterogeneous population of multipotent cells that can be isolated from any adult and fetal tissues and exhibit a capacity to give rise to cells of multiple cell lineages. Adipose-derived MSC (AD-MSC) are a good cell source alternative that offers promising potential in organ-specific tissue engineering and tissue reconstruction. AD-MSC have the ability to differentiate into mesodermal cells, such as adipocytes, fibroblasts, myocytes, osteocytes, and cartilagocytes, a process known as lineage-specific differentiation (Lin et al. 2006). Inducing specific cell differentiation of AD-MSC into chrondogenic or osteogenic cells provides a possible clinical application for the treatment of several diseases in humans, including inherited, traumatic, or degenerative bone, joint, and soft tissue defects.

In cats, the functional capacity of bone-marrow-derived MSC (BM-MSC) was measured by their capacity to give rise to cells of multiple cell lineages, including adipocyte, osteocyte, and neural-like cells (Martin et al 2002; Zhang et al 2011). Similarly, cat AD-MSC isolated from subcutaneous adipose tissue differentiated into mesodermal cells when stimulated with inductive factors (Quimby et al. 2011). AD-MSC offer several advantages for use in endangered species compared to BM-MSC, because adipose tissue is easier to access, simpler enzyme-based protocols for cell isolation have been described, and seemingly their potential to differentiate into osteoblasts, chrondocytes, and adipocytes is similar to that of BM-MSC (Schäffler and Büchler 2007).

4.1.1 Characteristics of Cat MSC

In the domestic cat, MSC isolated from BM had morphological characteristics that were similar to those reported in human and mice. Cat BM-MSC grew as isolated colonies between 7 and 10 days after initial plating, but after further expansion, cells were fibroblast-like, appearing polygonal or spindle-shaped with long processes (Martin et al. 2002). Cat BM-derived MSC do not grow well when cultured at low cell densities; however, when they were cultured with an initial concentration of 1.5×10^7 BM mononuclear cells, frequency of colony formation was increased (Martin et al. 2002). Cat AD-MSC were plastic adherent, spindle-shaped cells that formed within 24–72 h of culture with a replication rate significantly greater than that of BM-MSC (Webb et al. 2012). These studies suggest that cat BM and AD-MSC shared similar morphological characteristics, but AD-MSC can be generated more quickly due to its higher proliferative rates (Webb et al. 2012).

4.1.2 Molecular Signature of Cat MSC

MSC isolated from cat BM have shown similar patterns of cell-surface antigens as those reported for other mammalian species, where cells are strongly positive for CD9, CD44, and MHC-1, and negative for CD45 (Martin et al. 2002). Moreover, cat AD-MSC express high levels of CD44 and CD105, and expression of the CD90 surface antigen was higher in AD-MSC than that of BM-MSC (Webb et al. 2012). Studies on MSC gene expression have led to identification of some genes associated with multipotency (Song et al. 2006; Mitchell et al. 2006). However, molecular mechanisms regulating proliferation and multipotency of MSC are yet to be identified. Expression of transcription factors involved in the maintenance of pluripotency in ESC (*OCT-4*, *SOX-2* and *NANOG*) has been proposed to play a similar role in adult multipotent cells, including MSC. In humans, expression of *OCT-4* and *SOX-2* transcripts has been reported in MSC isolated from different tissues (Kolf et al. 2007; Riekstina et al. 2009; Greco et al. 2007). Similarly, expression of the three transcription factors *OCT-4*, *SOX-2*, and *NANOG* has been reported in cat BM-MSC (Zhang et al. 2011).

4.1.3 Induced Differentiation of Cat MSCs

Cat AD-MSC isolated from subcutaneous adipose tissue differentiated into mesodermal cells (osteoblasts, chrondocytes, and adipocytes) when stimulated with inductive factors (Quimby et al. 2011). The original anatomical location of the adipose tissue may influence both the proliferation rate and the differentiation capacity of the cells. For instance, in rabbits, the osteogenic potential of AD-MSC isolated from abdominal adipose tissue was greater than that of AD-MSC isolated from subcutaneous adipose tissue (Peptan et al. 2006). Conversely, in humans, osteogenesis was more robust in adipose-derived stromal cells from the flank and thigh, as compared with cells from the abdomen (Levi et al. 2010). It is not clear why MSC isolated from different adipose tissues are committing to differentiate into a specific cell lineage, but is known that certain transcriptional and molecular events activate cell differentiation (Hong et al. 2005; Hong and Yaffe 2006). Although, cat AD-MSC differentiate into mesodermal cells, it is important for future clinical applications to determine if expression of transcription factors between AD-MSC isolated from different anatomical locations may favor one cell lineage, rather than another.

4.1.4 Current and Future Applications of Stem-Cell-Based Therapies Using Cat Adipose-Derived MSC

CKD is a major cause of morbidity and mortality in cats. At present, the only definitive treatment option for cats with CKD is renal transplantation (Adin et al. 2001; Schmiedt et al. 2008). Similarly, the black-footed cat (*Felis nigripes*), one of the smallest wild cats and listed as threatened, suffers a high mortality rate due to kidney failure resulting from a condition known as amyloidosis (Terio et al. 2008). At our institute, we have black-footed cats, which are part of our assisted reproduction research program, that have been dying at 3–5 years of age due to amyloidosis. Similarly, San Diego and other zoos also reported the deaths of black-footed cats at early ages due to the same condition. Amyloidosis is a progressive, incurable, metabolic disease characterized by abnormal deposits of protein in one or more organs, in this instance, the kidneys. Even though, supportive care and treatment were provided to black-footed cats to stabilize renal function and reverse metabolic complications, the disease is progressive and animals die prematurely. Recently, BM and AD-MSC were transplanted into domestic cats for treatment of CKD (Quimby et al. 2011). Although the pilot study did not report a cure for the disease, the intrarenal injection of autologous MSC was well tolerated and induced a mild improvement in renal function. Thus, MSC transplantation is a promising tool for cell replacement therapy and regenerative medicine in endangered species and possibly an effective treatment for amyloidosis in black-footed cats.

Many rare and endangered animals held in zoological institutions are predisposed to chronic joint and ligament problems. This may be due to a combination of their captive environment and the longer life-span they usually achieve therein. Some treatments are not curative and, instead, are palliative for treating pain and offer a certain level of well-being to the animals. MSC have promising application for tissue repair, treatment of autoimmune diseases, and regenerative medicine (Huang et al. 2011). MSC transplantation for regenerative medicine has become an important area of research for treating human diseases and recently adopted in veterinary medicine. Currently, veterinarians are using MSC for treatment of tendon, ligament, or joint/cartilage injuries, mainly in dogs and horses, but also in domestic cats. Recently (2011), a Florida panther at the Tallahassee Museum of Natural History, was treated with AD-MSC to treat chronic arthritis in his elbow. Over a period of 7 weeks, the panther recovered the complete use of his elbow without any apparent pain. Thus, it is possible that the cells transdifferentiated into several types of lineages that may be beneficial when transplanted into endangered felids as a treatment to regenerate skeletal elements and repair cartilage.

4.2 *Spermatogonial Stem Cells*

In domestic cats, SSC are single cells localized in the basal membrane of the seminiferous tubules (Powell et al. 2013) that are capable of differentiating into mature spermatozoa after testicular transplantation (Silva et al. 2012). Domestic cat SSC have been isolated from prepubertal and adult testis (Kim et al. 2006; Powell et al. 2012, 2013; Silva et al. 2012; Vansandt et al. 2012; Tiptanavattana et al. 2013). SSC constitute only a small fraction of the adult testicular germ cells. Identification of germ-cell-specific markers that are expressed uniquely on SSC, but not on other somatic cells or differentiated spermatogenic cells, facilitates enrichment of a pure population of SSC for further studies or spermatogonial cell transplantation (SCT).

4.2.1 Molecular Signature SSC

Cat SSC have been characterized phenotypically by the expression of spermatogonial markers previously identified in rodents. Expression of cell-surface markers *GFRα1* (glial cell-derived neurotrophic factor (GDNF) receptor) which is a signaling molecule involved in SSC self-renewal, *GPR125* (G protein-coupled receptor 125) which is expressed in SSC, and *C-Kit* (which is a receptor for stem cell factor and an early marker for germ cell differentiation) was detected by immunohistochemistry of adult and prepubertal cat testes (Powell et al. 2012). Moreover, it also was demonstrated that *GFRα1* and *GPR125* cell-surface markers were localized both at the basement membrane of the seminiferous tubules and also across the seminiferous tubules section (Powell et al. 2013); a clear indication that these two cell-surface markers are not only expressed by SSC but also by more differentiated spermatogonial cells, and may not be used for isolation of a purified population of cat SSC.

Expression of the major pluripotent ESC surface markers *SSEA-4*, *SSEA-1*, *TRA1-81*, *TRA-1-60*, and transcription factor *OCT-4* also were detected in cat SSC, but all of them were located only along the basement membrane of the seminiferous tubules (Powell et al. 2013). On the other hand, expression of cell-surface markers (*GFRα1, GPR125*), and thymus cell antigen-1 (*THY1*), and intracellular markers (ubiquitin carboxi-terminal hydrolase I—*UCHL,*; promyelocytic leukemia zinc finger—PLZF, and *OCT-4*), previously identified in rodents, were detected at the mRNA level in a pool of spermatogonial cells isolated from prepubertal and adult cat testes (Vansandt et al. 2012). Overall, these studies confirm that SSC of prepubertal and adult cats express cell-surface and intracellular germ-cell-specific markers in common with SSC of other mammalian species, as well as pluripotent ESC markers. The location of cat SSC along the basement membranes of the seminiferous tubules, and the positive expression of pluripotent ESC markers provide further support for classifying the cells as SSC.

Further characterization by flow cytometry revealed that cat testes contain distinct cell populations expressing *GFRα1* (47 %), *GPR125* (52 %), *TRA-1* (18 %), and *TRA-1-81* (16 %), and smaller populations expressing *SSEA-1* (7 %) and *SSEA-4* (3 %). Dual staining of germ cells also revealed the presence of three distinct cell populations that either express *GFRα1* only (23 %), are positive for both *GFRα1* and *SSEA-4* (6 %) or are positive only for *SSEA-4* (1 %; Powell et al. 2013). Recently, immunohistochemistry in sections of adult cat testis showed a distinct cell population of SSC, located at the basement membrane of the seminiferous tubules, showing positive expression for *GFRα1* and a germ cell DEAD (Asp-Glu-Ala-Asp) box polypeptide 4 (DDX-4) marker (Tiptanavattana et al. 2013). These results demonstrated that cat SSC can be purified into different subpopulations by determining the expression of specific markers and can be isolated from more differentiated spermatogonial cells.

4.2.2 Transplantation of Spermatogonial Cells in Felids

SCT has been carried out in the domestic cat. Kim et al. (2006) investigated the ability of domestic cat germ cells to colonize the seminiferous tubules of mice by xenogenic transplantation. Successful colonization was observed, but spermatogenesis was not supported by the mouse testes. The taxonomic distance between the donor cat and recipient mice may have influenced the results. In contrast, in a recent study, it was demonstrated that xenogenic transplantation of ocelot spermatogonial cells, selected by the expression of the *GFRα1* marker, into domestic cat testes, a taxonomically similar species, resulted in both colonization and spermatogenesis (Silva et al. 2012). Nonetheless the production of live offspring from spermatozoa generated from SCT has not been accomplished.

Several steps during the process of SCT may affect the success of the technique. One critical step is preparation of recipient testes. Endogenous differentiated germ cells in the recipient male must be depleted to provide additional space at the basal compartment of seminiferous tubules for colonization and expansion of the transplanted cells (Brinster et al. 2003). The approaches applied to deplete germ cells in the domestic cat were similar to those reported in mice. The first approach involved focal irradiation with a total of 9–10 Gy (Kim et al. 2006; Silva et al. 2012), and in the second method the cat was treated with busulfan (Silva et al. 2012). The ability of both methods to deplete male germ cells was demonstrated, but treatment with busulfan at doses that depleted germ cells was toxic and resulted in animal mortality (Silva et al. 2012), Thus, in the domestic cat, focal irradiation is the recommended method to induce germ cell depletion.

Another critical step that may affect the success of SCT is the age of the recipient. Transplantation of SC into the testis of prepubertal mice and goats resulted in higher rates of colonization and spermatogenesis than was seen in older animals (Shinohara et al. 2001; McLean et al. 2003; Honaramooz et al. 2003). However, spermatogenesis of ocelot germ cells was supported after transplantation into testis of adult domestic cats. With the limited number of animals in which spermatogenesis was restored after SCT into domestic cat testes, it is not possible to conclude that adult cats were better recipients than younger cats. Moreover, depletion of germ cells in domestic cats before SCT may not be needed. Most of the original studies in SCT were performed in the mouse. Possible species-specific differences, including the dissimilarity between the mouse and the cat in size of the testes, may be suggestive evidence that depletion of germ cells before SCT may not be necessary. Further studies are required to better understand the factors that affect SCT success in domestic cats.

5 Conclusions

Understanding the transcriptional network of the signaling pathway(s) for self-renewal in pluripotent and multipotent cat stem cells is essential for optimization of the in vitro culture conditions required to produce robust cat stem cells. In these studies, it was shown that “core” transcription factors *OCT-4*, *SOX-2*, and *NANOG* are expressed by pluripotent and multipotent cat stem cells. At the transcriptional level, in cat ESC, LIF and bFGF signaling is able to induce activation of the downstream transcription factor STAT3 and protein kinase ERK1/2, and possibly by activation of other signaling pathways, up-regulate expression of the core of transcription factors *OCT-4*, *SOX-2*, and *NANOG*. Even though, cat ESC use the JAK/STAT and MEK/ERK signaling pathways for self-renewal ESC were not maintained in an undifferentiated state for prolonged periods of time. Additional studies are required to define if other signaling pathways, such as Wnt or Activin/nodal signaling, which sustain pluripotency in mouse and human ESC, may also regulate self-renewal in cat ESC. Similar studies are needed to determine if self-renewal pathways are conserved in multipotent cat stem cells.

Lastly, for application in regenerative medicine, it is relevant to elucidate the mechanisms that regulate lineage commitment in pluripotent and multipotent cat stem cells at the single-cell level to improve the efficacy of differentiation into specific cell types.

Acknowledgments We would like to thank Monica Biancardi for the RT-PCR analysis of signaling for self-renewal in cat ESC, Dr. Qian Qin for the neuronal induced differentiation of cat ESC, and Dr. Qin and Jason Galiguis for the design of figures.

References

Adin CA, Gregory CR, Kyles AE, Cowgill L (2001) Diagnostic predictors of complications and survival after renal transplantation in cats. Vet Surg 30:515–521. doi:10.1053/jvet.2001.28418

Amit M, Shariki C, Margulets V, Itskovit-Eldor J (2004) Feeder layer- and serum-free culture of human embryonic stem cells. Biol Reprod 70:837–845. doi:10.1095/biolreprod.103.021147

Bettiol E, Sartiani L, Chicha L et al (2007) Fetal bovine serum enables cardiac differentiation of human embryonic stem cells. Differentiation 75:669–681. doi:10.1111/j.1432-0436.2007.00174.x

Brevini TAL, Antonini S, Cillo F et al (2007) Porcine embryonic stem cells: facts, challenges and hopes. Theriogenology 68:206–213. doi:10.1016/j.theriogenology.2007.05.043

Brevini TAL, Pennarosa G, Gandolfi F (2010) No shortcuts to pig embryonic stem cells. Theriogenology 74:544–550. doi:10.1016/j.theriogenology.2010.04.020

Brinster RJ, Ryu BY, Avarbock MR et al (2003) Restoration of fertility by germ cell transplantation requires effective recipient preparation. Biol Reprod 69:412–420. doi:10.1095/biolreprod.103.016519

Burdon T, Stracey C, Chambers I et al (1999) Suppression of SHP-2 and ERK signaling promotes self-renewal of mouse embryonic stem cells. Dev Biol 210:30–43

Cyranoski D (2013) Stem cells boom in vet clinics. Nature 496:148–149. doi:10.1038/496148a

Eastham AM, Spencer H, Soncin F et al (2007) Epithelial-mesenchymal transition events during human embryonic stem cell differentiation. Cancer Res 67:11254–11262. doi:10.1158/0008-5472.CAN-07225

Evans MJ, Kaufman MH (1981) Establishment in culture of pluripotent cells from mouse embryos. Nature 292:154–156. doi:10.1039/292154a0

Fortier LA, Travis AJ (2011) Stem cells in veterinary medicine. Stem Cell Res Ther 2:9. doi:10.1186/scrt50

Garcia-Gonzalo FR, Izpisua JC (2008) Albumin-associated lipids regulate human embryonic stem cell self-renewal. PLoS One 3:e1384. doi:10.1371/journal.pone.0001384

Gómez MC, Pope CE, Giraldo AM et al (2004) Birth of African wildcat cloned kittens born from domestic cats. Cloning Stem Cells 6:217–228. doi:10.1089/clo.2004.6.247

Gómez MC, Pope CE, López M (2006) Chromosomal aneuploidy in African wildcat somatic cells and cloned embryos. Cloning Stem Cells 8:69–78

Gómez MC, Pope CE, Kutner RH et al (2008) Nuclear transfer of sand cat cells into enucleated domestic cat oocytes is affected by cryopreservation of donor cells. Cloning Stem Cells 10:469–483. doi:10.1089/clo.2008.0021

Gómez MC, Pope CE, Kutner RH et al (2009) Generation of domestic transgenic cloned kittens using lentivirus vectors. Cloning Stem Cells 11:167–176. doi:10.1089/clo.2008.0054

Gómez MC, Serrano MA, Pope CE et al (2010) Derivation of cat embryonic stem-like cells from in vitro produced blastocysts on homologous and heterologous feeder cells. Theriogenology 74:498–515. doi:10.1016/j.theriogenology.2010.05.023

Greco SJ, Liu H, Rameshwar P (2007) Functional similarities among genes regulated by OCT4 in human mesenchymal and embryonic stem cells. Stem Cells 25:3143–3154. doi:10.1634/stemcells.2007-0351

Gunn-Moore DA, McVee J, Bradshaw JM et al (2006) Ageing changes in cat brains demonstrated by beta-amyloid and AT8-immunoreactive phosphorylated tau deposits. J Feline Med Surg 8:234–242. doi:10.1016/j.jfms.2006.01.003

Gunn-Moore D, Moffat K, Christie LA, Head E (2007) Cognitive dysfunction and the neurobiology of ageing in cats. J Small Anim Pract 48:546–553. doi:10.1111/j.1748-5827.2007.00386.x

Honaramooz A, Bekboodi E, Blash S et al (2003) Germ cell transplantation in goats. Mol Reprod Dev 64:422–428. doi:10.1002/mrd.10205

Hong JH, Yaffe MB (2006) TAZ: a beta-catenin-like molecule that regulates mesenchymal stem cell differentiation. Cell Cycle 5:176–179

Hong JH, Hwang ES, McManus ET et al (2005) TAZ, a transcriptional modular of mesenchymal stem cell differentiation. Science 309:1074–1078. doi:10.1126/science.1110955

Hsieh YC, Intawicha P, Lee KH et al (2011) LIF and FGF cooperatively support stemness of rabbit stem cells derived from parthenogenetically activated embryos. Cell Reprogram 13:241–255. doi:10.1089/cell.2010.0097

Huang S, Leung V, Peng S et al (2011) Developmental definition of MSCs: new insights into pending questions. Cell Reprogram 13:465–472. doi:10.1089/cell.2011.0045

Jaenish R, Eggan K, Humpherys D et al (2002) Nuclear cloning, stem cells and genomic reprogramming. Cloning Stem Cells 4:389–396. doi:10.1089/153623002321025069

Kee K, Angeles VT, Flores M et al (2009) Human DAZL, DAZ and BOULE genes modulate primordial germ-cell and haploid gamete formation. Nature 462:222–225. doi:10.1038/nature08562

Kim Y, Selvaraj V, Dobrinski I et al (2006) Recipient preparation and mixed germ cell isolation for spermatogonial stem cell transplantation in domestic cats. J Androl 27:248–256. doi:10.2164/jandrol.05034

Kolf CM, Cho E, Tuan RS (2007) Mesenchymal stromal cells. Biology of adult mesenchymal stem cells: regulation of niche, self-renewal and differentiation. Arthritis Res Ther 9:204

Laping NJ, Grygielko A, Mathur S et al (2002) Inhibition of transforming growth factor (TGF)-1-induced extracellular matrix with a novel inhibitor of the TGF-Type I receptor kinase activity: SB-431542. Mol Pharm 62:58–64. doi:10.1124/mol.62.1.58

Leung T, Chen XQ, Manser E, Lim L (1996) The p160 Rho A-binding kinase ROK alpha is a member of a kinase family and is involved in the reorganization of the cytoskeleton. Mol Cell Biol 116:5313–5327

Levi B, James AW, Glotzbach JP et al (2010) Depot-specific variation in the osteogenic and adipogenic potential of human adipose-derived stromal cells. Plast Reconstr Surg 126:822–834. doi:10.1097/PRS.0b013e3181e5f892

Lin Y, Chen X, Yan Z et al (2006) Multilineage differentiation of adipose derived stromal cells from GFP transgenic mice. Mol Cell Biochem 285:69–78. doi:10.1007/s110-005-9056-8

Lyssiotis CA, Lairson LL, Boitano AE et al (2011) Chemical control of stem cell fate and development potential. Angew Chem Int Ed 50:200–242. doi:10.1002/anie.20114284

Mann MD (1979) Sets of neurons in somatic cerebral cortex of the cat and their ontogeny. Brain Res 180:3–45. doi:10.1016/0165-0173(79)90015-8

Martin DR, Cox NC, Hathcock GP (2002) Isolation and characterization of multipotential mesenchymal stem cells from feline bone marrow. Exp Hematol 30:879–886

McLean DJ, Friel PJ, Johnston DS, Griswold MD (2003) Characterization of spermatogonial stem cell maturation and differentiation in neonatal mice. Biol Reprod 69:2085–2091. doi:10.1095/biolreprod.103.017020

Mitchell JB, McIntosh K, Zvonic S et al (2006) Immunophenotype of human adipose-derived cells: temporal changes in stromal-associated and stem cell-associated markers. Stem Cells 24:376–385. doi:10.1634/stemcells.2005-0234

Murry CE, Keller G (2008) Differentiation of embryonic stem cells to clinically relevant populations: Lessons from embryonic development. Cell 132:661–680. doi:10.1016/j.cell.2008.02.008

Nayernia K, Nolte J, Michelmann HW (2006) In vitro-differentiated embryonic stem cells give rise to male gametes that can generate offspring mice. Dev Cell 11:125–132. doi:10.1016/j.devcel.2006.05.010

Niwa H, Burdon T, Chambers I, Smith A (1998) Self-renewal of pluripotent embryonic stem cells is mediated via activation of STAT3. Genes Dev 12:2048–2060. doi:10.1101/gad.12.13.2048

Paling NR, Wheadon H, Bone HK et al (2004) Regulation of embryonic stem cell self-renewal by phosphoinositidine 3-kinase-dependent signaling. J Biol Chem 279:48063–48070. doi:10.1074/jbc.M406467200

Peptan IA, Hong L, Mao JJ (2006) Comparison of osteogenic potentials of visceral and subcutaneous adipose-derived cells of rabbits. Plast Reconstr Surg 117:1462–1470. doi:10.1097/01.prs.0000206319.80719.74

Pontius JU, Mullikin JC, Smith DR et al (2007) Initial sequence and comparative analysis of the cat genome. Genome Res 17:1675–1689. doi:10.1101/gr.6380007

Pope CE, Keller GL, Dresser BL (1993) In vitro fertilization in domestic and nondomestic cats, including sequences of early nuclear events, in vitro development, cryopreservation and successful intra- and interspecies embryo transfer. J Reprod Fertil Suppl 47:189–201

Pope CE, Gómez MC, Dresser BL (2006a) In vitro production and transfer of cat embryos in the 21st century. Theriogenology 66:59–71. doi:10.1016/j.theriogenology.2006.03.014

Pope CE, Gómez MC, Dresser BL (2006b) In vitro embryo production and embryo transfer in domestic and non-domestic cats. Theriogenology 66:1518–1524. doi:10.1016/j.theriogenology.2006.01.026

Pope CE, Gómez MC, Galiguis J, Dresser BL (2012) Applying embryo cryopreservation technologies to the production of domestic and black-footed cats. Reprod Domest Anim 47:125–129. doi:10.1111/rda.12053

Powell RH, Biancardi MN, Pope CE et al (2012) Isolation and characterization of domestic cat spermatogonial cells. Reprod Fertil Dev 24:221–222 [abstr]

Powell RH, Biancardi MN, Galiguis JL et al (2013) Expression of pluripotent stem cells markers in domestic cat spermatogonial cells. Reprod Fertil Dev 25:290–291 [abstr]

Qi X, Li TG, Hao J (2004) BMP4 supports self-renewal of embryonic stem cells by inhibiting mitogen-activated protein kinase pathways. Proc Natl Acad Sci U S A 101:6027–6032. doi:10.1073/pnas.0401367101

Quimby JM, Webb TL, Gibbons DS, Dow SW (2011) Evaluation of intrarenal mesenchymal stem cell injection for treatment of chronic kidney disease in cats: a pilot study. J Feline Med Surg 13:418–426. doi:10.1186/scrt198

Rajala K, Vaajasaari H, Suuronen R (2011) Effects of the physiochemical culture environment on the stemness and pluripotency of human embryonic stem cells. Stem Cell Stud 1:e3. doi:10.4081/or.2011.e3

Ribitsch I, Burk J, Delling U et al (2010) Basic science and clinical applications of stem cells in veterinary medicine. Adv Biochem Eng Biotechnol 123:219–263. doi:10.1007/10_2010_66

Richards M, Tan S, Fong CF et al (2003) Comparative evaluation of various human feeders for prolonged undifferentiated growth of human embryonic stem cells. Stem Cells 21:546–556. doi:10.1634/stemcells.21-5-546

Riekstina U, Cakstina I, Parfejevs V et al (2009) Embryonic stem cell marker expression pattern in human mesenchymal stem cells derived from bone marrow, adipose tissue, heart and dermis. Stem Cell Rev 5:378–386. doi:10.1007/s12015-009-9094-9

Schäffler A, Büchler C (2007) Concise review: adipose tissue-derived stromal cells-basic and clinical implications for novel cell-based therapies. Stem Cells 25:818–827. doi:10.1634/stemcells.2006-0589

Schmiedt CW, Holzman G, Schwarz T, McAnulty JF (2008) Survival, complications, and analysis of risk factors after renal transplantation in cats. Vet Surg 37:683–695. doi:10.1111/j.1532-950X.2008.00435.x

Schuldiner M, Yanuka O, Itskovitz-Eldor J et al (2000) From the cover: effect of eight growth factors on differentiation of cells derived from human embryonic stem cells. Proc Natl Acad Sci U S A 97:11307–11312. doi:10.1073/pnas.97.21.11307

Shinohara T, Orwig KE, Avarbock MR, Brinster RL (2001) Remodeling of the post natal mouse testis is accompanied by dramatic changes in stem cell number and niche accessibility. Proc Natl Acad Sci U S A 98:6186–6191. doi:10.1073/pnas.111158198

Siebelink KH, Chu IH, Guss F et al (1990) Feline immunodeficiency virus (FIV) infection in the cat as a model for HIV infection in man: FIV-induced impairment of immune function. AIDS Res Hum Retroviruses 6:1373–1378. doi:10.1089/aid.1990.6.1373

Silva RC, Costa GM, Lacerda SM (2012) Germ cell transplantation in felids: a potential approach to preserving endangered species. J Androl 33:64–76. doi:10.2164/jandrol.110.012898

Song L, Webb NE, Song Y, Tuan RS (2006) Identification and functional analysis of candidate genes regulating mesenchymal stem cell self-renewal and multipotency. Stem Cells 24:1707–1718. doi:10.1634/stemcells.2005-0604

Talbot NC, Powell AM, Rexroard CE Jr (1995) In vitro pluripotency of epiblasts derived from bovine blastocysts. Mol Reprod Dev 42:35–52. doi:10.1002/mrd.1080420106

Terio KA, O'Brien T, Lamberski N et al (2008) Amyloidosis in black-footed cats (*Felis nigripes*). Vet Pathol 45:393–400. doi:10.1354/vp.45-3-393

Thomson JA, Marshall VS (1998) Primate embryonic stem cells. Curr Top Dev Biol 38:133–165. doi:10.1016/S0070-2153(08)60246-X

Thomson JA, Kalishman J, Golos TG et al (1995) Isolation of a primate embryonic stem cell line. Proc Natl Acad Sci U S A 92:7844–7848

Thomson JA, Itskovitz-Eldor J, Shapiro SS (1998) Embryonic stem cell lines derived from human blastocysts. Science 282:1145–1147. doi:10.1126/science.282.5391.1145

Tiptanavattana N, Thongkittidilok C, Techakumphu M, Tharasanit T (2013) Characterization and in vitro culture of putative spermatogonial stem cells derived from feline testicular tissue. J Reprod Dev 59:189–195

Travis AJ, Kim Y, Meyers-Wallen V (2009) Development of new stem cell-based technologies for carnivore reproduction research. Reprod Domest Anim 44:22–28. doi:10.1111/j.1439-0531.2009.01396.x

Vansandt LM, Pukazhenthi BS, Keefer CL (2012) Molecular makers of spermatogonial stem cells in the domestic cat. Reprod Domest Anim 47:256–260. doi:10.1111/rda.12079

Verma R, Holland MK, Temple-Smith P, Verma PJ (2012) Inducing pluripotency in somatic cells from the snow leopard (Panthera uncia), an endangered felid. Theriogenology 77:220–225. doi:10.1016/j.theriogenology.2011.09.022

Verma R, Liu J, Holland MK et al (2013) Nanog is an essential factor for induction of pluripotency in somatic cells from endangered felids. Biores Open Access 2:72–76. doi:10.1089/biores.2012.0297

Villa-Diaz LG, Pacut C, Slawny NA (2009) Analysis of the factors that limit the ability of feeder cells to maintain the undifferentiated state of human embryonic stem cells. Stem Cells Dev 18:641–651. doi:10.1089/scd.2008.0010

Watanabe K, Uneo M, Kamiya D et al (2007) A ROCK inhibitor permits survival of dissociated human embryonic stem cells. Nat Biotechnol 25:681–686. doi:10.1038/nbt1310

Webb TL, Quimby JM, Dow SW (2012) In vitro comparison of feline bone marrow-derived and adipose tissue-derived mesenchymal stem cells. J Feline Med Surg 14:165–168. doi:10.1177/1098612X11429224

Weissman IL, Anderson DJ, Gge F (2001) Stem and progenitor cells: origins, phenotypes, lineage commitments, and transdifferentiations. Annu Rev Cell Dev Biol 17:387–403. doi:10.1146/annurev.cellbio.17.1.387

Willet BJ, Flynn JN, Hosie MJ (1997) FIV infection of the domestic cat: an animal model for AIDS. Immunol Today 18:182–189. doi:10.1016/S0167-5699(97)84665-8

Wongsrikeao P, Saenz D, Rinkoski T et al (2011) Antiviral restriction factor transgenesis in the domestic cat. Nat Methods 8:853–859. doi:10.1038/nmeth.1703

Xu C, Inokuma MS, Denham J (2001) Feeder-free growth of undifferentiated human embryonic stem cells. Nat Biotechnol 19:971–974. doi:10.1038/nbt1001-971

Xu Y, Zhu X, Hahm HS et al (2010) Revealing a core signaling regulatory mechanism for pluripotent stem cell survival and self-renewal by small molecules. Proc Natl Acad Sci U S A 107:8129–8134. doi:10.1073/pnas.1002024107

Yin XJ, Lee HS, Yu XF et al (2008) Generation of cloned transgenic cats expressing red fluorescence protein. Biol Reprod 78:425–431. doi:10.1095/biolreprod.107.065185

Ying QL, Nichols J, Chambers I, Smith A (2003) BMP induction of id proteins suppresses differentiation and sustains embryonic stem cell self-renewal in collaboration with STAT3. Cell 115:281–292. doi:10.1016/S0092-8674(03)00847-X

Young RA (2011) Control of the embryonic stem cell state. Cell 144:940–954. doi:10.1016/j.cell.2011.01.032

Yu X, Jin G, Yin X et al (2008) Isolation and characterization of embryonic stem-like cells derived from in vivo-produced cat blastocysts. Mol Reprod Dev 75:1426–1432. doi:10.1002/mrd.20867

Yu X, Kim JH, Jung EJ et al (2009a) Cloning and characterization of cat *POUEF1* and for identification of embryonic stem cells. J Reprod Dev 55:361–366

Yu Z, Ping J, Cao J et al (2009b) DAZL promotes germ cell differentiation from embryonic stem cells. J Mol Cell Biol 1:93–103. doi:10.1093/jmcb/mjp026

Zhang Z, Maiman DJ, Kurpad SN et al (2011) Feline bone marrow-derived mesenchymal stem cells express several pluripotent and neural markers and easily turned into neural-like cells by manipulation with chromatin modifying agents and neural inducing factors. Cell Reprogram 13:385–390. doi:10.1089/cell.2011.0007

Chapter 7
Brief Introduction to Coral Cryopreservation: An Attempt to Prevent Underwater Life Extinction

Tiziana A.L. Brevini, Sara Maffei, and Fulvio Gandolfi

In recent decades, the coupling of climate change and anthropogenic stressors has caused a widespread and well-recognized reef crisis (Hughes et al. 2003; Veron et al. 2009) (Table 7.1). Throughout the world coral reefs are being degraded at unprecedented rates. The increased greenhouse gases make corals more susceptible to stress, bleaching, and newly emerging diseases. Locally, reefs are damaged by pollution, nutrients, and sedimentation from outdated land use, fishing, and mining practices (Hughes et al. 2003).

Although in situ conservation practices, such as marine protected areas, reduce these stressors and may help slow the loss of genetic diversity on reefs, the global effects of climate change will continue to cause population declines.

Gamete cryopreservation has already acted as an effective insurance policy to maintain the genetic diversity of many wildlife species, but has only just begun to be explored for coral.

So far Mary Hagedorn reported having a great deal of success with cryopreserving sperm and larval cells from a variety of coral species. Building on this success, she has now begun to establish genetic banks using frozen samples (Hagedorn et al. 2012a, b). She demonstrated that cells that are cryopreserved and banked properly can retain viability for years, or even centuries, without DNA damage (Hagedorn et al. 2012b).

Of all the reefs, Caribbean reefs are suffering the most severe declines, and their fate may predict the future of corals throughout the world. To build new tools for the continued protection and propagation of coral from the Great Barrier Reef, an international group of coral and cryopreservation scientists known as the Reef Recovery Initiative joined forces during the November 2011 mass-spawning event.

T.A.L. Brevini (✉) • S. Maffei • F. Gandolfi
Laboratory of Biomedical Embryology, UniStem, Centre for Stem Cell Research, Università degli Studi di Milano, Via Celoria 10, 20133 Milan, Italy
e-mail: tiziana.brevini@unimi.it

T.A.L. Brevini (ed.), *Stem Cells in Animal Species: From Pre-clinic to Biodiversity*, Stem Cell Biology and Regenerative Medicine, DOI 10.1007/978-3-319-03572-7_7,

Table 7.1 Extinction risk of different coral species

Species	Extinction risk
Acropora austera	++
Acropora cervicornis	+++
Acropora Florida	++
Acropora formosa	++
Acropora humilis	++
Acropora hyacinthus	++
Acropora millepora	++
Acropora nasuta	++
Acropora palmata	+++
Acropora paniculata	+
Acropora tenuis	++
Acropora verweyi	+

+ Low risk, ++ moderate risk, +++ high risk

Storage of important coral and related cells through cryopreservation will profoundly advance basic research in embryology, genetics, systematics and molecular biology, as well as enhance management strategies for reef restoration.

Current protocols for coral cryopreservation and associated organisms are not fully developed, and so the associated programs that could employ these important genetic resources have not reached their full potential. In the past 9 years some of the fundamental cryobiology for coral sperm, larvae, and associated symbionts have been characterized (Hagedorn et al. 2010; Hagedorn et al. 2006a, b). Additionally, sperm have been successfully cryopreserved and used to create new coral in vitro, and have been added, along with dissociated larval cells, to frozen repositories in the US and Australia (Hagedorn et al. 2006a, b).

Eight individual *Acropora tenuis* (Fig. 7.1) and eight *Acropora millepora* (Fig. 7.2) colonies were collected from western Pelorus Island (146_29.3040 E, 18_33.0010 S) and were transported in seawater to Australian Institute of Marine Science (AIMS) within 24 h. Each colony was labeled and maintained in 1-lm filtered seawater in 1,000 L tanks for the duration of the spawning event. Each night, colonies were examined for evidence of pre-spawning behavior or "setting" (Hagedorn et al. 2006a) and then isolated into individual bins.

A. tenuis and *A. millepora* are hermaphrodites, the sperm and buoyant eggs from individual colonies were allowed to separate and were then isolated using pipettes. The coral sperm cells were successfully cryopreserved, and after thawing, samples were used to fertilize eggs, resulting in functioning larvae.

In addition, developing larvae were also dissociated, and these samples with pluripotent cells (Heyward and Negri 2012) were cryopreserved and alive after thawing.

With advances in stem cell technologies, these frozen embryonic cells are likely to respond to specific media be coaxed to form developing larvae.

The outcome of these studies was the creation of the first frozen bank for Australian coral from two important reef-building species, *A. tenuis* and *A. millepora*.

Fig. 7.1 *Acropora tenuis*

Fig. 7.2 *Acropora millepora*

Moreover, the successful results from these studies indicated that cryomethods previously devised for coral specific (Hagedorn et al. 2006a, b) are widely applicable to other coral species.

The data described in these experiments also demonstrated that using frozen sperm, creating new coral and increasing genetic diversity in small populations is now possible.

However, the possibility to collect germplasm from coral is extremely limited by the short reproductive season (3 consecutive days each year) of most coral.

Besides reproducing sexually, a second way that coral can reproduce is through asexual fragmentation. These fragments consist of a host-symbiotic partnership of coral tissue and intracellular symbiotic *Symbiodinium* sp., commonly called zooxanthellae.

If it could cryopreserve these asexual fragments the speed of conservation efforts would be increased.

Since the fragment is so complex, the studies of Mary Hagedorn et al. had to take into account both the cryosensitivities of the coral tissue and those of their endosymbionts.

These studies used small fragments of *Pocillopora damicornis* of a size that would make them suitable for long-term cryopreservation storage in cryovials, yet still large enough to outplant and form a new colony. *P. damicornis* colonies were collected from various shallow reef flats around Coconut Island in Kaneohe Bay, Hawaii, and then maintained at the Hawaii Institute for Marine Biology, University of Hawaii. All fragments were labeled and followed individually throughout the treatments.

During the post-treatment recovery identified fragments were returned to water tables for visual monitoring over a 3-week period. The number of intact and missing polyps was determined by visually examining each fragment under a microscope at the end of the 3-week period. The presence of intracellular zooxanthellae was critical to the health of the coral polyp. Therefore, zooxanthellae number and the robustness of their photosystem were monitored over time to examine relative levels of photosynthesis in photosystem I and II between the treatments and controls. If a fragment had been affected by chilling or the toxicity of a cryoprotectant, it may lose tissue, zooxanthellae or both.

To understand the fragments' sensitivity to cryoprotectants, *P. damicornis* fragments were exposed to dimethyl sulfoxide solutions at different concentrations (1.0, 1.5, and 2.0 M) for different durations (1.0, 1.5 or 2.0 h) and monitored.

Coral embryos are extremely sensitive to chilling (Hagedorn et al. 2006a, b), therefore assessing coral fragments' sensitivity to chilling injury was a very important point.

Coral fragments chilled in the winter demonstrated a robust coral tissue with little loss to chilling, whereas fragments chilled in late spring were far more sensitive. In contrast regardless of the season, the zooxanthellae were consistently chill sensitive, demonstrating significant loss of zooxanthellae numbers even after a 5 min chilling event (the fragments appeared pale after 24 h).

These seasonal sensitivities have been demonstrated in other corals as well.

Successful cryopreservation of cells, germplasm, and tissues must address intrinsic biophysical properties (e.g., water and cryoprotectant permeability, osmotic tolerance limits, and intracellular ice nucleation) to maximize survival.

A similar systematic approach is vital to improving post-thaw survival of coral and its associated organisms.

Coral fragment cryopreservation is an entirely new field of research with great potential opportunities because the process would store the valuable whole, diploid fragments of the coral. Coral fragment storage is important because a whole fragment can grow much more quickly than coral re-derived from germplasm.

Being able to cryopreserve coral fragments would rapidly move cryoconservation efforts forward avoiding the limits deriving from the present dependence yearly spawning events. Nevertheless, all types of tissues, including sexually derived germplasm and asexual fragments, should be part of a comprehensive national plan for maintaining reef diversity.

References

Hagedorn M, Carter VL, Steyn RA, Krupp D, Leong JC, Lang RP, Tiersch TR (2006a) Preliminary studies of sperm cryopreservation in the mushroom coral, Fungia scutaria. Cryobiology 52: 454–458

Hagedorn M, Pan R, Cox EF, Hollingsworth L, Krupp D, Lewis TD, Leong JC, Mazur P, Rall WF, MacFarlane DR et al (2006b) Coral larvae conservation: physiology and reproduction. Cryobiology 52:33–47

Hagedorn M, Carter VL, Leong JC, Kleinhans FW (2010) Physiology and cryosensitivity of coral endosymbiotic algae (Symbiodinium). Cryobiology 60:147–158

Hagedorn M, Carter V, Martorana K, Paresa MK, Acker J, Baums IB, Borneman E, Brittsan M, Byers M, Henley M et al (2012a) Preserving and using germplasm and dissociated embryonic cells for conserving Caribbean and Pacific coral. PLoS One 7:e33354

Hagedorn M, van Oppen MJ, Carter V, Henley M, Abrego D, Puill-Stephan E, Negri A, Heyward A, MacFarlane D, Spindler R (2012b) First frozen repository for the Great Barrier Reef coral created. Cryobiology 65:157–158

Heyward AJ, Negri AP (2012) Turbulence, cleavage, and the naked embryo: a case for coral clones. Science 335:1064

Hughes TP, Baird AH, Bellwood DR, Card M, Connolly SR, Folke C, Grosberg R, Hoegh-Guldberg O, Jackson JB, Kleypas J et al (2003) Climate change, human impacts, and the resilience of coral reefs. Science 301:929–933

Veron JE, Hoegh-Guldberg O, Lenton TM, Lough JM, Obura DO, Pearce-Kelly P, Sheppard CR, Spalding M, Stafford-Smith MG, Rogers AD (2009) The coral reef crisis: the critical importance of<350 ppm CO2. Mar Pollut Bull 58:1428–1436

Part III
Stem Cell Banking for the Future

Chapter 8
Fundamental Principles of a Stem Cell Biobank

Ida Biunno and Pasquale DeBlasio

1 Introduction

Alternatives to transplantation medical practice are of utmost importance since, dramatically successful in some cases, remains an option for a limited number of patients due to the shortage of donated organs, to the epidemic of chronic diseases and to the high costs associated with patient transplantation and follow-up (Nelson et al. 2008; Dalay and Scadden 2008). Recent scientific advances in stem cell-based regeneration provides enormous potential to the replacement of damaged tissues with "new" ones, indeed pluripotent (PSCs) stem cells are becoming promising tools for replacing, repairing, regenerating, and rejuvenating dead, degenerating or trauma injured affected tissues. PSCs essentially are of two types: embryonic (ESCs), derived from embryos, and induced pluripotent (PSCs) generated by molecular reprogramming.

2 Stem Cells Are Pillar to the "Bioeconomy" in Many Countries

The bioindustry requests for stem cell products of human PSCs has increased over the last few years, the global market was \$3.8 billion in 2011 and expected to reach \$6.6 billion by 2016, increasing at a compound annual growth rate (CAGR) of

I. Biunno (✉)
Institute for Genetic and Biomedical Research of the National Research Council (IRGB-CNR), Via Fantoli 16/15, Milan, Italy

IRCCS Multimedica, Via Fantoli 16/15, Milan, Italy
e-mail: ida.biunno@irgb.cnr.it

P. DeBlasio
Integrated Systems Engineering SRL, Via Fantoli 16/15, Milan, Italy

T.A.L. Brevini (ed.), *Stem Cells in Animal Species: From Pre-clinic to Biodiversity*, Stem Cell Biology and Regenerative Medicine, DOI 10.1007/978-3-319-03572-7_8,

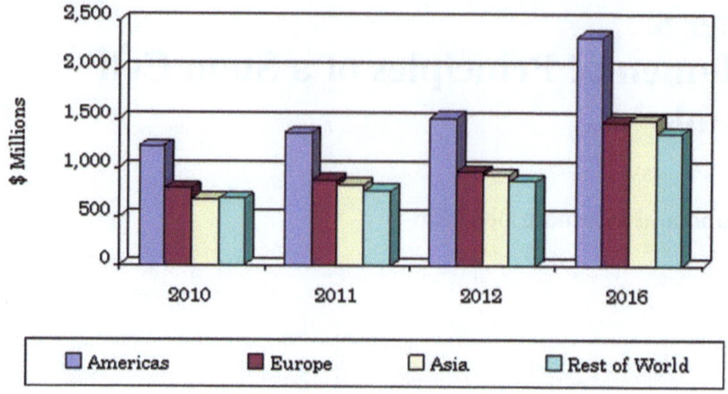

Fig. 8.1 Global stem cell market and forecasts, 2010–2016 (*Source*: BCC Research)

11.7 % from 2011 to 2016. The American market for stem cell products was $1.3 billion in 2011. This sector is expected to rise at a CAGR of 11.5 % and reach nearly $2.3 billion by 2016. The European market for stem cell products was $872 million in 2011 and is expected to reach nearly $1.5 billion by 2016, a CAGR of 10.9 % (bcc/research Market forecasting. Global Market for Stem Cells, Report code BI0035D, July 2012) (Fig. 8.1).

Neurological applications is forecasted to exert the highest supporters of this growth with a CAGR (Compound annual growth rate) of 24.4 % during 2010–2015, hematology and diabetes will closely follow the expected rate during the same time frame (Research and Markets May 11, 2011). North America and Europe are expected to have a lead position in stem cell market growth although Asian economies probably will maintain a large CAGR.

3 Stem Cells Types

Totipotent stem cells are found only in embryos. Each cell can form a complete organism (e.g., identical twins). Pluripotent stem cells exist in the undifferentiated inner cell mass of the blastocyst and can form any of the over 200 different cell types making up a living organism. Multipotent stem cells are derived from fetal tissue, cord blood, and adult stem cells. Although their ability to differentiate is more limited than pluripotent stem cells, they already have a track record of success in cell-based therapies (Bongo and Lee 2005) (Fig. 8.2).

Here is a current list of stem cell sources and basic characteristics:

1. *Embryonic stem cells*—are harvested from the inner cell mass of the blastocyst (Riazi et al. 2009) 7–10 days after fertilization. These cells have no natural environment as they exist only in vitro. When implanted, interact minimally with the surrounding tissues, form teratomas or structures reminiscent of in vitro embryoid bodies.

Stem Cells Types

Type	Description	Examples
Totipotent	Each cell can develop into new individual	Early embryos cells (1-3 days)
Pluripotent	able to form any of the 200 cell types that make the organism	Blastocyst cells
Multipotent	differentiated cells able to form several tissues	Fetal tissues, cord blood and adult stem cells

Fig. 8.2 Stem cells can be classified into three broad categories, based on their ability to differentiate

2. *Fetal stem cells*—are taken from the germline tissues that make up the gonads of aborted fetuses. These cells overall are the most plastic since they must adapt constantly to their evolving environment hence their potency must vary over time as the organs develop.
3. *Umbilical cord stem cells*—Umbilical cord blood contains stem cells similar to those found in bone marrow.
4. *Placenta derived stem cells*—up to ten times as many stem cells can be harvested from a placenta as from cord blood.
5. *Adult stem cells*—Many adult tissues contain stem cells that can be isolated. These cells are multipotent capable of persistent self-renew, live in a quiescent state and stable environment, in the body throughout life.
6. *Reprogrammed stem cells*, iPS, or retro-differentiation of tissue-restricted stem cells are obtained by the conversion of somatic donor cells (harvested from organs) into pluripotent progenitors to derive patient-specific progenitors.

Particular interest is being placed on induced Pluripotent Stem (iPS) cells by virtue of their enormous potential to model a variety of human cell types and human conditions, in particular can model: (1) in vitro diseases, particularly for those disorders for which no animal model system is available; (2) to implement drug discovery, assess its efficacy and safety; (3) in preclinical and clinical toxicity studies; (4) in vitro stratification of patients based upon the phenotype; (5) can become in vitro reference control stem cell lines particularly useful in those genetic diseases where the pathophysiology is still controversial; (6) in iPS-derived organotypic cell lines as model of organ response to drug metabolism based upon individual genotype and phenotype. The successful use of these cells in drug discovery, toxicology (Pistollato 2012) studies, and product safety assessments will also lead to a reduction in the use of animals for laboratory research and commercial testing (Baker 2007; Takahashi and Yamanaka 2006; Nobel Foundation Retrieve 2012).

4 Stem Cells in Translational Medicine

The clinical application of stem cells depends on the fundamental principle that the final cell phenotype rely on a combination of two factors: the starting cell population and the environment in which the cells are eventually implanted (Hipp and Atala 2008). This combination is fundamental and can be obtained by implanting: (i) raw tissue-restricted stem cells, (ii) progenitors capable to differentiate only after transplantation or (iii) pre-differentiating stem cells. The choice of the strategy depends on the tissue "niche" where the cells will ultimately be placed and essentially whether the "niche" has the appropriate differentiation program. Traumatic or ischemic disorders probably would benefit from implantation of raw tissue restricted progenitor cells, on the other hand, degenerative type of disorders would probably benefit from in vitro differentiated cells (Simara et al. 2013; Baker 2007; Takahashi et al. 2007).

Stem cell research and clinical translation constitute fundamental and indivisible modules catalyzed through the biobanking activity and eventually will generate a return of investment since banking through long-term storage represents a fundamental source to preserve the cells' original features (Bardelli 2010).

5 Stem Cell Specialized "Repository"

Clinics around the world offer stem-cell-based therapies to desperate patients for an array of intractable medical conditions and, in most cases, the benefits are overpriced and the risks are ignored (Gorski 2013). While the derivation and use of ES and iPS cells are increasing rapidly for biomedical research purposes, information on the actual number, provenience, characteristics, and accessability of these cells remain "*SCARSE*". The use of poorly defined stem cells hampers: disease modelling, toxicology studies, drug discoveries, and most of all affects research results credibility. To decrease the use of poorly characterized cells in "stem-cell-based therapies" it is auspicable to encourage the establishment of STEM CELL BANK(S) with a CENTRALIZED DATABASE and discourage "collegue-supply-routes."

There are a number of key elements required to support the development of high quality PSC-based research and human applications of stem cells of which the most fundamental are (Day and Stacey 2008; Diaferia et al. 2012; Inamdar et al. 2012):

- The establishment of pristine early passage stocks of cell lines to enable accurate and independent verification of research data and to sustain these tools for use in future research developments.
- Availability of well characterized and quality controlled sources of PSC lines established through implementation of up-to-date scientific norms, to ensure that research efforts are not wasted on cells that are unauthenticated, contaminated, or inadequately reprogrammed.
- Direct linkage of PSC stocks with a well documented and accessible database of donor and PSC characteristics.

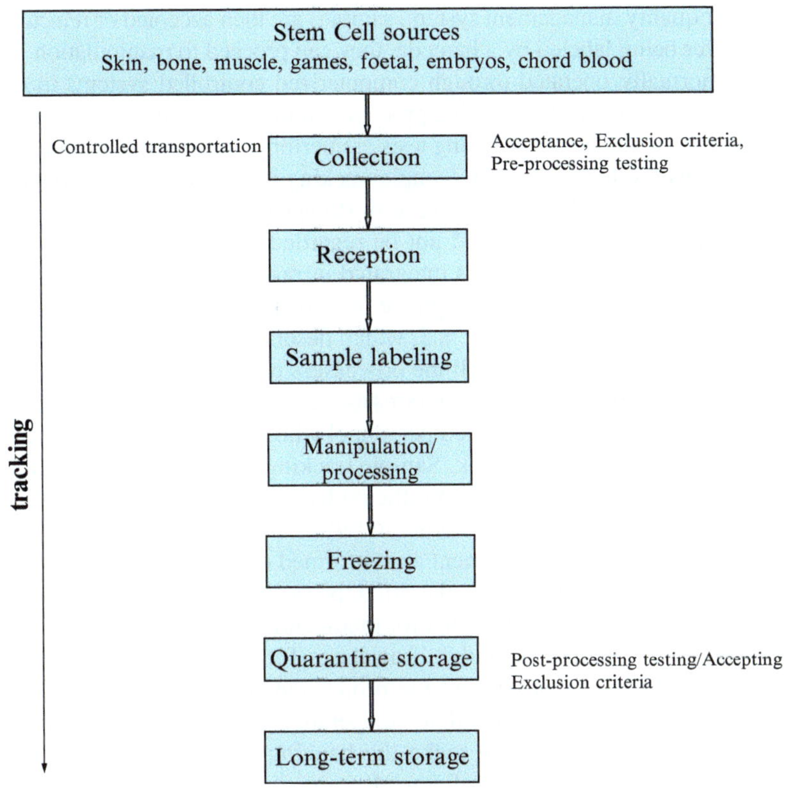

Fig. 8.3 Database and tracking system

- Assurance that cells intended for human application have appropriate traceability (Knoppers and Isasi 2010), and processing/storage history to meet regulatory demands and that they can be made available at the right time and in the right condition for therapeutic use.
- Provision of specific cell types, that can be produced at a sufficient scale with minimized batch-to-batch variability.

In general, to be operative the stem bank must implement a quality management system to make the banking facility function properly in every step of its activity; the standardization of procedures and protocols for sample collection (Murdock 2012), transport, processing, testing and storage, and release (Fig. 8.3). Standardization will guarantee that all the samples will share the same conditions and same quality controls. There must be sample traceability, meaning proper sample coding and identification; computer management and paper copies of all applications and processes to which samples and associated data have been submitted; employment of qualified personnel who will exclusively be dedicated to the different operations of the biobank.

Stem cell samples, collected upon informed consent should be sent to the biobank in a special transportation kit in order to preserve their viability and quality. According to the proper quality management system, samples are then accepted or rejected and if accepted, after being labeled by a barcode, they can proceed to manipulation. Sample freezing is normally operated through computerized controlled systems in order to retain a traceable graph of the freezing process. Samples are then stored in quarantine containers until all the post-processing tests are performed and samples undergo thorough acceptance/exclusion criteria. Sample tracking is maintained all throughout the process, starting from collection to long-term storage (Healy et al. 2011).

Stem-Cell-specific banks should not be regarded as separate academic entities with limited activities but rather, as integrated infrastructural modules with experienced workforce. An essential prerequisite for biobanking is a robust sample tracking system, as described in the Fig. 8.3, which describes how to handle the sample, its codification or identification. All the processess performed on the samples must be managed by computer assisted programs as well as archives of paper copies. This requires the employment of qualified personnel who is exclusively dedicated to the overall management of the biobank. Sample tracking must be maintained throughout the entire processess, starting from the collection to the long-term storage and distribution and must be efficient in order to follow the many phases of the biobanking process starting from: recruitment and informed consent of cell donation; sample collection and transportation; registration, processing, freezing, and testing; and the potential release of the stored units for therapeutical purposes. This necessitates the application of standard operative procedures for sample collection, transport, identification, processing, storage at the proper temperature, and informatic data collection and processing. In general, to be operative the stem bank must implement a quality management system to make the banking facility function properly in every step of its activity by standardizing all the procedures including lines release.

6 Stem Cells Cryopreservation

Whilst effective methods exist for the cryopreservation of a wide variety of lines, many cell types (including few stem cells) still prove to be refractory to freezing, or yield unacceptably low rates of recovery. Effective cryopreservation requires an understanding of the basic physical principles that underlie in the freeeze/thaw process and the cells' response to the imposition of such a process (Pegg 2007; Maryman 2007). In order to guarantee that all the samples share the same conditions and the same quality controls it is auspicable to employ qualified personnel who will exclusively be dedicated to the different biobanking operations.

7 Freezing and Thawing Process of Human Neural Stem Cells

We determined the best procedure, equipment, and velocity of NS cell freezing and thawing. Freezing media, conditions, and equipment for human glioblastoma NS cells (GliNS2, G179, G144, G166) and human fetal NS cell lines (CB660, CB660SP,

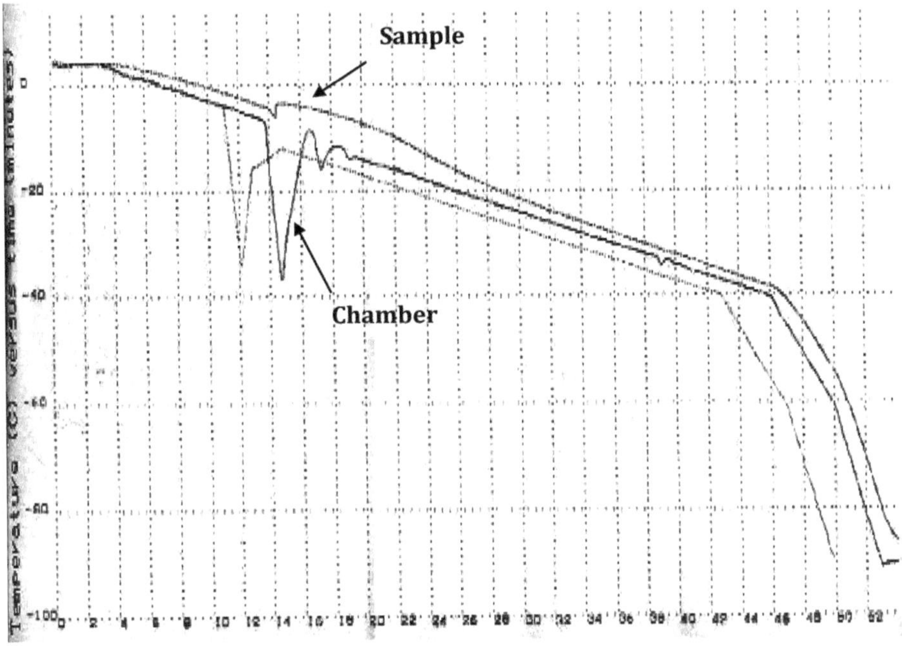

Fig. 8.4 Cooling rate profile recorded with a probe in the chamber along with the sample

CB541) were optimized. Using a rate freezer (Planer 560-16) with a programmable cooling rate, equipped with probes to measure and control the cooling chamber and the sample temperature, we were able to obtain a cooling rate of ~1 °C/min following these steps tested to control ice crystal formation: step 1, hold 3 min at 4.0 °C; step 2, −1.0 °C/min to −4.0 °C; step 3, −28.0 °C/min to −35.0 °C; step 4, +25.0 °C/min to −16.0 °C; step 5, +2.0 °C/min to −12.0 °C; step 6, −1.0 °C/min to −40.0 °C; step 7, −5.0 °C/min to −60.0 °C; step 8, −10.0°/min to −90.0 °C (Fig. 8.4).

To assure a better temperature exchange and minimum chance of contamination, cells were dispensed and sealed in glass ampoules and then stored in liquid nitrogen tanks. As cryopreservation media was used a protein-free, GMP manufacturing, freezing solution formulated with 10 % DMSO (CryoStor™, BioLife Solution) that proved to better control ionic and hydraulic balance in cells during hypothermia. After rapid thawing in water bath (for 1–2 min) cells were put back in culture and assessed for viability and apoptosis by flow cytometry. The use of a rate freezer with a programmable cooling rate and an intracellular-like freezing solution improved NS cell survival and reduced apoptosis at 24 h after thawing when compared with the manual step-down procedure using the isopropanol chamber and culture medium supplemented with 10 % DMSO as freezing medium (Fig. 8.5). Several studies showed that the freezing procedure displays its detrimental effects mainly at 24 h after thawing, due to failure to recover from the freezing damages, so we focused on this time point to evaluate survival and apoptosis.

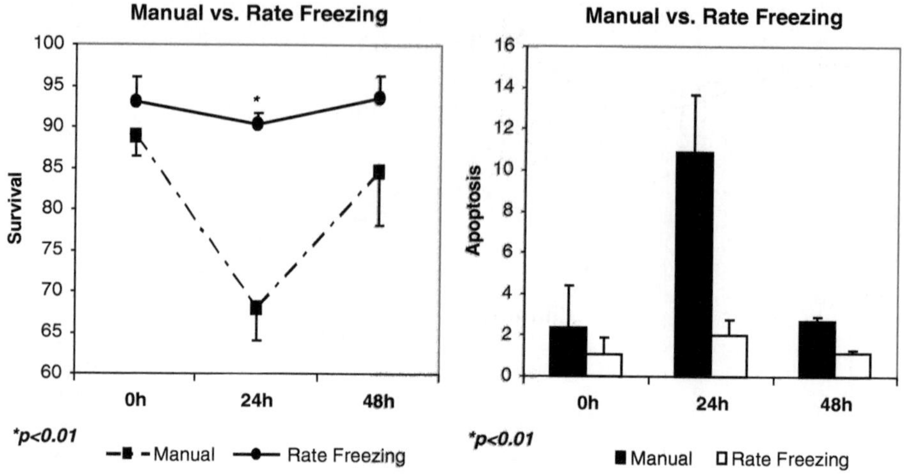

Fig. 8.5 Quantification of the survival (*left panel*) and apoptosis (*right panel*) rate of mouse neural stem cells embryo-derived (NS46C) comparing standard manual freezing procedure vs. rate freezing in CryoStor™

The same parameters used to freeze mouse neural stem cells were successfully used to cryopreserve other human cell types (human fetal neural stem cells and glioma neural stem cells) obtaining an optimal cooling rate profile of ~1 °C/min (Fig. 8.6).

Local import regulations with the recipient (helpful information is available from the world federation of culture collection, www.wfcc.org) and include a "Material Transfer Agreement" (MTA) that needs to be signed by the recipient scientist before distribution. In this document all the terms and conditions for the use of the biomaterial are explained to assure intellectual rights to the scientists that generated the cell lines. Furthermore a custom declaration form and a material datasheet are included in the shipment to assure proper clearance and handling of the material upon delivery.

8 Stem Cell Registry

Along the development of a high quality management system, a database is essential to caputre scientific and other relevant data that are directly linked to the physical stocks of cells. This will be achieved by implementing a registry of all the cell lines including hESC and a comprehensive and fully relational database providing access to details of the cell lines and other information held in the biobanking network facilitating sourcing and shipments directly to the researchers under standardized and non contentious transfer agreements. The database should contain informations regarding the origin of the cell donor (anonymous identity),

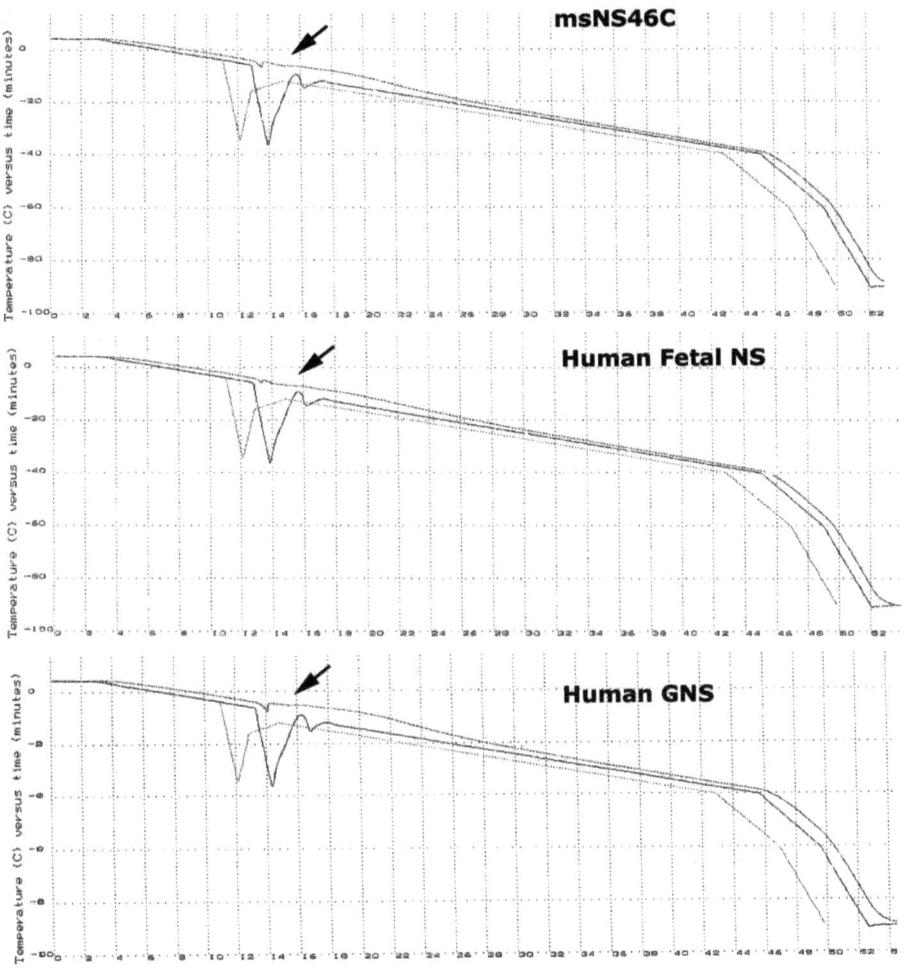

Fig. 8.6 Cooling rate profile recorded with a probe in the chamber along with the sample

phenotype (healthy controls as reference line), diseases, susceptibility to certain toxicities, HLA diversity, drug response, age, sex, gender, genotype, original tissue source, epigenetic signature, pluripotency status, gene expression profiles, epigenetic changes, and stability upon passages. The database allows the downloading of individual datasets, gives support to queries across multiple experiments and links together different types of data obtained from other ongoing projects. Added value to the database will be the addition of protocols in SOPs format, primers sequences, and experimental conditions.

(*Stem Cell Information: NIH guidelines (Young et al. 2010) July 7, 2009*)

9 Quality Control (QC) Stem Cell Pipeline

Whilst the quality and data management resources are essential to the work of the biobank, the delivery of sustainable supplies of lines that are scientifically and ethically sourced (based on solid scientific methodologies) is both a practical and legal prerequisite for the continued progression of stem cell research towards clinical therapies. Stem cell banks are indeed invaluable resources and will provide access to high quality controlled and characterized lines that meet the needs of current research and safety standards. The QC stem cell workflow pipeline of Fig. 8.7 describes in the workflow which was designed following *ISCBI* (*International Stem*

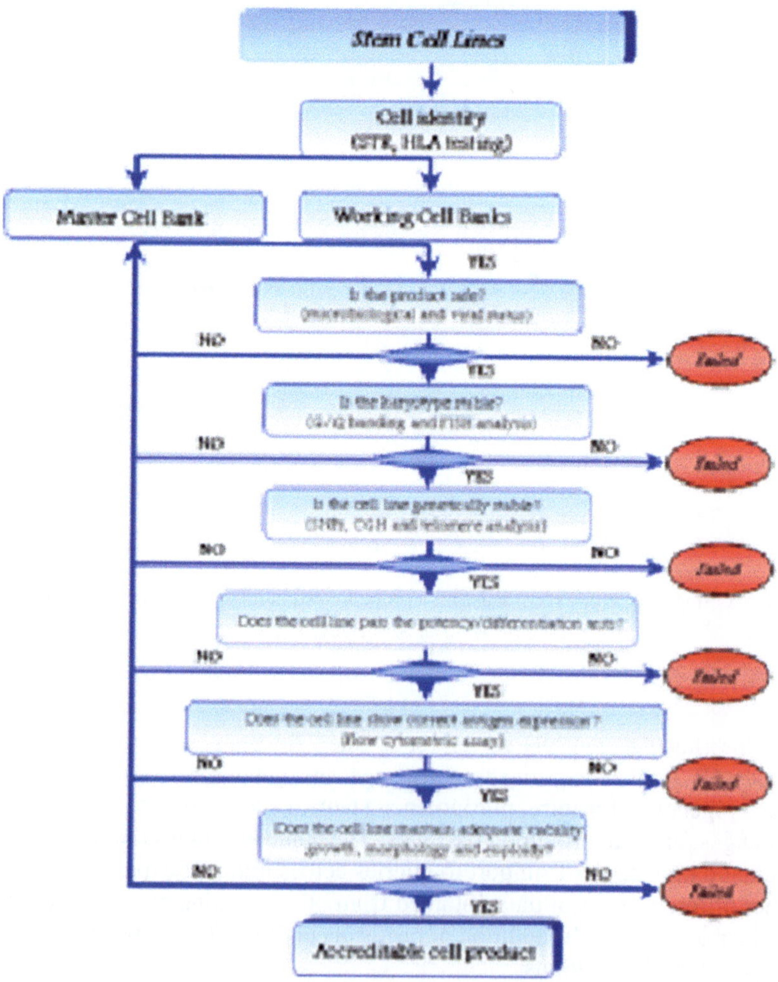

Fig. 8.7 Pluripotent stem cells bank workflow designed to reduce the use of uncharacterized lines and distribution using "collegues supply route"

Cell Banking Initiative) (Adewumi et al. 2007; Crook et al. 2010) best practice document applicable only for research-grade stem cells supply.

The aim of the workflow is to reduce the number of stem cell lines that are being produced, reduce the use of unproven stem cells, and substitute the uncharacterized lines for those that are well characterized. The successful use of stem cells and their derivatives in drug discovery, toxicology, and safety assessments requires (1) their production at a sufficient amount, (2) reduce batch-to-batch variability, (3) use of appropriate in vitro QC assays, and (4) animal testing drastic reduction. The QC pipeline consists in performing sequentially and systematically a number of activities:

1. Identify the iPS and derived cell line products following the nomenclature committee specifications. This step is fundamental to ensure that research efforts are not wasted on cells that are unauthenticated, contaminated, or inadequately reprogrammed. Cell identity can be obtained by various means, *NSCB* (National Stem Cell Bank) profiles the NIH registry of hESC lines using PowerPlexSTR testing (16 STR alleles) using the WiCellPopwerPlex-16. STR analysis must be performed on DNA extracted upon cell arrival to the bank (master bank) and the working bank line. The same tests need to be re-performed after cell lines cryostorage and reseeding to confirm cell identity. The following STR are usually evaluated: Fluorescent D8S1179; D21S11; D7S820; CSF1PO; D3S1358; TH01(chr.11p15.5); D13S317; D16S539; D2S1338; D19S433; vWA(chr.12p13.31); TPOX(chr.2p25.3); D18S51; Amel (sex-typing); D5S818, and FGA (chr. 4q28). HLA typing is another form of cell authentication.
2. Establish pristine early passage stocks of cell lines defined as a master cell bank to enable accurate and independent verification of the working cell bank
3. Establish and standardise contamination quality control biological assays including second line (cross-contamination) and microorganism contamination. Two lines of quality controls can be identified: (i) first line, contamination by pathogens and cross-cell contamination and (ii) chromosomal stability.

   (i) First line QC: (a) cross-cell contamination and (b) pathogens detection

      (a) Cross-cell contamination may be explained by the presence of an additional cell population within a cell line that makes it "impure", this may be caused by the overgrowth of a second (human or animal) cell line during the initial establishment or maintenance, by mislabeling or poor tissue culture practices.
      (b) Among the microorganisms (bacteria, fungi, viruses) contaminations particular attention must be given to the presence of mycoplasma. These pathogens are the smallest self-replicating organisms (0.3–0.8 μm), difficult to detect and their contamination is frequent in laboratories. It has been reported that mycoplasm contamination occurs in 1 % of primary and 5 % of early passage cultures, increasing to 15–35 % in continuous long-term culturing cell lines. The effects of the contamination on the cells vary and depend on the type of mycoplasm specie present. Several species which are known to occur in 98 % of the laboratory infections of cell cultures: *Mycoplasma hyorhinis*, *Mycoplasma arginini*, *Mycoplasma*

Table 8.1 lists three mycoplasma detection assays and detection limits generally used in cell biology labs

MycoAlert®	PCR	Enrichment + DAPI staining
Most species	All TC related	Many species
10^4–10^5 cfu/mL	10^2–10^3 cfu/mL	1–10 cfu/mL

orale, *Mycoplasma fermentans*, *Mycoplasma salivarium*, *Mycoplasma hominis*, and *Acholeplasma laidawii*. The effects of mycoplasma infection are more insidious than those produced by other bacterias, fungi, and yeast. They may remain undetected by microscopic observation but can cause a reduction of growth rate, morphological changes, chromosomal aberrations, induction or suppression of cytokine expression, changes in membrane composition and alterations in the amino acid, and nucleic acid metabolism. A variety of tests for mycoplasma detection are available, and it is usually recommended to use at least two techniques for testing cell banks to ensure optimum sensitivity and specificity. The kits usually used in biobanks are: Detection of Mycoplasma using indirect DNA staining (Hoechst 33258); Mycoplasma detection using PCR and Detection of mycoplasma using the GenProbe mycoplasma tissue culture non-isotopic detection system (Cobo et al. 2005) (Table 8.1).

Contamination from adventitious agents
As part of the quality control process it is necessary to test for Adventitious Agents. Following the blood transfusion guidelines and the geographical areas of the incoming cells or tissues, blood-borne viruses such as Hepatitis B Virus (HBV), Human Immunodeficiency Virus (HIV), and others such as Hepatitis C virus (HCV) easily transmissible must be assayed. Viral contaminations may be due to the operator, cell or tissue culture reagents of animal or human origin and viral latency. Assays for a defined set of viruses must include: HCV, HBV, EBV, HCMV, HSV1-2, HPV, and HIV using the sensitive NAT technology with a detection efficiency ranging from 1 × 10^3 to 30 × 10^3 copies/mL.

(ii) Genetic/genomic stability is a fundamental process in stem cell monitoring Fig. 8.8 (Diaferia et al. 2011). Characterize the genetic stability of the cell lines using G/Q-banded chromosome spreads able to detect deletions of greater than 10 Megabases in addition to comparative genomic hybridization (CGH) by microarray to look for micro chromosomal aberrations. This allows for an immediate evaluation of the population dynamic with regard to chromosomal stability, other "omics" technologies can be used to monitor the consistency of the results. To further validate the genomic stability of these cells, telomere length and telomerase activity can be analyzed. Telomere maintenance appears to be essential for chromosomal integrity and the prolonged persistence of stem cell function. hTERT (human telomerase reverse transcriptase) mRNA expression, relative telomerase activity by

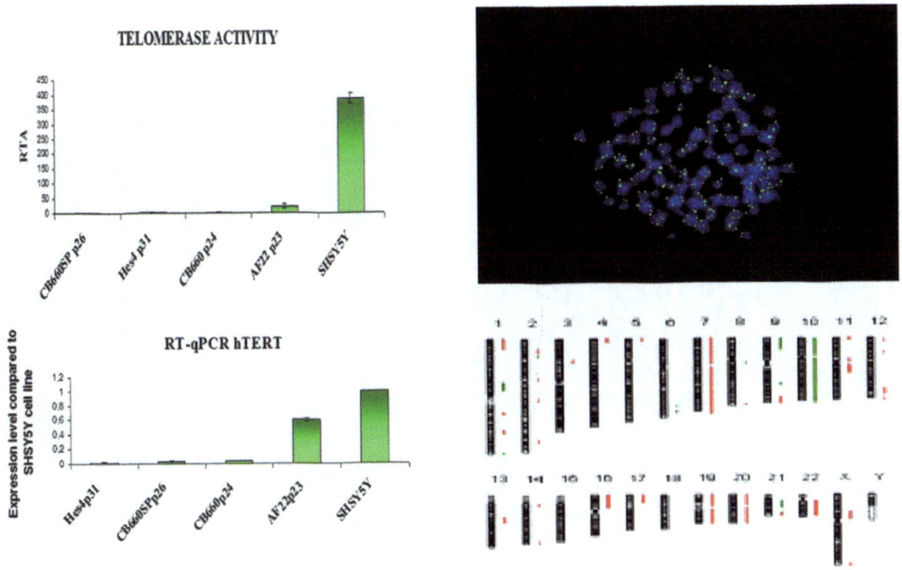

Fig. 8.8 Genetic stability

TRAP assay, average telomere length by fluorescent in situ hybridization (FISH), and telomere restriction fragment (TRF) length analysis and evaluation of γ-H2ax focus formation as a sign of DNA damage response (DDR) can all be determined as a means to follow the safety of the cell lines.

9.1 Epigenetic/Epigenomics Analysis

A process which can be performed in stem cell biobanks but not necessarily be part of the QC pipeline is to assess the epigenetic/epigenomic stability of the cells during extended passages. Emerging studies uncovered genetic and epigenetic alterations both in hES and hiPS whose origin can be attributed to culture adaptation, in vitro differentiation, and in reprogramming of iPS to preexisting parental somatic cell modifications. The genomic and epigenomic differences often involve cancer related genes and onco-microRNAs with negative effects on the function, stability, differentiation potential, and safety of the cells (Baronchelli et al. 2003).

Stem Cell lines pluripotency and differentiation characterization is to be assayed before cryopreservation and after thawing. Pluripotency markers (referring to those markers that indicate the ability to differentiate the cells into the three germ layers) following a combination of qRT-PCR assay (looking at the expression of at least 12 genes) and protein level using flow cytometry and immune-cytochemistry (Figs. 8.9 and 8.10).

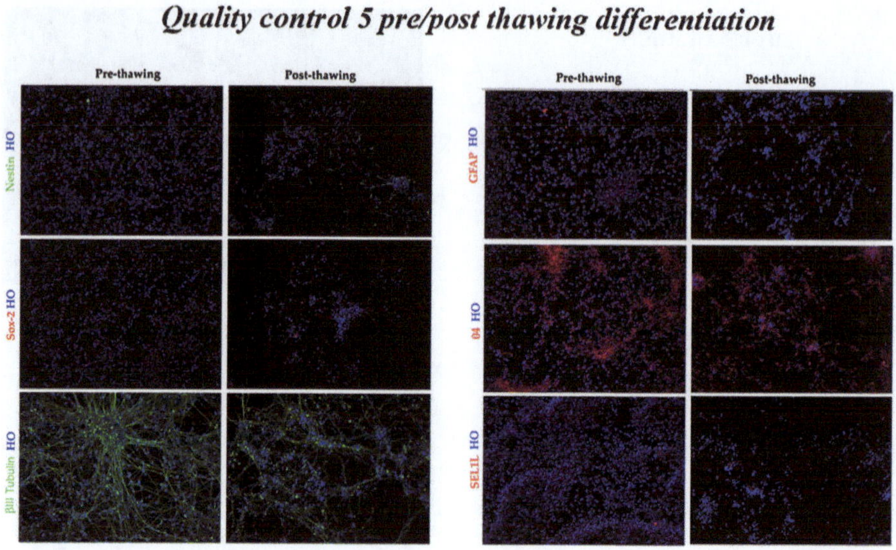

Fig. 8.9 Immunofluorescence analysis of neuronal derived human stem cells showing pluripotency and positivity to differentiation markers. The same cells were also assayed by q-PCR (not shown)

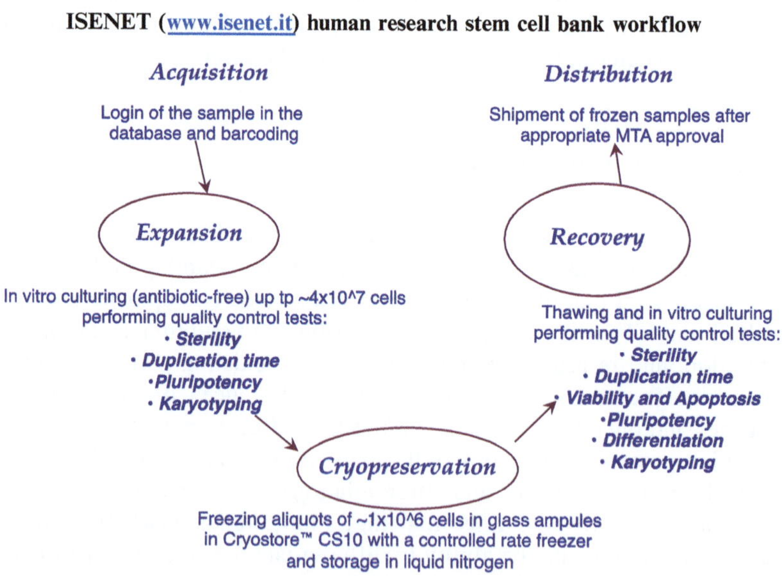

Fig. 8.10 ISENET cultures the stem cell lines in antibiotic free media and controls the sterility and stability of all the stem cells for research purposes before and after cryopreservation. Cells are cryopreserved in 10 % DMSO and/or in Cryostore CS10 (BioLife solution), a GMP cryoreagent containing 10 % DMSO but free of animal proteins

10 Important Elements to Include in the Material Transfer Agreement (MTA)

The MTA needs to be signed by the recipient scientist before product distribution. The RECIPIENT SCIENTIST must agree that the MATERIAL (Isasi et al. 2012):

- Is to be used solely for teaching and academic *research purposes* and for the Project indicated in the implementing letter annexed to the Agreement
- Will not be used in human subjects, in clinical trials, or for diagnostic purposes involving human subjects
- Is to be used only at the RECIPIENT organization and only in the RECIPIENT SCIENTIST's laboratory under the direction of the RECIPIENT SCIENTIST or others working under his/her direct supervision
- Will not be transferred to anyone else within the RECIPIENT organization without the prior written consent of the PROVIDER

Acknowledgment Ministero della salute# RF-MUL-2008-1248034.

References

Adewumi O et al (2007) Characterization of human embryonic stem cell lines by the International Stem Cell Initiative. Nat Biotechnol 25(7):803–816
Baker M (2007) Adult Cells reprogrammed to pluripotency, without tumors. Nat Rep Stem Cells 6 December 2007 doi:10.1038/stemcells.2007.124
Bardelli S (2010) Stem cell biobanks. J Cardiovasc Transl Res 3(2):128–134
Baronchelli S, Bentivegna A, Redaelli S, Riva G, Butta V, Paoletta L, Isimbaldi G, Miozzo M, Tabano S, Daga A, Marubbi D, Cattaneo M, Biunno I, Dalprà L (2003) Delineating the cytogenomic and epigenomic landscapes of glioma stem cell lines. PLoS ONE 8(2):e57462. doi:10.1371/journal.pone.0057462
Bongo A, Lee EH (2005) Stem cells—from bench to bedside. World Scientific Publishing Co Pte. Ltd, Singapore
Cobo F, Stacey GN, Hunt C, Cabrera C, Nieto A, Montes R, Cortés JL, Catalina P, Barnie A, Concha A (2005) Microbiological control in stem cell banks: approaches to standardization. Appl Microbiol Biotechnol 68(4):456–466
Crook JM et al (2010) Consensus guidance for banking and supply of human embryonic stem cell lines for research purposes. In Vitro Cell Dev Biol Anim 46(3–4):169–172
Dalay GQ, Scadden DJ (2008) Prospects for stem cell based therapy. Cell 132(4):544–548
Day JG, Stacey GN (2008) Biobanking. Mol Biotechnol 40(2):202–213
Diaferia GR, Conti L, Redaelli S, Cattaneo M, Mutti C, DeBlasio P, Dalprà L, Cattaneo E, Biunno I (2011) Systematic chromosomal analysis of cultured mouse neural stem cell lines. Stem Cells Dev 20(8):1411–1423
Diaferia G et al (2012) Stem cell biobanking: investing in the future. J Cell Physiol 227(1):14–19
Gorski D (2013) Science-based medicine: exploring issues and controversies in the relationship between science and medicine
Healy L et al (2011) Banking stem cells for research and clinical applications. Methods Mol Biol 767:15–27
Hipp S, Atala A (2008) Stem cells for regenerative medicine. Stem Cell Rev 4(1):3–11

Inamdar MS et al (2012) Global solutions to the challenges of setting up and managing a stem cell laboratory. Stem Cell Rev 8(3):830–843
International Stem Cell Banking Initiative (2009) Stem cell banks: preserving cell lines, maintaining genetic integrity, and advancing research. Stem Cell Rev 5(4):301–314
International Stem Cell Initiative (2011) Screening ethnically diverse human embryonic stem cells identifies a chromosome 20 minimal amplicon conferring growth advantage. Nat Biotechnol 29(12):1132–1144
Isasi R et al (2012) Disclosure and management of research findings in stem cell research and banking: policy statement. Regen Med 7(3):439–448
Knoppers BM, Isasi R (2010) Stem cell banking: between traceability and identifiability. Genome Med 2(10):73
Maryman HT (2007) Cryopreservation of living cells: principles and practice. Transfusion 47(5): 935–945
Murdock A et al (2012) The procurement of cells for the derivation of human embryonic stem cell lines for therapeutic use: recommendations for good practice. Stem Cell Rev 8(1):91–99
Nelson TJ, Behfar A, Terzic A (2008) Strategies for therapeutic repair: the "R(3)" regenerative medicine paradigm. Clin Transl Sci 1(2):168–171
Pegg DE (2007) Principles of cryopreservation. Methods Mol Biol 368:39–57
Pistollato F (2012) Standardization of pluripotent stem cell cultures for toxicity testing. Expert Opin Drug Metab Toxicol 8(2):239–257
Riazi AM et al (2009) Stem cell sources for regenerative medicine. Methods Mol Biol 482:55–90
Simara P, Motl JA, Kaufman DS (2013) Pluripotent stem cells and gene therapy. Transl Res 161(4):284–293
Stacey GN (2011) Detection of mycoplasma in cell cultures. Methods Mol Biol 731:79–91
Takahashi K, Tanae K, Oriuki M, Narita M, Ichisaka T, Tomoda N, Yamanaka S (2007) Cell 131(5):861–872
Takahashi K, Yamanaka S (2006) Induction of pluripotent stem cells from mouse embryonic and adult fibroblasts cell lines by defined factors. Cell 126(4):663–676
Young L et al (2010) The International Stem Cell Banking Initiative (ISCBI): raising standards to bank on. Nat Protoc 5(5):929–934

Chapter 9
Freezing and Freeze-Drying: The Future Perspective of Organ and Cell Preservation

Sara Maffei, Tiziana A.L. Brevini, and Fulvio Gandolfi

1 Cryobiology

1.1 Historical Background

Cryopreservation originates from the Greek word "kryos," which means "cold or frost," it indicates storage of cells or tissue, usually in liquid nitrogen, at temperatures below −130 °C. The lowest natural temperature on earth is −80 °C. Under normal pressure, the inert gas nitrogen, which is commonly used in cryopreservation, becomes a liquid at −196 °C.

A small number of species like various fish, frogs, and insects can survive at low temperature using two different biological principles or a combination of these. The first process is cell dehydration, which reduces the freezing point secondary to a concentration of solutes. The other mechanism is by endogenous production of special antifreeze molecules, like sugars, which prevent formation of large ice crystal. However, recent studies provide the first demonstration that freeze tolerance can also occur amongst vertebrate species. A freeze tolerance vertebrate would make an excellent model system for studies of the medical cryopreservation of tissues and organs.

The main goal of the cryopreservation procedure is to minimize tissue injury from low, subzero temperatures (Shaw and Jones 2003). The storage at a low temperature can continue for decades and the only theoretical limitation of storage time is influence by cosmic radiation, which over several thousand years would degrade the genome of the cryopreserved material. This factor can be neglected in any practical work of cryopreservation in our society. In other words, by usage of cryopreservation the biological clock can be halted for an unlimited time (Kuwayama 2007).

S. Maffei (✉) • T.A.L. Brevini • F. Gandolfi
Laboratory of Biomedical Embryology, UniStem, Centre for Stem Cell Research, Università degli Studi di Milano, Via Celoria 10, 20133 Milan, Italy
e-mail: sara.maffei@unimi.it

T.A.L. Brevini (ed.), *Stem Cells in Animal Species: From Pre-clinic to Biodiversity*, Stem Cell Biology and Regenerative Medicine, DOI 10.1007/978-3-319-03572-7_9,

The term cryobiology refers to the knowledge and understanding of the effects of low temperature on cellular system and the utilization of this information to develop improved cryopreservation protocols. The science of cryobiology can be considered to have its starting point about 70 years ago (Luyet and Gehenio 1940). At that time, Luyet tried to achieve cryopreservation by cooling epidermal plant cells quickly and published a monograph about his pioneer work in 1940 (Luyet and Gehenio 1940). Three years later a scientist from England succeeded in cryopreservation of human and fowl spermatozoa, using glycerol as the agent that would protect against freezing injury (Polge et al. 1949).

The studies of cells and organ freezing have increased considerably since Prof. Polge's early attempts and it may well be that the technique of cryopreservation will be routinely used in the future to store cells and organs.

1.2 Fundamental Cryobiology

Therefore cells are mostly constituted of water, which makes up 60–85 % of the cell volume (Mazur 2004). In addition to free water, biological systems also contain "bound" water molecules. The "bound" water molecules that are intimately hydrogen-bonded to the atoms within molecules are such as proteins, RNA, DNA, or membrane phospholipids head groups. These water molecules form a substantial coat around the biological molecule and are essential to maintain cells structure and function (Shaw and Jones 2003). The "bound" water molecules are incapable of freezing (Sun 1999). Thus, water plays an important role in cryobiology since at cooling to low subzero temperature, 90 % of water in a cell will convert into ice (Mazur 2004) and the 10 % of bound water will not freeze.

Cell is in osmotic equilibrium, which means that concentrations of any solution inside and outside the cell are the same, since the cell membrane is semipermeable. In other words, water moves in or out of the cell depending on changes in solution concentration outside the cell and permeability of the cell membrane, which vary for each specific cell type. In addition to passive diffusion through the membrane lipid bilayer (Solomon 1968), water moves through water transport pores known as aquaporins (Verkman et al. 1996). These active transport proteins can transport water up to 100-fold more efficient than passive diffusion. Also, permeability is temperature dependant (Elmoazzen et al. 2002), so that permeability is higher with increasing temperature. Osmotic stress is defined as shrinkage and swelling of a cell due to osmotic differences between the inside and outside of the cell. When these changes are large it may lead to major damage of the cell and in the worst scenario to cell death.

In pure water, the highest temperature at which ice can form at normal pressure is 0 °C, but ice nuclei form only very reluctantly at this temperature, with the result that nuclei do not usually form until the temperature falls below 0 °C. In fact, ice in general forms at temperatures between −5 and −15 °C; through spontaneous or induced (by seeding) ice nucleation. Once an ice nucleus has formed, further water molecules can very rapidly bond onto this frozen surface, allowing it to grow in size.

When ice forms there is a release of heat (the latent heat of fusion or crystallization), as the change from liquid to crystal releases energy. The amount of energy released may be considerable and commonly brings the temperature of the whole sample back up to 0 °C. With pure water the temperature will remain at 0 °C until the freezable water has formed ice, after which it will re-equilibrate with the ambient temperature. Following the crystallization of pure water, very little remains unfrozen even at temperatures as high as ±5 °C (>80 % ice) or ±10 °C (>90 % ice) state (Shaw and Jones 2003).

To thaw all this ice the temperature has to rise above the freezing point. Pure water therefore melts at 0 °C, the same as the highest temperature at which ice crystals can first form. During the thawing process, the change from crystal to liquid requires energy input. The solution which remains free on ice in a temperature below freezing point is in a supercooled state (Shaw and Jones 2003).

It is well established that cell survival rate during cryopreservation depends mostly on the cooling rate (Mazur 1970). In extensive studies of survival of yeast cells and human red blood cells, as a function of cooling rate, it was showed that a curve of survival versus cooling rate exhibited the shape of an inverted U. This curve can be interpreted as the resultant of two different mechanisms of which one damages living cells at high cooling rates and the other damages at low cooling rates. If the rate is too fast or too slow, cells do not survive and balancing these two cryoinjuries is the key to success (Mazur 1970).

1.3 Cryoprotectant Solutions

Adding salts and other solutes to pure water lowers the temperature at which ice forms/melts. There is no limit to how low the freezing/melting point can be depressed by solutes, with the result that solutions with very high concentrations may never form ice, rather they will solidify into an ice-free (vitreous) state. Even when salts and solutes are present, they rarely become incorporated into the ice crystals themselves, because of the very specifc shape and bonding requirements of the growing ice crystals. Molecules other than water molecules are instead excluded from the growing ice crystals, and tend to accumulate between crystals or are physically pushed ahead of an advancing ice front. The formation and growth of ice crystals can, however, be modified by compounds such as antifreeze proteins (O'Neil et al. 1998), ice blocking agents (Wowk et al. 2000), and cryoprotectants.

Many chemicals have been recognized as having a cryoprotective function and these are known as "cryoprotectant solutions" (CPAs). During the freezing process, CPAs can serve several functions including lowering the freezing point, binding water to prevent it freezing at zero degree and decreasing membrane damages (Farrant 1980). According to their ability to transport through cell membrane, CPAs can be divided into two categories, permeable and non-permeable. Permeable CPAs can diffuse through cell membrane and non-permeable CPAs do not enter the cytoplasm.

Permeable CPAs are small, nonionic molecules with low toxicity and high solubility in water. The rate of CPA permeation and dilution is determined by the species, cell type, and stage of development, solution composition, temperature, and hydrostatic pressure (Liebermann et al. 2002; Shaw and Jones 2003). The most commonly used permeable CPAs are DMSO, propylene glycol (PROH), and ethylene glycol (EG). The exact mechanism by which these permeable CPAs protect living cells from cryoinjury is not completely understood. However, their general mechanisms seem to be by lowering the freezing point by replacement of some of the bound water molecules in and around proteins, deoxyribonucleic acid and head groups of phospholipids (Shaw and Jones 2003). Moreover, the permeable CPAs stabilize cellular proteins in the cytoplasm as well as in the cell membrane (Karlsson and Toner 1996). At entrance into the cell, they also reduce the concentration of electrolytes by lowering the amount of ice formed at a given temperature (Pegg 1984).

Non-permeable CPAs are usually long-chain polymers that are soluble in water and increase the osmolality of the solution. The most frequently used non-permeable CPAs are disaccharides (sucrose, glucose, fructose, sorbitol, saccharose, trehalose), some macromolecules (polyvinilpyrrolidone, polyvinyl alcohol, Ficoll), and proteins (bovine serum albumin; BSA). They contribute to cell dehydration, counteract osmotic stress and reduce the toxicity of permeable CPAs (Muldrew 2004).

1.4 Cryopreservation Techniques

There are two principally different cryopreservation procedures: vitrification and slow freezing. Recently, a new directional freezing technology was introduced (Arav and Natan 2009; Revel et al. 2004) which provides identical cooling rates through the specimen being frozen. All three procedures include four common steps: exposure of the samples to CPA, freezing/cooling to the storage temperature (−196 °C), thawing/warming, and CPA removal. The terms "freezing and thawing" relate to the slow and directional freezing procedures whereas "cooling and warming" are more correct in relation to the vitrification procedure.

1.4.1 Vitrification

The term vitrification originated from the Greek word "vitri," which means "glass." Vitrification is a procedure in which solution/specimen solidifies to form a glasslike, or vitreous, state without any ice crystal formation during cooling and remains in this state throughout the warming step (Shaw and Jones 2003).

The main idea of ultrarapid cooling is to pass rapidly through the critical temperature zone where the cells are most sensitive for chilling injury (Liebermann et al. 2002). To achieve this glasslike solidification, high cooling rates in combination with high concentrations of CPA that interact strongly with water preventing water molecules from interaction to form ice, are used. In general, the rate of cooling/warming and the concentration of the cryoprotectant required to achieve vitrification are inversely related. Typically, the temperature is reduced directly from 0 °C to −130 °C by plunging the sample into LN. Cooling rate at vitrification varies in the range of 2,500–30,000 °C/min or greater. However, one has to take into account that every cell seems to require its own optimal cooling rate. Also, the natural state of liquid water inside living cells is retained during vitrification (Sformo et al. 2010), leading to minimal disturbance within the cryopreserved sample. A sample that will be vitrified is exposed to CPA at the same manner as during slow freezing, but the vitrification procedure requires higher concentration of CPAs to achieve high cooling rate. Warming after vitrification is usually also fast. A major potential drawback of vitrification is the use of high concentration of cryoprotectant, and an unintentional negative impact of these cryoprotectants in turn can be their toxicity, which may affect the cell or tissue.

Vitrification has been recommended as the method of choice for oocytes or embryos. All developmental stages of human embryos cultured in vitro have been successfully vitrified and warmed, with the generation of offsprings. In 1999 and 2000 successful pregnancies and deliveries after vitrification and warming of human oocytes were reported (Kuleshova et al. 1999). Since that time, and because it seems to be that both entities appear to be especially chill-sensitive cells in assisted reproductive technology (ART), oocytes and blastocysts seem to receive a potentially significant boost in survival rates by avoiding ice crystallization using vitrification.

Vitrification is very simple, requires no expensive programmable freezing equipment, and relies especially on the placement of the embryo in a very small volume of vitrification medium (referred also as "minimal volume approach") that must be cooled at extreme rates not obtainable in traditional enclosed cryostorage devices such as straws and vials. The importance of the use of a small volume, also referred as "minimal volume approach" was described and published in 2005 (Kuwayama et al. 2005) (Kuwayama 2007). In addition, recent publications have shown the dominance of warming rate over cooling rates in the survival of oocytes subjected to a vitrification procedure (Seki and Mazur 2009) (Mazur and Seki 2011). Vitrification has also been proposed for cryopreserving whole organs. However, there is evidence that a large tissue or whole organ will require slow cooling rates (Hubel et al. 1991). In fact, fractures of the organ could be caused by vitrification and warming procedures, and devitrification if the storage temperature is above the glass transition temperature (Fahy et al. 1990). The large problem is eliminating or sufficiently limiting ice formation throughout the large organs without inducing unacceptable toxicity caused by high doses of CPA (Fahy et al. 1990; Gavish et al. 2008)

1.4.2 Conventional Freezing

Convectional slow freezing is a method which aims to have a protocol that is slow enough to dehydrate the cells in order to prevent intracellular crystallization and fast enough to minimize osmotic stress to the cells (Mazur 1990). The convectional cooling procedure is usually ended when the temperature lies between −30 °C and

−80 °C and after the sample can be plunged into LN (Shaw and Jones 2003). When applied on cells the convectional freezing method reduces the likelihood of intracellular ice formation by initiation of extracellular ice crystal formation at a high subzero temperature. The freezing rate is then responsible for how fast the extracellular ice crystals will grow. The extracellular ice draws water out of the cell until little amount of free water remains and only small (nonlethal) ice crystal has been formed (Shaw et al. 2000). The best outcome of this cryopreservation method is obtained when the rate of freezing allows equilibrium between cell dehydration and the rate at which water is integrated into extracellular ice crystals.

The rate of dehydration, which is the movement of water outward across the cell membrane, depends on the cooling rate that is determined by temperature drop in relation to time. Thus, the movement of water through the cell membrane decreases as a consequence of lower temperature (Mazur 1984).

It is important to emphasize that in this procedure, the cooling rate is dependent on the size and permeability of the specific cells/tissue to be cryopreserved. When trying to freeze a large tissue or a whole organ with convectional freezing, one of the major problems is that cooling rates will vary across the samples depending on the geometry (surface, depth, and volume) and heat conductivity. In fact, the conventional slow-freezing method involves lowering the temperature of the chamber in a controlled stepwise manner. This method is based on using multidirectional (equiaxial) heat transfer to achieve a rate of temperature change in the sample that depends on the thermal conductivity and geometrical shape of the container and of the biological material within it (Armitage 1987) (Bischof 2000; Karlsson and Toner 1996; Maas et al. 2000). The thermal gradient within the sample is determined implicitly by the temperature of the chamber and the thermal conductivity of the materials of the sample, and is not directly controllable. Furthermore, the ambient temperature gradients within the freezing chamber and the unreliability of temperature recording measurements add to the difficulty of achieving the optimal cooling rate.

One of the major problems that occurs during cryopreservation of large tissue using convectional freezing is the release of latent heat when crystallization occurs. There are two problems regarding latent heat release. The first problem is that the heat is transferred to colder areas in which the ice is just formed and so it could possibly re-warm the ice and cause local melting (Fig. 9.1).

The second problem is the isothermal period. When samples of large volume with relatively surface/volume are frozen, the release of latent heat may cause a long isothermal period in the material being frozen. At the same time, the temperature in the freezing chamber or on the surrounding medium is lowered, increasing the temperature different between the sample and its surrounding (Fig. 9.1). Consequently, since the thermal conductivity of water (0.6) is lower than that of ice (1.6), when latent heat is no longer released, the temperature in the material being frozen will drop very quickly to a temperature close to the temperature of the surrounding environment. This might lead to a non-optimal cooling rate and possibly to cellular damage through intracellular crystallization (Petersen et al. 2006) (Armitage 1987) (Bischof 2000; Maas et al. 2000).

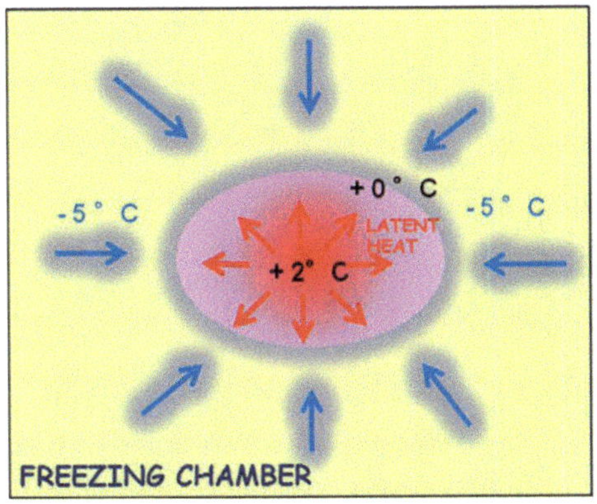

Fig. 9.1 Release of latent heat. See the temperature differences between the core and the surface of the sample and between sample and the freezing chamber

The current approach to overcome the problem of cell damage due to release of latent heat is the removal of the excessive heat through the adjustment of the cooling rate at specific time points and ensuring uniform freezing by maintaining a high ratio of surface to volume. In other words, the sample to be frozen is made as thin as possible, whereupon heat from the inner part of the sample, which is being released through the surface of the sample, will be removed faster due to the steep temperature gradient. In this way, it is possible to apply the optimal cooling rates for each sample while providing a heat sink for rapid absorption of the released latent heat.

1.4.3 Directional Freezing

Recently, a new directional freezing technology (MTG) was introduced (Arav and Natan 2009; Revel et al. 2004) which provides identical cooling rates through the specimen being frozen and it is aimed at cryopreserving whole organs (Fig. 9.2).

This technology is based on a series of heat conductive block (usually built of brass or aluminum) arranged in a line, with a straight track running through the blocks. Along the blocks, different temperatures (T1, T2, and T3, Fig. 9.2) can be set, in order to obtain a temperature gradient along the blocks (G1, G2, and G3). The blocks are separated by gap and the temperature of the block on one side of the gap (T1) is above the freezing point temperature and on the other side of the gap (T2) is below the freezing point temperature, thereby imposing a temperature gradient across the gap (G1). Biological samples to be frozen are placed inside test tube and are moved along the track at a certain velocity (V). The samples are frozen at rates according to the specific protocols of the samples; the rate of freezing is given by the temperature gradient multiplied by the velocity (CR1,2,3 = G1,2,3*V).

The MTG can generate a linear thermal gradient in an organ/tissue to be cooled. It is possible to change heat transfer in a directional watt according to the heat transfer in the sample. The movement of the sample in the thermal gradient does not exceed this velocity in a way that heat will remain directional. In this way, heat

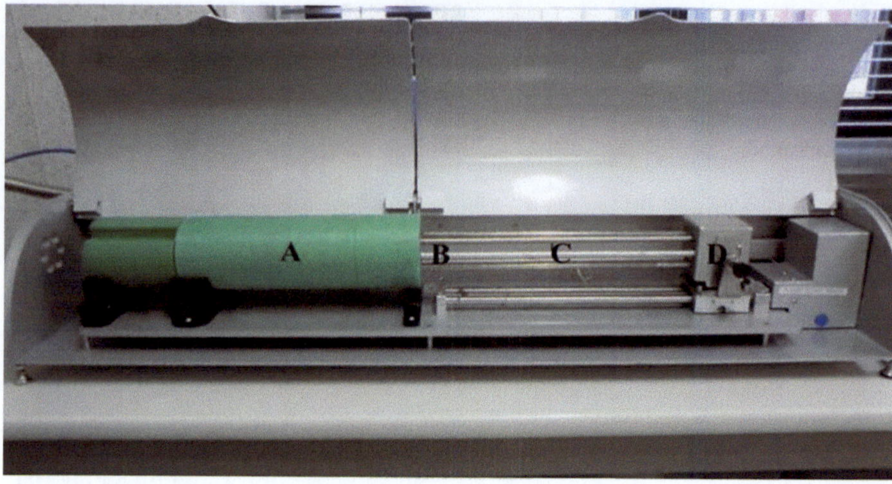

Fig. 9.2 The Multi-Thermal-Gradient (MTG) freezing device. (*A*) cooling block; (*B*) entrance of the sample's glass tube; (*C*) metal rod, which advances the sample's glass tube forward into the cooling channel in the cooling block; and (*D*) electric motor

transfer will propagate in the tissue in a linear way and the cooling rate for each point in the organ will propagate the same.

Heat transfer in directional solidification is opposed to the direction of sample movement. This is when the velocity is slower than the speed at which the heat is removed from the center of the sample toward the direction of the liquid and toward the conductive material of the device. In directional freezing the ice formation occurs according to the solution's freezing point temperature. Therefore supercooling, a non-equilibrium thermodynamic state in which a solution is cooled below its freezing point without crystallization, is avoided completely.

The freezing technology is based on directional freezing in which the biological material is transferred through a linear temperature gradient so that the cooling rate and ice front propagation is precisely controlled. Thus maximizing the survival rate of cells subjected to freezing and thawing requires the careful control of the freezing process. Although this method is rather new, results showed that directional freezing is a promising method for cell, tissue, and entire organ cryopreservation (Arav et al. 2005; Revel et al. 2004; Saragusty and Arav 2011).

1.4.4 Thawing

The thawing process and the removal of cryoprotectant are also important steps for a successful procedure. There is an opinion that the ice formed during thawing is less dangerous (Shaw et al. 1991), although some authors recognize slow thawing as a prime destructive factor (Arav and Natan 2009). The longer the duration of the thawing period, the greater the damage occurring to the cells. The increase of the solute effects and the maximal growth of ice crystals are responsible for this damage. The large crystals have an abrasive action producing a mechanical disruption of cells. Crystal growth is maximal at the range of 0 °C and −40 °C. The disruptive ultrastructural changes in cells are increased with recrystallization. The thawing is most destructive if it is done completely, so that all of the frozen tissue is thawed, which takes full advantage of recrystallization. For these authors, basically two types of damage occur during slow thawing. The first is recrystallization, which is the growth of ice that was formed during the freezing process, and the second is the refreezing of the solution that occurs after complete thawing.

1.4.5 Cryoinjury

All cells and tissues that are subjected to freezing may be damaged during the freezing procedure or during thawing. This is referred to as cryoinjury and includes several types and causes of injuries. It may be a result of excessive cell dehydration (Mazur et al. 1972), intracellular ice formation (Acker et al. 2001), CPA toxicity (Fahy 2010), or a combination of these. It is well known that formation of ice inside the cell is lethal (Muldrew and McGann 1990).

This event occurs when a cell is unable to maintain equilibrium with the extracellular space during slow freezing or when the critical cooling rate is not obtained by vitrification methods. A recently introduced factor is the possible protective effect of the so-called "innocuous" intracellular ice formation (Acker and McGann 2003), which may reduce osmotic stress. The survival of a cell at freezing can also be affected by the specific forms of ice crystals that are present and mechanical deformation of cells caused by large extracellular ice crystals have been reported (Hubel et al. 1992). Moreover, a phenomenon known as devitrification results in recrystallization (Rall et al. 1984), that is a formation of ice crystals during warming. Despite the intended protective effect of CPAs, they may also be toxic and their presence may contribute to osmotic stress (Pedro et al. 1997). However, decreasing the time and temperature of cell exposure to CPAs can reduce the toxicity while stepwise addition and removal of CPAs decreases osmotic stress and excessive volume change of the cells (Wowk 2010). It should be pointed out that the exact mechanism of cell damage during cryopreservation and CPA toxicity has not yet been elucidated (Fahy 2010; Karlsson and Toner 1996).

2 Organ Banking

The narrow time frame between harvesting of organs and transplanting the donated organs area and the ever-growing need for organs, has motivated intensive research in the organs preservation. At present, solid organs, such as liver, are preserved ex vivo under hypothermic conditions during which the organ remains viable for only a short period of time that is limited to 12–24 h.

Because hypothermic preservation failed to provide a solution for the problem of long-term preservation of organs, alternative approaches were sought, among them storage at subzero conditions.

There are also numerous problems associated with freezing and thawing large vascularized organs, including the difficulty of generating heat transfer within a large thermal mass with a geometry that has little malleability, the packing density of cells within the organ, and the presence of many different cell types, each with its own requirements for optimal freezing and thawing conditions. Another problem associated with freezing large biological samples is the isothermal period caused by the massive release of latent heat during the crystallization stage of the process.

2.1 Cryopreservation of Whole Ovary

Current cancer treatments improve the long-term survival rates of young women suffering from malignancies. However, many of these therapies have lasting effects on fertility because they cause severe injuries to the ovarian reserve, which may lead to consequent premature ovarian failure, with a negative impact on the life quality of young cancer survivors (Maffei et al. 2013a, b).

At the present, one of the future strategies to preserve fertility is the transplantation of the cryopreserved whole intact ovary. Whole ovary transplantation with vascular anastomosis was proposed as a mechanism to reduce ischemic damage and, in theory, prolong the longevity of the graft (Jeremias et al. 2002). In this technique, the whole ovary with its vascular pedicle, is removed, cryopreserved, thawed, and then transplanted using a microvascular anastomosis into a heterotopic or orthotopic site (Jadoul et al. 2007).

Whole frozen ovary transplantation with microvascular anastomosis was first described in rats by Wang et al. (Wang et al. 2002). They described successful vascular transplantation of frozen–thawed rat ovaries and reproductive tract in four of seven (57 %) transplants, which survived for ≥60 days, were ovulatory and resulted in one pregnancy. Chen et al. (Chen et al. 2006) showed that frozen–thawed rabbit ovaries remained functional for at least 7 months after microvascular transplantation in 13 of 15 (86.7 %) animals.

The challenge of whole ovary cryopreservation and transplantation technology is not only the surgical technique but the cryopreservation protocol for an entire organ. It appears that, in large mammals and humans, cryopreserving such a large-sized intact ovary may prove more problematic than in small animals, largely due to the physical constraints that limit an appropriate heat transfer between the core and the periphery of the organ (Arav and Natan 2009). In addition, the large volume of the whole organ poses some limitation to the perfusion and diffusion of cryoprotectants (Falcone et al. 2004; Torre et al. 2013) (see Chap. 1.1.4.2). Both are essential for preventing intravascular ice formation which would irreversibly compromise a rapid and efficient resumption of the blood supply (Pegg 2010).

Nevertheless, data obtained from experiments mostly performed in sheep show some positive results (Grazul-Bilska et al. 2008) and pregnancies (Salle et al. 2002) but also highlight stromal and vascular damage following either slow freezing (Onions et al. 2009; Wallin et al. 2009) or vitrification (Courbiere et al. 2009; Salle et al. 2003).

Imhof et al. reported that 18 months after transplantation of whole ovary frozen with slow freezing, the follicular survival rate was less than an 8 % (Imhof et al. 2006). Other authors reported an even lower follicular survival rate (6 %) and the depletion of the entire follicular population after whole ovary vitrification and transplantation (Courbiere et al. 2009). Bedaiwy et al. (2003) reported the restoration of ovarian function after autotransplantation of intact frozen–thawed sheep ovaries with microvascular anastomosis, but it should be noted that 8 of 11 ovaries were lost due to thrombotic events in the reanastomosed vascular pedicle.

Although transplantation of the whole cryopreserved-thawed ovary was not performed in humans, cryopreservation of a whole ovary using a slow freezing protocol has been successfully attempted.

Recently, Martinez-Madrid et al. (Martinez-Madrid et al. 2007) described a cryopreservation protocol for intact human ovary with its vascular pedicle and proved high survival rates of follicles (75.1 %), small vessels and stroma, and a normal histological structure in all the ovarian components after thawing.

Most recently, a multigradient freezing device was used with promising results (see Chap. 1.1.4.3). Very recently Maffei et al. compared this new freezing method with the conventional freezing demonstrating that directional freezing allows good preservation of whole ovaries (Maffei et al. 2013a, b). In the follow-up study, they also showed that directional freezing improves the viability of cryopreserved ovarian tissue not only when used with whole organs but also with ovarian fragment (Maffei et al. 2013a, b). Furthermore, Arav et al. (Arav et al. 2005) reported progesterone activity 36 months after vascular transplantation of frozen–thawed sheep ovaries in three of eight transplants, and retrieval of six oocytes, resulting in embryonic development up to the 8-cell stage after parthenogenic activation. Even after 6 years the cryopreserved and transplanted tissue appeared functional as was indicated by histological examination. Normal tissue architecture, healthy blood vessels, and follicles at different stages were all observed in these ovaries. Even after 6 years the cryopreserved and transplanted tissue appeared functional as was indicated by histological examination. Normal tissue architecture, healthy blood vessels, and follicles at different stages were all observed in these ovaries (Arav et al. 2010). These results are promising in finding an optimal method for the cryopreservation of whole ovaries, so that in the future it will be possible to restore fertility long-term and to prevent premature menopause in young female cancer patients.

2.2 *Cryopreservation of Kidney and Liver*

Currently there are over 100,000 patients awaiting liver transplants with transplant rates of only 38 %. There is clearly a need to improve both preservation of the organs and to expand the donor pool.

Hypothermia was employed for organ preservation to reduce the kinetics of metabolic activities that would otherwise lead to cellular degradation when oxygen is removed from the donor organ. Simple cold storage (SCS) is a process by which the cryoprotectant is infused into the organ and then stored at hypothermic temperatures.

Currently there are two clinical methods for preserving livers: SCS and hypothermic machine perfusion (HMP). All other methods, such as normothermic machine perfusion (NMP) and oxygen persufflation (OP), are in various stages of preclinical and early clinical studies.

Hypothermia was employed for organ preservation to reduce the kinetics of metabolic activities that would otherwise lead to cellular degradation when oxygen is removed from the donor organ. SCS is a process by which the preservation solution is infused into the organ and then stored statically at hypothermic temperatures.

For the liver, prior to the discovery of the University of Wisconsin (UW) solution, preservation by SCS was limited to 6 h. With the UW solution, preservation improved to 16 h and allowed long distance procurement of the donor organ and this UW solution is currently used for other donor organs including kidneys, hearts, pancreas, intestine, and lungs.

The other method for preserving livers is the HMP. This method of preservation was developed for kidney to extend both preservation time and quality. HMP is able to supply oxygen to the tissue for ATP synthesis via perfusion of fluids that can carry oxygen. The oxygen requirements of cold tissues are low so the demand for oxygen is also low. This allows for slow flow rates during hypothermia and the relatively low oxygen carrying capacity of most crystalloid perfusates are adequate at low temperatures. However, perfusing organs at low temperatures cause side effects from both hypothermia and perfusion. Mitigating these side effects is a main function of most modern day perfusion solutions. HMP preservation has proven to be a reliable method for preserving good renal function in explanted cadaveric kidneys for transplantation (Moers et al. 2009). This method was used also for the liver preservation. In large animal studies, Piennar (Pienaar et al. 1990) showed that HMP with the UW solution could be used for 72 h preservation of the liver in the canine transplant model. In human, the first clinical trial was described in 2010 (Guarrera et al. 2010) demonstrating that HMP for donor livers provided safe and reliable preservation method.

3 Preservation of Blood Cells

3.1 Freeze-Drying

Water is essential to life, providing a universal solvent supporting bio-chemical activities within cells, which enables metabolism to continue and sustains all living processes. In the absence of water, life will cease, resulting in a state of death or dormancy in live cells. Water also plays a major role in the degradation of stored material, providing conditions that potentiate autolysis, or promote the growth of spoilage organisms.

In order to stabilize labile products, it is therefore necessary to immobilize or reduce the water content of stored samples.

Vaccines, other biological materials, cells, and tissues can be stabilized by chilling or freezing. However, maintaining and transporting samples in the frozen state is costly, whereas freezer breakdown may result in the complete loss of valuable products (Fanget and Francon 1996).

Alternatively, bioproducts can be dried in air using high processing temperatures. Freeze-drying or lyophilization describe precisely the same process.

Freeze-drying can be technically defined as a controllable method of dehydrating labile products by vacuum desiccation and can be summarized in two steps:

1. Cooling of the liquid sample, followed by the conversion of freezable solution water into ice
2. "Evaporation" of water from the amorphous matrix

Freeze-drying has been used in the preparation of pharmaceuticals and vaccines as one of the most important processes for the preservation of heat-sensitive biological materials. Compared to other techniques, freeze-drying has some well-known advantages, including sample stability at room temperature, defined porous product structure, easy reconstitution by the addition of water or aqueous solution, and easy transportation (Xiao et al. 2004). Freeze-drying of cells used to be limited to prokaryotes. Over recent years, a few types of mammalian cells have been successfully freeze-dried, such as human erythrocytes and human platelets (Han et al. 2005; Wolkers et al. 2001). A study carried out in mouse demonstrated that freeze-dried platelets can be stored at room temperature for several months while their physiological viabilities remain (Wolkers et al. 2002).

3.2 *Freeze-Drying of Human Blood Cells*

Hematopoietic as well as other somatic cells are currently cryopreserved and stored in liquid nitrogen tanks, in nitrogen vapor phase or in 280 °C freezers (Choi et al. 2001; Halle et al. 2001; Rogers et al. 2001).

This conventional mode of cryopreservation is prone to transient warming events and various other hazards, such as cross-contamination (Bielanski et al. 2003; Morris 2005), and results in cell losses of 20–30 % (Laroche et al. 2005).

The possibility of storing blood cells and other cell lines in a dry state through the process of lyophilization could be of immense benefit for clinical application especially when long-term preservation is required.

In addition, freeze-drying does not involve the thawing process to which cell damage is attributed, particularly at large volumes (e.g. from recrystallization). However, the use of freeze-dried blood is still far from clinical application.

In order to freeze-dry the cells, researchers had to develop a method that would enable freezing in the absence of permeating cryoprotectant agents (CPAs) such as DMSO, glycerol or ethylene glycol, to name a few, and would enable the use of additives that have a high glass transition temperature (Tg) and that are solid at room temperature.

The major damaging factor associated with freeze-drying liposomes is lipid-phase transition, it is the main obstacle for successful cryopreservation of many cell types, including sperm (Drobnis et al. 1993) and oocytes (Arav et al. 1996). Chilling injury can be overcome by stabilizing the membrane phospholipids using disaccharides such as sucrose or trehalose (Crowe and Crowe 1991). Other approaches have involved changing the lipid composition of the membrane by using liposomes in vitro (Zeron et al. 2002a, b) or dietary additives in vivo (Zeron et al. 2002a, b).

The second factor is membrane fusion, liposomes stored above the high glass transition temperature have been shown to rapidly fuse and become damaged, and it was therefore concluded that glass transition or vitrification is an important factor in decreasing the chances of fusion upon drying (Crowe et al. 1998).

As described above, the freeze-drying of cells can be divided into two steps: the first is the freezing process during which large ice crystals are formed, pushing the cells into an area defined as the unfrozen fraction. The unfrozen fraction incorporates the cells that have dehydrated and vitrified in an amorphous matrix during the freezing process. The second step is sublimation of the ice crystals, which occurs in two stages, termed primary and secondary drying.

Recently, an Israeli group tried to freeze human red blood with a new device (Arav and Natan 2011). In this study, human blood freezing was performed by the MTG-1314 apparatus (see chapter 1.4.2). Lyophilization was performed by putting the frozen samples in the commercial lyophilizer near the condenser (Freezone Plus 6, Labconco), which reaches a temperature of −80 °C, for 3.5 days. Thawing was performed by immersing the frozen samples in a water bath heated to 37 °C. At the end of the lyophilization process, samples were immediately rehydrated by adding 2.4 mL double-distilled water that had been pre-warmed to 37 °C.

Currently, Arav and his group working on determining the optimal conditions for sublimation and storage which they believe will result in further increasing the survival of cells after freeze-drying.

Acknowledgments This work was supported by Associazione Italiana per la Ricerca sul Canro (AIRC) and Carraresi Foundation.

References

Acker JP, McGann LE (2003) Protective effect of intracellular ice during freezing? Cryobiology 46:197–202

Acker JP, Elliott JA, McGann LE (2001) Intercellular ice propagation: experimental evidence for ice growth through membrane pores. Biophys J 81:1389–1397

Arav A, Natan Y (2009) Directional freezing: a solution to the methodological challenges to preserve large organs. Semin Reprod Med 27:438–442

Arav A, Natan D (2011) Freeze drying (lyophilization) of red blood cells. J Trauma 70:S61–S64
Arav A, Zeron Y, Leslie SB, Behboodi E, Anderson GB, Crowe JH (1996) Phase transition temperature and chilling sensitivity of bovine oocytes. Cryobiology 33:589–599
Arav A, Revel A, Nathan Y, Bor A, Gacitua H, Yavin S, Gavish Z, Uri M, Elami A (2005) Oocyte recovery, embryo development and ovarian function after cryopreservation and transplantation of whole sheep ovary. Hum Reprod 20:3554–3559
Arav A, Gavish Z, Elami A, Natan Y, Revel A, Silber S, Gosden RG, Patrizio P (2010) Ovarian function 6 years after cryopreservation and transplantation of whole sheep ovaries. Reprod Biomed Online 20:48–52
Armitage WJ (1987) Cryopreservation of animal cells. Symp Soc Exp Biol 41:379–393
Bedaiwy MA, Jeremias E, Gurunluoglu R, Hussein MR, Siemianow M, Biscotti C, Falcone T (2003) Restoration of ovarian function after autotransplantation of intact frozen-thawed sheep ovaries with microvascular anastomosis. Fertil Steril 79:594–602
Bielanski A, Bergeron H, Lau PC, Devenish J (2003) Microbial contamination of embryos and semen during long term banking in liquid nitrogen. Cryobiology 46:146–152
Bischof JC (2000) Quantitative measurement and prediction of biophysical response during freezing in tissues. Annu Rev Biomed Eng 2:257–288
Chen CH, Chen SG, Wu GJ, Wang J, Yu CP, Liu JY (2006) Autologous heterotopic transplantation of intact rabbit ovary after frozen banking at -196 degrees C. Fertil Steril 86:1059–1066
Choi CW, Kim BS, Seo JH, Shin SW, Kim YH, Kim JS (2001) Long-term engraftment stability of peripheral blood stem cells cryopreserved using the dump-freezing method in a -80 degrees C mechanical freezer with 10 % dimethyl sulfoxide. Int J Hematol 73:245–250
Courbiere B, Caquant L, Mazoyer C, Franck M, Lornage J, Salle B (2009) Difficulties improving ovarian functional recovery by microvascular transplantation and whole ovary vitrification. Fertil Steril 91:2697–2706
Crowe LM, Crowe JH (1991) Solution effects on the thermotropic phase transition of unilamellar liposomes. Biochim Biophys Acta 1064:267–274
Crowe JH, Carpenter JF, Crowe LM (1998) The role of vitrification in anhydrobiosis. Annu Rev Physiol 60:73–103
Drobnis EZ, Crowe LM, Berger T, Anchordoguy TJ, Overstreet JW, Crowe JH (1993) Cold shock damage is due to lipid phase transitions in cell membranes: a demonstration using sperm as a model. J Exp Zool 265:432–437
Elmoazzen HY, Elliott JA, McGann LE (2002) The effect of temperature on membrane hydraulic conductivity. Cryobiology 45:68–79
Fahy GM (2010) Cryoprotectant toxicity neutralization. Cryobiology 60:S45–S53
Fahy GM, Saur J, Williams RJ (1990) Physical problems with the vitrification of large biological systems. Cryobiology 27:492–510
Falcone T, Attaran M, Bedaiwy MA, Goldberg JM (2004) Ovarian function preservation in the cancer patient. Fertil Steril 81:243–257
Fanget B, Francon A (1996) A varicella vaccine stable at 5 degrees C. Dev Biol Stand 87: 167–171
Farrant J (1980) General observations on cell preservation. Ashwood-Smith MJ, Farrant J (eds) Low temperature preservation in medicine and biology. Pitman Medical Limited, Kent, pp 1–18
Gavish Z, Ben-Haim M, Arav A (2008) Cryopreservation of whole murine and porcine livers. Rejuvenation Res 11:765–772
Grazul-Bilska AT, Banerjee J, Yazici I, Borowczyk E, Bilski JJ, Sharma RK, Siemionov M, Falcone T (2008) Morphology and function of cryopreserved whole ovine ovaries after heterotopic autotransplantation. Reprod Biol Endocrinol 6:16
Guarrera JV, Henry SD, Samstein B, Odeh-Ramadan R, Kinkhabwala M, Goldstein MJ, Ratner LE, Renz JF, Lee HT, Brown RS Jr et al (2010) Hypothermic machine preservation in human liver transplantation: the first clinical series. Am J Transplant 10:372–381
Halle P, Tournilhac O, Knopinska-Posluszny W, Kanold J, Gembara P, Boiret N, Rapatel C, Berger M, Travade P, Angielski S et al (2001) Uncontrolled-rate freezing and storage at -80 degrees C, with only 3.5-percent DMSO in cryoprotective solution for 109 autologous peripheral blood progenitor cell transplantations. Transfusion 41:667–673

Han DW, Kim HH, Lee MH, Baek HS, Lee KY, Hyon SH, Park JC (2005) Protection of osteoblastic cells from freeze/thaw cycle-induced oxidative stress by green tea polyphenol. Biotechnol Lett 27:655–660

Hubel A, Toner M, Cravalho EG, Yarmush ML, Tompkins RG (1991) Intracellular ice formation during the freezing of hepatocytes cultured in a double collagen gel. Biotechnol Prog 7: 554–559

Hubel A, Cravalho EG, Nunner B, Korber C (1992) Survival of directionally solidified B-lymphoblasts under various crystal growth conditions. Cryobiology 29:183–198

Imhof M, Bergmeister H, Lipovac M, Rudas M, Hofstetter G, Huber J (2006) Orthotopic microvascular reanastomosis of whole cryopreserved ovine ovaries resulting in pregnancy and live birth. Fertil Steril 85(Suppl 1):1208–1215

Jadoul P, Donnez J, Dolmans MM, Squifflet J, Lengele B, Martinez-Madrid B (2007) Laparoscopic ovariectomy for whole human ovary cryopreservation: technical aspects. Fertil Steril 87: 971–975

Jeremias E, Bedaiwy MA, Gurunluoglu R, Biscotti CV, Siemionow M, Falcone T (2002) Heterotopic autotransplantation of the ovary with microvascular anastomosis: a novel surgical technique. Fertil Steril 77:1278–1282

Karlsson JO, Toner M (1996) Long-term storage of tissues by cryopreservation: critical issues. Biomaterials 17:243–256

Kuleshova L, Gianaroli L, Magli C, Ferraretti A, Trounson A (1999) Birth following vitrification of a small number of human oocytes: case report. Hum Reprod 14:3077–3079

Kuwayama M (2007) Highly efficient vitrification for cryopreservation of human oocytes and embryos: the Cryotop method. Theriogenology 67:73–80

Kuwayama M, Vajta G, Ieda S, Kato O (2005) Comparison of open and closed methods for vitrification of human embryos and the elimination of potential contamination. Reprod Biomed Online 11:608–614

Laroche V, McKenna DH, Moroff G, Schierman T, Kadidlo D, McCullough J (2005) Cell loss and recovery in umbilical cord blood processing: a comparison of postthaw and postwash samples. Transfusion 45:1909–1916

Liebermann J, Nawroth F, Isachenko V, Isachenko E, Rahimi G, Tucker MJ (2002) Potential importance of vitrification in reproductive medicine. Biol Reprod 67:1671–1680

Luyet BJ, Gehenio PM (1940) Life and death at low temperatures. Normandy, Missouri Byodynamica

Maas WJ, de Graaf IA, Schoen ED, Koster HJ, van de Sandt JJ, Groten JP (2000) Assessment of some critical factors in the freezing technique for the cryopreservation of precision-cut rat liver slices. Cryobiology 40:250–263

Maffei S, Hanenberg M, Pennarossa G, Silva JR, Brevini TA, Arav A, Gandolfi F (2013a) Direct comparative analysis of conventional and directional freezing for the cryopreservation of whole ovaries. Fertil Steril 100:1122–1131

Maffei S, Pennarossa G, Brevini TA, Arav A, Gandolfi F (2013b) Beneficial effect of directional freezing on in vitro viability of cryopreserved sheep whole ovaries and ovarian cortical slices. Hum Reprod 29:114–124

Martinez-Madrid B, Camboni A, Dolmans MM, Nottola S, Van Langendonckt A, Donnez J (2007) Apoptosis and ultrastructural assessment after cryopreservation of whole human ovaries with their vascular pedicle. Fertil Steril 87:1153–1165

Mazur P (1970) Cryobiology: the freezing of biological systems. Science 168:939–949

Mazur P (1984) Freezing of living cells: mechanisms and implications. Am J Physiol 247: C125–C142

Mazur P (1990) Equilibrium, quasi-equilibrium, and nonequilibrium freezing of mammalian embryos. Cell Biophys 17:53–92

Mazur J (2004) Principles of cryobiology. Life in the frozen state. CRC Press, Boca Raton

Mazur P, Seki S (2011) Survival of mouse oocytes after being cooled in a vitrification solution to -196 degrees C at 95 degrees to 70,000 degrees C/min and warmed at 610 degrees to 118,000 degrees C/min: A new paradigm for cryopreservation by vitrification. Cryobiology 62:1–7

Mazur P, Leibo SP, Chu EH (1972) A two-factor hypothesis of freezing injury. Evidence from Chinese hamster tissue-culture cells. Exp Cell Res 71:345–355
Moers C, Smits JM, Maathuis MH, Treckmann J, van Gelder F, Napieralski BP, van Kasterop-Kutz M, van der Heide JJ, Squifflet JP, van Heurn E et al (2009) Machine perfusion or cold storage in deceased-donor kidney transplantation. N Engl J Med 360:7–19
Morris GJ (2005) The origin, ultrastructure, and microbiology of the sediment accumulating in liquid nitrogen storage vessels. Cryobiology 50:231–238
Muldrew K (2004) The water to ice transition: implications for living cells. Life in the frozen state. CRC Press, Boca Raton, pp 67–103
Muldrew K, McGann LE (1990) Mechanisms of intracellular ice formation. Biophys J 57: 525–532
O'Neil L, Paynter SJ, Fuller BJ, Shaw RW, DeVries AL (1998) Vitrification of mature mouse oocytes in a 6 M Me2SO solution supplemented with antifreeze glycoproteins: the effect of temperature. Cryobiology 37:59–66
Onions VJ, Webb R, McNeilly AS, Campbell BK (2009) Ovarian endocrine profile and long-term vascular patency following heterotopic autotransplantation of cryopreserved whole ovine ovaries. Hum Reprod 24:2845–2855
Pedro PB, Zhu SE, Makino N, Sakurai T, Edashige K, Kasai M (1997) Effects of hypotonic stress on the survival of mouse oocytes and embryos at various stages. Cryobiology 35:150–158
Pegg DE (1984) Red cell volume in glycerol/sodium chloride/water mixtures. Cryobiology 21:234–239
Pegg DE (2010) The relevance of ice crystal formation for the cryopreservation of tissues and organs. Cryobiology 60:S36–S44
Petersen A, Schneider H, Rau G, Glasmacher B (2006) A new approach for freezing of aqueous solutions under active control of the nucleation temperature. Cryobiology 53:248–257
Pienaar BH, Lindell SL, Van Gulik T, Southard JH, Belzer FO (1990) Seventy-two-hour preservation of the canine liver by machine perfusion. Transplantation 49:258–260
Polge C, Smith AU, Parkes AS (1949) Revival of spermatozoa after vitrification and dehydration at low temperatures. Nature 164:666
Rall WF, Reid DS, Polge C (1984) Analysis of slow-warming injury of mouse embryos by cryomicroscopical and physiochemical methods. Cryobiology 21:106–121
Revel A, Elami A, Bor A, Yavin S, Natan Y, Arav A (2004) Whole sheep ovary cryopreservation and transplantation. Fertil Steril 82:1714–1715
Rogers I, Sutherland, Holt D, Macpate F, Lains A, Hollowell S, Cruickshank B, Casper RF (2001) Human UC-blood banking: impact of blood volume, cell separation and cryopreservation on leukocyte and CD34(+) cell recovery. Cytotherapy 3:269–276
Salle B, Demirci B, Franck M, Rudigoz RC, Guerin JF, Lornage J (2002) Normal pregnancies and live births after autograft of frozen-thawed hemi-ovaries into ewes. Fertil Steril 77:403–408
Salle B, Demirci B, Franck M, Berthollet C, Lornage J (2003) Long-term follow-up of cryopreserved hemi-ovary autografts in ewes: pregnancies, births, and histologic assessment. Fertil Steril 80:172–177
Saragusty J, Arav A (2011) Current progress in oocyte and embryo cryopreservation by slow freezing and vitrification. Reproduction 141:1–19
Seki S, Mazur P (2009) The dominance of warming rate over cooling rate in the survival of mouse oocytes subjected to a vitrification procedure. Cryobiology 59:75–82
Sformo T, Walters K, Jeannet K, Wowk B, Fahy GM, Barnes BM, Duman JG (2010) Deep supercooling, vitrification and limited survival to -100°C in the Alaskan beetle Cucujus clavipes puniceus (Coleoptera: Cucujidae) larvae. J Exp Biol 213:502–509
Shaw JM, Jones GM (2003) Terminology associated with vitrification and other cryopreservation procedures for oocytes and embryos. Hum Reprod Update 9:583–605
Shaw JM, Kola I, MacFarlane DR, Trounson AO (1991) An association between chromosomal abnormalities in rapidly frozen 2-cell mouse embryos and the ice-forming properties of the cryoprotective solution. J Reprod Fertil 91:9–18

Shaw JM, Oranratnachai A, Trounson AO (2000) Fundamental cryobiology of mammalian oocytes and ovarian tissue. Theriogenology 53:59–72

Solomon AK (1968) Characterization of biological membranes by equivalent pores. J Gen Physiol 51:335–364

Sun WQ (1999) State and phase transition behaviors of quercus rubra seed axes and cotyledonary tissues: relevance to the desiccation sensitivity and cryopreservation of recalcitrant seeds. Cryobiology 38:372–385

Torre A, Ben Brahim F, Popowski T, Boudjenah R, Salle B, Lornage J (2013) Factors related to unstained areas in whole ewe ovaries perfused with a metabolic marker. Hum Reprod 28: 423–429

Verkman AS, van Hoek AN, Ma T, Frigeri A, Skach WR, Mitra A, Tamarappoo BK, Farinas J (1996) Water transport across mammalian cell membranes. Am J Physiol 270:C12–C30

Wallin A, Ghahremani M, Dahm-Kahler P, Brannstrom M (2009) Viability and function of the cryopreserved whole ovary: in vitro studies in the sheep. Hum Reprod 24:1684–1694

Wang X, Chen H, Yin H, Kim SS, Lin Tan S, Gosden RG (2002) Fertility after intact ovary transplantation. Nature 415:385

Wolkers WF, Walker NJ, Tablin F, Crowe JH (2001) Human platelets loaded with trehalose survive freeze-drying. Cryobiology 42:79–87

Wolkers WF, Looper SA, McKiernan AE, Tsvetkova NM, Tablin F, Crowe JH (2002) Membrane and protein properties of freeze-dried mouse platelets. Mol Membr Biol 19:201–210

Wowk B (2010) Thermodynamic aspects of vitrification. Cryobiology 60:11–22

Wowk B, Leitl E, Rasch CM, Mesbah-Karimi N, Harris SB, Fahy GM (2000) Vitrification enhancement by synthetic ice blocking agents. Cryobiology 40:228–236

Xiao HH, Hua TC, Li J, Gu XL, Wang X, Wu ZJ, Meng LR, Gao QR, Chen J, Gong ZP (2004) Freeze-drying of mononuclear cells and whole blood of human cord blood. Cryo Letters 25:111–120

Zeron Y, Sklan D, Arav A (2002a) Effect of polyunsaturated fatty acid supplementation on biophysical parameters and chilling sensitivity of ewe oocytes. Mol Reprod Dev 61:271–278

Zeron Y, Tomczak M, Crowe J, Arav A (2002b) The effect of liposomes on thermotropic membrane phase transitions of bovine spermatozoa and oocytes: implications for reducing chilling sensitivity. Cryobiology 45:143–152

Index

T.A.L. Brevini (ed.), *Stem Cells in Animal Species: From Pre-clinic to Biodiversity*, Stem Cell Biology and Regenerative Medicine, DOI 10.1007/978-3-319-03572-7,

FSC
www.fsc.org

MIX
Papier aus verantwortungsvollen Quellen
Paper from responsible sources
FSC® C105338

If you have any concerns about our products,
you can contact us on
ProductSafety@springernature.com

In case Publisher is established outside the EU,
the EU authorized representative is:
Springer Nature Customer Service Center GmbH
Europaplatz 3, 69115 Heidelberg, Germany

Printed by Libri Plureos GmbH
in Hamburg, Germany